SISTER CHROMATID EXCHANGES

25 Years of Experimental Research

Part B

Genetic Toxicology and Human Studies

BASIC LIFE SCIENCES

Alexander Hollaender, General Editor

Council for Research Planning in Biological Sciences, Inc., Washington, D.C.

Recent volumes in the series:

Volume 18 TRENDS IN THE BIOLOGY OF FERMENTATIONS FOR FUELS AND CHEMICALS
Edited by Alexander Hollaender, Robert Rabson, Palmer Rogers, Anthony San Pietro, Raymond Valentine, and Ralph Wolfe

Volume 19 GENETIC ENGINEERING OF MICROORGANISMS FOR CHEMICALS
Edited by Alexander Hollaender, Ralph D. DeMoss, Samuel Kaplan, Jordan Konisky, Dwayne Savage, and Ralph S. Wolfe

Volume 20 MOLECULAR AND CELLULAR MECHANISMS OF MUTAGENESIS
Edited by J. F. Lemontt and W. M. Generoso

Volume 21 GENETIC TOXICOLOGY: An Agricultural Perspective
Edited by Raymond A. Fleck and Alexander Hollaender

Volume 22 PLANT CELL CULTURE IN CROP IMPROVEMENT
Edited by S. K. Sen and Kenneth L. Giles

Volume 23 INDUCED MUTAGENESIS: Molecular Mechanisms and Their Implications for Environmental Protection
Edited by Christopher W. Lawrence

Volume 24 ORGAN AND SPECIES SPECIFICITY IN CHEMICAL CARCINOGENESIS
Edited by Robert Langenbach, Stephen Nesnow, and Jerry M. Rice

Volume 25 BASIC BIOLOGY OF NEW DEVELOPMENTS IN BIOTECHNOLOGY
Edited by Alexander Hollaender, Allen I. Laskin, and Palmer Rogers

Volume 26 GENETIC ENGINEERING OF PLANTS: An Agricultural Perspective
Edited by Tsune Kosuge, Carole P. Meredith, and Alexander Hollaender

Volume 27 NEUTRONS IN BIOLOGY
Edited by Benno P. Schoenborn

Volume 28 GENETIC CONTROL OF ENVIRONMENTAL POLLUTANTS
Edited by Gilbert S. Omenn and Alexander Hollaender

Volume 29 SISTER CHROMATID EXCHANGES: 25 Years of Experimental Research
Part A: The Nature of SCEs
Part B: Genetic Toxicology and Human Studies
Edited by Raymond R. Tice and Alexander Hollaender

A Continuation Order Plan is available for this series. A continuation order will bring delivery of each new volume immediately upon publication. Volumes are billed only upon actual shipment. For further information please contact the publisher.

SISTER CHROMATID EXCHANGES

25 Years of Experimental Research

Part B
Genetic Toxicology and Human Studies

Edited by

Raymond R. Tice
Brookhaven National Laboratory
Upton, New York

and

Alexander Hollaender
Council for Research Planning in Biological Sciences, Inc.
Washington, D.C.

Associate Editors

Bo Lambert
Karolinska Hospital
Stockholm, Sweden

and

Kanehisa Morimoto
University of Tokyo
Tokyo, Japan

Technical Editor

Claire M. Wilson
Council for Research Planning in Biological Sciences, Inc.
Washington, D.C.

Plenum Press • New York and London

Library of Congress Cataloging in Publication Data

Main entry under title:

Sister chromatid exchanges.

(Basic life sciences; v. 29, pt. A-B)

"Proceedings of a symposium held in December 1983 at Brookhaven National Laboratory, Upton, New York"—T.p. verso.

Includes bibliographies and indexes.

Contents: pt. A. The nature of SCEs—pt. B. Genetic toxicology and human studies.

1. Sister chromatid exchange—Congresses. 2. Genetic toxicology—Congresses. I. Tice, Raymond R. II. Hollaender, Alexander, Date- . III. Brookhaven National Laboratory. IV. Series. [DNLM: 1. Crossing Over (Genetics)—congresses. W3 BA255 v.29 pt.A-B / QH 445 S6236]

QH445.S573 1984 599′.0873282 84-18083

ISBN 0-306-41882-7 (pt. B)

Proceedings of a symposium held in December 1983 at Brookhaven National Laboratory, Upton, New York

A Division of Plenum Publishing Corporation
233 Spring Street, New York, N.Y. 10013

Printed in the United States of America

FOREWORD

Chromosomes, being well-defined structures that are easily visible under the optical microscope, readily lend themselves to intense physical and biochemical study. The understanding of the structure and function of this most critical genetic material has progressed through a number of interesting stages. Often connected with the development of new techniques in staining and photography, using the standard microscope and the electron microscope.

It is interesting to look back at the history of cytogenetics. I would like especially to emphasize the work of Karl Sax and many of his students. Work with Tradescantia became feasible after Edgar Anderson straightened out the ecology and Sax took advantage of the small number of chromosomes easily visible under the microscope. As a matter of fact, this development is seen as the foundation for the quantitative analysis of radiation effects on chromosomes. During the 50 years since then, more refined studies have been initiated. The study of cytogenetic mechanisms has become an important tool for the recognition of the effects of environmental factors on all living systems and has made SCE studies possible.

One of the most important stages in chromosome research was the development, in radiation biology, of radiolabeling the chromosome with tritiated thymidine. This technique, published in 1957 by Dr. J. Herbert Taylor and his colleagues, demonstrated the easy visibility of sister chromatid exchanges (SCEs), thus opening a broad field for experimental investigation. The research continued to advance with the discovery of new staining techniques, many of which will be discussed at the present symposium.

This symposium will outline the present status of the very prolific study of the SCE phenomena, making possible a discussion of the interesting developments which have helped us to interpret the significance of chromosome structure, and to validate this significance through experimental work, especially through the use of radiation and chemicals.

The anniversary of the important development that Dr. Taylor initiated over 25 years ago gives us an opportunity to determine where we stand in this field, and to outline the possible directions for future research.

Alexander Hollaender

PREFACE

In 1957, J. Herbert Taylor and his colleagues in the Biology and Medical Departments at Brookhaven National Laboratory published their first contribution on the cytological demonstration of sister chromatid exchanges (SCEs) in plant root-tip cells following the incorporation of tritiated thymidine into replicating DNA. The following year, the first contribution concerned with examining the nature of SCEs was published, also by Taylor. Twenty-five years later, scientific interest in the nature, significance, and utility of this cytogenetically observed phenomenon continues to expand. The purpose of this symposium, organized by the Biology and Medical Departments at Brookhaven National Laboratory, was to honor this initial research by bringing together internationally recognized leaders in the fields of genetics, cytogenetics, carcinogenesis, mutagenesis, radiation biology, toxicology, and environmental health into an open forum to present and discuss: (i) current knowledge of the induction and formation of SCEs and their relationship to other biological endpoints, including carcinogenesis, mutagenesis, transformation, clastogenesis, DNA damage and repair, and cellular toxicity; (ii) the optimal strategies for the utilization of SCEs in genetic toxicology testing schemes involving in vitro and in vivo exposure situations; (iii) the most valid statistical methods for analyzing SCE data obtained from cells in culture, from cells in intact organisms, and from cells in humans; (iv) the relevance of SCEs as an indicator of human disease states, both inherited and acquired, and of progress in disease treatment; and (v) the use of SCEs as an indicator of human exposure to genotoxic agents and their relevance as a prognosticator of future adverse health outcome.

To ensure a truly international program, three program committees representing the U.S., Japan, and Europe were formed, chaired by R.R. Tice and A. Hollaender, K. Morimoto, and B. Lambert, respectively, to invite scientists, expert in different aspects of SCE research and utilization, to make presentations and/or to contribute posters. This approach was extremely successful as demonstrated by the participation of 175 scientists from 20 countries. The active participation of scientists from academia, industry, research laboratories, and regulatory agencies, all with a common interest in

SCEs, provided for an extremely useful interchange of knowledge, and facilitated the prioritization of future research directions.

The symposium program was divided into three areas of interest: The Nature of SCEs, SCEs and Genetic Toxicology, and SCEs in Human Studies. Each area included also in appropriate poster session. The symposium was highlighted by keynote addresses by J. Herbert Taylor and Samuel A. Latt. A general discussion on the relevance of SCE studies to public health, prevention, and intervention closed the meeting.

This symposium and the preparation of the proceedings which followed were successful due to the enthusiastic support of many individuals. We especially appreciate the efforts of the following: to the members of the program committees, for their assistance in planning a scientifically exciting and rewarding program; to Ms. H. Kondratuk (Biology), for her cheerful and efficient preparation and management of the symposium; to Ms. K. Kissel (Biology) and Ms. K. DiPierro (Medical) for their able assistance during the symposium; to Ms. C. Wilson, assisted by Ms. C. von Dohlen (the Council for Research Planning in Biological Sciences), who are responsible for the technical quality of the proceedings volume; and not least of all, to the Department of Energy, the Environmental Protection Agency, the National Institute of Environmental Health Sciences, the National Institute of Occupational Safety and Health, and the Occupational Safety and Health Administration for their financial support of the symposium and these proceedings; and to the institutions or governmental agencies which provided travel support for many of the attendees.

R.R. Tice
A. Hollaender
B. Lambert
K. Morimoto

CONTENTS OF PART B

SCE AND GENETIC TOXICOLOGY

Nonmammalian Systems

In Vitro Mammalian Systems

In Vivo Mammalian Systems

HUMAN STUDIES

The Human Lymphocyte System

Genetic Disease

Human Variability and Cancer

GENERAL DISCUSSION

CONTENTS OF PART A

HISTORICAL PERSPECTIVE

THE NATURE OF SCEs

Spontaneous and Halogenated Pyrimidines

Modulation of SCE Induction

Correlations

Statistical Analysis

SISTER CHROMATID EXCHANGE ANALYSIS IN CULTURED PERIPHERAL BLOOD LEUKOCYTES OF THE COLDWATER MARINE FISH, PACIFIC STAGHORN SCULPIN (LEPTOCOTTUS ARMATUS): A FEASIBLE SYSTEM FOR ASSESSING GENOTOXIC MARINE POLLUTANTS

Helen R. Zakour, Marsha L. Landolt,
and Richard M. Kocan

School of Fisheries, WH-10
College of Ocean and Fishery Sciences
University of Washington
Seattle, Washington 98195

ABSTRACT

The genotoxicity of environmental contaminants and test compounds to aquatic and marine fish has primarily been assessed by in vivo techniques that require sacrifice of the test organism for analysis. The major objective of this research was to develop an in vitro sister chromatid exchange (SCE) assay which would utilize cultured peripheral blood leukocytes (PBLs) of a coldwater marine fish species. Use of PBLs in cytogenetic genotoxicity tests has several advantages, the major one being that the experimental fish need not be sacrificed for sample collection. In addition, this nondestructive method of tissue collection permits the investigator to take multiple samples from a single individual and thereby allows the use of an individual as its own control and to monitor its SCE frequency over time.

A suitable in vitro culture method for fish PBLs was a prerequisite for cytogenetic analysis of this tissue. The in vitro culture conditions necessary to provide a sufficient number of dividing cells for performance of the SCE assay were established in our laboratory for the PBLs of the Pacific staghorn sculpin (Leptocottus armatus), a common bottom-dwelling Puget Sound fish. The major components of this culture system are heparinized whole blood, fetal bovine serum-supplemented enriched tissue culture medium (RPMI

1640), purified protein derivative of tuberculin as a mitogen, and an incubation temperature of 13.5°C. This in vitro PBL culture system is unique because it involves cultured blood cells from a coldwater marine fish species.

Using this culture method, SCE induction was investigated in Pacific staghorn sculpin PBLs which had been exposed in vitro to N-methyl-N'-nitro-N-nitrosoguanidine (MNNG), a known direct-acting inducer of SCEs. Cultured cells exposed in vitro responded to MNNG in a dose-related manner in regard to SCE induction, and the frequency of "outlier" cells increased at the higher concentrations of MNNG.

With further development, this technique may be adaptable for use with in vivo genotoxicity studies and provide information concerning the induction and persistence of chemically induced SCEs in fish. This PBL/SCE assay may also be a feasible assessment tool for detecting exposure of marine fish to genotoxic environmental contaminants in laboratory and field situations.

INTRODUCTION

Genotoxic Environmental Contaminants

Numerous environmental contaminants enter the marine and freshwater environments from a variety of sources including industrial and municipal discharges, agricultural runoff and other man-made sources. Many of these pollutants contain known or suspected mutagens and carcinogens that can accumulate in water and sediment, and thus, may affect the well-being of aquatic life as well as present a potential hazard for human health (1,2). The long-range goal of our research is to develop a reliable short-term in vitro method for detecting the presence of genotoxic contaminants in the marine environment. The establishment of a genotoxicity test system which utilizes the SCE assay and cultured PBLs from a marine fish provides a practical and feasible approach to achieve this goal.

The SCE Assay

The SCE assay measures the interchange of DNA at apparently homologous loci between sister chromatids of a chromosome at metaphase. SCEs occur spontaneously at a low frequency in normal cells. An increase in the frequency of SCEs can be induced in a dose-related manner by different types of DNA-damaging agents including many environmental contaminants (3,4). For this reason, the SCE assay is gaining acceptance as a sensitive indicator of genotoxic damage and has been used to monitor the impact of genotoxic agents on various biological systems (3,4), including fish (5-13). Other aspects of this cytogenetic assay are reviewed and discussed in this volume.

Use of PBLs

The use of PBLs in genotoxicity tests which measure cytogenetic endpoints is advantageous since the experimental animal is not sacrificed for tissue collection and analysis. Thus, multiple samples can be obtained from a single individual; also, each organism can serve as its own control and be monitored over time. The ability to use each organism as its own control is particularly important for genotoxicity testing of fish because a large degree of variability is inherent in feral fish populations. The use of PBLs for these analyses is also advantageous since these cells circulate throughout the entire body of an organism and are distributed in various tissues.

In Vitro Culture of PBLs

The characteristic of PBLs which makes them suitable for use with in vitro genotoxicity assays is the ability of this normally nondividing cell population to be stimulated to divide in vitro when grown under the appropriate conditions (14). This has enabled SCEs in cultured PBLs to be used to assess in vivo and in vitro exposure to genotoxic agents for several mammals (3,4,15-18).

In the past, the major difficulty in applying the in vitro PBL/SCE technique to fish and aquatic pollution studies has been the paucity of reliable in vitro culture methods for fish PBLs. Most attempts to culture fish PBLs in vitro have been unsuccessful or have yielded inconsistent results (reviewed by 17,19). However, culture techniques for the PBLs of marine (19) and freshwater (20) fish species have recently been developed specifically for use with in vitro cytogenetic genotoxicity tests.

Objectives

The aim of this study is to adapt and apply the techniques for SCE analysis to cultured fish PBLs, to evaluate the effects of a known mutagen on these cells, and to examine the feasibility of using this in vitro fish PBL/SCE assay to detect the presence of genotoxic contaminants in the marine environment and to assess the effect of these compounds on resident marine fish.

EXPERIMENTAL PROCEDURE AND RESULTS

The Organism

The Pacific staghorn sculpin (*Leptocottus armatus*), a common bottom-dwelling Puget Sound fish, was selected as the test organism. Most important, an in vitro culture technique has been developed for the PBLs of this species (19), and this fish possesses a karyotype that is suitable for SCE analysis (2n = 42; see Fig. 1). In addition, this coldwater marine fish is hardy, adaptable to laboratory conditions, and available throughout most of the year. A number of other Puget Sound fish were considered but were not appropriate due to unsuitable karyotypes, lack of availability and/or difficulty in keeping them alive.

In Vitro PBL Culture and Genotoxic Exposure

Sculpin PBLs were grown in vitro using the culture conditions established for this species by Zakour et al. (19). Briefly, the major components of this culture system are heparinized whole blood (0.1 ml), fetal bovine serum-supplemented RPMI 1640 culture medium (1.0 ml), and purified protein derivative (PPD) of tuberculin (100 μg/ml) as a mitogen. Cells were cultured for 5-6 da at 13.5°C which is typical of the environmental temperatures for this coldwater marine fish. For genotoxicity studies and SCE analysis, bromodeoxyuridine (BrdUrd) (7.5 μM) was added to established PBL cultures on da 1 to achieve sister chromatid differentiation (SCD) (Zakour, unpublished data); and MNNG (0.15-7.36 μg/ml), a known direct-acting SCE inducer was added on da 3. All solutions and cultures containing BrdUrd were handled in subdued light or in the dark to minimize photodegradation. MNNG was dissolved in dimethylsulfoxide (DMSO) and equivalent amounts of DMSO were added to all cultures except untreated controls which received a similar amount of culture medium.

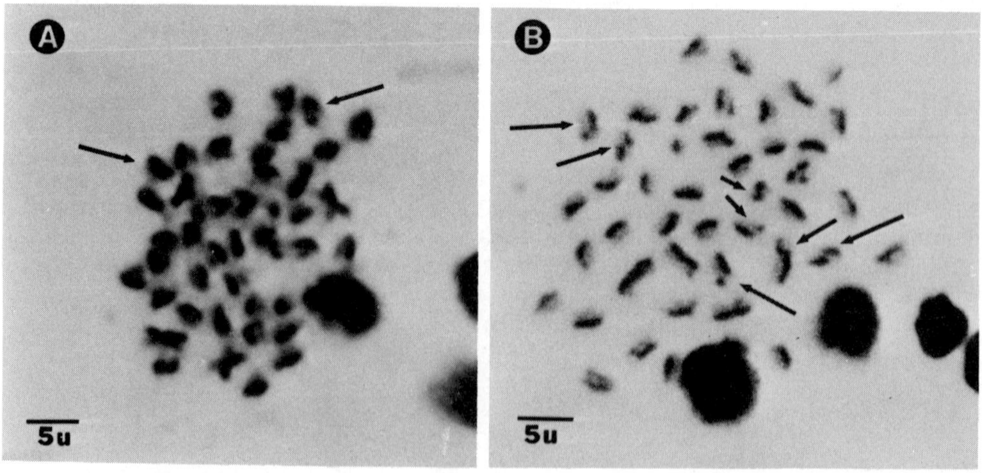

Fig. 1. Metaphase spreads from Pacific staghorn sculpin cultured PBLs showing SCD and SCEs. A. DMSO control; B. In vitro exposure to MNNG (1.47 μg/ml). Arrows indicate some of the SCEs observed.

SCE Analysis

Colchicine (5.5 μg/ml) was added to cultures 6 hr prior to harvest, and metaphase spreads were prepared according to standard cytological techniques (19). The direct Giemsa staining method of Alves and Jonasson (21) was used to visualize SCD, and thus SCEs. This staining procedure was the most reliable method for obtaining SCD in the chromosomes of Puget Sound marine fish (5). Prior to visual examination, all microscope slides were masked and coded. SCEs were counted in second-division metaphases that exhibited SCD in all chromosomes and contained at least 40 chromosomes. Twenty, or all suitable metaphases per replicate per treatment, were analyzed for the presence of SCEs. A green broad-pass substage filter was used to enhance SCD staining. Tests for significant differences between the means of treated and control groups were conducted using Student's t-test (22). To determine whether "outlier" (23) or "high frequency" (24) cells were present, histograms of the number of cells exhibiting each SCE frequency were constructed.

SCE Response of Sculpin PBLs to MNNG In Vitro

Using the method that we developed for the in vitro culture of fish PBLs (19), SCEs could be detected and measured in blood cells of the Pacific staghorn sculpin. SCE induction occurred in sculpin PBLs exposed to MNNG, and representative metaphases showing SCD and SCEs are shown in Fig. 1.

Sufficient cells that met the criteria for SCE analysis were obtained from cultures exposed to 0-2.50 μg MNNG/ml (Tab. 1), even though the frequency of dividing cells was reduced in cells exposed to higher mutagen concentrations. It should be noted that sculpin chromosomes are smaller than many mammalian chromosomes, and thus some SCEs may go undetected, especially very small ones. Yet, for sculpin PBLs, a concentration-related increase in SCE frequency was observed in response to MNNG exposure (Fig. 2). For the lowest concentration tested in that trial (0.25 μg MNNG/ml, Experiment 2, Tab. 1), a 2.5-fold increase in SCE frequency, relative to control values, was observed, and a 4.4-fold increase for the highest concentration at which cell division was evident (5.0 μg MNNG/ml, Experiment 2, Tab. 1). Significant differences ($p \leq 0.05$) were evident between treated and control cultures at exposure concentrations of 0.25-5.00 μg MNNG/ml (Tab. 1). An alternative way to examine the SCE response is shown in Fig. 3. Here the histograms indicate that the precentage of cells exhibiting SCEs and the frequency of outlier (23) or high frequency (24) cells (for our study, cells exhibiting $\geq$ 5 SCEs/cell) increased with exposure to greater amounts of MNNG. Note that for control samples (untreated and DMSO), no outliers were present compared to 11 outliers at 1.25 μg MNNG/ml.

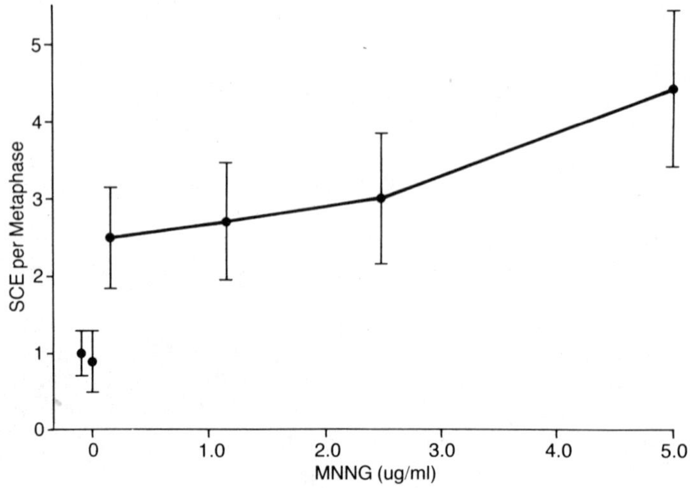

Fig. 2. Dose-response curve for SCE frequencies in Pacific staghorn sculpin cultured PBLs exposed in vitro to MNNG (X ± S.D.). (Also see Tab. 1, Experiment 2).

DISCUSSION

PBLs and the SCE Assay

Cultured PBLs and the SCE assay have been used to monitor in vivo and in vitro exposure to mutagenic and carcinogenic compounds for humans (25,26), rabbits (15,16), rats (17,24), and mice (18). In our study it is shown that the SCE assay can also be adapted for use with PBLs cultured in vitro from the Pacific staghorn sculpin. PBLs of the oyster toadfish (Opanus tau), another marine fish, have also been reported to be suitable for in vitro SCE analysis (12). It is likely that the in vitro PBL/SCE assay which has routinely been used to assess environmental exposure of humans to genotoxic compounds (26) may also provide a feasible method for detecting the presence of mutagenic and carcinogenic contaminants in the marine environment.

In Vitro Sensitivity of Fish PBLs

The sensitivity of sculpin cultured PBLs to the effects of genotoxic compounds was demonstrated by the induction of SCEs in a concentration-related manner following in vitro exposure to a known SCE inducer, MNNG. Also, the percentage of cells exhibiting a relatively high number of SCEs, that is, outlier (23) or high frequency (24) cells, increased in PBL cultures exposed to the higher genotoxicant concentrations. Harrison and Jones (23) report that this parameter may prove to be a more sensitive measure of SCE induction,

Tab. 1. The effects of in vitro exposure to MNNG on SCE frequencies of sculpin cultured PBLs.

CONCENTRATION	NUMBER OF SCEs/NUMBER OF METAPHASES		MEAN SCE METAPHASE ± S.D.	
	Experiment 1[a]	Experiment 2[b]	Experiment 1[a]	Experiment 2[b]
Control	50/40	113/120	1.3 ± 0.9	0.9 ± 0.8
DMSO	47/40	121/120	1.2 ± 0.8	1.0 ± 0.6
MNNG (ug/ml)				
0.15	95/40	---	2.4 ± 1.7	---
0.25	---	220/93	---	2.5 ± 1.3*
1.25	---	214/72	---	2.7 ± 1.5*
1.47	127/40	---	3.2 ± 2.1*	---
2.50	---	201/67	---	3.0 ± 1.7*
5.00	---	43/10	---	4.4 ± 2.0*

[a]Data from one fish.

[b]Data from three fish.

*Significantly different from control; Student's t test, $p \leq 0.05$.

and thus genotoxic exposure, than the mean SCE frequency. Cultured blood cells from the oyster toadfish exposed in vitro to another alkylating agent, ethylmethanesulfonate (EMS), responded with increased SCE frequencies (12). These 2 studies indicate that the SCE assay can be adapted for use with cultured PBLs from marine fish, and that this cytogenetic assay can be used to measure in vitro exposure of fish blood cells to known genotoxic compounds. These studies differ, however, in the number of cells in which SCEs could be analyzed, an important parameter for the statistical evaluation of SCE data. The present study of sculpin PBLs employed culture conditions which had specifically been developed for this species and in vitro cytogenetic assays (19). This culture technique provided a sufficient number of cells that exhibited SCD and an ample sample size for statistical treatment of the data. In contrast, the culture method reported for oyster toadfish PBLs did not supply an adequate number of dividing cells for statistical analysis, since SCEs could only be analyzed in a small number of cells (a total of 50 cells from 13 cultures) (12). This demonstrates the importance of selecting a reliable in vitro culture system for in vitro cytogenetic analysis of fish PBLs.

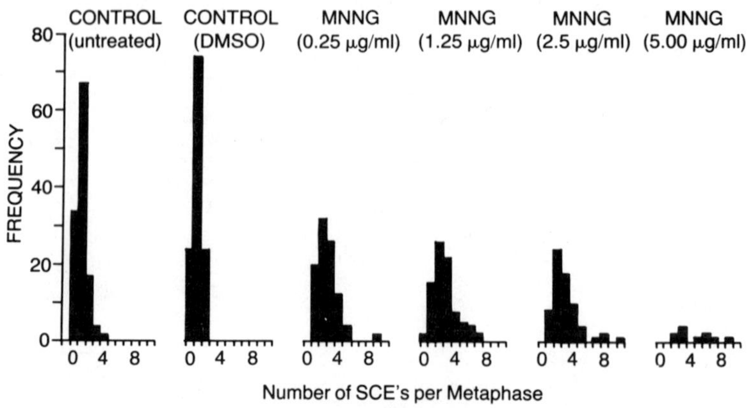

Fig. 3. Histograms showing distribution of metaphases with the indicated number of SCEs after in vitro exposure to the indicated concentration of MNNG. Note the presence of "outliers" (cells with $\geq$ 5 SCEs for this study) at the higher mutagen concentrations. Data from 3 fish (see Tab. 1, Experiment 2).

In Vitro Culture of Fish PBLs

A general approach has not yet been achieved for the in vitro culture of fish PBLs. Most attempts to culture these cells have been unsuccessful or have yielded inconsistent results (17). These culture methods have recently been reviewed by Zakour et al. (19). Foremost, fish PBLs do not appear to respond to mitogenic stimulation as readily as mammalian PBLs (17,19). Nevertheless, in vitro culture of fish PBLs has been used to obtain cells for karyotype analysis (27-30) and to investigate the ontogeny of fish leukocytes (31-34). However, it is uncertain whether the culture methods used for these studies which do not require many dividing cells could also provide a sufficient number of dividing cells for performance of in vitro cytogenetic assays such as the SCE assay. Therefore, a reliable culture method for fish PBLs that can adequately stimulate cell division is a prerequisite for in vitro cytogenetic analysis of this tissue.

To establish the in vitro culture technique mentioned above for sculpin PBLs, a variety of culture parameters were investigated for their ability to stimulate or enhance in vitro cell division in fish blood cells (19). The 2 major factors that affected the division and growth of sculpin PBLs were the type of mitogen and culture medium used. Furthermore, there were interactions between these 2 components of the culture system. Purified protein derivative of tuberculin, which is considered to be a mammalian B-cell mitogen

(35), stimulated sculpin PBLs to divide in vitro (19). Maximal cell division was observed at ∿100 μg PPD/ml, but individual fish varied in the responsiveness of their PBLs to this mitogen. Of the other compounds tested which included phytohemagglutinin, pokeweed mitogen, concanavalin A and lipopolysaccharide, none was stimulatory for sculpin PBLs (19), even though some of these compounds are mitogenic for PBLs of other fish species (see 19) and various mammals (13). The type of culture medium in which the cells were grown influenced the mitogenic activity of PPD for sculpin PBLs (19). RPMI 1640, an enriched medium, supported PPD-stimulated cell division in sculpin PBLs at a level that was sufficient for conducting the SCE assay (present study). The response of these fish cells to PPD was less evident when Leibovitz's L-15 medium, a less enriched culture medium, was used (19). Also, minimal essential medium failed to support cell division and growth at any of the PPD concentrations tested (19). In addition, other culture parameters such as type and amount of serum supplement, cell preparation and incubation temperature affected the division and growth of these fish cells (19). The culture conditions established for sculpin PBLs were also stimulatory for blood cells from the starry flounder (Platichythys stellatus), a marine fish, and the goldfish (Carassius auratus), a freshwater species, but not for 2 other marine fish, the Pacific sanddab (Citharichthys sordidus) and the speckled sanddab (C. stigmeaus) (Zakour, unpublished data).

Recently, another culture method which is suitable for investigating cytogenetic endpoints in vitro has been developed for PBLs of the central mudminnow (Umbra limi), a freshwater species (20). Chromosome aberrations were scored in blood cells grown in the presence of TC-199 culture medium, phytohemagglutinin and an atmosphere of 100% oxygen. Sister chromatid differentiation was achieved in these cultured cells (20), and it is likely that this culture technique could be used with the SCE assay. However, the effects of various culture parameters on the division and growth of mudminnow PBLs in vitro were not examined.

Sensitivity of Fish Cells to Genotoxic Agents

The sensitivity of fish cells to genotoxic compounds has been demonstrated in both in vitro and in vivo studies; selected data from these studies are summarized in Tab. 2. Alkylating agents such as MNNG and EMS induced SCEs in vitro in sculpin PBLs (present study), oyster toadfish PBLs (12), and transformed cells derived from embryonic tissue of Ameca splendens, a freshwater fish (13). Fish cells exposed in vitro may be less sensitive to the effects of genotoxic compounds than many cultured mammalian cells (Ref. 13 and present study). However, because of the paucity of data for in vitro fish systems, such comparisons with mammalian systems are limited.

Tab. 2. SCE investigations of fish.

ORGANISM	CELLS ASSAYED	EXPOSURE CONDITIONS	SCE/CELL	REFERENCE
I. In Vitro Exposure:				
Pacific staghorn sculpin	PBLs	Control	1.0	Present study
		MNNG-0.25 ug/ml	2.5	
		MNNG-5.00 ug/ml	4.4	
Oyster toadfish	PBLs	Control	7.0	12
		EMS-62 ug/ml	21.3	
		EMS-124 ug/ml	35.1	
Ameca splendens	fibroblast?[a]	control	4.8	13
		MNNG-0.15 ug/ml	11.8	
		MNNG 1.47 ug/ml	30.2	
II. In Vivo Exposure:				
English sole	anterior kidney[b]	control	2.2	5
		BaP-0.5 ug/g wet wt	3.0	
		BaP-5.0 ug/g wet wt	5.9	
		Puget Sound-"pristine"[c]	2.2	
		"polluted"[c]	3.3	
central mudminnow	gill	control	3	8
		MMS-10 ug/ml	4	
		MMS-65 ug/ml	10	
Eastern mudminnow	gill	control	1.4	11
		Rhine River water-3 da	2.8	
		Rhine River water-11 da	3.4	
Eastern mudminnow	gill	control	2.2	9
		Rhine River water-8 da	8.4	
Notobranchius rachowi	gill	control	1.6	10
		EMS-120 ug/ml	11.6	

[a]transformed cells derived from embryonic tissue

[b]in fish this tissue is hematopoietic

[c]"pristine" site received little industrial input, whereas the "polluted" site was near a heavily industrialized area (the Duwamish River)

Fish are also sensitive to the effects of genotoxic agents in vivo. Kligerman (18) demonstrated SCE induction in the freshwater fish, the central mudminnow, following in vivo exposure to the known genotoxic compounds--EMS, cyclophosphamide (CP) and waterborne neutral red dye. Other SCE studies based on in vivo exposures have also shown that freshwater (9-11) and marine (5) fish are sensitive to various genotoxic compounds such as benzo(a)pyrene, EMS, and CP.

Most important, fish are sensitive to known and unknown environmental contaminants (5,9-11). Environmental exposure of English sole (Paraphrys vetulus) to unknown genotoxic compounds was monitored by measuring SCE induction in hematopoietic tissue (anterior kidney) of this marine fish (5). This study indicates that SCE induction in resident fish could be used to detect the presence of genotoxic contaminants in the marine environment. English sole collected from areas that received industrial effluents and other man-made pollutants exhibited SCE frequencies that were significantly greater than those of fish collected from relatively unpolluted areas (5). Likewise, increased SCE frequencies have been reported for Eastern mudminnows (Umbra pygmaea) experimentally exposed to Rhine River water that contained various organic pollutants (9,11). Also, the degree of SCE induction was related to the time of exposure to Rhine River water (11).

Advantages of Using the PBL/SCE Test System

The PBL/SCE assay has a number of advantages compared to other methods that have been used to study the induction of SCEs in fish. The first advantage is that the experimental animal is not sacrificed for tissue collection since PBLs can be obtained by nondestructive means. Therefore, multiple samples can be collected from a single individual, each organism can serve as its own control, and the temporal sequence of the induction and persistence of SCE formation can be investigated in individuals and the sample population. Using each organism as its own control should reduce the effect of genetic differences between fish (Ref. 12 and present study), an aspect that is especially important when examining feral populations that inherently exhibit a large degree of variation. Second, since fish are exposed to the genotoxic agent in vivo, the uptake, metabolic and elimination capabilities of the organism are maintained intact. Third, PBLs circulate throughout the body; thus, these cells may be exposed to a variety of metabolite intermediates as well as to the parent compound. In addition, the exposure method can be selected to correspond to a known or suspected route of exposure; environmental exposure could also be monitored. Therefore, the presence of unidentified contaminants, especially those from low-level and/or nonpoint sources, may be detected. Finally, the conditions needed to achieve SCD, and therefore SCE detection, can be controlled more easily in vitro than in vivo.

Difficulties in Applying the PBL/SCE Assay to Fish

There are two major difficulties encountered in applying the PBL/SCE assay to fish. These are the lack of reliable PBL culture methods for many fish species and the typical fish karyotype which in many cases is unsatisfactory for SCE analysis. Culture methods that are suitable for in vitro cytogenetic analysis of fish PBLs are currently available for only 2 species, the Pacific staghorn sculpin (Ref. 19 and present study), a marine fish, and the central mudminnow (20), a freshwater species. However, both of these methods have been developed recently and the results of these studies should stimulate further research in this area. It should be possible to apply this assay to any fish species whose blood cells can successfully be cultured in vitro. Secondly, fish tend to have karyotypes that consist of a large number of small chromosomes. This makes enumeration of SCEs tedious and time consuming, and small exchanges may not be detected. However, a number of fish possess chromosomes that are few in number and relatively large in size, and these species are the ones most suitable for SCE studies. Among these fish are marine species such as the Pacific tomcod (Microgadus proximus) (2n = 24), the Pacific sanddab (2n = 20), and the speckled sanddab (2n = 20) (Zakour, unpublished data), and freshwater fish including Ameca splendens (2n = 26), the central (2n = 22) and Eastern (2n = 22) mudminnows, and several other species noted by Kligerman (7).

Method Validation

Presently, the PBL/SCE assay has not been validated for fish cells. In order to establish the sensitivity and reliability of this method and to determine whether environmental genotoxic contaminants would be detected by this technique, a variety of compounds including ones that induce and do not induce SCEs in other systems should be tested with fish PBLs. Nevertheless, the demonstration of the sensitivity of fish cells to genotoxic compounds, both in vivo and in vitro, and the ability to measure SCE induction in fish cultured PBLs strongly suggest that studies be undertaken to determine the validity of this test system. Fish PBL/SCE analysis may provide an important indicator of genotoxic contaminants in the aquatic environment.

Applications for the Fish PBL/SCE Test System

The fish PBL/SCE test system may provide a practical method for detecting genotoxic contaminants in marine and freshwater environments. Human consumption or contact with water supplies contaminated by mutagenic and carcinogenic substances from various sources, such as chemical waste dumps and industrial effluents, could pose a serious health problem. Since fish are in direct contact with water- and sediment-borne contaminants, the detection of genetic damage in these organisms by the SCE assay could indicate the presence of genotoxic compounds in their environment. The in vivo exposure-in vitro assay system for fish PBLs could possibly be used to detect pollutants that might otherwise go undetected. Among these substances are trace contaminants, genotoxic compounds from nonpoint sources, and mutagens and carcinogens that are unknown or unsuspected, including genotoxic products that are formed in the environment.

Thus, this nonselective assay could function as a monitoring system for the detection of potential mutagenic and carcinogenic contamination in public and private water supplies. Once the presence of genotoxic substances has been recognized, studies to identify and locate the source of the contaminants could be initiated. Notably, the capability to measure genetic damage in the same individual at different times is an important aspect of this test system, for it provides a means to assess the impact of changes in the environment, such as dredging and effluent treatment, on fish and water quality.

SUMMARY

A cytogenetic method that utilizes the SCE assay and a newly developed in vitro culture method for marine fish PBLs has been described for detecting exposure of marine fish to genotoxic agents. Sculpin cultured PBLs responded to in vitro exposure to MNNG with a concentration-related increase in SCE frequencies. Also the number of "outlier" (23) or "high frequency" (24) cells increased with mutagen exposure. Using this PBL/SCE test system, both in vivo and in vitro exposure to experimental and environmental substances may be monitored. However, selection of an appropriate culture method for fish PBLs is fundamental to the usefulness and success of this technique. The PBL/SCE assay described herein has several advantages, when compared to other SCE methods used to study genotoxic damage in fish, that enable one to monitor the induction and persistence of cytogenetic damage in individual fish under different exposure conditions across time. This test system is also potentially useful for determining the presence of genotoxic contaminants in the marine and freshwater environments including those water supplies used by man.

ACKNOWLEDGEMENTS

This research was supported in part by grants from NIEHS (5-T32-ES07032-05, training grant, HRZ; and 5-P30-ES-02190-02) and the Office of Marine Pollution Assessment, NOAA (NA80RAD00053).

REFERENCES

1. Beardmore, J., C. Barker, B. Battaglia, R. Berry, A. Longwell, J. Payne, and A. Rosenfield (1980) The use of genetic approaches to monitoring biological effects of pollution. Rapp. P.-V. Reun. Cons. Int. Explor. Mer. 179:299-305.
2. Payne, J., and I. Martins (1980) Monitoring for mutagenic compounds in the marine environment. Rapp. P.-V. Reun. Cons. Int. Explor. Mer. 179:292-298.

3. Latt, S., R. Schreck, K. Loveday, and C. Shuler (1979) In vitro and in vivo analysis of sister chromatid exchange. Pharmacol. Rev. 30:501-535.
4. Perry, P. (1980) Chemical mutagens and sister chromatid exchange. In Chemical Mutagens: Principles and Methods for Their Detection, F. de Serres and A. Hollaender, eds. Plenum Press, New York, Vol. 6, pp.1-39.
5. Stromberg, P., M. Landolt, and R. Kocan (1981) Alterations in the frequency of sister chromatid exchanges in flatfish from Puget Sound, Washington, following experimental and natural exposure to mutagenic chemicals. NOAA Technical Memorandum, Office of Marine Pollution Assessment, Boulder, Colorado. 43 pp.
6. NCI/EPA Collaborative Symposium: The Use of Small Fish Species in Carcinogenicity Testing. Washington, D.C., December, 1981 (unpublished).
7. Kligerman, A. (1980) The use of aquatic organisms to detect mutagens that cause cytogenetic damage. In Radiation Effects on Aquatic Organisms, E. Egami, ed. Japan Science Soc. Press, Tokyo, pp. 241-252.
8. Kligerman, A. (1979) Induction of sister chromatid exchanges in the central mudminnow following in vivo exposure to mutagenic agents. Mutat. Res. 64:205-217.
9. Hooftman, R., and G. Vink (1981) Cytogenetic effects on the Eastern mudminnow, Umbra pygmaea, exposed to ethyl methansulfonate, benzo(a)pyrene and river water. Ecotoxicol. Environ. Safety 5:261-269.
10. van der Hoevan, J., I. Bruggeman, G. Alink, and J. Koeman (1982) The killifish Notobranchius rachowi, a new animal in genetic toxicology. Mutat. Res. 97:35-42.
11. Alink, G., E. Frederix-Wolters, A. van der Gaag, J. van der Kerkhoff, and C. Poels (1980) Induction of sister-chromatid exchanges in fish exposed to Rhine water. Mutat. Res. 78:369-374.
12. Maddock, M., and J. Kelly (1980) A sister chromatid exchange assay for detecting genetic damage to marine fish exposed to mutagens and carcinogens. In Water Chlorination: Environmental Impact and Health Effects, R. Jolley et al., eds. Ann Arbor Sci. Publ., Inc., Ann Arbor, Michigan, pp. 835-844.
13. Barker, C., and B. Rackham (1979) The induction of sister-chromatid exchanges in cultured fish cells (Ameca splendens) by carcinogenic mutagens. Mutat. Res. 68:381-387.
14. Loeb, L. (1974) Molecular analysis of lymphocyte transformation. In Developments in Lymphoid Cell Biology, A. Gottlieb, ed. CRC Press, Cleveland, Ohio, pp. 104-131.
15. Stetka, D., and S. Wolff (1976) Sister chromatid exchange as an assay for genetic damage induced by mutagens-carcinogens. I. In vivo test for compounds requiring metabolic activation. Mutat. Res. 41:333-342.

16. Stetka, D., and S. Wolff (1976) Sister chromatid exchange as an assay for genetic damage induced by mutagens-carcinogens. II. In vitro test for compounds requiring metabolic activation. Mutat. Res. 41:343-350.
17. Kligerman, A., J. Wilmer, and G. Erexson (1981) Characterization of a rat lymphocyte culture system for assessing sister chromatid exchange after in vivo exposure to genotoxic agents. Environ. Mut. 3:531-543.
18. Erexson, G., J. Wilmer, and A. Kligerman (1983) Analyses of sister-chromatid exchange and cell-cycle kinetics in mouse T- and B-lymphocytes from peripheral blood cultures. Mutat. Res. 109:271-281.
19. Zakour, H., M. Landolt, and R. Kocan. An in vitro culture technique for the peripheral blood leukocytes of a coldwater marine fish, the Pacific staghorn sculpin (Leptocottus armatus). (Submitted for publication).
20. Suyama, I., and H. Etoh (1983) X-ray-induced dicentric yields in lymphocytes of the teleost, Umbra limi. Mutat. Res. 107:111-118.
21. Alves, P., and J. Jonasson (1978) New staining method for the detection of sister-chromatid exchanges in BrdU-labelled chromosomes. J. Cell Sci. 32:185-195.
22. Snedecor, G., and W. Cochran (1967) Statistical Methods. The Iowa State University Press, Ames, Iowa, 593 pp.
23. Harrison, F., and I. Jones (1982) An in vivo sister chromatid exchange assay in the larvae of the mussel Mytilus edulis: Response to 3 mutagens. Mutat. Res. 105:235-242.
24. Kligerman, A., G. Erexson, M. Phelps, and J. Wilmer (1983) Sister-chromatid exchange induction in peripheral blood lymphocytes exposed to ethylene oxide by inhalation. Mutat. Res. 120:37-44.
25. Lambert, B., and A. Lindblad (1980) Sister chromatid exchange and chromosome aberrations in lymphocytes of laboratory personnel. J. Toxicol. Environ. Health 6:1237-1243.
26. Buckton, K., and H. Evans (1982) Human peripheral blood lymphocytes. In Cytogenetic Assays of Environmental Mutagens, T. Hsu, ed. Allanheld, Osmun and Co., Totowa, New Jersey. pp. 183-202.
27. Ojima, Y., S. Hitotsumachi, and M. Hayashi (1970) A blood culture method for fish chromosomes. Japan. J. Genet. 45:161-162.
28. Heckman, J., F. Allendorf, and J. Wright (1971) Trout leukocytes: growth in oxygenated cultures. Science 173:246.
29. Kang, Y., and E. Park (1975) Leukocyte culture of the eel without autologous serum. Japan. J. Genet. 50:159-161.
30. Thorgaard, G. (1976) Robertsonian polymorphism and constitutive heterochromatin distribution in chromosomes of the rainbow trout (Salmo gairdneri). Cytogenet. Cell Genet. 17:174-184.
31. Etlinger, H., H. Hodgins, and J. Chiller (1976) Evolution of the lymphoid system. I. Evidence for lymphocyte heterogeneity

in rainbow trout revealed by organ distribution of mitogenic responses. J. Immunol. 166:1547-1553.
32. Chilmonczyk, S. (1978) In vitro stimulation by mitogens of peripheral blood lymphocytes from rainbow trout (Salmo gairdneri). Annal Immunol. 129:3-12.
33. Lopez, D., M. Sigel, and J. Lee (1974) Phylogenetic studies on T-cells. I. Lymphocytes of the shark with differential response to phytohemagglutinin and concanavalin A. Cell. Immunol. 10:287-293.
34. McKinney, E., G. Ortiz, J. Lee, M. Sigel, D. Lopez, R. Epstein, and T. McLeod (1976) Lymphocytes of fish: multipotential or specialized? In Phylogeny of Thymus and Bone Marrow-Bursa Cells, R. Wright and E. Cooper, eds. Elsevier/North Holland Biomedical Press, Amsterdam, The Netherlands, pp. 73-82.
35. Sultzer, B., and B. Nilsson (1972) PPD-tuberculin -- a B-cell mitogen. Nature (New Biol.) 240:198-200.

SISTER CHROMATID EXCHANGE STUDIES IN THE CHICK EMBRYO AND NEONATE: ACTIONS OF MUTAGENS IN A DEVELOPING SYSTEM

Stephen E. Bloom

Institute for Comparative and Environmental Toxicology
and Department of Poultry and Avian Sciences
Cornell University
Ithaca, New York 14853

ABSTRACT

The embryonic and neonatal periods represent times when the disease process may be initiated as a result of exposure to environmental mutagens and teratogens. We are using the chick embryo/neonate as an experimental system to detect and study the genotoxicity of environmental chemicals in developing tissues and the resultant biological alterations in survivors of perinatal chemical exposure. In vivo bromodeoxyuridine (BrdUrd) labeling of replicating DNA has been employed to measure basal and induced sister chromatid exchanges (SCEs), a candidate cytogenetic endpoint in genetic toxicology testing. Additionally, SCE induction studies with model promutagens have permitted the detection and study of components of the developing mixed-function oxidase (MFO) system of the liver and other organs. The relationship between specific MFO enzyme induction and SCE generation by promutagens has been studied in ovo and using in vitro assays.

The in vivo SCE induction potential of 53 compounds, including known mutagens and nonmutagens, was evaluated in the early chick embryo. About 90% of the mutagens induced SCEs; all nonmutagens failed to induce SCE above baseline. Clastogens such as bleomycin did not induce SCE but did cause massive chromosome damage that was easily detected. Gentian violet (GV) is a direct-acting clastogen that did not show any SCE induction. This agrees with the findings from in vitro mutation assays that incorporate rat liver S-9 preparations. Potency for inducing SCE in the chick embryo correlates well with true mutagenic potency, DNA inhibition, and to some extent with carcinogenic activity.

Indirect-acting mutagen-carcinogens induced SCEs and also unscheduled DNA synthesis (UDS) in embryonic cells. Biochemical studies revealed that aryl hydrocarbon hydroxylase (AHH) activity develops in the early embryonic liver as it is first formed at 4-5 da of incubation. The level of AHH activity is sufficient to account for the dramatic SCE response. Enhanced SCE induction occurred in older stage embryos, correlating with the increased basal AHH level and enhanced induction capacity of the liver. Modulation of the MFO enzyme system with specific inducers resulted in altered SCE and UDS responses in vivo and in vitro using a chick microsome/chinese hamster ovary (CHO) mammalian cell assay.

Viable embryos and neonates have been obtained following exposure to SCE-inducing levels of the mutagen-carcinogen aflatoxin B_1 applied at either 6 da or 12 da of development. Phenotypically "normal" chicks demonstrated deficiencies in the hematolymphoid system and in postnatal growth potential. The postnatal biological outcome was correlated to some degree with the developmental/differentiation status at the time of toxicant exposure. Thus, baseline and induced SCEs (if they occur in the absence of the BrdUrd probe) in the above instance are not associated with gross disturbances of development. Rather, any induced cellular alterations are subtle and expression is delayed.

The chick embryo should be useful for the further study of SCEs in a developing system, the effects of chemicals on tissue-specific SCE induction, the role of MFO enzymes in mediating the production of DNA-damaging and SCE-inducing metabolites, and for studying the cellular and developmental consequences of embryo exposure to genotoxic environmental chemicals.

INTRODUCTION

Genetic alteration underlies much of animal and human disease. This thesis continues to be central in investigations of the causes and cellular mechanisms involved in the development of cancer and birth defects (1-6). Additionally, there is increasing evidence that the genome changes leading to various disease states are initially induced either in the embryonic or neonatal stages. In certain animal model systems, specific cancers can be induced in high frequency following prenatal exposure to chemical mutagens (7,8).

Malignant cancer represents the second leading cause of mortality in children. About 80% of these cancers are leukoses, lymphomas, tumors of the nervous system, and kidney neoplasias (9). Such cancers are most likely initiated in the perinatal period, since latency periods of expression of cancers in man are typically long (10-40 yr). A proportion of childhood cancers can be attributed to inherited factors often including structural chromosome aberrations

(10-13). However, a significant proportion of these same childhood malignancies (e.g., 40% of retinoblastoma cases) arises from newly generated genetic aberrations. The de novo production of genetic aberrations is also seen in the context of the prenatal and neonatal genetic screening. In a more general sense, human disease is seen that cannot be readily attributed to any particular cause. One question is, then, are de novo forms of cancer, birth defects, and other diseases linked to specific factors present in the environment or are they truly "spontaneous"? Are some of the chemicals prevalent in our environment responsible for a portion of de novo genetically based disease? If so, when in the life cycle are the hazards the greatest?

At present there is a dearth of information concerning the periods (and cellular events) in the life cycle of an organism when exposure to a chemical is most apt to cause the form(s) of cellular damage, such as mutation, that result(s) in cancer and birth defects. There is certainly ample evidence to suggest that the embryo and neonate are sensitive to the immediate damaging effects of chemicals, but less known and explored is the study of phenotypically "normal" survivors. That is, some chemicals (at certain dosages) are not highly teratogenic. The biological response is manifest later in life. At present, it is not clear how initial cellular damage (or what forms of damage) remains largely unexpressed until the postnatal period. These and other questions concerning the biological hazards of environmental mutagens on developing systems need to be addressed. In addition, there is a need for assays that can be used to screen compounds for mutagenic and carcinogenic potential in developing systems. Although knowledge of cellular and molecular mechanisms leading to cancer and other diseases is still incomplete, judgements still have to be made concerning potential/probable hazards of chemicals to man and animal. In this context, SCEs have emerged as an important component of assays for cytogenetic damage (14). Even though SCEs do not entail wholesale breakage of the chromosome, they nevertheless indicate the interaction of a chemical, or its activated metabolites, with chromosomal DNA. The power of the SCE endpoint is in its ready application to a wide variety of in vitro and in vivo genetic assay systems (15). We have used BrdUrd-based SCE detection to study mutagen-induced DNA damage in developing chick embryos. As I will discuss, SCE analysis has allowed: 1) the detection of actions of mutagens in embryos at defined periods of development, 2) the discrimination of mutagens from nonmutagens in vivo, 3) the detection of the embryonic MFO system that can activate and detoxify chemicals, and 4) a quantitative measure of genome attack in studies to determine the long-term biological hazards of prenatal chemical insult. Studies with the chick embryo cytogenetic system have provided new information on the nature of SCEs and the use of this chromosomal endpoint in genetic toxicology testing (16-18). There is also the suggestion of the possible use of this BrdUrd-based SCE in vivo system to further investigate the role of induced cellular damage in embryonic cells in the production of diseases (such as cancer and birth defects) expressed in the neonatal period or even later in life (19).

DETECTION OF SCE IN EMBRYONIC CELLS IN VIVO

The first in vivo demonstration of SCEs was in allantoic and limb bud cells from the 4-da chick embryo (20). The 2-round pattern of differentiation of sister chromatids (SCD) was achieved following a single application of BrdUrd onto the inner shell membrane (ISM) and incubation for 26 hr, which permitted the completion of 2 full cell cycles. SCEs were also demonstrated in various tissues of the central mudminnow (*Umbra limi*) after a 5-da in vivo exposure to BrdUrd (21). Results with these 2 systems indicated the possible general use of BrdUrd labeling for studying SCE in vivo as well as showing the low baseline rate of SCE in various tissues. The concept that SCE could be studied at the embryonic stage as well as in adults was applied to other species including mammals (22-26).

Since our first report in 1975, the chick embryo system (CES) has been developed further and used extensively in studying the genotoxic activity of different types of chemicals (16-18). Since detailed procedures have been reported previously, only a brief outline will be given here.

Sources and Handling of Chick Embryos

The Cornell K-strain (27) is used as the source of fertile eggs. Uniform eggs are produced that average 59 gm in weight. Fertility and hatchability are both over 90%, and this genetic strain is highly resistant to Marek's disease.

Eggs are set in a Robbins incubator with temperatures at 37.5°C and humidity at 85%. Eggs are mechanically rotated hourly in this forced-air incubation system. Prior to incubation, fertile eggs are stored at 15°C for not more than 7 da. The general procedure is to initiate incubation of a batch of eggs (usually 90) at a given time for the experiment. Incubation time is counted from the moment of setting in the incubator to the time of removal. Details of the course of development are well-known for the chick.

Schedules and Administration of Chemical Reagents

Chemical compounds are applied to the chick embryo at 3-4 da of incubation (DI) (Fig. 1). At this stage, normal embryos are easily identified by candling, and this material is excellent for cytogenetic analysis. Also, dramatic changes in embryo size and form occur between da 3 and 4, facilitating the detection of toxic effects.

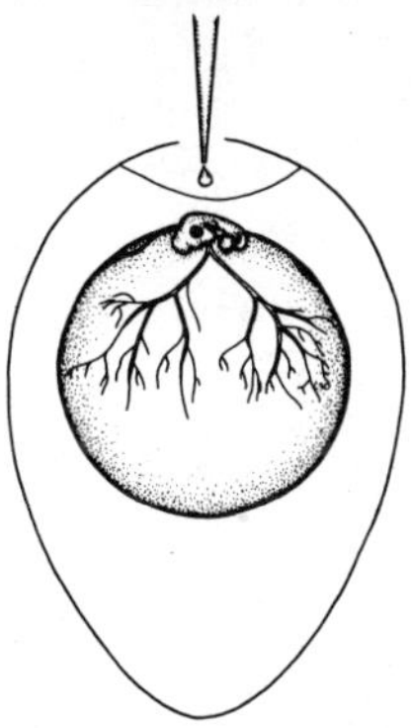

Fig. 1. Air-cell method for applying chemicals to the chick embryo. Solutions are pipetted onto the inner shell membrane.

In the most recent studies, older embryos (6-7 DI) were used to test the theory that, with a more developed organ system, a greater genotoxic response to indirect-acting mutagens would be obtained.

Previously, we reported 2 methods of application of chemical solutions to chick embryos: the air-cell method and the window method (16,20). In our evaluation of chemicals, we chose the former method to avoid unnecessary trauma to the CES. In the air-cell method, applied to the early and late CES, a 0.5-cm^2 portion of shell overlying the air cell is removed with forceps and the test solution is dropped onto the ISM from an Eppendorf micropipette (Fig. 1). The hole is covered with 1/2"-wide Johnson & Johnson waterproof adhesive tape. Amounts in the range of 10-100 μl are used. Since the embryo is positioned just below the ISM (vertical position), solutions are rapidly absorbed by the embryo proper, the surrounding vascular network, or both.

Dosing the embryo. The early embryo can tolerate small amounts of various solvents such as water, Hanks balanced salt solution (HBSS), 15-30% ethanol, and acetone. For most work 50-100 μl injections are given, but with acetone 10 μl is the maximum amount. These volumes do not interfere with growth over the experimental period, and successive injections can be given without inducing undue trauma.

A series of chemical treatments is applied to the embryo sequentially and at precise times (Fig. 2) (17,18). For chromosome-breakage studies, the test chemical is given to the 74-hr embryo, and 2 hr prior to sacrifice (hr 94) colcemid is applied to arrest

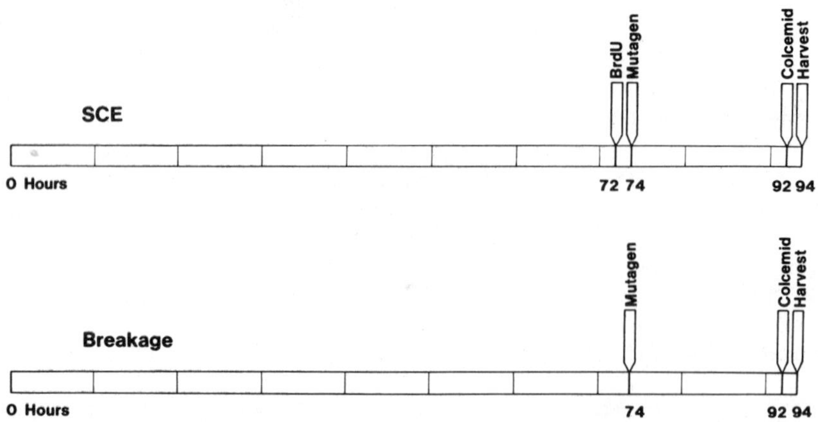

Fig. 2. Schedules for performing SCE and chromosome breakage assays in the chick embryo. Scale is in hr of development.

cells at metaphase. For SCE studies, BrdUrd is injected before the mutagen to produce sister chromatid differentiation (SCD).

Test chemicals. Some 53 chemical compounds have been evaluated in the CES. Among these, 21 are known mutagens, 18 are nonmutagens, and 14 are miscellaneous but probable nonmutagens (18). Of the mutagens, 13 are known carcinogens. All compounds were evaluated in the 3-4 da CES and others, including ethylmethanesulfonate (EMS), the flame retardants tris-BP (28) and Fyrol FR2 (29), 2-acetylaminofluorene (2-AAF), and aflatoxin B_1 (AFB_1), were also tested at the 6- to 7-da stage. All solutions were made fresh before use and compounds were dissolved in the appropriate solvents at various concentrations (18). A wide variety of types of chemicals was selected, representing different modes of action on DNA including alkylation, intercalation, and base substitution. Both direct-acting compounds and those requiring metabolic activation for action were included. Among the nonmutagens, solvents, tissue culture components, antibiotics, buffers, and stains were tested (18).

Use of BrdUrd to Reveal SCEs In Vivo

The interaction of chemical mutagens with the eukaryotic genome can be monitored through quantitative analysis of SCEs (30, 31). If mitotic cells are exposed to BrdUrd for 2 cell cycles, the chromatids show SCD after staining with fluorochromes (32-34) or special Giemsa methods (35-39). With these procedures, SCEs are clearly demarcated. Chemical mutagens are known to cause significant increases in the rate of SCEs (15,32,40). Advantages of the BrdUrd-SCD technique include: 1) ease in identifying and scoring SCEs, 2) relatively small number of cells are needed to detect a response to mutagens, 3) simultaneous measurement of SCE rate and cell cycle time, and 4) sensitivity of the BrdUrd technique to low levels of chemicals.

BrdUrd is applied to the chick embryo by the air-cell method (Fig. 1). An optimal injection of 100 µg is given to deliver BrdUrd concentrations of 150 µl and 1 mg for the 3-4 DI and 6-7 DI systems, respectively. Very dramatic SCD is obtained at 150 µg BrdUrd for both the fluorescence and Geimsa methods. Higher levels do not improve contrast between chromatids. The typical SCD pattern will occur only if cells are exposed to BrdUrd for 2 cell cycles and hence 2 complete S periods. Shorter exposure produces replication patterns, and if a third S period is included, a typical 3-cycle pattern is revealed. If the cell population under study is mostly synchronous, numerous metaphases with SCD will be recovered. This facilitates scoring of SCEs. Interestingly enough, we find a high proportion (about 70%) of metaphases with SCD in limb bud and allantoic cells from 3-da chick embryos exposed to BrdUrd for 22 hr. Chromosome preparations are made by a modified solid tissue technique and stained with either 33258 Hoechst or by a Giemsa protocol (18).

<u>Analysis of Chromosome Preparations</u>

The diploid complement of <u>Gallus domesticus</u> contains 78 chromosomes, ranging in size from about 8 µm to barely visible by light microscopy (41-45). The sex-determining mechanism is ZZ(♂)-ZW(♀). The first 5 pairs (macrochromosomes) are large and distinctive enough to permit identification and detection of aberrations without the aid of a photokaryotype (16,20,46,47). Pairs 1-5 (including the Z sex chromosome) contain about 50% of the nuclear DNA per cell. In photographs, pairs 6-10 can also be identified. The W sex chromosome, present only in females, is similar in size and shape to chromosomes in pair 9. When the C-banding technique is used, however, the W chromosome is differentially stained.

Scoring of SCEs is accomplished by counting the number of exchange points on each macrochromosome (pairs 1-5) in the metaphase. For routine screening studies, we calculate SCEs per metaphase and incorporate equal numbers of male (ZZ) and female (ZW) embryos in the control and treated classes.

One problem is scoring SCEs at the centromere. Many such exchanges turn out to be subcentromeric on careful examination. In other cases, it is clear that each chromatid is continuous on one side through the centromere. In still other cases, it is not clear whether there is a centromeric SCE or a twisting of the chromatids. When there is such doubt, we do not score an SCE.

<u>Baseline rate of SCE.</u> The spontaneous rate of SCE has been studied in embryos from 5 Cornell experimental chicken strains (C,

K, S, N, and P) and one commercial source (Rich-Glo, El Campo, Texas). The mean background rate of SCE measured in various experiments is 1.2 exchanges per metaphase (range: 0.8-3.0 SCEs/metaphase). No strain differences in SCE are apparent.

SCE DISCRIMINATES MUTAGENS FROM NONMUTAGENS

Using the SCE and breakage cytogenetic assay, chemical mutagens were discriminated from nonmutagens with a high degree of accuracy. However, SCE analysis was far more reliable an indicator of mutagenicity than chromosome breakage. SCEs were easy to score, and the presence of mutagens could usually be determined from examination of just a few cells. Such cells were characterized by the presence of multiple exchanges within each macrochromosome. Of the 21 mutagens, only 6 were classified as definite clastogens, 5 as weak clastogens, 5 as nonclastogens, and 5 not determined (17). With SCE analysis, 80% of the mutagens were detected (Tab. 1). With both SCE and breakage, 90% of the mutagens were detected. All 18 nonmutagens and 14 miscellaneous (probably nonmutagenic) chemicals failed to induce SCEs or breakage in our study (18).

Of the 3 mutagens that failed to induce SCE, bleomycin (BLM) caused chromosome breakage, Fyrol FR2 is a weak mutagen in the Ames Salmonella test, and GV is a clastogen in vitro but is converted to a nonomutagenic, but cytotoxic, form after metabolism (48,49). Whether Fyrol FR2 is a mutagen and carcinogen in higher animals and man is not known. Our test suggests that it probably is not. Gentian violet must be considered a potent toxicant and direct-acting mutagen.

Direct-acting chemical mutagens were effective SCE inducers. Neutral red (NR) was the most potent of these, followed by methyl methanesulfonate (MMS), quinacrine mustard (QM), ethidium bromide (EB), EMS, acridine orange (AO), BrdUrd, and trimethylphosphate (TMP). This order of mutagens represents a dosage range of 5 μg (NR) to 1,500 μg (TMP). The indirect-acting chemical mutagens or promutagens were the most potent SCE inducers, with AFB_1, mitomycin C (MMC), cyclophosphamide (CP), and 2-aminoanthracene (2-AA) active at dosages as low as 0.05-1.0 μg per embryo (Tab. 1). Active in the range of 1.3-4.0 μg were sterigmatocystin (SMC), 3,4-benzo-(a)pyrene (BP), 2-AAF, and dimethylbenzanthracene (DMBA). O-tolidine (TD) and tris-BP caused increased SCE response beginning at 10.0 and 25 μg, respectively.

If SCEs are a quantitative indicator of DNA damage, a dose-response relationship, linear at least for the direct-acting mutagens, is expected. Such was the case for these compounds, except for EB and BrdUrd, which gave only a minimal SCE increase. These 2 compounds, therefore, cause statistically significant but borderline

Tab. 1. Induction of SCE in chick embryos by chemical mutagens.

Chemical	Lowest dosage of chemical giving significant SCE increase[a] (μg)	Rate of SCE/cell at highest dose[b] (mean ± S.D.)	Carcinogenicity[c]
1. Aflatoxin B_1	0.05	12.9 ± 2.5	+
2. Mitomycin C	0.05	5.2 ± 1.7	+
3. Cyclophosphamide	0.5	13.6 ± 1.2	+
4. 2-Aminoanthracene	1.0	13.0 ± 2.4	+
5. Sterigmatocystin	1.3	10.7 ± 3.2	+
6. 3,4-Benzo(a)pyrene	1.8	7.0 ± 2.2	+
7. 2-Acetylaminofluorene	2.5	9.7 ± 1.6	+
8. 9,10-Dimethyl-1,2-benzathracene	4.0	16.5 ± 2.5	+
9. Neutral red	5	18 ± 1.7	ND
10. O-Tolidine	10	6.3 ± 1.0	+
11. MMS	12.5	9.5 ± 0.3	+
12. Quinacrine mustard	25	8.1 ± 4.8	ND
13. Tris-BP	25	8.5 ± 1.4	+
14. Ethidium bromide	50	4.3 ± 0.1	ND
15. EMS	100	8.6 ± 1.8	+
16. Acridine orange	100	7.9 ± 0.5	+
17. 5-Bromodeoxyuridine	500	4.6 ± 0.1	NC
18. TMP	1,500	7.5 ± 1.4	+
- Bleomycin	NS	2.2 ± 0.8	NC
- Fyrol FR2	NS	2.2 ± 0.5	ND
- Gentian violet	NS	2.4 ± 0.1	ND

[a] A one-sided Student t-test was performed on all data. Dosages shown give a statistically significant increase in SCE ($P<0.01$). NS = no significant increase in SCE at any dosage.

[b] This dose refers to the high doses given for each chemical (17).

[c] + = a definite positive; NC = not carcinogenic; ND = not determined.

increases in exchange frequency. In general, as doses increased, so did the SCE frequency for the class of promutagens.

SCE INDUCTION BY PROMUTAGENS: METABOLIC ACTIVATION BY EMBRYOS

The SCE assay was applied as a probe for the metabolic activation of promutagen-carcinogens by specific components of the embryonic MFO system and for determining the reliability of the SCE endpoint for assessing true mutagenic potency. Seven promutagens, known to be metabolically activated by the different major enzyme families of the MFO system, were applied to the chick embryo during the 3-4 da and 6-7 da developmental periods (Tab. 2). The liver differentiates at about 4.5 da. Compounds such as AFB_1 are known to be activated by the cytochrome P-450-associated MFO enzymes; BP and other compounds are activated by the P-448 enzymes.

All 7 promutagens induced SCE above baseline in a dose-dependent manner. In addition, SCE induction was observed in both the 3- to 4-da and 6- to 7-da developing embryos. The compounds known to be activated by the P-450 MFO enzymes were the most potent inducers of SCEs (Tab. 2). In addition, promutagens were detected at very low dosages. For example, AFB_1 genotoxicity was detected at 3.2×10^{-6} moles/Kg.

These results show that the DNA-damaging effects of promutagens of various types can be detected in the early chick embryo using the SCE endpoint (50). This was true for both developmental periods studied. Since liver is not yet present as a differentiated organ at 3 da, it is hypothesized the MFO activity is present in the embryo proper and perhaps in the embryonic membranes. Later, it is entirely possible that MFO enzymes are present in the liver in sufficient quantities to perform biotransformation. We addressed

Tab. 2. SCE induction in chick embryos using promutagens.

Promutagen-Carcinogen	Activation by MFO System[a]	SCE Induction[b]
Aflatoxin B_1	PB	3.2×10^{-6}
Cyclophosphamide	PB	3.6×10^{-5}
Sterigmatocystin	PB	8.0×10^{-5}
2-Aminoanthracene	PB	1.0×10^{-4}
3,4-Benzo(a)pyrene	AR	1.4×10^{-4}
2-Acetylaminofluorene	AR	2.2×10^{-4}
9,10-Dimethyl-1,2,-benzanthracene	AR	3.2×10^{-4}

[a] The promutagens are metabolically activated by the enzyme families shown. Phenobarbital-induced (PB) or P-450 mixed-function (MFO) enzymes; aroclor-induced (AR) or the P-448 enzymes.

[b] Lowest dosages in moles/kg giving significant SCE induction above baseline.

these points in subsequent work that will be considered in this chapter.

The relationship between mutagenic potency and SCE induction capacity of promutagens, as well as a few direct-acting compounds, was also considered in these studies. Data from the Ames Salmonella mutation test were compared with results from the chick embryo SCE test. The potency ranking for 8 compounds is identical for both assays (Tab. 3).

DEVELOPMENTAL CONTROL OF SCE INDUCTION BY PROMUTAGEN-CARCINOGENS

The previous studies suggested that promutagens were metabolically activated in the chick embryo at both early and later developmental times. However, since liver is present at 6 da and not at 3 da, it is not unreasonable to expect a difference in the extent of promutagen activation at these 2 time points. To examine this aspect in greater detail, 2 promutagens and 1 direct-acting mutagen were administered in identical dosages to 3-da and 6-da embryos, respectively. BrdUrd was also administered to label cells for SCE detection. Embryos were harvested 24 hr post-treatment. All 3 mutagens (AFB_1, 2-AAF, and EMS) increased the frequency of SCE above the control level of 1.8 SCE/cell (51). However, the SCE frequency for the 6-da embryos was significantly higher than that observed for the 3-da embryos for each of the dosages tested. In contrast, a reduced

Tab. 3. Potency of chemicals for inducing mutation and SCE.

	Ames Assay[a]		Chick SCE Assay[b]	
Chemical	Rank	Rev/nM	Rank	E.D. (moles/kg)
$AF\text{-}B_1$	1	7057	1	3.2×10^{-6}
ST	2	915	2	8.0×10^{-5}
2-AA	3	510	3	1.0×10^{-4}
BP	4	121	4	1.4×10^{-4}
2-AAF	5	108	5	2.2×10^{-4}
tris-BP	6	21	6	7.2×10^{-4}
MMS	7	0.63	7	2.2×10^{-3}
EMS	8	0.16	8	1.6×10^{-2}

[a] Data from McCann et al. (4).

[b] E.D. is the effective dosage, i.e., lowest dose giving significant SCE induction above baseline.

level of SCE was observed in the older embryos for all dosages of EMS.

Even though the mass of the embryo increased 15-fold from 3-6 da, an enhanced SCE induction response was observed for the promutagens (Tab. 4). This was interpreted to indicate the presence of a well-developed liver MFO system that might also be inducible. The reduced EMS response could be due to either increased dilution of the given dosage in the older embryo or the enhanced detoxification of this compound.

DEMONSTRATION OF AHH ACTIVITY IN THE CHICK EMBRYO: DEVELOPMENTAL PATTERN AND INDUCIBILITY

The development of the hepatic microsomal MFO system was studied to determine the basal level of embryonic enzyme activity and the inducibility of this system throughout growth and differentiation (52). Chick embryo livers were assayed for basal and inducible

Tab. 4. Influence of development on the MFO system and the SCE-inducing activity of promutagens.

Developmental and Experimental Parameters	Embryos at: 3 Days	6 Days
Embryo weight	20 mg	290 mg (15x)
Liver	Absent	Present
Liver weight	-	5.1 mg
AHH activity detected in:		
Embryo	0.5 nmol[a]	-
Liver	-	0.5 nmol[a]
AHH inducible in:		
Embryo	No	No
Liver	-	Yes (2x)
SCE induction by promutagens,		
$AF\text{-}B_1$ 0.1 μg	6.2 SCE[b]	11.0 SCE[b]

[a] nmol of benzo(a)pyrene metabolized per minute per mg of protein in a fluorometric assay.

[b] The values are SCE/cell and they are significantly different in a t-test at $P < 0.01$.

hepatic AHH activity from the first appearance of the liver as a discrete organ at 5 DI through 10 da post-hatching. In addition, whole embryo and viscera preparations were assayed at 3 and 4 DI. Basal AHH activity was equal to, or greater than, adult levels from 3 DI through hatching in all preparations (approximately 0.3-0.5 nmol/min/mg). A 3-fold increase in basal activity above adult values occurred at hatching. The onset of inducibility in chick embryo liver between 5 and 6 DI was concomitant with hepatocyte differentiation (Fig. 3). A developmental profile of 24-hr 3,4,3',4'-tetrachlorobiphenyl (TCB)-induced AHH activity showed a 15- to 30-fold induction over controls from 7 DI through 10 da posthatching, with a maximum of 15 nmol/min/mg at 14 DI and 1 da posthatching. This specific activity was more than 50% greater than the maximum induction level in the adult. Embryonic AHH activity was also induced by 2,3,7,8-tetrachlorodibenzo-ρ-dioxin, 3- methylcholanthrene, β-naphthoflavone, and sodium phenobarbital (PB). Induction kinetics throughout embryonic development were similar to those reported for the adult chicken and other animals. These findings demonstrate development of a MFO system in very early embryogenesis and then in the liver as it differentiates. Liver AHH activity is inducible throughout development and perinatally but such activity is under strict developmental regulation. The chick embryo possesses adult levels of AHH activity which would be sufficient to achieve metabolic activation of promutagens/carcinogens before and after hepatocyte

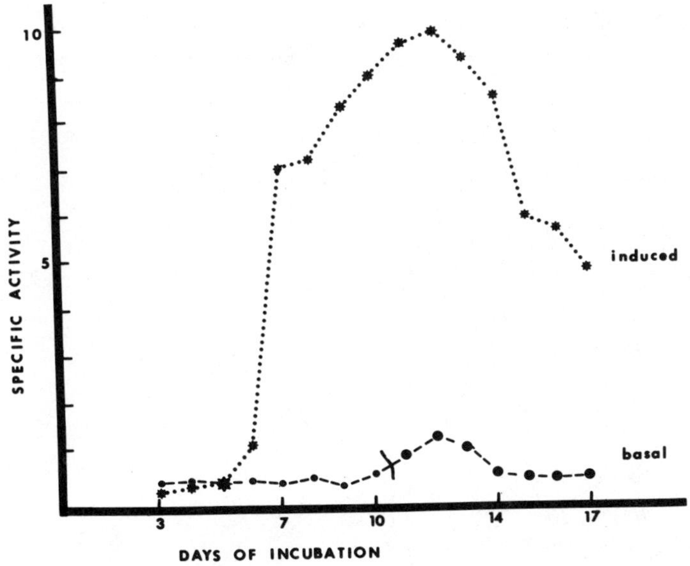

Fig. 3. Developmental pattern of basal and induced AHH activity in the developing chicken embryo. Enzyme activity is given in nmol of BP metabolized/min/mg of protein.

differentiation (Tab. 4). These results provide the biochemical basis for the previous observation on promutagen SCE induction in embryos. In addition, the high level of MFO activity detected in embryos explains why such low levels of promutagens were detected by the SCE technique.

FORMATION AND MODULATION OF SCE-INDUCING PRODUCTS IN CHICK EMBRYO MICROSOMES

The formation of SCE-inducing metabolic products has been further investigated using the model promutagens AFB_1 and BP and specific induction of enzyme components of the MFO system (53). To facilitate the study of genotoxic activity of the primary metabolic products, microsomal preparations were used in conjunction with the CHO SCE test. That is, CHO cells were exposed to the products of primary chick embryo metabolism in an in vitro assay. Additionally, metabolism was altered by specific induction of the P-450 (PB) vs. P-448 (TCB) systems. If the embryo MFO system resembles the adult, PB induction should enhance AFB_1 activation and TCB should enhance BP activation.

Livers from control, TCB, and PB-induced 14-da chick embryos were homogenized and centrifuged at 16,000 x g (S-16) to obtain the appropriate postmitochondrial supernatant fractions. Levels of AHH activities in the microsomal fractions were about 0.5 in control and

Tab. 5. SCE induction in CHO cells exposed to promutagens activated using chick embryo microsomal preparations (S-16).

Treatment		SCE/Metaphase[a]
Acetone Control		5.38
Benzo(a)pyrene at 10^{-4} M		
Heat-inactivated S16		7.38
Non-induced S16		9.14
Benzo(a)pyrene with TCB-induced		
Chick S16 preparations	10^{-6} M	12.50
	10^{-5} M	17.40
	10^{-4} M	22.70
Aflatoxin B_1 with phenobarbital-induced		
Chick S16	10^{-5} M	14.20

[a] All SCE values are significantly higher than the acetone solvent control and significantly different from each other at $p < 0.01$.

6.0 nmol/mg of protein in the induced fractions. Chinese hamster ovary cells were exposed to graded dosages of promutagen, together with the S-16 fraction and appropriate cofactors. Heat-inactivated S-16 preparations were also included as controls.

Small but significant SCE induction was observed when CHO cells were exposed to BP and either heat-inactivated or noninduced S-16 (Tab. 5). A greatly enhanced SCE induction pattern was seen when TCB-induced S-16 preparations were used. With AFB_1 as the promutagen, the best SCE induction was seen with PB-induced S-16 (Tab. 5). Preparations of TCB did not give any better SCE response than did noninduced samples.

The above results are consistent with the notion that the chick embryo MFO system is well-developed and can be induced. Such induction yields either a higher level of activation products or ones that are more highly genotoxic, or both. Specific induction of components of the embryonic MFO system gave the expected genotoxic responses with the respective model promutagens.

MUTAGEN-INDUCED SCE AND DEVELOPMENTAL POTENTIAL

The chick embryo was used to study the short- and long-term biological consequences of embryonic exposure to SCE-inducing levels of a promutagen.

Some 2,000 Cornell K-strain and N-line embryos were exposed to graded dosages (0.01-1.0 μg) of the promutagen/carcinogen aflatoxin B_1 at either 6 or 12 DI. Successful hatches of chicks were achieved at AFB_1 dosages of 0.01-0.20 μg per embryo. The corresponding percentage hatchability percentage varied from 83% to 50% in the K-strain and 81% to 46% in the N-line eggs. At dosages from 0.5 to 1.0 μg, hatching rate was reduced to 1%. Acetone controls, on the average, hatched at a level of 70%.

Aflatoxin B_1 induced 10.6 SCE/cell at 0.1 μg dosage/embryo over a control rate of 1.8 SCE. This level of AFB_1 was used for studies of long-term effects on growth and development.

Postnatal Monitoring of the Erythroid System

The influence of embryonic exposure to AFB_1 upon the erythroid system of the developing chicken was examined (19). Embryos were exposed to the mutagen or solvent at either 6 or 12 DI. Yolk sac-derived red blood cells (RBCs) are present at da 6 but a new embryo-derived cell population is present at da 12 that becomes the source of cells in postnatal life. Chickens surviving the prenatal mutagen exposure exhibited erythroid anemia in the postnatal period when compared to acetone controls. Red blood cell count, hematocrit, and hemoglobin level were all significantly reduced in the 12-da AFB_1

treatment group. In contrast, control and treated values for these same RBC parameters did not differ in the 6-da exposed group. Other blood parameters such as mean cell volume, percentage of circulating reticulocytes and incidence of an erythroid differentiation marker (chicken fetal antigen) were essentially unaltered by the embryonic treatment. These results suggest that mutagenic damage to embryonically derived RBCs may reduce the population size or growth potential of cells in the embryonic and post-hatch bone marrow.

Postnatal Studies of the Developing Immune System

The influence of embryonic exposure to AFB_1 on parameters important to overall health was investigated using our avian model. In ovo exposure of 6- or 12-da Cornell K-strain chicken embryos to either acetone (10 µl) or acetone + AFB_1 (0.1 µg) was followed by assessment of the comparative development of control and treatment groups. This dosage of AFB_1 induced DNA damage and SCE in embryonic cells. Assays emphasized evaluation of the status of hematolymphoid development, particularly immune capabilities. Only chickens surviving both chronic and acute embryonic toxicity were available for long-term post-hatch analysis. Age of embryonic exposure appeared to influence the spectrum of hematolymphoid effects. Aflatoxin B_1-treated 6-da embryos exhibited a marked reduction in post-hatch cell-mediated immune function compared to the acetone controls. In contrast, exposure of 12-da embryos to AFB_1 produced only slight immune depression but resulted in significant erythroid anemia. These findings suggest that a single embryonic exposure to AFB_1 can produce long-term alteration of the hematolymphoid system that could lead to impaired immune capabilities, but the outcome is dependent on the developmental status of the embryo at the time of toxicant exposure.

Comparison of the Chick Cytogenetic System to Other Assays: Mutation, Cancer, and SCE

Our preliminary comparison of SCE data from nonmammalian systems, particularly the avian embryo, shows their utility in 1) detecting mutagens, 2) detecting nonmutagens, 3) determining mutagenic potential, and 4) establishing a data base to assign a risk factor (for mutation and cancer) to any newly evaluated compound.

Using the SCE technique, 86% of mutagens were detected in the chick. When chromosome aberration data were included, accuracy was increased to 90%. The chick test also gave a negative result for compounds rendered genetically inactive in vivo (e.g., GV). No false positives were found (17).

A comparison of SCE data to other genetic assays reveals that a potency value can be assigned to a compound from in vivo testing in the chick embryo (Tabs. 3 and 6). For example, we compared the chick embryo results with those of the Ames Salmonella test (4) and in a DNA inhibition assay (55). Chemicals causing point mutations and DNA inhibition also induce SCE. Ten of the Ames-tested mutagens were ranked according to the number of revertants/plate/1 μg and then compared with the ranking in the chick SCE test determined from both dose-response data and the number of SCEs/1 μg of mutagen (Tab. 6). In both these test systems, AFB_1 ranked as the most potent mutagen. The 5 top-ranked mutagens were identical for both tests; these are AFB_1, SMC, 2-AA, 2-AAF, and BP. These highly genotoxic compounds include the mold toxins and the polycyclic aromatic hydrocarbons, also known for their carcinogenic potency. Acridine orange, EB, tris-BP, MMS, and EMS were in the lower ranking in the Ames system, the DNA inhibition assay, and in the SCE test. When data are expressed on a molar basis, the potency ranking of 8 carcinogens is identical in the Ames and chick tests (Tab. 3). The chick embryo SCE assay appears to be more sensitive than the mouse SCE assay for detecting mutagens in vivo (Tab. 7). These results show that the induction of SCE in the chick embryo in vivo test is related to the ability of a chemical to induce point mutations in DNA. This is true for both direct-acting mutagens and those requiring metabolic activation for mutagenesis and carcinogenesis.

Tab. 6. Potency of chemicals for inducing SCE, causing point mutations and inhibiting DNA synthesis.

Chemical	Rank by Ames test	Dose giving 1,000 rev/plate[a] (μg)	Rank in chick SCE test	Dose giving significant SCE (μg)[b]	Rank in DNA inhibition assay	Dose giving 40% inhibition in 2.5 hr[c] (M)
Aflatoxin B_1	1	0.04	1	0.05	1	5×10^{-8}
Sterigmatocystin	2	0.36	3	1.3	-	
2-Aminoanthracene	3	0.38	2	1.0	-	
2-Acetylaminofluorene	4	2.08	5	2.5	2	1×10^{-7}[d]
3,4-Benzo(a)pyrene	5	2.08	4	1.8	3	7×10^{-6}
Acridine orange	6	4.57	9	100	-	
Ethidium bromide	7	4.95	8	50	-	
Tris-BP	8	30.03	7	25	-	
MMS	9	175.44	6	12.5	4	5×10^{-4}
EMS	10	769.23	10	100	5	1×10^{-2}

[a] Calculated from McCann et al. (4)
[b] Data from Table 1.
[c] Calculated from Painter (56).
[d] N-acetoxy-2-acetylaminofluorene.

Tab. 7. Comparative detection power of avian and mammalian SCE systems. [Mouse data calculated from Latt et al. (57).]

	Effective Dose (moles/kg)[a]	
Compound	Mouse	Chick
$AF\text{-}B_1$	7×10^{-4}	3.2×10^{-6}
CP	2×10^{-4}	3.6×10^{-5}
BP	6×10^{-4}	1.4×10^{-4}
2-AAF	2×10^{-5}	2.2×10^{-4}
DMBA	6×10^{-4}	3.2×10^{-4}
Tris-BP	2.2×10^{-3}	7.2×10^{-4}
MMS	10^{-3}	2.2×10^{-3}
EMS	1.6×10^{-3}	1.6×10^{-2}

[a] Dosage giving significant SCE induction above baseline.

Using the current data base, it is possible to assign a mutagenic risk factor and to indicate the potential for inducing tumors. Unknown compounds inducing SCE in the range of 0.05 to 2 µg (10^{-6} to 2×10^{-4} moles/kg) per embryo are expected to be potent mutagens and carcinogens, while those active at 500-1,500 µg would be weak ones (Tab. 8). In any event, compounds inducing SCE in the chick embryo system should be considered at least potentially carcinogenic.

Tab. 8. Potency of chemicals for inducing cancer and SCE.

Cancer Ranking	Daily Dosage[a]	SCE Ranking[b]	(total dose moles/kg)
1. Aflatoxin B_1	1 µg	$AF\text{-}B_1$	3.2×10^{-6}
2. Sterigmatocystin	400 µg	SMC	8.0×10^{-5}
3. 2-Acetylaminofluorene	1,000 µg	2-AAF	2.2×10^{-4}
4. Tris-BP	100,000 µg	Tris-BP	7.2×10^{-4}
5. Methyl methanesulfonate	500,000 µg	MMS	2.2×10^{-3}

[a] Gives 50% tumor incidence in rats.

[b] Data are from the chick embryo in vivo assay. Lowest doses giving significant induction of SCE above baseline.

Since the study of SCE induction in the chick embryo incorporated a broad spectrum of chemical types, one could ask whether the SCE response is related to a particular mode of interaction with chromosomal DNA. Known alkylating agents, cross-linkers, intercalating dyes, and those chemicals that form bulky adducts in DNA all produce increases in SCEs. It appears then that SCEs reflect multiple forms of damage to chromosomes. In terms of the utility of SCE as a screening device for chemical mutagens, this finding of wide-spectrum SCE induction is fortunate. In addition, since increasing amounts of BrdUrd cause increased rates of SCE, base substitution is presumably another mode of action leading to exchange. Perhaps many forms of damage to DNA lead to SCE formation or else agents produce exchanges by a common lesion. This does not mean that the chromosome is necessarily replicated or repaired without mistakes. Small changes at the nucleotide level would obviously not be visualized at the cytological level. Further studies are needed to elucidate the mechanism(s) leading to SCE formation and the mutational changes, if any, that occur as a result.

Applications of the Chick Embryo Cytogenetic Test System in Genetic Toxicology Testing and Environmental Monitoring

The avian embryo SCE system has features which make it attractive for use in various aspects of genetic toxicology testing and environmental monitoring. Nonmammalian species are generally easy to use, economical, and permit in vivo analysis without the need for fancy surgical apparatus.

The CES can have important uses in agriculture, industry, and medicine. In the agricultural sphere, this test can be useful for assessing the genetic toxicity of poultry health products such as antibiotics used in feed, medications for the treatment of diseases, and pesticides used to control insects in the poultry house and on the birds. If a chemical is found to be genotoxic in the CES, this would warn of the possible induction of mutations in breeder flocks. In addition, a positive result in the chick test would signal a warning of potential human hazard.

In the industrial setting, the avian embryo test is attractive as a rapid screen of chemicals for potential mutagenicity, embryo toxicity, and also teratogenicity. The test is relatively fast and definitely not as expensive as in vivo mammalian tests. The CES should be amenable to automation similar to that achieved with the manufacture of vaccines. In that case, large numbers of eggs are handled by machine, including the injection of solutions.

In the area of medicine, the CES might be a very useful screen for the in vivo effects of drugs such as chemotherapeutic agents, anesthetics, sedatives, and other materials included with primary medications.

In the broad scheme of things, the CES fits in the category of tier 2 testing. The Ames test, at tier 1, establishes mutagenic potential of chemicals with and without activation. At the second tier, the CES would extend the information to effects in a eukaryote (and a vertebrate). Additional information on mutagenic activity of the test compound as effected by metabolism could be obtained quite rapidly by testing in early- vs. late-stage embryos. From both the Ames test and the CES, the mutagenic potency would be reasonably well established. The data would be invaluable in determining the need for, and type of, tier 3 testing. The CES also complements in vitro cytogenetic tests, which are very valuable in determining any direct effects of the test compound.

Finally, the avian egg may be a very useful monitor of the environment in 2 respects. First, embryos from bird populations such as herring gulls and ringdoves may be examined cytogenetically and analytically to determine the presence of mutagenic materials accumulated in the wild. Second, environmental contaminants or materials containing such pollutants can be applied to avian embryos for assessment of genetic hazard (18).

SUMMARY AND IMPORTANCE OF FINDINGS

This work demonstrates the feasibility and utility of performing an assay of DNA/chromosome damage and repair in vivo in the developing avian embryo. Specifically, we have been able to study SCE and chromosome breakage in embryos exposed to environmental mutagens and carcinogens as well as to nonmutagenic toxicants. Indirect-acting mutagen-carcinogens induced DNA damage in developing embryos and this finding led to more extensive studies into the nature and activity of the embryonic MFO system. Adult AHH enzyme levels are present in the embryo and neonate, and the liver is highly inducible just after it forms in early embryogenesis. Metabolites formed from the activation of promutagens by the embryo MFO systems are highly genotoxic in vivo and to mammalian cells in vitro. Thus, the compact in vivo chick system is ideal for studying DNA damage and repair induced by indirect- and direct-acting mutagens during the course of development.

We have assayed the in vivo SCE induction potential of 53 chemicals and find a highly positive correlation ($\sim$90%) between mutagenic activity and SCE formation. Nonmutagens did not induce SCEs in vivo. Potency for inducing SCEs is related to true mutagenic potency, DNA inhibition, and to carcinogenic activity. Some clastogens such as bleomycin do not induce SCE but cause massive chromosome damage that is easily noticed in the chick assay. One clastogen, GV, failed to induce SCE or cause breakage. This was because GV is a direct-acting clastogen that is converted to a nonmutagenic but cytotoxic compound upon metabolism. Thus, the chick embryo cytogenetic system gives the predictable responses for an in vivo assay. The CES was found to be quite useful in screens for the mutagenic activities of drugs, dyes, and pesticides.

The work on SCE induction in embryos by promutagens led to an investigation of the developing MFO system. This direction, which was not originally planned, proved to be very exciting and yielded a great deal of new information on the nature of the embryo MFO system, including its developmental pattern in tissues, induction capacity, and activity of the AHH component, which activates promutagens such as BP.

We were successful in obtaining viable embryos and neonates following early and late prenatal exposures to toxicants including AFB_1, acetone, and other compounds such as MFO inducers. Of interest was the study of the developmental potential of survivors of embryonic mutagenesis. Phenotypically normal chicks demonstrated deficiencies of both the erythroid and lymphoid systems in postnatal development. The developmental time when toxicant exposure occurred determined, in part, the postnatal alteration. The erythroid system was affected more by the 12-da insult than by the 6-da exposure. This correlated with the difference in the erythroid cell populations at these 2 stages. The 12-da RBC population is the progenitor of the adult population; the 6-da RBC primitive (yolk-sac derived) cells disappear at mid-gestation. These findings warrant follow-up studies on developmental control of the biological response to toxicants.

Sister chromatid exchange analysis has proven a most useful monitor for in vivo and in vitro studies. Sister chromatid exchanges are an indicator of DNA damage on the entire genome (chromosomal level) and "nontoxic" responses that correlate with the true mutagenic activity of chemicals. Linear dose-response relationships can be obtained in vivo using SCEs. Sister chromatid exchanges are also useful as a probe for the metabolic activation/detoxification of compounds. We initially discovered the chick embryo AHH activity by this means. This was subsequently confirmed by our biochemical studies. Cells in vivo and in vitro can replicate and differentiate in the presence of induced SCEs (assuming SCEs occur in the absence of BrdUrd).

Lesions leading to SCE formation are removed after 2 or 3 cell cycles, although some may persist to 10 (56). However, our in vivo evidence indicates that differentiation/growth patterns of embryo cells exposed to certain mutagens can be affected in subtle ways and expression is delayed until the postnatal period. We are not sure about the mechanism of this delay. Certain point mutations may not be expressed until the change from embryo to adult cell (gene expression) occurs. Prenatal damage may reduce embryo cell populations available for additional growth/differentiation postnatally. This situation will require further investigation.

The research with the chick embryo has opened up a number of new approaches for investigating the causes and mechanisms of disease formation and may, in itself, provide a useful system for preventing future disease related to exposure to environmental agents.

ACKNOWLEDGEMENTS

The author wishes to acknowledge the assistance and collaborative efforts with Drs. T. C. Hsu, A. D. Kligerman, C. Goodpasture, D. Lisk, R. R. Dietert, and C. F. Wilkinson. The following graduate students have contributed greatly to studies of the embryo MFO system including L. A. Todd, J. W. Hamilton, and M. Denison. The technical assistance of E. G. Polakoff, P. Shalit, M. Qureshi, U. Nanna, O. P. Schaeffer, and D. Muscarella is greatly appreciated. The author further thanks Diane Colf for editorial assistance. This research was supported by grants from the United States Department of Agriculture and National Cancer Institute (CA 28953).

REFERENCES

1. Ames, B.N. (1979) Identifying environmental chemicals causing mutations and cancer. Science 204:587-593.
2. Drake, J.W., S. Abrahamson, J.F. Crow, A. Hollaender, S. Lederberg, M.S. Legator, J.V. Neel, M.W. Shaw, H.E. Sutton, R.C. von Borstel, and S. Zimmering (1975) Environmental mutagenic hazards. Science 187:503-514.
3. Hollaender, A., and F.J. de Serres (1971-1982) Chemical Mutagens: Principles and Methods for Their Detection, Vols. 1-7. Plenum Press, New York.
4. McCann, J., E. Choi, E. Yamasaki, and B.N. Ames (1975) Detection of carcinogens as mutagens in the Salmonella/microsome test: Assay of 300 chemicals. Proc. Natl. Acad. Sci., USA 72:5135-5139.
5. Nagao, M., and T. Sugimura (1978) Environmental mutagens and carcinogens. Ann. Rev. Genet. 12:117-159.
6. Vogel, F., and G. Röhrborn (1970) Chemical Mutagenesis in Mammals and Man. Springer-Verlag, New York.
7. Perinatal Carcinogenesis (1979) National Cancer Institute Monograph 51, DHEW Publication No. (NIH) 79-1633, 282 pp.
8. Rice, J.M. (1973) An overview of transplacental chemical carcinogenesis. Teratology 8:113-126.
9. Ivankovic, S. (1979) Teratogenic and carcinogenic effects of some chemicals during prenatal life in rats, Syrian golden hamsters, and minipigs. In Perinatal Carcinogenesis, NCI Monograph 51, DHEW Publication No. 79-1633, pp. 103-115.
10. Knudson, Jr., A.G. (1979) Mutagenesis and embryonal carcinogenesis. In Perinatal Carcinogenesis, NCI Monograph 51, DHEW publication No. 79-1633, pp. 19-23.

11. Nichols, W.W. (1982) Status of prezygotic chromosome lesions in relation to cancer. Cytogenet. Cell Genet. 33:179-184.
12. Riccardi, V.M., E. Sujansky, A.C. Smith, and U. Francke (1978) Chromosomal imbalance in the Aniridia-Wilms' tumor association: 11p interstitial deletion. Pediatrics 61:604-610.
13. Strong, L.C., V.M. Riccardi, R.E. Ferrell, and R.S. Sparkes (1981) Familial retinoblastoma and chromosome 13 deletion transmitted via an insertional translocation. Science 213: 1501-1503.
14. Sandberg, A.A., ed. (1982) Progress and Topics in Cytogenetics. II. Sister Chromatid Exchange. Alan R. Liss, Inc., New York.
15. Latt, S.A., R.R. Schreck, K.S. Loveday, C.P. Dougherty, and C.F. Shuler (1980) Sister chromatid exchanges. In Advances in Human Genetics, Vol. 10, H. Harris and K. Hirschhorn, eds. Plenum Press, New York, pp. 267-331.
16. Bloom, S.E. (1978) Chick embryos for detecting environmental mutagens. In Chemical Mutagens: Principles and Methods for Their Detection, Vol. 5, A. Hollaender and F.J. de Serres, eds. Plenum Press, New York, pp. 203-232.
17. Bloom, S.E. (1982) Detection of sister chromatid exchanges in vivo using avian embryos. In Cytogenetic Assays of Environmental Mutagens, T.C. Hsu, ed. Allanheld, Osmun & Co., Montclair, New Jersey, pp. 137-159.
18. Bloom, S.E. (1982) Avian and aquatic systems for in vivo detection of sister chromatid exchanges. In Progress and Topics in Cytogenetics. II. Sister Chromatid Exchange, A.A. Sandberg, ed. Alan R. Liss, New York, pp. 249-277.
19. Dietert, R.R., S.E. Bloom, M.A. Qureshi, and U.C. Nanna (1983) Hematological toxicology following embryonic exposure to aflatoxin-B_1. Proc. Soc. Exp. Biol. Med. 173:481-485.
20. Bloom, S.E., and T.C. Hsu (1975) Differential fluorescence of sister chromatids in chicken embryos exposed to 5-bromodeoxyuridine. Chromosoma 51:261-267.
21. Kligerman, A.D., and S.E. Bloom (1976) Sister chromatid differentiation and exchanges in adult mudminnows (Umbra limi) after in vivo exposure to 5-bromodeoxyuridine. Chromosoma 56:101-109.
22. Allen, J.W., and S.A. Latt (1976) Analysis of sister chromatid exchange formation in vivo in mouse spermatogonia as a new test system for environmental mutagens. Nature (Lond.) 260:449-451.
23. Pera, F., and P. Mattias (1976) Labelling of DNA and differential sister chromatid staining after BrdU treatment in vivo. Chromosoma 57:13-18.
24. Schneider, E.L., J.R. Chaillet, and R.R. Tice (1976) In vivo BUdR labeling of mammalian chromosomes. Exp. Cell Res. 100: 396-399.
25. Vogel, W., and T. Bauknecht (1976) Differential chromatid staining by in vivo treatment as a mutagenicity test system. Nature 260:448-449.

26. Tice, R., J. Chaillet, and E.L. Schneider (1976) Demonstration of spontaneous sister chromatid exchanges in vivo. Exp. Cell Res. 102:426-429.

27. Cole, R.K. (1967) Leukosis control through genetics. In Proc. Poult. Health Conf., University of New Hampshire, Durham, pp. 59-72.

28. Blum, A., and B.N. Ames (1977) Flame-retardant additives as possible cancer hazards. Science 195:17-23.

29. Gold, M.D., A. Blum, and B.N. Ames (1978) Another flame retardant, tris(1,3-dichloro-2-propyl)-phosphate, and its expected metabolites are mutagens. Science 200:785-787.

30. Kato, H. (1977) Spontaneous and induced sister chromatid exchanges as revealed by the BUdR-labelling method. Int. Rev. Cytol. 49:55-97.

31. Wolff, S. (1977) Sister chromatid exchange. Ann. Rev. Genet. 11:183-201.

32. Latt, S.A. (1974) Sister chromatid exchanges, indices of human chromosome damage and repair: Detection of fluorescence and induction by mitomycin C. Proc. Natl. Acad. Sci., USA 71:3162-3166.

33. Dutrillaux, B., A.M. Fosse, M. Prieur, J. Lejeune (1974) Analyse des échanges de chromatides dans les cellules somatiques humaines: Traitment au BUdR (5-bromdeoxyuridine) et fluorescence bicolore par l'acridine orange. Chromosoma 48: 327-340.

34. Kato, H. (1974) Spontaneous sister chromatid exchanges detected by a BUdR-labelling method. Nature 251:70-72.

35. Kihlman, B.A., and D. Kronberg (1975) Sister chromatid exchanges in Vicia faba. I. Demonstration by a modified fluorescent plus Giemsa (FPG) technique. Chromosoma 51:1-10.

36. Korenberg, J.R., and E.R. Freedlender (1974) Giemsa technique for the detection of sister chromatid exchanges. Chromosoma 48:355-360.

37. Perry, P., and S. Wolff (1974) New Giemsa method for the differential staining of sister chromatids. Nature 251:156-158.

38. Wolff, S., and P. Perry (1974) Differential Giemsa staining of sister chromatids and the study of sister chromatid exchanges without autoradiography. Chromosoma 48:341-353.

39. Zakharov, A.F., and N.A. Egolina (1972) Differential spiralization along mammalian mitotic chromosomes. I. BUdR-revealed differentiation in Chinese hamster chromosomes. Chromosoma 38:341-365.

40. Perry, P., and H.J. Evans (1975) Cytological detection of mutagen-carcinogen exposure by sister chromatid exchange. Nature 258:121-125.

41. Owen, J.J.T. (1965) Karyotype studies on Gallus domesticus. Chromosoma 16:601-608.

42. Bloom, S.E. (1974) Current knowledge about the avian W chromosome. BioScience 24:340-344.

43. Ohno, S. (1961) Sex chromosomes and microchromosomes of Gallus domesticus. Chromosoma 11:484-498.

44. Van Brink, J.M., and G.A. Ubbels (1956) La question des heterochromosomes chez les sauropsides: Oiseaux. Experientia 12: 162-164.
45. Yamashina, M.Y. (1944) Karyotype studies in birds. I. Comparative morphology of chromosomes in seventeen races of domestic fowl. Cytologia 13:270-296.
46. Bloom, S.E. (1970) Marek's disease: Chromosome studies of resistant and susceptible strains. Avian Dis. 14:478-490.
47. Cormier, J.M., and S.E. Bloom (1973) An in vivo study of the effects of X-radiation on the chromosomes of chick allantoic membranes. Mutat. Res. 20:77-85.
48. Au, W., S. Pathak, C.J. Collie, T.C. Hsu (1978) Cytogenetic toxicity of gentian violet and crystal violet on mammalian cells in vitro. Mutat. Res. 58:269-276.
49. Au, W., M.A. Butler, S.E. Bloom, and T.S. Matney (1979) Further study of the genetic toxicity of gentian violet. Mutat. Res. 66:103-112.
50. Bloom, S.E. (1982) Sister chromatid exchange as an indicator of promutagen activation and mutagenic potency in early embryonic development. Environ. Mutagen. 4:361-362.
51. Todd, L.A., and S.E. Bloom (1980) Differential induction of sister chromatid exchanges by indirect-acting mutagen-carcinogens at early and late stages of embryonic development. Environ. Mutagen. 2:435-445.
52. Hamilton, J.W., M.S. Denison, and S.E. Bloom (1983) Development of basal and induced aryl hydrocarbon [benzo(a)pyrene] hydroxylase activity in the chick embryo in ovo. Proc. Natl. Acad. Sci., USA 80:3372-3376.
53. Bloom, S.E., O.P. Schaefer, and J.W. Hamilton (1983) Avian embryonic liver homogenates for activating promutagens in the Chinese hamster cell sister chromatid exchange assay. Environ. Mutagen. 5:386-387.
54. Hamilton, J.W., and S.E. Bloom (1984) Correlation between mixed-function oxidase enzyme induction and aflatoxin-B_1 induced unscheduled DNA synthesis in the chick embryo in vivo. Environ. Mutagen. (6:41-48).
55. Painter, R.B. (1981) DNA synthesis inhibition in mammalian cells as a test for mutagenic carcinogens. In Short-Term Tests for Chemical Carcinogens, H.R. Stich and R.H.C. San, eds. Springer-Verlag, New York, pp. 59-64.
56. Muscarella, D.E., and S.E. Bloom (1982) The longevity of chemically induced sister chromatid exchanges in Chinese hamster ovary cells. Environ. Mutagen. 4:647-655.
57. Latt, S.A., J. Allen, S.E. Bloom, A. Carrano, E. Falke, D. Kram, E. Schneider, R. Schreck, R. Tice, B. Whitfield, and S. Wolff (1981) Sister chromatid exchanges: A report of the Gene-Tox program. Mutat. Res. 87:17-62.

SCE INDUCTION BY INDIRECT MUTAGENS/CARCINOGENS IN METABOLICALLY ACTIVE CULTURED MAMMALIAN CELL LINES

Syuiti Abe

Chromosome Research Unit
Faculty of Science
Hokkaido University
North 10, West 8, Sapporo 060, Japan

SUMMARY

Sister chromatid exchange (SCE) analysis in a total of 20 cultured mammalian cell lines of various types exposed to a variety of indirect mutagens/carcinogens suggested that, among these, several human and rat tumor cell lines are intrinsically capable of metabolizing a wide range of mutagens/carcinogens into genotoxic forms to induce SCEs. These cells may therefore serve as simple in vitro SCE test systems for screening of activation-dependent genotoxins without the use of an exogenous activating system. In addition, a comparison of the relevant metabolizing enzyme activity and inducibility of SCEs by mutagens/carcinogens in such cell systems appears to provide clues about the metabolic pathways related to SCE induction. Based on our own data and those in the literature, practical and theoretical advantages of the use of a direct-activating SCE test system in genetic toxicology are discussed.

INTRODUCTION

The sister chromatid exchange (SCE) test in mammalian cell cultures has been widely used for the detection of genotoxic mutagens/carcinogens. Most chemical mutagens/carcinogens, however, require metabolic activation to become ultimate genotoxic substances (e.g., Ref. 18). The current in vitro SCE test for such indirect mutagens/carcinogens generally uses metabolically less active or almost inactive cells and, therefore, needs the addition of an exogenous activation system such as the liver microsomal S-9 fraction or metabolizing feeder cells (16). However, such microsome- or cell-mediated systems are sometimes accompanied by technical complications.

Recently, attention has been focused on the possibility of developing in vitro test systems which use cells intrinsically capable of metabolic activation (e.g., Refs. 1,7), since the use of such metabolically active cells may provide a promising approach with less technical complications than the conventional S-9 pretreatment or cocultivation with feeder cells. During the past six years, we have attempted to detect metabolically active cultured cell systems by SCE and/or chromosome aberration tests and have found several human and rat tumor cell lines retaining intrinsic metabolizing ability (1,3,4,13). In this chapter, I will describe the results of our SCE studies in these cell lines, paying particular attention to the relevant metabolizing enzyme activity and SCE induction by mutagens/carcinogens.

SCE INDUCTION AS A MEASURE OF GENOTOXICITY OF MUTAGENS/CARCINOGENS

Most DNA-damaging agents can induce SCEs, although the types of DNA damage resulting in an SCE response may be different (14). Alkylating agents, certain DNA-binding dyes, and some base analogs including 5-bromodeoxyuridine (BrdUrd) are known to be effective inducers of SCEs and also to be mutagenic and/or carcinogenic (2,15).

We attempted to evaluate the SCE test by a comparison of existing data on SCE induction and the mutagenicity or carcinogenicity of environmental chemicals (2). The extensive literature survey, although provisional, revealed a high degree of concordance between SCE induction and mutagenicity and/or carcinogenicity for various kinds of chemicals. This concordance indicated the potential usefulness of SCE analysis for the detection of genotoxic mutagens/carcinogens. The quantitative correlation between inducibility of SCEs and mutagenic or carcinogenic potential of chemicals remains to be explored, although such an attempt has successfully been made on several aflatoxins, including 3 known liver carcinogens and 1 noncarcinogen (6).

SCE INDUCTION BY MUTAGENS/CARCINOGENS IN CULTURED CELL LINES

Current SCE testing is primarily conducted in vitro using cultured mammalian cell systems (2). However, the yields of SCEs are not necessarily similar among different cell types when exposed to mutagens/carcinogens. For example, human dermal fibroblasts tend to exhibit a greater SCE increase than do peripheral lymphocytes (14). In addition, Chinese hamster lung or ovary cells exposed to a variety of environmental agents including mutagens/carcinogens seem to yield a higher frequency of SCEs than human lymphocytes or embryonic fibroblasts (e.g., Refs. 19,24). The exact reason for such differences is unknown at present, although SCE induction may be influenced by various factors, such as the inherent DNA repair characteristics of cells (28) and the metabolic requirements of the chemicals tested (14), the latter being particularly important for detecting activation-dependent genotoxins.

METABOLICALLY ACTIVE CULTURED MAMMALIAN CELL LINES AS DETECTED BY THE SCE TEST

The metabolic activation of indirect mutagens/carcinogens to their genotoxic forms is detectable by the induction of SCEs, as shown by the microsome- or cell-mediated SCE assays (e.g., Ref. 30). Seeking for metabolically active cultured cell systems, we investigated the production of SCEs in a total of 20 mammalian cell lines after treatment with some known indirect mutagens/carcinogens, including benzo(a)pyrene (BP), cyclophosphamide (CP), and dimethylnitrosamine (DMN) (Tab. 1) (1,3,4,13). Several human and rodent cell lines exhibited a high level of SCE response to more than 2 chemicals tested, each in the absence of exogenous activation. Among these cell lines, human hepatomas C-HC-4 and C-HC-20, rat ascites hepatoma AH66-B, and rat esophageal tumor R1 were selected to further examine whether they exhibit SCE sensitivity to a wider spectrum of mutagens/carcinogens. As shown in Tab. 2, all of these cell lines, in terms of an induction of SCEs, were highly sensitive to a variety of indirect mutagens/carcinogens. In addition, AH66-B and R1 cells were able to enhance the yields of SCEs in cocultivated, metabolically deficient Chinese hamster Don-6 cells, when exposed to some mutagens/carcinogens (1). These results suggest that these cell lines are intrinsically capable of metabolizing a wide range of activation-dependent genotoxins into active forms capable of inducing SCEs, and that they therefore offer a useful in vitro test system for detecting genotoxic agents without the use of an exogenous activation system (1,4,13).

Similar attempts to obtain metabolically competent cells in vitro have also been conducted using cultured human and rat hepatoma or hepatoblastoma cells or neonatal or adult rat-liver cells (7,8,17,29). Usually, the metabolically active cell lines detected thus far by the SCE test are of hepatic or hepatoma origin, although other types of established cell lines, e.g., Syrian hamster BHK cells, are also known to possess some metabolic ability (22).

METABOLIZING ENZYME ACTIVITY AND INDUCIBILITY OF SCES BY MUTAGENS/CARCINOGENS

Many indirect mutagens/carcinogens are metabolically activated by a range of mixed-function oxidases (MFOs) referred to as the

Tab. 1. Frequency of SCEs in various mammalian cell lines exposed to BP, CP, or DMN.

Cell line	Origin	Relative frequency of SCEs (dose mM)		
		BP	CP	DMN
AH66-B	Rat ascites hepatoma	1.2 (0.5)	6.6 (3.0)	2.5 (100.0)
AH66-C	Rat ascites hepatoma	NT	2.1 (5.0)	NT
AH70-B	Rat ascites hepatoma	NT	8.1 (3.0)	NT
AH109-A	Rat ascites hepatoma	NT	4.0 (1.0)	NT
YS	Yoshida rat ascites hepatoma	NT	3.0 (3.0)	1.1 (30.0)
R1	Rat esophageal tumor	2.5 (0.5)	8.3 (3.0)	1.9 (100.0)
R2	Rat esophageal tumor	NT	5.5 (1.0)	NT
R3	Rat esophageal tumor	NT	6.2 (1.0)	1.9 (100.0)
B-13	Chinese hamster embryo	3.3 (0.1)	3.8 (3.0)	1.1 (100.0)
CHL	Chinese hamster lung	NT	2.8 (3.0)	NT
CHO-K1	Chinese hamster ovary	NT	2.1 (3.0)	NT
Don-6	Chinese hamster lung	1.8 (0.5)	1.6 (3.0)	2.1 (100.0)
GHE	Golden hamster embryo	3.8 (0.3)	3.8 (3.0)	NT
C-HC-4	Human hepatoma	4.3 (0.4)	3.5 (0.6)	NT
C-HC-20	Human hepatoma	3.6 (0.4)	4.6 (0.6)	NT
C-HC-32	Human hepatoma	3.0 (0.4)	NT	NT
HE2144	Human embryo	1.8 (0.1)	3.0 (5.0)	NT
HE2236	Human embryo	NT	5.5 (5.0)	NT
P3HR-1	Burkitt's lymphoma	NT	2.9 (5.0)	0.9 (100.0)
TH	Human esophageal tumor	NT	3.1 (3.0)	NT

Maximum relative frequency of SCEs was given. Induced SCE frequencies were compared with the respective control values of the cell types used. Data from Abe & Sasaki (1, 3), Abe et al. (4) and Ikeuchi & Sasaki (13). NT, not tested.

P-450 cytochromes. For example, aflatoxins and CP are activated by cytochrome P-450 (10,11,27) and polycyclic aromatic hydrocarbons (PAHs, e.g., BP) by P-448 (11) or arylhydrocarbon hydroxylase (AHH) (9). These enzyme systems are known to be present or inducible in a variety of cultured mammalian cells, e.g., early passage Syrian hamster embryonic cell cultures (20). However, little is known about such activation enzymes and their relation to SCE induction in the aforementioned human and rat cell lines and those reported in the literature.

We compared the AHH activity induced by 1,2-benz(a)anthracene (BA) and the induction of SCEs by several PAHs including BP, 7,12-dimethylbenz(a)anthracene (DMBA) and 3-methylcholanthrene (MCA) in

Tab. 2. SCE frequency in several human and rat cell lines at the maximum dose of 12 indirect mutagens/carcinogens tested.

Chemical	Relative frequency of SCEs (dose mM)			
	AH66-B	R1	C-HC-4	C-HC-20
2-Acetylaminofluorene	2.1 (4.5)	2.0 (0.45)	NT	NT
Aflatoxin B_1	NT	NT	3.0 (0.001)	3.8 (0.001)
4-Aminoazobenzene	1.7 (0.05)	1.5 (0.025)	NT	NT
Butylbutanolnitros-amine	2.5 (1.5)	2.3 (1.5)	NT	NT
n-Dibutylnitrosamine	1.4 (0.63)	1.2 (0.25)	NT	NT
Diethylstilbestrol	2.4 (0.01)	1.4 (0.01)	NT	NT
7,12-Dimethylbenz[α]-anthracene	3.9 (0.2)	2.1 (0.2)	4.0 (0.04)	4.2 (0.04)
3-Methylcholanthrene	1.6 (0.38)	1.8 (0.38)	4.5 (0.07)	2.9 (0.07)
2-Methyl-4-dimethyl-aminoazobenzene	2.4 (0.2)	2.0 (0.4)	NT	NT
3'-Methyl-4-dimethyl-aminoazobenzene	2.7 (34.0)	2.0 (17.0)	NT	NT
4-*o*-Tolylazo-*o*-toluidine	2.2 (0.09)	3.2 (0.09)	NT	NT
Urethane	2.5 (1.0)	1.1 (1.0)	NT	NT

Relative frequency of SCEs, induced SCE frequencies were compared with the respective control values of the cell types used. Data from Abe & Sasaki (1) and Abe et al. (4). NT, not tested. Underlined numbers are values over twice the control SCE frequency showing an apparent dose-related increase for at least three different doses examined, which are tentatively considered to be positive.

some human, rat, and Chinese hamster cell lines (Fig. 1) (5). After induction by BA, 2 human hepatoma cell lines (C-HC-4 and C-HC-20) showed a dramatic elevation of AHH activity, reaching levels about 140 and 64 times their respective basal levels, whereas none of 3 rodent cell lines (AH66-B, Don-6, and R1) exhibited an appreciable increase in the enzymic activity, i.e., less than 1.08 of the basal values (Fig. 1B). When compared in terms of SCE yields and the minimal effective dose of the chemicals needed to induce SCEs, the 2 AHH-inducible human hepatoma cell lines were highly sensitive to all PAHs tested, whereas the 3 noninducible rodent cell lines were moderately or only slightly sensitive to these chemicals (Fig. 1A). Therefore, SCE induction by PAHs in the 2 AHH-inducible human cell lines parallels their metabolism by these cells, particularly since both cell lines, as shown by a biochemical analysis, liberate various metabolites of BP including a proximate form BP-7,8-dihydrodiol

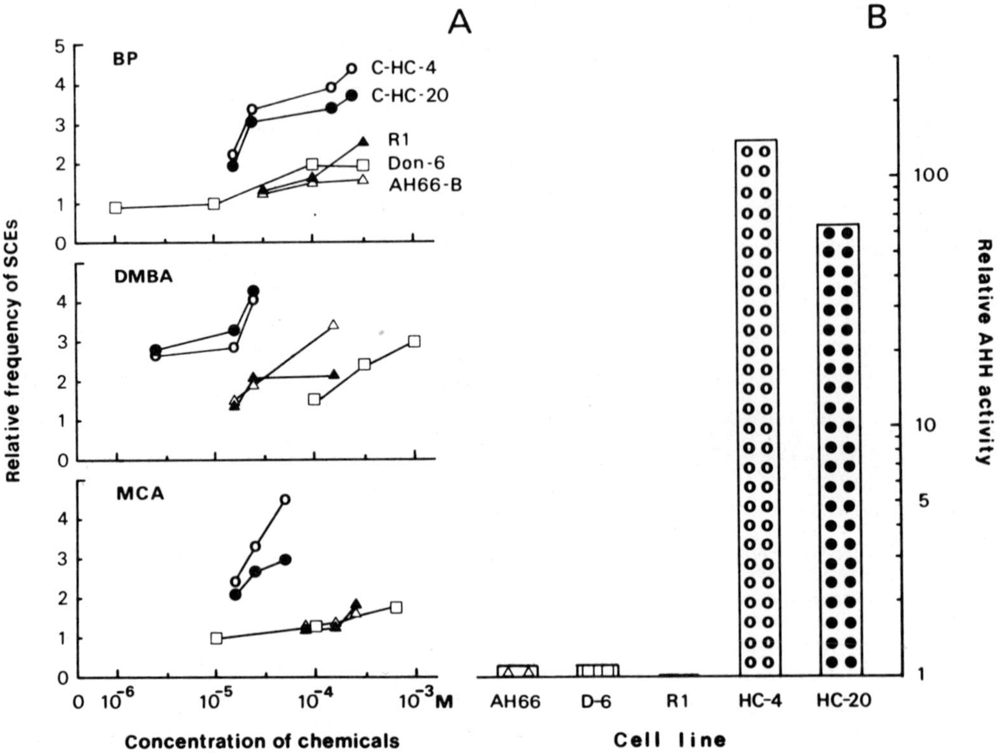

Fig. 1. A. Relative frequencies of SCEs in various mammalian cell lines exposed to BP, DMBA, and MCA. Induced SCE frequencies were compared with the respective control values of the cell types used (5). B. Relative activities of AHH in various mammalian cell lines treated with 13 μM of 1,2-benz(a)anthracene. Induced enzymic activities were compared with respective control values of the cell types used (5).

(4) which is the most potent SCE inducer among the BP metabolites examined (21).

Although a similar observation has also been reported for another AHH-inducible human hepatoma cell line capable of activating BP (12), the results on the 3 noninducible rodent lines suggest that the induction of SCEs by PAHs is not necessarily dependent on the inducibility of AHH. Similar discrepancies were also observed in BP-exposed cultured human lymphocytes (23) and in bone marrow and regenerating liver cells of mice given BP (25,26). In addition, a human hepatoblastoma cell line in which P-450 associated enzyme activities were not detectable exhibited a highly increased SCE frequency when exposed to CP (8).

Even though the available data on the relevant enzyme activity and SCE induction by mutagens/carcinogens are still insufficient to warrant the correlation, such comparative studies may provide clues about the metabolic pathways related to SCE induction. For example, SCE induction by PAHs in the above 3 noninducible rodent lines suggests a possible involvement of enzymes other than AHH in the activation of such chemicals (5).

Another approach to assess the metabolic pathways related to SCE induction may be the use of inducers or inhibitors of the relevant activation enzymes. Such studies, however, are still few and provide conflicting findings (8,23). The effects of metabolic modulators on SCE induction should be further explored.

METABOLICALLY ACTIVE MAMMALIAN CELL LINES AS DIRECT-ACTIVATING SCE TEST SYSTEM

As described earlier, metabolically active cultured cells seem to provide a simple in vitro test system in which a simultaneous analysis of drug activation and SCE induction in the same cell system is possible. Such direct-activating SCE assays may further be advantageous in genetic toxicology, in that a closer approximation to in vivo conditions, longer exposure times for liberating the active metabolites of test chemicals, and an increased sensitivity to chemicals almost incapable of exerting their genotoxicity in the conventional microsome- or cell-mediated assays can be expected to occur. Human cell lines with metabolizing ability, e.g., C-HC-4 and C-HC-20, deserve particular attention in this context because they allow a more direct estimation of the genotoxic effects of mutagens/carcinogens on man.

The known metabolically active cell lines exhibit considerably different sensitivities to indirect mutagens/carcinogens, as defined by SCE yields or the minimal effective dose of a chemical needed to induce SCEs (Tab. 3). For example, human cell lines C-HC-4 and C-HC-20 are far more sensitive to PAHs than rat cell lines AH66-B and Rl. Such differences may essentially be accounted for by the presence or absence of relevant activation enzymes including known or unknown ones in a given cell system, although other factors affecting SCE formation, such as DNA repair capacity, should also be taken into consideration. In practice, the combined use of different cell types may be needed for identifying a wide variety of genotoxic mutagens/carcinogens, since not all cell lines seem to retain similar metabolic functions.

CONCLUSION

SCE analysis appears to be useful both for detecting genotoxic mutagens/carcinogens and for testing the metabolic ability to

Tab. 3. Comparison of effective dose of several indirect mutagens/carcinogens to induce SCEs in various mammalian cell lines.

Cell line	Origin	Least effective dose (μM) of AFB$_1$	BP	CP	DMBA	Reference
AH66-B	Rat ascites hepatoma	NT	—	< 300	< 200	Abe & Sasaki (1)
ARL 18	Adult rat liver	NT	< 10	NT	1	Tong et al. (29)
H4-II-E	Rat hepatoma	NT	NT	< 960	NT	Dearfield et al. (8)
HTC	Rat hepatoma	0.01	NT	< 100	1	Dean et al. (7)
R1	Rat esophageal tumor	NT	< 50	< 300	40	Abe & Sasaki (1)
RL$_1$	Suckling rat liver	NT	NT	NT	0.2	Meyer & Dean (17)
BHK-21	Newborn Syrian hamster kidney	NT	NT	10000	(1)*	Perry (22)
C-HC-4	Human hepatoma	< 0.5	< 40	< 10	< 4	Abe et al. (4)
C-HC-20	Human hepatoma	< 0.5	< 40	10	< 4	Abe et al. (4)
HepG2	Human hepatoblastoma	NT	NT	< 960	NT	Dearfield et al. (8)

Least effective dose, the lowest dose at which the control SCE frequency was doubled showing a dose-related increase for at least three different doses examined. AFB$_1$, aflatoxin B$_1$. BP, benzo[α]pyrene. CP, cyclophosphamide. DMBA, 7,12-dimethylbenz[α]anthracene. NT, not tested. —, induced SCE frequencies less than twice the control value. * only two doses were tested.

activate such chemicals in a given cell system (5). Several metabolically active cultured human and rat cell lines, as detected by the SCE test, are expected to provide a useful in vitro screening system for the detection of genotoxic agents without the use of an exogenous activating system. These cell lines also seem to be useful for studying the metabolic pathways related to SCE induction by indirect mutagens/carcinogens. However, a definite validation of a direct-activating SCE test system utilizing these cells should be reserved until more SCE data on a large number of activation-dependent genotoxins are accumulated by using such cell systems.

ACKNOWLEDGMENTS

The author wishes to thank Professor Motomichi Sasaki for valuable advice and for a critical reading of the manuscript. Our study was supported by grants from the Ministry of Education, Science and Culture, and the Ministry of Health and Welfare, Japan.

REFERENCES

1. Abe, S., and M. Sasaki (1982a) Induction of sister-chromatid exchanges by indirect mutagens/carcinogens in cultured rat hepatoma and esophageal tumor cells and in Chinese hamster Don cells co-cultivated with rat cells. Mutat. Res. 93:409-418.
2. Abe, S., and M. Sasaki (1982b) SCE as an index of mutagenesis and/or carcinogenesis. In Sister Chromatid Exchange, A.A. Sandberg, ed., Alan R. Liss, New York, pp. 461-514.
3. Abe, S., and M. Sasaki (unpublished data).
4. Abe, S., N. Nemoto, and M. Sasaki (1983a) Sister-chromatid exchange induction by indirect mutagens/carcinogens, aryl hydrocarbon hydroxylase activity and benzo(a)pyrene metabolism in cultured human hepatoma cells. Mutat. Res. 109:83-90.
5. Abe, S., N. Nemoto, and M. Sasaki (1983b) Comparison of aryl hydrocarbon hydroxylase activity and inducibility of sister-chromatid exchanges by polycyclic aromatic hydrocarbons in mammalian cell lines. Mutat. Res. 122:47-51.
6. Batt, T.R., J.L. Hsueh, H.H. Chen, and C.C. Huang (1980) Sister chromatid exchanges and chromosome aberrations in V79 cells induced by aflatoxin B_1, B_2, G_1, and G_2 with or without metabolic activation. Carcinogenesis 1:759-763.
7. Dean, R., G. Bynum, D. Kram, and E.L. Schneider (1980) Sister-chromatid exchange induction by carcinogens in HTC cells. An in vitro system which does not require addition of activating factors. Mutat. Res. 74:477-483.
8. Dearfield, K.L., D. Jacobson-Kram, N.A. Brown, and J.R. Williams (1983) Evaluation of a human hepatoma cell line as a target cell in genetic toxicology. Mutat. Res. 108:437-449.
9. Gelboin, H.V. (1980) Benzo(a)pyrene metabolism, activation and carcinogenesis: Role and regulation of mixed-function oxidases and related enzymes. Physiol. Rev. 60:1107-1166.
10. Gurtoo, H.L., and N. Bejba (1974) Hepatic microsomal mixed function oxygenase: Enzyme multiplicity for the metabolism of carcinogens to DNA-binding metabolites. Biochem. Biophys. Res. Commun. 61:735-742.
11. Gurtoo, H.L., and C.V. Dave (1975) In vitro metabolic conversion of aflatoxins and benzo(a)pyrene to nucleic acid-binding metabolites. Cancer Res. 35:382-389.
12. Huh, N., N. Nemoto, and T. Utakoji (1982) Metabolic activation of benzo(a)pyrene, aflatoxin B_1, and diethylnitrosamine by a human hepatoma cell line. Mutat. Res. 94:339-348.
13. Ikeuchi, T., and M. Sasaki (1981) Differential inducibility of chromosome aberrations and sister-chromatid exchanges by indirect mutagens in various mammalian cell lines. Mutat. Res. 90:149-161.
14. Latt, S.A. (1981) Sister chromatid exchange formation. Ann. Rev. Genet. 15:11-55.
15. Latt, S.A., J. Allen, S.E. Bloom, A. Carrano, E. Falke, D. Kram, E. Schneider, R. Schreck, R. Tice, B. Whitefield, and S.

Wolff (1981) Sister-chromatid exchanges: A report of the Gene-Tox program. Mutat. Res. 87:17-62.

16. Madle, S., and G. Obe (1980) Methods for analysis of the mutagenicity of indirect mutagens/carcinogens in eukaryotic cells. Human Genet. 56:7-20.
17. Meyer, A.L., and B.J. Dean (1981) Induction of sister-chromatid exchanges in a rat-liver cell line with chemical carcinogens. Mutat. Res. 91:47-50.
18. Miller, J.A. (1970) Carcinogenesis by chemicals: An overview-G.H.A. Clowes memorial lecture. Cancer Res. 30:559-576.
19. Morgan, W.F., and P.E. Crossen (1982) A comparison of induced sister chromatid exchange levels in Chinese hamster ovary cells and cultured human lymphocytes. Environ. Mutagen. 4:65-71.
20. Nebert, D.W., and H.V. Gelboin (1968) Substrate-inducible microsomal aryl hydroxylase in mammalian cell culture. 1. Assay and properties of induced enzyme. J. Biol. Chem. 243:6242-6249.
21. Pal, K., B. Tierney, P.L. Grover, and P. Sims (1978) Induction of sister-chromatid exchanges in Chinese hamster ovary cells treated in vitro with non-K-region dihydrodiols of 7-methylbenz(a)anthrancene and benzo(a)pyrene. Mutat. Res. 50:367-375.
22. Perry, P.E. (1980) Chemical mutagens and sister-chromatid exchange. In Chemical Mutagens: Principals and Methods for Their Detection, F.J. de Serres and A. Hollaender, eds., Vol. 6, Plenum Press, New York, pp. 1-39.
23. Rüdiger, H.W., F. Kohl, W. Mangels, P. von Wichert, C.R. Bartram, W. Wöhler, and E. Passarge (1976) Benzpyrene induces sister chromatid exchanges in cultured human lymphocytes. Nature (Lond.) 262:290-292.
24. Sasaki, M., K. Sugimura, M.A. Yoshida, and S. Abe (1980) Cytogenetic effects of 60 chemicals on cultured human and Chinese hamster cells. La Kromosomo II-20:574-584.
25. Schreck, R.R., and S.A. Latt (1980) Comparison of benzo(a)pyrene metabolism and sister chromatid exchange induction in mice. Nature (Lond.) 288:407-408.
26. Schreck, R.R., I.J. Paika, and S.A. Latt (1982) Differences in murine procarcinogen activation enzymes are not accompanied by parallel differences in procarcinogen-induced sister-chromatid exchange. Mutat. Res. 94:143-153.
27. Sladek, N.E. (1972) Therapeutic efficacy of cyclophosphamide as a function of its metabolism. Cancer Res. 32:535-542.
28. Tomkins, D.J., S.E. Kwok, G.R. Douglas, and D. Biggs (1982) Sister chromatid exchange response of human diploid fibroblasts and Chinese hamster ovary cells to dimethylnitrosamine and benzo(a)pyrene. Environ. Mutagen. 4:203-214.
29. Tong, C., S. Ved Brat, and G.M. Williams (1981) Sister-chromatid exchange induction by polycyclic aromatic hydrocarbons in an intact cell system of adult rat-liver epithelial cells. Mutat. Res. 91:467-473.

30. Wojciechowski, J.P., P. Kaur, and P.S. Sabharwal (1981) Comparison of metabolic systems required to activate promutagens/carcinogens in vitro for sister-chromatid exchange studies. Mutat. Res. 88:89-97.

ERYTHROCYTE-MEDIATED METABOLIC ACTIVATION DETECTED BY SCE

Hannu Norppa[1] and Francesco Tursi[2]

[1]Institute of Occupational Health
Haartmaninkatu 1
SF-00290 Helsinki 29, Finland

INTRODUCTION

Microsomal metabolism has proved to be essential for the activation of many chemical carcinogens. For this reason in vitro mutagenicity tests include a metabolic activation system. As a rule, the metabolizing mixtures are prepared from the postmitochondrial fraction of rodent liver. While it is generally accepted that the liver is the primary organ of xenobiotic metabolism for a number of chemicals, it is also known that other tissues can contribute both to the activating and inactivating reactions. The ratio between activation and inactivation may be an important parameter in determining the target tissue of a precarcinogen. However, S-9 mixes from sources other than the liver have only seldom been used in mutagenicity tests.

Exogenous metabolizing systems are especially important when established cell lines with poor metabolic activation capacity are used. On the other hand, some types of cultured cells are capable of activating premutagens. Examples of such cells are certain liver-derived cell lines, some embryonic cells, and stimulated lymphocytes. Human lymphocytes and other leukocytes contain mixed-function oxidases able to activate many premutagenic chemicals (1, 2). Consequently, such classical premutagens as cyclophosphamide (CP) and benzo(a)pyrene induce sister chromatid exchanges (SCEs) in cultures of human lymphocytes without any exogenous metabolizing systems (3,4).

---

[2]Istituto di Ricerche Farmacologiche "Mario Negri," Via Eritrea 62, 20157 Milan, Italy.

The whole blood lymphocyte culture system, employed often in cytogenetic tests, differs from most in vitro assays, because the targets of the analysis, the lymphocytes, represent only a small portion of the cells present in the cultures. The predominant cell type is the erythrocyte: the ratio of leukocytes to erythrocytes is usually about 1:1,000. The presence of erythrocytes may greatly influence the results obtained with the test system. Erythrocytes are additional targets for reactive molecules and may thus reduce the cytogenetic effects observed after an in vitro treatment. Another possibility is that erythrocytes are able to activate some premutagens. This was first demonstrated by Ray and Altenburg (5,6), who showed that the induction of SCEs by sodium selenite in cultured human lymphocytes and in a human lymphoblastoid cell line is dependent on the presence of erythrocytes or erythrocyte lysate. Accordingly, some other premutagens, previously thought to be activated in whole blood lymphocyte cultures by leukocytes, might actually be dependent on erythrocytes.

We shall discuss the role of erythrocyte-mediated metabolic activation of premutagens in lymphocyte cultures and other in vitro assays. The data shown mainly derive from our studies on SCE induction by styrene, styrene analogs, and other vinyl compounds.

STYRENE

Several mutagenicity tests have shown that styrene is mutagenic after metabolic activation. The primary reactive metabolite seems to be an epoxide, styrene-7,8-oxide, which is a well-known mutagen (see Ref. 7). Recent studies indicate that styrene also has other mutagenic metabolites, such as arene oxides, but the importance of these intermediates is presently unclear (8).

In human whole blood lymphocyte cultures, styrene is able to induce structural chromosome aberrations, SCEs, and micronuclei without exogenous metabolic activation systems (9,10). Gas chromatographic analyses have shown that styrene-7,8-oxide is formed as a function of incubation time in such cultures, which indicates that the cultured cells are capable of oxidizing styrene (10).

The effect of styrene on SCEs in cultured human lymphocytes is dependent on the presence and number of erythrocytes (11). In cultures of isolated lymphocytes, which contain only 0.01% of the erythrocytes present in whole blood cultures, the clear dose-dependent effect of styrene on SCEs almost completely disappears (Fig. 1). Furthermore, the addition of different amounts of erythrocytes in purified lymphocyte cultures results in a gradual increase in styrene-induced SCEs (Tab. 1). These findings suggest that styrene is metabolically activated by erythrocytes in this in vitro system and that the reactive intermediate is styrene-7, 8-oxide. Cyclophosphamide, a well-known premutagen tested for comparison, does not require erythrocytes to induce SCEs (Fig. 1).

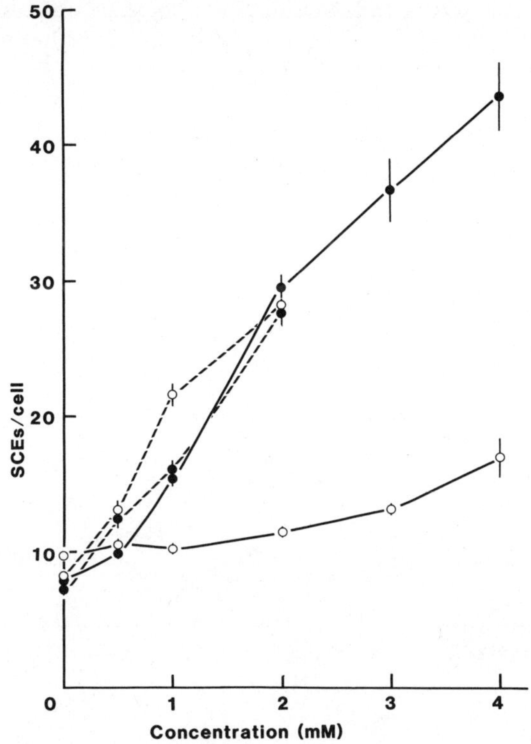

Fig. 1. The induction of SCEs by styrene (solid line; combined results from cultures of 2 persons) and CP (broken line) in whole blood lymphocyte cultures (with 2 x 10^8 erythrocytes/ml; black symbols) and in lymphocytes isolated with Ficoll-Paque^R (with less than 2 x 10^4 erythrocytes/ml; open symbols). The compounds were present in the 72 hr cultures for the last 48 hr. The symbols represent means, bars are standard errors. (Based on data in Ref. 11.)

The oxidation of styrene to styrene-7,8-oxide by human erythrocytes has also been demonstrated in gas chromatographic studies based on the indirect analysis of styrene-7,8-oxide (12). In whole blood, the formation of styrene-7,8-oxide is entirely explained by the action of the erythrocytes. The metabolizing capacity of erythrocytes is not restricted only to human cells--rat red blood cells are equally efficient in styrene oxidation (13). Styrene increases SCE also in whole blood lymphocyte cultures of rat (14).

Styrene has given contradictory results in mutagenicity assays using liver microsomal fractions for metabolic activation (7). The main reason for this appears to be the better effectivity of styrene-7,8-oxide inactivation as compared to styrene oxidation (15).

Tab. 1. SCEs in purified human lymphocyte cultures (72 hr) after treatment with styrene (2 mM, 48 hr) in the presence of different amounts of erythrocytes (11).

Treatment	Number of erythrocytes/ml	Number of metaphases analyzed	SCEs/cell ± S.E.
Control (acetone)	2×10^4	50	9.3 ± 0.5
Styrene	2×10^4	50	9.5 ± 0.5
	2×10^5	50	11.0^a ± 0.5
	2×10^6	50	9.6 ± 0.4
	2×10^7	50	12.4^b ± 0.6
	2×10^8	50	21.5^b ± 0.8
	2×10^9	10	38.0^b ± 2.3

Probabilities counted according to 1-tailed t-test:

^{a}P < 0.05

^{b}P < 0.001

S.E. = standard error

The negative result with styrene in the conventional SCE test with CHO cells with or without S-9 mix is an example (Ref. 16 and Tab. 2). However, the addition of erythrocytes to cultures of Chinese hamster ovary (CHO) cells results in SCE induction if long treatment times are used (Tab. 2).

Results from gas chromatographic studies suggest that the oxidation of styrene in human erythrocytes is catalyzed by oxyhemoglobin. The formation of styrene-7,8-oxide by erythrocytes requires oxygen and depends on the molar fraction of oxyhemoglobin (17). Furthermore, in erythrocytes where oxyhemoglobin has been oxidized to methemoglobin by sodium nitrite, the oxidation of styrene correlates with the disappearance of methemoglobin if methylene blue is used to reduce methemoglobin. Styrene-7,8-oxide is similarly formed in a reconstituted system containing methemoglobin, NADPH, glucose, and rat liver cytochrome c reductase--methylene blue stimulates this reaction (18). The oxidation of styrene also takes place in another artificial system consisting of methemoglobin and either hydrogen peroxide or cumene hydroperoxide. Methemoglobin is probably converted into oxyhemoglobin in the presence of these peroxides (19). Styrene oxidation is not the only example of the catalytic activity

Tab. 2. Mean SCE frequencies (± S.E.) in CHO cells after styrene treatment in the presence and absence of erythrocytes (RBCs) (2×10^8/ml) or Wistar rat liver S-9 mix (induced with Clophen A50). If not otherwise shown, 60 harlequin stained metaphases per treatment were analyzed for SCEs (14,31).

Treatment	Without S-9 mix	With S-9 mix	Without RBCs		With RBCs	
Concentration (mM)	(4 h)	(4 h)	24 h	34 h	24 h	34 h
Control (no treatment)	8.8 ± 0.4	10.0 ± 0.5				
Control (acetone)	8.1 ± 0.4	10.8 ± 0.5	9.9 ± 0.4	9.5 ± 0.5	8.9 ± 0.3[c]	9.4 ± 0.5
Cyclophosphamide						
0.01	9.6 ± 0.5	33.9[b] ± 0.8				
Styrene						
1	9.2 ± 0.4	10.6 ± 0.5				
2			10.4 ± 0.5		8.5 ± 0.5	
4			9.6 ± 0.4		9.3 ± 0.5	
5	10.2[a] ± 0.4	10.4 ± 0.5				
8			9.9 ± 0.5		14.6[b] ± 0.5[c]	
10	9.7 ± 0.4	11.5 ± 0.6	9.9 ± 0.5[d]		17.0[b] ± 0.7	
12			9.5 ± 0.5[e]	10.9[a] ± 0.5	17.5[b] ± 0.6	23.5[b] ± 0.6
15	9.5 ± 0.5	11.4 ± 0.4	No 2nd div. cells	13.7[b] ± 0.5	No 2nd div. cells	27.3[b] ± 0.9
20	No 2nd div. cells	11.3 ± 0.4				

[a] $p < 0.05$, 1-tailed t-test; [b] $p < 0.001$, 1-tailed t-test; [c] 120 cells analyzed; [d] 57 cells analyzed; [e] 30 cells analyzed

of hemoglobin. Such reactions as the *p*-hydroxylation of aniline and the decarboxylation of 3,4-dihydroxyphenylalanine are supported by hemoglobin (20,21).

The above in vitro data quite conclusively show that styrene can be activated in cell cultures by erythrocyte metabolism. Styrene is converted into styrene-7,8-oxide probably through the action of oxyhemoglobin. Human blood contains 10-25 times more erythrocytes than ordinary whole blood lymphocyte cultures--one could thus expect that some activation of styrene might also take place in erythrocytes in vivo. Actually, chromosome aberrations and SCEs have been reported in lymphocytes of styrene-exposed workers of the reinforced plastics industry (22-24), and styrene-7,8-oxide has been detected in blood circulation of men and mice after styrene exposure (25). However, no direct evidence for the importance of erythrocytes in styrene metabolism exists in vivo. In comparison with hepatic metabolism, erythrocyte-dependent oxidation of styrene is probably negligible, but may be important locally in blood circulation, especially for the formation of lesions seen as chromosome damage in cultured lymphocytes.

OTHER VINYL COMPOUNDS

Several analogs of styrene, differing from the parent compound by 1 or 2 substitutions, are able to induce SCEs in human whole blood lymphocyte cultures without exogenous metabolic activation (26-28). In analogy with styrene, these derivatives are probably converted into reactive metabolites by the cultured cells, as they are not known to be direct mutagens. The presence of the vinyl side chain appears to be important for the SCE-inducing capacity of the compounds, as related chemicals--lacking the double bond in the side chain--are negative (2-phenylethanol) or weak (ethylbenzene and phenylacetaldehyde) in SCE induction. Furthermore, substitution of the vinyl chain seems to modify the SCE-inducing potency of the analogues, so that α-methylstyrene and *cis*-β-methylstyrene are less effective than *trans*-β-methylstyrene and 4-methoxy-*trans*-β-chlorostyrene. The substitution of the ring has less effect on SCE induction (26-28).

In a recent experiment (28), we have studied the role of erythrocytes in the induction of SCEs by the previously tested styrene analogs. The results, summarized in Tab. 3, indicate that the ring-methylated and *trans*-β-substituted derivatives are almost completely (2-, 3-, and 4-methylstyrene and 4-methoxy-*trans*-β-chlorostyrene) or largely (*trans*-β-methylstyrene) dependent on erythrocytes. The presence of erythrocytes is not important for SCE induction by α-methylstyrene and *cis*-β-methylstyrene.

These findings suggest that styrene analogs with intact or *trans*-β-substituted vinyl chain are activated by erythrocytes in

Tab. 3. SCE induction by styrene analogs in human lymphocytes cultured (72 hr) with whole blood (2×10^8 erythrocytes/ml) or isolated with Ficoll-Paque[R] (less than 1,000 erythrocytes/ml). The chemicals were present in the cultures for the last 48 hr (28).

Treatment	Whole blood		Isolated lymphocytes	
(mM)	Cells analyzed	SCEs/cell ± S.E.	Cells analyzed	SCEs/cell ± S.E.
Control I	50	7.6 ± 0.4	50	10.1 ± 0.5
α-methylstyrene				
1	50	8.9[a] ± 0.4	50	11.1 ± 0.6
2	50	11.4[c] ± 0.6	33	12.1[b] ± 0.6
cis-β-methylstyrene				
1	50	11.6[c] ± 0.5	50	16.4[c] ± 0.6
2	50	13.7[c] ± 0.7	50	21.7[c] ± 0.6
trans-β-methylstyrene				
0.5	50	25.8[c] ± 0.8	50	16.9[c] ± 0.6
1	50	37.8[c] ± 1.1	50	25.3[c] ± 0.6
4-methoxy-trans-β-chlorostyrene				
0.1	50	18.8[c] ± 0.8	50	13.9[c] ± 0.6
0.2	50	30.9[c] ± 1.3	50	14.2[c] ± 0.5
Control II	50	8.3 ± 0.4	50	10.1 ± 0.5
2-methylstyrene				
1	25	11.9[c] ± 0.8	50	10.8 ± 0.4
2	50	16.5[c] ± 0.8	50	11.9[b] ± 0.6
3-methylstyrene				
1	50	13.9[c] ± 0.6	50	11.1 ± 0.5
2	50	20.4[c] ± 0.7	28	13.2[c] ± 0.8
4-methylstyrene				
1	50	22.8[c] ± 1.0	50	12.3[c] ± 0.5
2	25	20.6[c] ± 1.1	50	12.5[c] ± 0.5

Probabilities counted according to 1-tailed t-test:
[a] $p < 0.05$
[b] $p < 0.01$
[c] $p < 0.001$

human lymphocyte cultures, while derivatives with α- or cis-β-substitution are not. Thus, the reaction would be very specific with respect to the structure of the compound.

Thus far, the compounds reported to be metabolized by hemoglobin have been aromatic (12,20,21). Does the activation of vinyl compounds in erythrocytes require the phenyl ring? To shed some light on this question, we have studies the induction of SCEs by vinyl compounds with diverse chemical structures: 3 of them were aliphatic (vinylacetate, acrylonitrile, and acrylamide), 2 were polyaromatic (4-vinylbiphenyl and 9-vinylanthracene), and 1 was alicyclic (N-vinyl-2-pyrrolidone). With the exception of vinylacetate, the effect of these compounds on SCEs in human whole blood lymphocyte cultures was weak (28-30). However, vinylacetate turned out to be an exceptionally potent inducer of SCEs. Vinylacetate, acrylonitrile, acrylamide, and N-vinyl-2-pyrrolidone were also tested in cultures of isolated lymphocytes. For the latter 3 chemicals the slight increase in SCEs remained also in this system, whereas vinylacetate elevated SCEs even more than in whole blood cultures. In other words, none of these chemicals seemed to be activated by the presence of erythrocytes (28-30). These results may suggest that, in addition to the vinyl group, the phenyl ring is important for the erythrocyte-mediated activation.

Vinylacetate has recently been shown to induce structural chromosome aberrations in human lymphocytes and SCEs, both with and without S-9 mix, in CHO cells (Fig. 2; Refs. 30,31). Vinylacetate does not seem to need microsomal metabolism to produce chromosome damage--in contrast, esterases may break this compound down to acetaldehyde, which is an efficient in vitro inducer of both SCEs and structural chromosome aberrations (30,32).

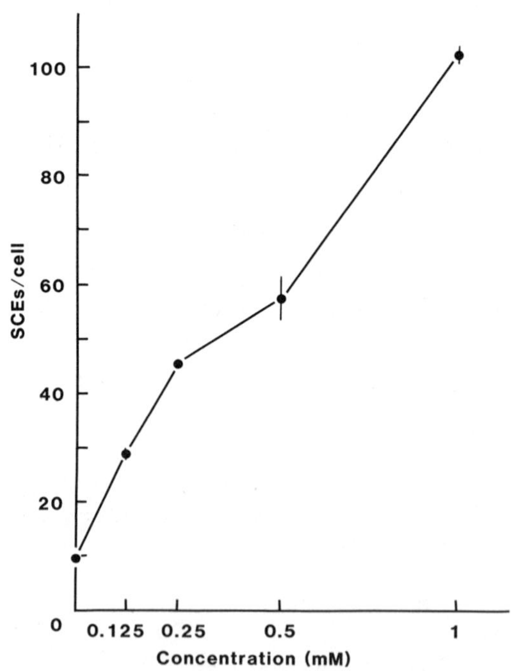

Fig. 2. The induction of SCEs in CHO cells by a 24-hr treatment with vinylacetate. Means ± S.E. are shown (31).

CONCLUSIONS

The list of compounds suggested to be metabolically activated by erythrocytes is not long (Tab. 4). On the other hand, the possibility for erythrocyte-mediated activation in mutagenicity tests was first pointed out only five years ago, and studies on the subject are few. At least for the oxidation of styrene (and possibly styrene analogs) the key molecule seems to be hemoglobin, which is normally not regarded as an enzyme, although it is structurally related to cytochrome P-450 and catalyzes a number of reactions.

Tab. 4. Compounds suggested to be dependent or independent on erythrocyte-mediated metabolic activation, as judged by the induction of SCEs in cultured human lymphocytes.

Compound	References
Erythrocyte-dependent	
Sodium selenite	5, 6
Styrene	11, 14
2-Methylstyrene	28
3-Methylstyrene	28
4-Methylstyrene	28
Trans-ß-methylstyrene	28
4-Methoxy-*trans*-ß-chlorostyrene	28
Aniline HCl	34
Erythrocyte-independent	
Cyclophosphamide	11
Benzo(a)pyrene[a]	4, 33
α-Methylstyrene	28
Cis-ß-methylstyrene	28
Vinylacetate	14, 29, 30
Acrylonitrile	29
Acrylamide	29
N-vinyl-2-pyrrolidone	28

[a]Reported to be an effective inducer of SCEs both in whole-blood cultures and in isolated lymphocytes.

The discovery of the metabolic activation capacity of the erythrocytes may lead to re-evaluation of some earlier results obtained with the lymphocyte assays. One should at least take into account the presence of erythrocytes when testing chemicals in whole blood lymphocyte cultures. If the erythrocyte-mediated activation were widely distributed among premutagens, erythrocytes could actually be used as a metabolic activation system for other in vitro mutagenicity tests as well. The isolation of erythrocytes from blood is much easier than the preparation of S-9 mix. The phenomenon may, however, be restricted to a quite limited number of chemicals, as the erythrocyte-dependent SCE induction by styrene analogs is closely related to the structure of the compound. As long as no direct information exists in vivo, the activation of premutagens by erythrocytes should be viewed as an in vitro phenomenon. There are, however, no reasons to believe that such reactions could not take place also in blood circulation.

ACKNOWLEDGEMENTS

This report was supported financially by the Swedish Work Environment Fund (H.N.: Grant 83-0349) and the European Science Foundation (F.T.). We are grateful to Leila Kuusela and Raija Marttala for typing the manuscript.

REFERENCES

1. Whitlock, J.P., H.L. Cooper, and H.V. Gelboin (1972) Aryl hydrocarbon (benzopyrene) hydroxylase is stimulated in human lymphocytes by mitogens and benz(a)anthracene. Science 177: 618-619.
2. Okano, P., H.N. Miller, R.C. Robinson, and H.V. Gelboin (1979) Comparison of benzo(a)pyrene and (-)-trans-7,8-dihydroxy-7,8-dihydrobenzo(a)pyrene metabolism in human blood monocytes and lymphocytes. Cancer Res. 39:3184-3193.
3. Waalkens, D.H., H.F.P. Joosten, R.D.F.M. Taalman, J.M.J.C. Scheres, T.D. Yih, and A. Hoekstra (1981) Sister-chromatid exchanges induced in vitro by cyclophosphamide without exogenous metabolic activation in lymphocytes from three mammalian species. Toxicol. Lett. 7:229-232.
4. Rüdiger, H.W., F. Kohl, W. Mangels, P. von Wichert, C.R. Bartram, W. Wöhler, and E. Passarge (1976) Benzypyrene induces sister chromatid exchanges in cultured human lymphocytes. Nature 262:290-292.
5. Ray, J.H., and L.C. Altenburg (1978) Sister-chromatid exchange induction by sodium selenite: Dependence on the presence of red blood cells or red blood cell lysate. Mutat. Res. 54:343-354.
6. Ray, J.H., and L.C. Altenburg (1982) Sister-chromatid exchange

induction by sodium selenite. Plasma protein-bound selenium is not the active SCE-inducing metabolite of Na_2SeO_3. Mutat. Res. 102:285-296.

7. Norppa, H., and H. Vainio (1983) Genetic toxicity of styrene and some of its derivatives. Scand. J. Work Environ. Health 9:108-114.
8. Watabe, T., A. Hiratsuka, T. Aizawa, T. Sawahata, N. Ozawa, M. Isobe, and E. Takabatake (1982) Studies on metabolism and toxicity of styrene IV. 1-Vinyl-benzene 3,4-oxide, a potent mutagen formed as a possible intermediate in the metabolism of styrene to 4-vinylphenol. Mutat. Res. 93:45-55.
9. Linnainmaa, K., T. Meretoja, M. Sorsa, and H. Vainio (1978) Cytogenetic effects of styrene and styrene oxide. Mutat. Res. 58:277-286.
10. Norppa, H., M. Sorsa, P. Pfäffli, and H. Vainio (1980) Styrene and styrene oxide induce SCEs and are metabolised in human lymphocyte cultures. Carcinogenesis 1:357-361.
11. Norppa, H., H. Vainio, and M. Sorsa (1983) Metabolic activation of styrene by erythrocytes detected as increased sister chromatid exchanges in cultured human lymphocytes. Cancer Res. 43: 3579-3582.
12. Belvedere, G., and F. Tursi (1981) Styrene oxidation to styrene oxide in human blood erythrocytes and lymphocytes. Res. Comm. Chem. Pathol. Pharmacol. 33:273-282.
13. Vainio, H., F. Tursi, and G. Belvedere (1982) What are the significant toxic metabolites of styrene? In Cytochrome P-450, Biochemistry, Biophysics and Environmental Implications, E. Hietanen, M. Laitinen, and O. Hänninen, eds. Elsevier Biomedical Press B.V., pp. 679-687.
14. Norppa, H., F. Tursi, and P. Einistö (1984) Erythrocytes as a metabolic activation system in mutagenicity tests. In Proceedings of the 13th Annual Meeting of the European Environmental Mutagen Society, September 5-9 1983, Montpellier, France. Inserm Symposia Series, Paris (in press).
15. Bauer, C., C. Leporini, G. Bronzetti, C. Corsi, R. Nieri, R. Del Carratore, and S. Tonarelli (1980) The problem of negative results for styrene in the in vitro mutagenesis test with metabolic activation (microsomal assay): Explanation by gas chromotographic analysis. Boll. Soc. Ital. Biol. Sper. 56:203-207.
16. de Raat, W.K. (1978) Induction of sister chromatid exchanges by styrene and its presumed metabolite styrene oxide in the presence of rat liver homogenate. Chem.-Biol. Interact. 20:163-170.
17. Tursi, F., M. Samaia, and G. Belvedere (1983) Styrene oxidation to styrene oxide in human erythrocytes is catalyzed by oxyhemoglobin. Experientia 39:593-594.
18. Belvedere, G., F. Tursi, and H. Vainio (1983) Non-microsomal activaton of styrene to styrene oxide. In Extrahepatic Drug Metabolism and Chemical Carcinogenesis, J. Rydström, J. Montelius, and M. Bengtsson, eds. Elsevier Science Publishers B.V., pp. 193-200.

19. Cantoni, L., D. Blezza, and G. Belvedere (1982) Effect of iron and hemoproteins on hydrogen peroxide-supported styrene oxidation to styrene oxide. Experientia 38:1192-1194.
20. Mieyal, J.J., R.S. Ackerman, J.L. Blumer, and L.S. Freeman (1976) Characterization of enzyme-like activity of human hemoglobin. Properties of the hemoglobin-P-450 reductase-coupled aniline hydroxylase system. J. Biol. Chem. 251:3436-3441.
21. Yamabe, H., and W. Lovenberg (1972) Decarboxylation of 3,4-dihydroxyphenylalanine by oxyhemoglobin. Biochem. Biophys. Res. Comm. 47:733-739.
22. Meretoja, T., H. Vainio, M. Sorsa, and H. Härkönen (1977) Occupational styrene exposure and chromosomal aberrations. Mutat. Res. 56:193-197.
23. Andersson, H.C., E.Å. Tranberg, A.H. Uggla, and G. Zetterberg (1980) Chromosomal abberations and sister-chromatid exchanges in lymphocytes of men occupationally exposed to styrene in a plastic boat factory. Mutat. Res. 73:387-401.
24. Camurri, L., S. Codeluppi, C. Pedroni, and L. Scarduelli (1983) Chromosomal aberrations and sister-chromatid exchanges in workers exposed to styrene. Mutat. Res. 119:361-369.
25. Byfält Nordqvist, M., E. Gullstrand, E. Lundgren, E.-M. Nydahl, A. Löf, and E. Wigaeus (1983) Förekomst av styren-7,8-oxid och styrenglykol i mus och människa efter administrering av styren. In 32:a Nordiska Yrkeshygieniska Mötet, September 19-21 1983, Stockholm, Sweden. Arbetarskyddsstyrelsen, Stockholm, pp. 77-78.
26. Norppa, H. (1981) The in vitro induction of sister chromatid exchanges and chromosome aberrations in human lymphocytes by styrene derivatives. Carcinogenesis 3:237-242.
27. Norppa, H., and H. Vainio (1983) Induction of sister-chromatid exchanges by styrene analogs in cultured human lymphocytes. Mutat. Res. 116:379-387.
28. Tursi, F., and H. Norppa (unpublished data).
29. Norppa, H., and F. Tursi (1983) Induction of sister chromatid exchanges by vinyl compounds in cultured human lymphocytes. In Nordic Symposium on Factors Affecting Mutagenicity and Evaluation of Mutagenicity Data. June 8-10 1983., Stockholm, Sweden, Abstracts, pp. 25.
30. Mäki-Paakkkanen, J., F. Tursi, H. Norppa, H. Järventaus, and M. Sorsa (1983) Vinylacetate is a potent inducer of chromosome damage in cultured mammalian cells. In 13th Annual Meeting of European Environmental Mutagen Society. September 5-9 1983, Montpellier, France, Abstracts, II-3-C-7.
31. Norppa, H., and H. Järventaus (unpublished data).
32. Böhlke, J.U., S. Singh, and H.W. Goedde (1983) Cytogenetic effects of acetaldehyde in lymphocytes of Germans and Japanese: SCE, clastogenic activity, and cell cycle delay. Human Genet. 63:285-289.
33. Hopkin, J.M., and P.E. Perry (1980) Benzo(a)pyrene does not contribute to the SCEs induced by cigarette smoke condensate. Mutat. Res. 77:377-381.

34. Wilmer, J.L., G.L. Erexson, and A.D. Kligerman (1984) The effect of erythrocytes and hemoglobin on sister chromatid exchange (SCE) induction in cultured human lymphocytes exposed to aniline HCl. In Sister Chromatid Exchange, R.R. Tice and A. Hollaender, eds. Plenum Press, New York.

THE EFFECT OF ERYTHROCYTES AND HEMOGLOBIN ON SISTER CHROMATID EXCHANGE INDUCTION IN CULTURED HUMAN LYMPHOCYTES EXPOSED TO ANILINE HCl

James L. Wilmer, Gregory L. Erexson,
and Andrew D. Kligerman

Chemical Industry Institute of Toxicology
Department of Genetic Toxicology
P.O. Box 12137
Research Triangle Park, North Carolina 27709

ABSTRACT

Erythrocytes [red blood cells (RBCs)] possess aniline hydroxylase activity. When aniline interacts with ferrohemoglobin in the presence of molecular oxygen, oxidation of nitrogen and ring carbons occurs. Thus, apart from the liver, RBCs may represent an important site of aniline metabolism. Because 2 metabolites of aniline, o-aminophenol and phenylhydroxylamine, can induce sister chromatid exchanges (SCEs), we examined the ability of RBCs and hemoglobin to activate aniline to genotoxic intermediates as evidenced by SCE induction in human lymphocytes. Aniline HCl (0.05-1.0 mM) induced significant concentration-related increases in the SCE frequency only in the whole blood cultures. Similarly, inhibition of cell cycle kinetics by aniline was observed only in the whole blood cultures, as shown by a concentration-dependent decrease in the percentage of third- (and later) division metaphases. Mitotic indices were not affected significantly at any concentration of aniline or hemoglobin. Hemoglobin (500 or 1,000 μg/ml) alone induced significant concentration-related increases in SCEs in the mononuclear leukocyte cultures. Therefore, human mononuclear leukocytes do not activate aniline to genotoxic intermediates capable of inducing SCEs during a 48-hr exposure. However, the inclusion of RBCs and granulocytes provides an activation system as demonstrated by a small, but statistically significant increase in the SCE frequency in the whole blood cultures. The weak genotoxicity of hemoglobin may be related to production of oxygen radicals during autoxidation to methemoglobin. Thus, a possible mechanism of aniline-induced splenic toxicity in rats may be the combined genotoxic effect of aniline metabolites generated or accumulated in the RBC-engorged spleen and hemoglobin released during phagocytosis of damaged RBCs.

INTRODUCTION

The spleen is the primary target organ for aniline-induced toxicity in the Fischer-344 (F-344) rat. This toxicity is manifested as increased splenic weight (12), elevated erythropoietic activity (12,20), increased hemosiderin content (12,21), sinusoidal engorgement (12), fibrous hyperplasia (9,16), and induction of hemangio- and fibroscarcomas (9,16). The exact mechanism for aniline-induced splenic toxicity is not known, but the RBC appears to play a major role. For example, RBCs can metabolize aniline to p-aminophenol (4) and possibly phenylhydroxylamine (11). Ferrohemoglobin ($HbFe^{2+}$) in the presence of molecular oxygen appears to be the primary component of the RBC responsible for the hydroxylation of aniline (4). $HbFe^{2+}$ is also readily oxidized to ferrihemoglobin ($HbFe^{3+}$) by hydroxylated aniline metabolites and oxygen (14). Heinz body formation (denatured hemoglobin) is increased during aniline treatment (12), but $HbFe^{3+}$ production is not necessarily the causative event (14). Also, covalently bound ^{14}C-aniline or metabolites accumulate preferentially in RBCs and spleens of male F-344 rats following either a 1- or 10-da exposure to orally administered ^{14}C-aniline HCl (7,8,22).

There are at least 3 possible mechanisms for the mediation of splenic damage by RBCs. First, aniline is metabolized primarily in the liver by cytochrome P-450 monoxygenases to the nitrogen- and ring-hydroxylated products (phenylhydroxylamine, o- and p-aminophenols) (3,14,18). Then, RBCs transport aniline and its metabolites to the spleen where macrophages scavenge damaged RBCs (24), subsequently leading to release of cytotoxic and genotoxic metabolites during erythroclasia (7,26). Second, if the phagocytic function of splenic macrophages is overtaxed or impaired, then sinusoidal congestion involving abnormal numbers of RBCs may result (24). Because RBCs possess aniline hydroxylase activity (4), RBCs or liberated $HbFe^{2+}$ may activate aniline directly in the spleen (7). Third, the release of $HbFe^{2+}$ in the spleen and accumulation of hemosiderin in macrophages may lead to iron toxicity due to autoxidative processes (6,7). The aim of this study was to determine if human mononuclear leukocytes, whole blood, or $HbFe^{2+}$ could activate aniline to genotoxic intermediates capable of inducing SCEs.

MATERIALS AND METHODS

Blood removal, blood processing, and cell culture were as described previously (25). Venous blood was drawn from a healthy male donor (33 yr old) and processed on a Ficoll-Paque^{R} (Pharmacia Fine Chemicals, Piscataway, New Jersey) density gradient to separate mononuclear leukocytes from RBCs and granulocytes (5). Duplicate whole blood and mononuclear leukocyte cultures were established by inoculating 10^6 nucleated cells into 1.9 ml of medium composed of RPMI 1640, 10% heat-inactivated fetal bovine serum, 100 units of penicillin and 100 μg of streptomycin sulfate/ml, and an additional 292 μg L-glutamine/ml. T lymphocytes were stimulated with either 30 or 4 μg concanavalin A/ml in the whole blood and mononuclear leukocyte cultures, respectively. The cultures were incubated at 37°C in a humidified CO_2 atmosphere for 24 hr at which time 5-bromo-2'-deoxyuridine (5 μM) and aniline HCl (0.05-1.0 mM) were added. Aniline HCl (Eastman Chemical Co., Rochester, New York) and human hemoglobin (Sigma Chemical Co., St. Louis, Missouri were dissolved in RPMI 1640 and sterilized through 0.22 μm Millex-GS filter units (Millipore Corp., Bedford, Massachusetts). Aliquots of 20.0 and 42.0 μl were used for aniline and hemoglobin, respectively. Hemoglobin (500 or 1,000 μg/ml) and graded concentrations of aniline HCl were added to the mononuclear leukocyte cultures at 24 hr after culture initiation. The cultures were harvested at 72 hr following a final 4-hr treatment with 1.35 μM demecolcine. The cytogenetic preparations were stained using a modified fluorescence-plus-Giemsa technique (26). A total of 50 second-division metaphases, 200 consecutive metaphases, and 2,000 nuclei were analyzed from each treatment for SCE frequency, cell cycle kinetics, and mitotic index, respectively. The SCE data were tested for equality of variance by the F-test for variance ratio and then subjected to a one-way analysis-of-variance with the level of significance chosen as 0.05 (21).

RESULTS

Aniline HCl induced statistically significant concentration-related increases in the SCE frequency only in the whole blood cultures (Tab. 1). The addition of 500 μg hemoglobin/ml to the mononuclear leukocyte cultures did not activate aniline. Although 1,000 μg hemoglobin/ml appeared to activate aniline at concentrations of 0.5 and 1.0 mM, the SCE increase was not statistically significant. Hemoglobin alone caused statistically significant concentration-related increases in the SCE frequency in the mononuclear leukocyte cultures. A statistically significant inhibition of cell cycle kinetics was observed only in the whole blood cultures treated with aniline as shown by a concentration-dependent decrease in the percentage of third-division (and later) metaphases (Tab. 1). Mitotic indices were not affected significantly at any concentration of aniline or hemoglobin (Tab. 1).

DISCUSSION

These results demonstrate that human mononuclear leukocytes do not activate aniline to genotoxic intermediates capable of inducing SCEs during a 48-hr exposure. However, the inclusion of RBCs and

Tab. 1. Effect of erythrocytes and hemoglobin on SCE induction in cultured human lymphocytes exposed to aniline HCl.

Aniline HCl Concentation	SCE Frequency (x ± S.D.)	Mitotic Index (x ± S.D.)	Cell Cycle Kinetics (x ± S.D.) 1st	2nd	≥3rd
Whole Blood					
Control	7.9 ± 0.4*	3.3 ± 1.2	15.5 ± 0.7	46.5 ± 2.1	38.0 ± 1.4**
0.05 mM	8.9 ± 0.2	2.3 ± 0.2	25.5 ± 2.1	45.0 ± 2.8	29.5 ± 4.9
0.10 mM	9.0 ± 1.0	2.5 ± 0.7	20.0 ± 2.8	48.5 ± 2.1	31.5 ± 4.9
0.50 mM	10.1 ± 0.1	1.9 ± 0.6	30.0 ± 4.5	42.0 ± 7.1	28.0 ± 2.8
1.00 mM	11.6 ± 0.6	2.1 ± 0.0	21.5 ± 2.1	55.5 ± 0.7	23.0 ± 1.4
Purified PBLs					
Control	8.6 ± 0.1†	11.5 ± 4.0	11.5 ± 9.2	34.0 ± 8.5	54.5 ± 17.7
0.05 mM	8.7 ± 0.3	9.2 ± 0.6	11.5 ± 0.7	37.5 ± 2.1	50.5 ± 0.7
0.10 mM	8.6 ± 1.5	8.4 ± 4.2	9.5 ± 0.7	25.0 ± 5.7	65.5 ± 0.7
0.50 mM	8.9 ± 0.2	5.1 ± 1.4	16.0 ± 4.2	35.5 ± 3.5	47.5 ± 6.4
1.00 mM	9.7 ± 0.4	8.0 ± 2.5	15.0 ± 4.2	31.0 ± 1.4	54.0 ± 5.7
Purified PBLs + 500 μg Hemoglobin/ml					
500 μg Hb/ml	10.4 ± 0.3†	8.2 ± 0.8	12.0 ± 2.8	36.5 ± 2.1	51.5 ± 4.9
0.05 mM	9.7 ± 0.3	6.5 ± 0.9	13.5 ± 3.5	34.5 ± 2.1	52.0 ± 1.4
0.10 mM	10.2 ± 0.5	7.9 ± 0.1	12.5 ± 3.5	37.5 ± 6.4	50.5 ± 10.6
0.50 mM	11.0 ± 0.8	7.1 ± 0.6	15.5 ± 0.7	35.0 ± 5.6	49.5 ± 16.3
1.00 mM	10.5 ± 0.5	3.7 ± 3.0	21.0 ± 12.7	31.5 ± 3.5	47.5 ± 16.3
Purified PBLs + 1000 μg Hemoglobin/ml					
1000 μg Hb/ml	11.6 ± 0.1†	6.3 ± 1.2	12.4 ± 4.2	40.0 ± 1.4	48.0 ± 5.7
0.05 mM	11.5 ± 1.8	6.8 ± 2.0	18.5 ± 7.8	29.0 ± 0.0	52.5 ± 7.8
0.10 mM	12.3 ± 1.0	6.4 ± 0.9	11.5 ± 0.7	38.0 ± 9.9	50.0 ± 8.5
0.50 mM	13.3 ± 0.4	7.7 ± 0.2	17.0 ± 8.5	38.0 ± 2.8	45.0 ± 5.7
1.00 mM	14.3 ± 0.6	5.1 ± 0.3	16.0 ± 11.3	40.0 ± 1.4	44.0 ± 12.7

* Significant increase in SCE frequency by aniline HCl in whole blood cultures using one-way ANOVA, $p < 0.05$

** Significant cell cycle delay induced by aniline HCl in whole blood cultures using one-way ANOVA, $p < 0.05$

† Significant increase in SCE frequency in purified PBLs (mononuclear leukocytes) exposed to hemoglobin alone using one-way ANOVA, $p < 0.05$.

granulocytes provides an activation system as shown by a small, but statistically significant increase in the SCE frequency and significant perturbation of the cell cycle in lymphocytes from whole blood cultures. These data are consistent with other studies that have shown that human RBCs can hydroxylate aniline to p-aminophenol (and other unidentified products) (4) and possibly phenylhydroxylamine (11). Phenylhydroxlamine and o-aminophenol are capable of inducing SCEs and inhibiting cell cycle kinetics at relatively low concentrations in human fibroblasts during a 2-hr exposure, whereas p-aminophenol inhibits only cell cycle kinetics (26). Similarly, o-aminophenol (15) and, to a much lesser extent, p-aminophenol (15,23) can induce SCEs in human lymphocytes from whole blood cultures exposed over 48-72 hr. Thus, it appears that certain cell culture systems including Chinese hamster lung fibroblasts (Don cell line) (1), the RL_4 rat liver epithelial cell line (10), and human whole blood can metabolize aniline to genotoxic products. Our results contrast with a previous study in which human whole blood lymphocytes stimulated with phytohemagglutinin did not exhibit a significant increase in SCEs when exposed to 1.0 mM aniline over a 48-hr period (23).

Although free and membrane-bound $HbFe^{2+}$ is thought to be responsible for oxidation of aniline (4), our data show that aniline activation to genotoxic intermediates was minor in mononuclear leukocyte cultures reconstituted with hemoglobin. This lower level of metabolism may be attributable to several factors. 1) The catalytically active $HbFe^{2+}$ was probably autoxidizing to $HbFe^{3+}$ under the culture conditions used (14), and a large proportion of the hemoglobin could have been oxidized during purification procedures. 2) Because an adult human RBC contains about 29 pg of hemoglobin (2) and the ratio of RBCs to leukocytes is about 1,000:1 (17), the total amount of hemoglobin in the whole blood cultures is 14.5-29.0 times the amount of purified hemoglobin added to the mononuclear leukocyte cultures. Therefore, the extent of aniline metabolism may have been lowered by the amount of enzyme present. 3) No cofactors such as NADPH, NADH, cytochrome b5 and NADH-cytochrome b5 reductase were added to the cultures (4,14), and, hence, methemoglobin ($HbFe^{3+}$) might have formed rapidly, especially in the presence of aniline and oxygen. These cofactors are normally present in the mature RBC and act as a methemoglobin reductase system. Thus, besides liver S-9 mix activation, it may be possible to use cell-free, reconstituted enzymes and cofactors from RBCs to activate or detoxify other xenobiotic compounds using mammalian cells as targets. Because RBCs can metabolize compounds such as styrene and its analogues (17) and sodium selenite (19) to potent genotoxic intermediates, the incorporation of this cell type into short-term assays should provide more insight into possible alternative mechanisms of activation (17).

Sister chromatid exchange induction by hemoglobin alone is consistent with another study showing that ferroproteins and ferric ions cause immunotoxic effects including inhibition of lymphocyte proliferation, colony formation, and cytotoxic lymphocyte generation in vitro (13). The genotoxic effect of hemoglobin may be mediated directly by oxygen radicals produced during autoxidation of $HbFe^{2+}$ (14) or indirectly by active oxygen interaction with membrane lipids to form peroxides (6).

Although higher aniline concentrations were used in the present study in vitro than would be expected in the in vivo situation, relatively high levels of aniline and its metabolites might have been achieved in the spleens of F-344 rats during a 2-yr carcinogenicity bioassay (9,16). In this regard, our data suggest that the mechanism of aniline-induced splenic toxicity in rats may result from the combined genotoxic effect of aniline metabolites generated or accumulated in the RBC-engorged spleen and hemoglobin released during phagocytosis of damaged RBCs.

ACKNOWLEDGMENTS

We thank James S. Bus and Michael Cox for helpful discussions on the mechanisms of hemoglobin-mediated metabolism of aniline, and Linda Smith and Joanne Quate for typing the manuscript.

REFERENCES

1. Abe, S., and M. Sasaki (1977) Chromosome aberrations and sister chromatid exchanges in Chinese hamster cells exposed to various chemicals. J. Natl. Cancer Inst. 58:1635-1641.
2. Altman, P.L., and D.S. Dittmer (1974) Biological Data Book, Vol. 3, Federation of American Societies for Experimental Biology, Bethesda, Maryland, 2123 pp.
3. Blaauboer, B.J., and C.W.M. van Holsteijn (1983) Formation and disposition of N-hydroxylated metabolites of aniline and nitrobenzene by isolated rat hepatocytes. Xenobiotica 13:295-302.
4. Blisard, K.S., J.J. Mieyal (1979) Characterization of the aniline hydroxylase activity of erythrocytes. J. Biol. Chem. 254:5104-5110.
5. Böyum, A. (1968) Isolation of mononuclear cells and granulocytes from human blood. Isolation of mononuclear cells by one centrifugation, and of granulocytes by combining centrifugation and sedimentation. Scand. J. Clin. Lab. Invest. 21 (Suppl. 97):77-89.
6. Bucher, J.R., M. Tien, L.A. Morehouse, and S.D. Aust (1983) Redox cycle and lipid peroxidation: The central role of iron chelates. Fund. Appl. Toxicol. 3:222-226.
7. Bus, J.S. (1983) Aniline and nitrobenzene: Erythrocyte and spleen toxicity. C.I.I.T. Activities 3(12):1,6. Chemical Industry Institute of Toxicology, Research Triangle Park, North Carolina.
8. Bus, J.S., and J. Sun (1979) Accumulation and covalent binding of radioactivity in rat spleen after ^{14}C-aniline HCl administration. Pharmacologist 21:221.
9. Chemical Industry Institute of Toxicology (1982) 104-Week Chronic Toxicity in Rats. Aniline Hydrochloride. CIIT Docket No. 11642.
10. Cunningham, M.L, and P.S. Ringrose (1983) Benzo(a)pyrene and aniline increase sister chromatid exchanges in cultured rat liver fibroblasts without addition of activating enzymes. Toxicol. Lett. 16:235-239.
11. Eyer, P., H. Kampffmeyer, H. Maister, and E. Rosch-Oehme (1980) Biotransformation of nitrosobenzene, phenylhydroxylamine, and aniline in the isolated perfused rat liver. Xenobiotica 10:499-516.
12. Jenkins, F.P., J.A. Robinson, J.B.M. Gellatly, and G.W.A. Salmond (1972) The no-effect dose of aniline in human subjects and a comparison of aniline toxicity in man and the rat. Fd. Cosmet. Toxicol. 10:671-679.

13. Keown, P., and B. Deschamps-Latscha (1983) In vitro suppression of cell-mediated immunity by ferroproteins and ferric salts. Cell. Immunol. 80:257-266.

14. Kiese, M. (1974) Methemoglobinemia: A Comprehensive Treatise. CRC Press, Inc., Cleveland, Ohio, 259 pp.

15. Kirchner, G., and U. Bayer (1982) Genotoxic activity of the aminophenols as evidenced by the induction of sister chromatid exchanges. Human Toxicol. 1:387-392.

16. National Cancer Institute (1978) Bioassay of aniline HCl for possible carcinogenicity, CAS No. 142-04-01. Carcinogenesis Technical Report No. 130, NTIS Report No. PB-287-539/AS. National Technical Information Service, Springfield, Virginia.

17. Norppa, H., H. Vainio, and M. Sorsa (1983) Metabolic activation of styrene by erythrocytes detected as increased sister chromatid exchanges in cultured human lymphocytes. Cancer Res. 43:3579-3582.

18. Parke, D.V. (1960) The metabolism of [^{14}C]aniline in the rabbit and other animals. Biochem. J. 77:493-503.

19. Ray, J.H., and L.C. Altenburg (1978) Sister-chromatid exchange induction by sodium selenite: Dependence on the presence of red blood cells or red blood cell lysate. Mutat. Res. 54:343-354.

20. Short, C.R., C. King, P.W. Sistrunk, and K.M. Kerr (1983) Subacute toxicity of several ring-substituted dialkyanilines in the rat. Fund. Appl. Toxiocol. 3:285-292.

21. Snedecor, G.W., and W.G. Cochran (1967) Statistical Methods. The Iowa State University Press, Ames, Iowa, 593 pp.

22. Sun, J.D., and J.S. Bus (1980) Comparison of covalent binding of ^{14}C-aniline HCl in red blood cells, spleen and liver of rats. Pharmacologist 22:247.

23. Takehisa, S., and N. Kanaya (1982) SCE induction in human lymphocytes by combined treatment with aniline and norharman. Mutat. Res. 101:165-172.

24. Vernon-Roberts, B. (1972) The Macrophage. Cambridge University Press, Cambridge, England, 242 pp.

25. Wilmer, J.L., G.L. Erexson, and A.D. Kligerman (1983) Implications of an elevated sister-chromatid exchange frequency in rat lymphocytes cultured in the absence of erythrocytes. Mutat. Res. 109:231-248.

26. Wilmer, J.L., A.D. Kligerman, and G.L. Erexson (1981) Sister chromatid exchange induction and cell cycle inhibition by aniline and its metabolites in human fibroblasts. Environ. Mutagen. 3:627-638.

DEVELOPMENT OF RODENT PERIPHERAL BLOOD LYMPHOCYTE CULTURE SYSTEMS TO DETECT CYTOGENETIC DAMAGE IN VIVO

A. D. Kligerman, G. L. Erexson, and
J. L. Wilmer

Chemical Industry Institute of Toxicology
Research Triangle Park, North Carolina 27709

ABSTRACT

Peripheral blood lymphocytes (PBLs) offer many advantages for in vivo cytogenetic studies. They can be removed nonlethally from the animal allowing a subject to serve as its own control, permitting the analysis of cytogenetic damage over time. Furthermore, mature PBLs normally do not divide, and some populations are long-lived. Thus, they have the potential to accumulate DNA lesions during chronic exposures to genotoxicants. We have developed standard methodologies for the whole blood culture of rat and mouse PBLs to serve as models for determining the sensitivity of PBLs to cytogenetic damage. The cultures obtained with these protocols give reproducible results with high mitotic indices, stable baseline sister chromatid exchange (SCE) frequencies, and ample numbers of first- and second-division methaphases for scoring both chromosomal aberrations and SCEs. The methodologies have been especially useful for examining cytogenetic damage after inhalation exposures to toxicants such as ethylene oxide, formaldehyde, benzene, and nitrobenzene. Of these compounds, only benzene and ethylene oxide were found to induce significant dose-dependent increases in SCEs in PBLs. Also, dose-response curves have been obtained for several carcinogens administered by ip injection. These studies show that PBLs are sensitive indicators of the genotoxic effects of the carcinogens benzo(a)pyrene, 2-acetylaminofluorene, cyclophosphamide (CP), N-nitrosomorpholine, and ethylmethanesulfonate (EMS). In addition, because subpopulations of lymphocytes can be stimulated to divide using different mitogens, it has been possible to compare the sensitivity of murine B and T lymphocytes following in vitro and in vivo cyclophosphamide exposure. Once the sensitivity and selectivity of rodent lymphocyte cultures are determined, these assays should be valuable not only as a means for predicting which environmental agents could lead to increases in human cytogenetic damage, but also as a way to corroborate human cytogenetic studies.

INTRODUCTION

With the advent of the 5-bromo-2'-deoxyuridine (BrdUrd) methodology for visualizing SCEs (1,2), a sensitive means for detecting the presence of genotoxicants as well as for studying their mechanisms of action became available. This is evidenced by the number of papers that have appeared on the subject of SCE, and its application to environmental monitoring studies using organisms ranging from the annelid, *Neanthes arenaceodentata* (3) to humans (4). Numerous studies of SCE induction in lymphocytes of iatrogenically or occupationally exposed individuals have appeared in the literature since 1976 (5-7).

Two factors that have led to the expansion of research in this area are the ease of obtaining and growing human lymphocytes in culture. This has many obvious advantages for studies on the effects of genotoxicant exposure, the primary one being that humans can be studied directly. However, there are some disadvantages. It is difficult to perform carefully controlled cytogenetic studies of humans because of inherent variability among individuals, which leads to the improbability of getting well-matched control groups. Furthermore, in retrospective epidemiological studies, determination of exposure level as well as confounding factors such as drug use, diet, and smoking habits are difficult to evaluate.

One way to help evaluate the various factors that may influence human cytogenetic studies is to develop laboratory animal models, which can be carefully manipulated in order to answer questions about experimental design, evaluation of data, and variability inherent in PBL studies. The laboratory rat and mouse are reasonable choices for such models because of their use in many toxicological investigations, as well as the availability of both hybrid and inbred strains with known genetic composition. Furthermore, their small size, modest housing requirements, and relatively low cost are other factors that support their use. This chapter describes the development and use of both a rat (8,9) and a mouse (10) PBL culture system for evaluating cytogenetic damage after in vivo exposure to known and suspected genotoxicants. For other PBL culture methodologies that have yielded varying degrees of success, see Refs. 11-16, and review by Erexson et al. (10).

ADVANTAGES AND DISADVANTAGES OF PERIPHERAL BLOOD LYMPHOCYTES FOR CYTOGENETIC STUDIES

Lymphocytes are found in the blood of all vertebrates and function in the processes of immune recognition and destruction of infectious agents, tumors, and "foreign substances." Peripheral blood lymphocytes can be broadly divided into 2 categories: T lymphocytes, which function in cell-mediated immunity, and B lymphocytes, which upon differentiation to plasma cells, secrete antibodies (17). T lymphocytes are produced in the bone marrow and differentiate in the thymus, and B lymphocytes originate in the bone marrow of higher vertebrates and are processed in the bursa or bursa equivalent. Both types of PBLs recirculate through the peripheral blood and lymphatic systems (17,18). The lifespan of PBLs is variable, depending upon the animal species and the subpopulation of lymphocytes examined. Human recirculating PBLs are estimated to have a mean lifespan of at least 3 yr (18-20), and PBLs of the rat and mouse appear to have an average lifespan of 1 mo or more (17,21,22).

The widespread distribution of lymphocytes throughout the body makes this cell type well-suited for in vivo studies of genotoxicants. Not only are PBLs likely to come in contact with the parent compound, but they also have the potential to encounter sitespecific short-lived metabolites during recirculation. Another major advantage of using PBLs for the study of induced cytogenetic damage is that PBLs normally do not divide after they enter circulation unless stimulated to do so by encounter with an antigen (17). Thus, unlike bone marrow and other rapidly cycling cells, the PBLs have the potential to accumulate lesions during chronic exposures because the damage will not be diluted through cell division or lost through cell death due to aberrant mitoses (23,24). In addition, PBLs can be removed nonlethally from the animals under study. This not only enables animals to serve as their own controls during the course of chronic studies, but also permits the analysis of persistence or removal of lesions over time after exposures have terminated.

Peripheral blood lymphocytes offer other advantages for cytogenetic studies. One advantage is that PBLs are in the G_0 stage of the cell cycle before they are induced to divide. In acute chemical exposure during in vivo studies, it is known at which stage of the cell cycle exposures occur. Furthermore, by being able to control the timing of blast transformation with mitogens, the investigator can have a modest degree of control over the cell cycle stage during in vitro exposures. In addition, the methodology used in the culture of rodent PBLs is comparable to that used with human PBLs. Therefore, it is technically possible to do direct in vitro (25) and in vivo comparisons between humans and rodent species and to determine the relative suitability of various species as models for human risk assessment.

The use of PBLs in cytogenetic studies also presents some disadvantages: PBLs are not a homogeneous cell population but are composed of various subpopulations that may display differences in sensitivity to geneotoxicants (18,26). Furthermore, the number of lymphocytes in the peripheral blood may vary due to age, sex, or health status of the animals under study (27), which may cause variability in mitotic activity and cell cycling time (9). Peripheral blood lymphocytes possess a limited ability to metabolize promutagens (28,29). Therefore, if a promutagen yields only short-lived metabolites in vivo, the PBLs may have to come into close proximity to the metabolically competent tissues for cytogenetic damage to be induced.

Even with these minor disadvantages, lymphocytes are still the only practical means to evaluate cytogenetic damage in humans. Therefore, it is important to have reliable models to verify human epidemiological studies and to predict effects before human exposures occur.

METHODOLOGY

Significant aspects of the rodent PBL culture methods will be discussed, which should give a better understanding of how to obtain rodent PBL cultures with high mitotic indices. No attempt will be made here to describe in detail the rodent lymphocyte culture methodologies; they are described in detail elsewhere (8-10,30) and outlined in Fig. 1.

Rat Lymphocyte Culture

One of the major factors in successful rat lymphocyte culture is the removal of the rat blood plasma before inoculation of the cultures with cells (8,11). Washing the blood 3 times with sterile phosphate buffered saline (pH 7.4) before inoculation increased the mitotic index by an order of magnitude in phytohemagglutinin (PHA)-stimulated cultures (8). The inhibitory factor in the plasma may account in part for some of the variability obtained in previous attempts to culture rat lymphocytes. Dilution may have reduced the inhibitory effect of the plasma with culture methods in which small volumes of whole blood were added to relatively large volumes of medium. Another major factor in obtaining rat lymphocyte cultures that yield ample numbers of dividing cells is the control of other cell culture variables such as the number of cells in the inoculum, inoculum volume, BrdUrd concentration, and the type and concentration of mitogen.

The number of cells in the inoculum has a significant effect on both mitotic activity and baseline SCE frequency (9). As the number of cells in the inoculum increases, there is a concomitant decrease in both mitotic activity and SCE frequency (Fig. 2). The majority of cells in the inoculum are erythrocytes. Thus, it was hypothesized that removal of the erythrocytes might increase the mitotic yields of the cultures. Cultures that were inoculated with mononuclear leukocytes separated on a Ficoll-Hypaque density gradient showed a substantial increase in mitotic activity; however, the PBLs

RAT PBL CULTURE	TIME (HOURS)	MOUSE PBL CULTURE
METHOXYFLURANE ANESTHESIA		METHOXYFLURANE ANESTHESIA
BLOOD REMOVAL		BLOOD REMOVAL
BLOOD WASHED 3x IN PBS		BLOOD WASHED 3x IN PBS
ADD WASHED BLOOD CONTAINING 1×10^6 LEUCOCYTES TO ABOUT 1.9 ml OF COMPLETE MEDIUM* (TOTAL VOLUME: 2 ml)	0	ADD WASHED BLOOD CONTAINING 5×10^5 LEUCOCYTES TO ABOUT 0.9 ml OF COMPLETE MEDIUM** (TOTAL VOLUME: 1 ml)
ADD BrdUrd (2 μM)	24	CHANGE MEDIUM; ADD COMPLETE MEDIUM WITHOUT 6% MOUSE PLASMA; ADD BrdUrd (2 μM)
ADD COLCEMID (0.5 μg/ml)	50-51	ADD COLCEMID (0.5 μg/ml)
HARVEST CULTURES	54-60	HARVEST CULTURES

Fig. 1. Rodent PBL culture methodology. *Medium for rat PBL culture consists of RPMI 1640 plus 25 mM HEPES buffer with L-glutamine, 10% heat-inactivated fetal bovine serum, 100 units penicillin/ml, 100 μg streptomycin/ml, an additional 292 μg L-glutamine/ml, and either 4 μg PHA/ml or 30 μg concanavalin A/ml. **Medium for mouse PBL culture consists of RPMI 1640 plus 25 mM HEPES buffer with L-glutamine, 20% heat-inactivated fetal bovine serum, 100 units penicillin/ml, 100 μg streptomycin/ml, an additional 292 μg L-glutamine/ml, and either 7 μg PHA/ml or 60 μg lipopolysaccharide/ml.

also displayed significantly elevated baseline SCE frequencies, and culture-to-culture SCE frequency variability was higher (31). A standard protocol was developed in which a volume of washed whole blood containing 10^6 leukocytes is added to the culture medium to give a final concentration of 5×10^5 leukocytes/ml. This procedure yields cultures with ample numbers of mitotic figures for cytogenetic analysis as well as cells with low and stable baseline SCE frequencies (9).

When the number of leukocytes in the inoculum is held constant, the inoculum volume is determined by the leukocyte counts of the animals. The use of animals with low leukocyte counts results in larger inoculum volumes. However, as the inoculum volume increases, the mitotic activity decreases significantly although there is no significant effect on SCE frequency (9). Cultures with optimal

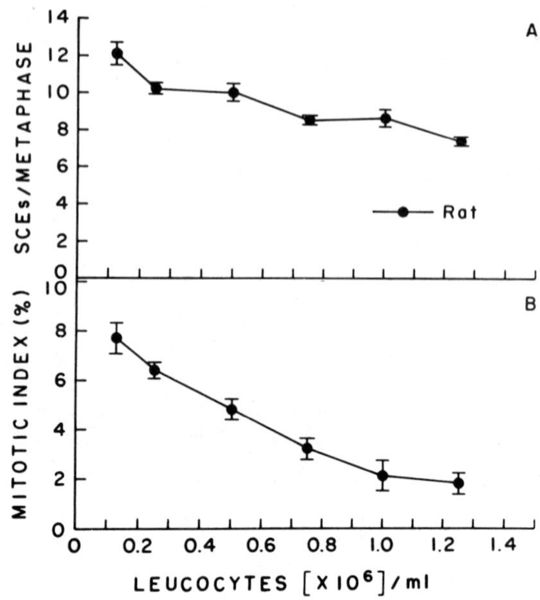

Fig. 2. Effect of the number of leukocytes in the inoculum on the SCE frequency (A) and the mitotic indices (B) of PHA-stimulated rat PBLs (mean ± S.E.M.). [Reproduced with permission from Kligerman et al. (30).]

mitotic indices were obtained from animals with leukocyte counts between 4 and 7 x 10^6 leukocytes/ml of blood. Also, animals should be kept as stress-free as possible because stress can have a significant inhibitory effect on leukocyte counts and mitogenic stimulation (32).

Rat lymphocytes are sensitive to the mitotic inhibitory effects of BrdUrd. As the concentration of BrdUrd in the culture medium increases, the mitotic activity decreases, and the cell cycle time increases (8,9). Adequate sister chromatid differentiation for SCE analysis can be obtained with a concentration of BrdUrd as low as 2 μM. Since high concentrations of BrdUrd do not stabilize the SCE frequency in rat lymphocyte cultures (9), BrdUrd concentrations from 2 to 5 μM can be used without seriously impairing the mitotic activity if the appropriate mitogen is used (see next paragraph). Also, addition of BrdUrd between 23 and 24 hr after culture initiation helps prevent inhibition of blast transformation (8). Few if any cycling cells will remain unlabeled by this time delay in addition of BrdUrd, because significant levels of DNA synthesis do not begin until 24 hr after culture initiation (31,33). Cultures are usually harvested between 54 and 56 hr after initiation in order to obtain a predominance of first- and second-division metaphases for scoring chromosome breakage and SCE, respectively. Later harvests can be used if significant cell cycle delay or lowered mitotic activity is expected after exposure to genotoxicants.

During the initial development of the rat lymphocyte culture assay, highly purified PHA (Burroughs-Wellcome HA-16) was the mitogen used (8,9,30). In subsequent studies concanavalin A (Sigma, Type IV) was found to yield cultures with higher mitotic activity (31,34). Whichever mitogen is used, it is important that each laboratory develop its own concentration-response curves because different lots of mitogen have variable mitogenic activity.

Mouse Lymphocyte Culture

Contrary to what is found when rat lymphocytes are cultured, washing mouse lymphocytes before inoculation decreases the mitotic activity (10). This response indicates that there are one or more "growth factors" in the mouse plasma, which are important for obtaining cultures with good mitotic yields. However, when blood is obtained from mutagen-treated animals, it is necessary to wash the blood cells before inoculation to prevent carry-over of genotoxicants into the culture medium (35), thereby resulting in loss of the plasma. Control mouse plasma was added to medium containing washed mouse blood cells in order to standardize the culture conditions. Experiments revealed (Fig. 3) that the addition of 6% control mouse plasma and 20% heat-inactivated fetal bovine serum to the supplemented RPMI 1640 culture medium (Fig. 1) produced conditions that consistently yielded ample numbers of metaphases for cytogenetic analysis. The results also indicated that some of the variability in success obtained in previous attempts by others to culture mouse lymphocytes may have been due to variations in the amount of mouse plasma that was inoculated along with the blood cells into the cultures.

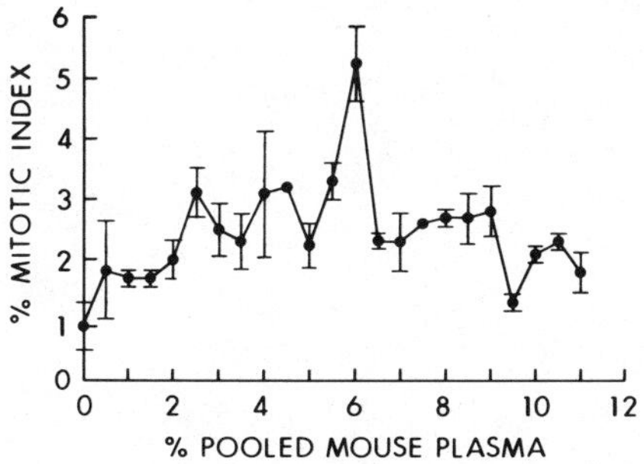

Fig. 3. Effect of pooled mouse plasma on mitotic indices of PHA-stimulated mouse PBLs. Each datum is the mean ± S.E.M. of 2 to 4 cultures from 2 independent experiments. [Reproduced with permission from Erexson et al. (10).]

Similar to what was found with rat lymphocyte culture, the number of cells in the inoculum has a significant effect on the mitotic activity (Fig. 4). When PHA (7 µg/ml) was used as the mitogen, peak mitotic activity occurred when whole blood containing 5 x 10^5 leukocytes was inoculated to obtain a total culture volume of 1 ml (10). The use of Ficoll-Hypaque-separated leukocytes instead of whole blood in the inoculum, increased the mitotic index, but this procedure also increased the SCE frequency (36). Mouse lymphocytes are not as sensitive to the mitotic inhibitory effects of BrdUrd as rat lymphocytes, and both species show similarly shaped SCE concentration-response curves when exposed to this DNA base analog (Fig. 5) (30).

The choice of mitogen is of particular importance in the culture of mouse lymphocytes. Again, it is recommended that each investigator develop a concentration-response curve for each mitogen. Phytohemagglutinin can be used to stimulate specifically T lymphocytes; lipopolysaccharide will cause B lymphocytes to divide (37). Similar to what had been found with human lymphocytes (38,39) (Tab. 1), mouse T cells have a higher baseline SCE frequency than mouse B cells (10).

USE OF RODENT LYMPHOCYTE CULTURE SYSTEMS TO INVESTIGATE GENOTOXICANTS

The determination of the sensitivity of PBLs to induction of cytogenetic damage after in vivo exposure to genotoxicants is an ongoing area of research. Table 2 summarizes the results obtained to date in our laboratory with both the mouse and rat PBL assays. Preliminary evidence suggests that the examination of SCEs in PBLs exposed in vivo is a more sensitive indicator of the presence of

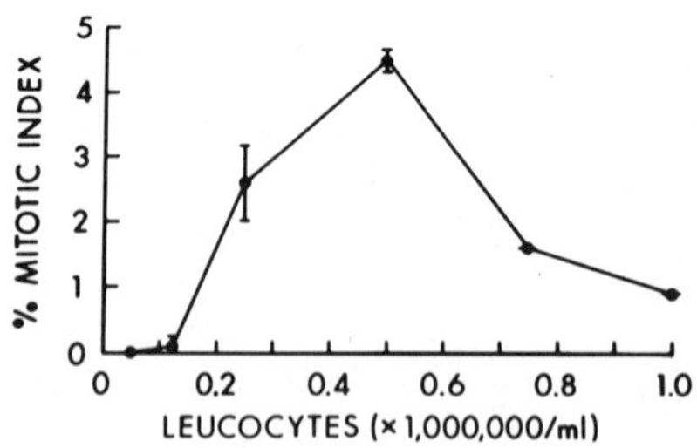

Fig. 4. Effect of number of PBLs in the inoculum on mitotic indices of PHA-stimulated mouse lymphocytes. Each datum is the mean ± S.E.M. of 4 cultures from 2 independent experiments. [Reproduced with permission from Erexson et al. (10).]

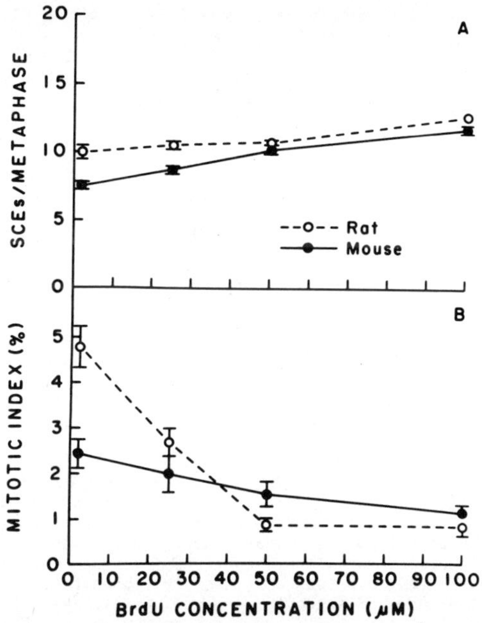

Fig. 5. Effect of BrdUrd concentrations on the SCE frequencies (A) and mitotic indices (B) of PHA-stimulated rat and mouse PBLs means ± S.E.M. [Reproduced with permission from Kligerman et al. (30).]

genotoxicants than is chromosome breakage, but more compounds must be studied to verify this finding. Furthermore, with PBL culture, both direct- and indirect-acting genotoxicants administered by several different routes can be detected by SCE analysis.

Because PBLs are normally nondividing and relatively longlived, repeated exposures can be given over a considerable length of time in an attempt to accumulate damage. This approach has been useful for studying suspected genotoxicants administered by inhalation. For example, male Fischer-344 (F-344) rats were exposed by inhalation to target concentrations of 50, 150, or 450 ppm ethylene oxide for 6 hr/da for 1 or 3 da. Ethylene oxide caused a concentration-dependent increase in SCE frequency, and the effects were additive over a 3-da period (Fig. 6)(42). However, no significant concentration-related increases in chromosome breakage could be detected. In 2 other experiments, male F-344 rats were exposed to target concentrations of 5, 16, or 50 ppm nitrobenzene for 6 hr/da, 5 da/wk, for a total of 21 da during a 29-da period (34), and male and female F-344 rats were exposed to formaldehyde target concentrations of 0.5, 6, or 15 ppm for 6 hr/da for 5 da (43). In both cases, no significant increase in SCE frequency or chromosome breakage could be detected in the PBLs (34,43). In experiments using DBA/2 male mice in inhalation studies, animals were exposed to

Tab. 1. Effects of mitogens and culture time on the SCE frequency of human T and B lymphocytes.

Mitogen[a]	Culture Duration [hours]	SCEs/Metaphase[b] T Lymphocytes	B Lymphocytes
PHA	72	13.3 ± 0.8 [120][c]	9.2 ± 3.0 [59][c]
	96	11.7 ± 0.5 [120]	7.7 ± 0.7 [57]
Con A	72	12.3 ± 0.8 [120]	7.6 ± 1.0** [35]
	96	12.5 ± 0.1 [111]	6.6* [7]
PWM	72	12.7 ± 1.0 [117]	7.7 ± 1.2 [35]
	96	13.1 ± 1.2 [77]	10.0* [10]

[a]Human lymphocytes were separated using a modification of the technique described by Saxon et al. (40). Blood was obtained from 3 male donors, separated on a Ficoll-Hypaque density gradient, and the B-lymphocyte fraction (non-rosetting cells) was separated from the T-lymphocyte (rosetting) fraction using sheep erythrocytes treated with 2-aminoethylisothiouronium bromide hydrobromide. This procedure was repeated for the B-lymphocyte fraction to ensure that this fraction was virtually devoid of contaminating T-lymphocytes. However, it did contain substantial numbers of monocytes and null cells. Replicate cultures were established for each treatment, and the lymphocytes were stimulated to divide with either 2 µg PHA/ml phytohemagglutinin (PHA), 8 µg concanavalin A/ml (Con A), or 10 µl pokeweed/ml mitogen (PWM). All cultures contained 10 µM BrdUrd. Where possible, 40 metaphases were scored for each of the 3 individuals, but due to the lack of T-lymphocytes in the B-lymphocyte fraction, cell division in the B-lymphocyte cultures was limited.

[b]mean ± s.d. among three subjects except where noted: *- 1 individual
**- 2 individuals

[c]total number of second division metaphases scored.

target concentrations of 10, 100, or 1000 ppm benzene for 6 hr. Eighteen hr later blood was removed by cardiac puncture and cultured for SCE analysis. Bone marrow preparations were made for analyzing the induction of micronuclei in polychromatic erythrocytes. A parallel concentration- dependent increase in both SCE and micronuclei induction was evident (44).

Alternatively, a single exposure can be given and animals sampled over an extended period of time. In initial experiments with the rat PBL culture system, male F-344 rats were injected ip with either 30, 100, or 300 mg EMS/kg. Blood was removed from animals sacrificed at 1, 3, 5, 7, 9 and 28 da after EMS administration. Although there was a precipitous drop in the SCE frequencies within 5 da of exposure, PBLs from treated animals still had an SCE frequency approximately 50% above the control frequency 28 da after exposure (8).

Cytogenetic damage in different subpopulations of PBLs can be examined in the mouse. As mentioned previously, specific mitogens

Tab. 2. SCE response in cultured PBLs after in vivo exposure to known and suspected genotoxicants.

Compound[a]	Species	Exposure Route	Total Exposure	SCE Response[b]	Comments[c] (Reference)
EMS	Rat	ip	30-300 mg/kg	+++	blood cultured either 1,3,5,7,9 or 28 days after exposure (8)
CP	Rat	ip	5 mg/kg	+++	blood cultured 23 hr after exposure (34)
EO	Rat	inhalation	288-7992 ppm x hr	++	50, 150,or 450 ppm for 1 or 3 days;*(42)
FORM	Rat	inhalation	14-443 ppm x hr	-	0.5, 6, or 15 ppm for 5 days;*(43)
NB	Rat	inhalation	60-7500 ppm x hr	-	10, 35, or 125 ppm for 1, 3 or 10 days;*(30)
NB	Rat	inhalation	626-30,173 ppm x hr	-	5, 16, or 50 ppm for 21 days during a 29 day period;*(34)
DNT	Rat	gavage	100 mg/kg	+	blood cultured 48 hr after exposure (30)
MMS	Rat	gavage	20 mg/kg	++	**(30)
TCDD	Rat	gavage	20, 200 ug/kg	-	blood cultured either 2, 12, or 48 hrs after exposure (30)
BZ	Mouse	inhalation	62-5947 ppm x hr	+++	10, 100, or 1000 ppm exposure for 6 hr; (44)
AAF	Mouse	ip	20-160 mg/kg	++	***(41)
BP	Mouse	ip	25-300 mg/kg	+++	***(41)
CP	Mouse	ip	0.5-5 mg/kg	+++	***(26)
DCV	Mouse	ip	5-25 mg/kg	-	***(41)
EMS	Mouse	ip	10-270 mg/kg	++	***(41)
NM	Mouse	ip	37.5-300 mg/kg	+++	***(41)

[a]Abbreviations: EMS, ethyl methanesulfonate; CP, cyclophosphamide; EO, ethylene oxide; FORM, formaldehyde; NB, nitrobenzene; AFB1, aflatoxin B1; DMBA, 7,12-dimethylbenz[a]anthracene; DNT, dinitrotoluene (technical grade); MMS, methyl methanesulfonate; MNNG, N-methyl-N'-nitro-N-nitrosoguanidine; TCDD, 2,3,7,8-tetrachlorodibenzo-p-dioxin; BZ, benzene; AAF, 2-acetylaminofluorene; BP, benzo[a]pyrene; DCV, dichlorvos; NM, N-nitrosomorpholine

[b]At highest concentration examined:
- \- no significant increase over control SCE frequency
- \+ < 50% increase over control SCE frequency
- ++ > 50% increase over control SCE frequency
- +++ >100% increase over control SCE frequency

[c] * blood removed and placed in culture within 1 hr after termination of exposure
** blood removed and placed in culture within 2 hr after termination of exposure
*** blood removed and placed in culture within 24 hr after termination of exposure

exist for stimulation of either T- or B-murine lymphocytes. This idea was incorporated in a study designed to examine the differential sensitivity of T and B lymphocytes to induction of cytogenetic damage by CP (26). Both lymphocyte types showed similar sensitivity to CP in vitro as measured by the induction of SCEs. However, after ip injection of mice with 5 mg CP/kg, B lymphocytes responded with a 3-fold increase in SCEs over baseline while T lymphocytes showed only a 2-fold increase over baseline SCE frequency. Whether this differential response is specific for CP, specific for promutagens, or if it is a generalized phenomenon is unknown at present.

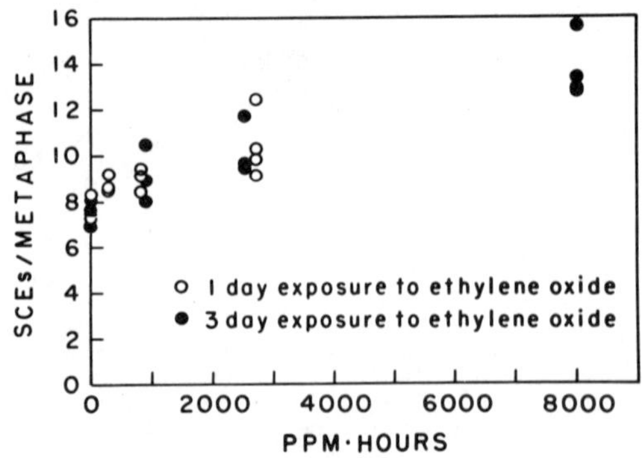

Fig. 6. Concentration × time plot showing induction of SCEs in PHA-stimulated PBLs of rats exposed to ethylene oxide by inhalation. Each datum is the mean SCE frequency from PBL cultures from 1 rat. Animals were exposed for 6 hr da for either 1 or 3 da to target concentrations of 50, 150, or 450 ppm ethylene oxide. [Reproduced with permission from Kligerman et al.] (42).

CONCLUSIONS

Reliable rat and mouse PBL culture systems are available and have proven useful for the study of genotoxicants administered by several different routes. However, a large number of studies remain to be conducted. More mutagens and nonmutagens need to be examined to determine the sensitivity and specificity of PBL culture systems for detecting cytogenetic damage after in vivo exposures. Time-course experiments should be conducted to determine the persistence of lesions that lead to cytogenetic damage and to evaluate the role of cell turnover and DNA repair in the disappearance of lesions. Furthermore, studies are needed to investigate the various cellular interactions that take place in vitro, which may influence the cytogenetic responses measured (31).

Although the culturing of PBLs is a reliable way to analyze human subjects for chromosomal damage, the meaning of such studies is often unclear because carefully controlled conditions are difficult to achieve. Investigations should be conducted to determine whether the mouse or the rat are good predictive models for determining human cytogenetic responses to genotoxic exposure. It is known already that some differences exist between rodent and human PBLs in culture with regard to baseline SCE frequencies (31), influence of erythrocytes on the SCE frequency (31), and DNA repair capabilities (33). If the rat or the mouse is proven to be a good model for human exposure, one can envision studies in which animals are exposed to suspect agents or to mixtures of agents that have been implicated as clastogens or SCE inducers in human epidemiological studies.

Then, a determination could be made if any of the agents do indeed have genotoxic potential. Alternatively, before an expensive cytogenetic epidemiological investigation is performed using a human population exposed to a suspected genotoxicant, the chemical could be tested in rodents to determine if it has the potential to cause cytogenetic damage in PBLs. If the agent is found to have no cytogenetic damaging potential even at high doses, it may be prudent to re-evaluate the proposed epidemiological study.

Future experiments will determine whether rodent PBL culture assays can serve as good cytogenetic models for understanding and predicting the cytogenetic consequences of genotoxicant exposure in human populations. Presently, these systems offer the investigator a valuable tool for conducting cytogenetic investigations as well as for studying cell interactions in vitro.

REFERENCES

1. Perry, P., and S. Wolff (1974) New Giemsa method for differential staining of sister chromatids. Nature (Lond.) 261:156-158.
2. Latt, S.A. (1973) Microflurometric detection of deoxyribonucleic acid replication in human metaphase chromosomes. Proc. Nat. Acad. Sci., USA 70:3395-3399.
3. Pesch, G.G., and C.E. Pesch (1980) Neanthes arenaceodentata (Polychaeta: Annelida), a proposed cytogenetic model for marine genetic toxicology. Can. J. Fish. Aquat. Sci. 37:1225-1228.
4. Daoud, C., M.W. Shaw, and A. Craig-Holmes, (1976) Sister chromatid exchange frequency among normal individuals and breast cancer patients. Mammalian Chromosome Newsletter 17:26.
5. Lambert, B., A. Lindblad, K. Holmberg, and D. Francesconi (1982) The use of sister chromatid exchange to monitor human populations for exposures to toxicologically harmful agents. In Sister chromatid Exchange, S. Wolff, ed. John Wiley and Sons, New York, pp. 149-182.
6. Brögger, A. (1982) Application of SCE to public health. In Progress and Topics in Cytogenetics, Vol. 2., Sister Chromatid Exchange, A.A. Sandberg, ed. Alan R. Liss, Inc., New York, pp. 655-673.
7. Raposa, T. (1978) Sister chromatid exchange studies for monitoring DNA damage and repair capacity after cytostatics in vitro and in lymphocytes of leukaemic patients under cytostatic therapy. Mutat. Res. 57:241-251.
8. Kligerman, A.D., J.L. Wilmer, and G.L. Erexson (1981) Characterization of a rat lymphocyte culture system for assessing sister chromatid exchange after in vivo exposure to genotoxic agents. Environ. Mutagen. 3:531-543.
9. Kligerman, A.D., J.L. Wilmer, and G.L. Erexson (1982) Characterization of a rat lymphocyte culture system for assessing

sister chromatid exchange. II. Effects of 5-bromodeoxyuridine concentration, number of white blood cells in the inoculum, and inoculum volume. Environ. Mutagen. 4:585-594.

10. Erexson, G.L., J.L. Wilmer, and A.D. Kligerman (1983) Analyses of sister-chromatid exchange and cell-cycle kinetics in mouse T- and B-lymphocytes from peripheral blood cultures. Mutat. Res. 109:271-281.

11. Metcalf, W.K. (1965) Some experiments on the phytohaemagglutinin culture of leucocytes from rats, and other mammals. Exp. Cell Res. 40:490-498.

12. Lilly, L.J., A.M. Cobon, and M.F. Newton (1979) Chromosomal aberations induced in rat lymphocytes by in vivo exposure to chemicals. A short term method of carcinogenicity screening. Meth. Find. Exptl. Clin. Pharmacol. 1:225-231.

13. Basler, A. (1982) SCEs in lymphocytes of rats after exposure in vivo to cigarette smoke or to cyclophosphamide. Mutat. Res. 102:137-143.

14. Beltz, P.A., and R.D. Benz (1980) Nondestructive cytogenetic toxicologic testing with rats. Environ. Mutagen. 2:297 (abstract).

15. Granberg-Ohman, I., S. Johansson, and A. Hjerpe (1980) Sister-chromatid exchanges and chromosome aberrations in rats treated with phenacetin, phenazone and caffeine. Mutat. Res. 79:13-18.

16. Tates, A.D., N. de Vogel, and I. Neuteboom (1980) Cytogenetic effects in hepatocytes, bone marrow cells and blood lymphocytes of rats exposed to ethanol in the drinking water. Mutat. Res. 79:285-288.

17. Sprent, J. (1977) Recirculating lymphocytes. In The Lymphocyte: Structure and Function, Part 1, J.J. Marchalonis, ed. Marcel Dekker, Inc., New York, pp. 43-112.

18. Obe, G., and B. Beek (1982) The human leucocyte test system. In Chemical Mutagens: Principles and Methods for their Detection, Vol. 7, F.J. de Serres and A. Hollaender, eds. Plenum Press, New York, pp. 337-400.

19. Buckton, K.E., P.G. Smith, and W.M. Court Brown, (1967) The estimation of lymphocyte lifespan from studies on males treated with x-rays for ankylosing spondylitis. In Human Radiation Cytogenetics, H.J. Evans, W.M. Court Brown, and A.S. McLean, eds. North Holland, Amsterdam, pp. 106-114.

20. Dolphin, G.W., D.C. Lloyd, and R.J. Purrott (1973) Chromosome aberration analysis as a dosimetric technique in radiological protection. Health Physics 27:7-15.

21. Robinson, S.H., G. Brecher, I.S. Lourie, and J.E. Haley (1965) Lymphocyte labeling in rats during and after continuous infusion of tritiated thymidine: Implications for lymphocyte longevity and DNA reutilization. Blood 26:281-295.

22. Röpke, C., and N.B. Everett (1973) Small lymphocyte populations in the mouse bone marrow. Cell Tissue Kinet. 6:499-507.

23. Carrano, A.V., and D.H. Moore II (1982) The rationale and methodology for quantifying sister chromatid exchange in humans.

In Mutagenicity: New Horizons in Genetic Toxicology, J.A. Heddle, ed. Academic Press, Inc., New York, pp. 267-304.

24. Evans, H.J. (1976) Cytological methods for detecting chemical mutagens. In Chemical mutagens: Principles and Methods for Their Detection, Vol. 4, A. Hollaender, ed. Plenum Press, New York, pp. 1-29.
25. Waalkens, D.H., H.F.P. Joosten, R.D.F.M. Taalman, J.M.J.C. Scheres, T.D. Yih, and A. Hoekstra (1981) Sister chromatid exchanges induced in vitro by cyclophosphamide without exogenous activation in lymphocytes from three mammalian species. Toxicol. Letters 7:229-232.
26. Wilmer, J.L., G.L. Erexson, and A.D. Kligerman (1984) Sister chromatid exchange induction in mouse B and T lymphocytes exposed to cyclophosphamide in vitro and in vivo. Cancer Res. (44:880-884).
27. Schalm, O.W., N.C. Jain, and E.J. Carroll (1975) Veterinary Hematology. Lea & Febiger, Philadelphia.
28. Busbee, D.L., C.R. Shaw, and E.T. Cantee II (1972) Aryl hydrocarbon hydroxylase induction in human leukocytes. Science 178:315-316.
29. Gurtoo, H.L., A.J. Marinella, H.J. Freedman, B. Paigen, and J. Minowada (1978) Cytochrome P-450 in a cultured human lymphocyte line. Biochem. Pharmacol. 27:2659-2662.
30. Kligerman, A.D., J.L. Wilmer, and G.L. Erexson (1982) The use of rat and mouse lymphocytes to study cytogenetic damage after in vivo exposure to genotoxic agents. In Banbury Report 13: Indicators of Genotoxic Exposure, B.A. Bridges, B.E. Butterworth, and I.B. Weinstein, eds. Cold Spring Harbor Laboratory, pp. 277-291.
31. Wilmer, J.L., G.L. Erexson, and A.D. Kligerman (1983) Implications of an elevated sister-chromatid exchange frequency in rat lymphocytes cultured in the absence of erythrocytes. Mutat. Res. 109:231-248.
32. Keller, S.E., J.M. Weiss, S.J. Schleifer, N.E. Miller, and M. Stein (1981) Suppression of immunity by stress: Effect of a graded series of stressors on lymphocyte stimulation in the rat. Science 213:1397-1400.
33. A.D. Kilgerman, and M. Cheng (unpublished data).
34. Kligerman, A.D., G.L. Erexson, J.L. Wilmer, and M.C. Phelps (1983) Analysis of cytogenetic damage in rat lymphocytes following in vivo exposure to nitrobenzene. Toxicol. Letters 18:219-226.
35. DuFrain, R.J., L.G. Littlefield, and J.L. Wilmer (1980) The effect of washing lymphocytes after in vivo treatment with streptonigrin on the yield of chromosome and chromatid aberrations in blood cultures. Mutat. Res. 69:101-105.
36. G.L. Erexson (unpublished data).
37. Greaves, M., and G. Janossy (1972) Elicitation of selective T and B lymphocyte responses by cell surface binding ligands. Transplant. Rev. 11:87-130.

38. Santesson, B., K. Lindahl-Kiessling, and A. Mattesson (1979) SCE in B and T lymphocytes. Possible implications for Bloom's syndrome. Clin. Genet. 16:133-135.
39. Lindblad, A., and B. Lambert (1981) Relation between sister chromatid exchange, cell proliferation and proportion of B and T cells in human lymphocyte cultures. Human. Genet. 57:31-34.
40. Saxon, A., J. Feldhaus, and R.A. Robins (1976) Single step separation of human T and B cells using AET treated SRBC rosettes. J. Immunol. Meth. 12:285-288.
41. Kligerman, A.D., J.L. Wilmer, and G.L. Erexson (unpublished data).
42. Kligerman, A.D., G.L. Erexson, M.E. Phelps, and J.L. Wilmer (1983) Sister-chromatid exchange induction in peripheral blood lymphocytes of rats exposed to ethylene oxide by inhalation. Mutat. Res. 120:37-44.
43. Kligerman, A.D., M.C. Phelps, and G.L. Erexson (1984) Cytogenetic analysis of lymphocytes from rats following formaldehyde inhalation. Toxicol. Letters (in press).
44. Erexson, G.L., J.L. Wilmer, C.D. Auman, and A.D. Kligerman (1984) Induction of sister chromatid exchanges and micronuclei in male DBA/2 mice after inhalation of benzene. Proceedings of the 15th Annual Meeting of the Environmental Mutagen Society, Montreal, Canada, February 19-23.

ACCUMULATION OF SCEs IN LYMPHOCYTES DURING CHRONIC INGESTION OF A MUTAGEN

A. F. McFee and M. N. Sherrill

Medical and Health Sciences Division
Oak Ridge Associated Universities
P.O. Box 117
Oak Ridge, Tennessee 37831

ABSTRACT

Studies in 2 large-animal species have shown that continued ingestion of a mutagen increases sister chromatid exchanges (SCE) in circulating lymphocytes at rates directly related to the accumulated dose but differing between species. Groups of 3 pigs received daily oral doses of 0, 1.25, or 2.50 mg/kg body weight of 7,12-dimethylbenzanthracene (DMBA) for 160 consecutive da; similar groups of sheep received 0, 0.625, or 1.25 mg/kg daily for up to 114 da. Lymphocytes taken at 3-wk intervals from pigs and every 2 wk from sheep were cultured, and 25 randomly selected second-division metaphases from each culture were scored for SCE. Swine lymphocytes had elevated numbers of SCE at the earliest sampling times after DMBA feeding was begun. Sister chromatid exchanges continued to increase gradually among pigs consuming 1.25 mg/kg/da for 160 da whereas those on 2.5 mg/kg tended to plateau at approximately 2.5 times their pretreatment level after some 80 da. Sheep treated with comparable levels of DMBA (1.25 mg/kg) exhibited a much more rapid rise in SCE than pigs; within 40 da their SCEs were at 4 times the control level and animals were exhibiting toxicity symptoms. Low-dose sheep also showed a rapid initial increase but plateaued at about 2.5 times the control level beyond 40 da.

INTRODUCTION

Numerous possibilities exist for the contamination of the environment or specific foodstuffs which could result in the chronic ingestion of mutagenic substances by humans as well as food-chain animals. No convenient biological dosimetry system has yet been developed to detect such exposure and to evaluate the cumulative effects of chemical build-up in the animal system. Since peripheral lymphocytes are nondividing cells, they might be expected to accumulate lesions which could be expressed as SCEs in vitro and thus, serve as a biological dosimeter. Stetka and Wolff (11) initially demonstrated that acute in vivo effects of mutagens could be detected by evaluating SCE induction in lymphocytes of treated animals, and elevated frequencies of SCE have been detected in human lymphocytes for extended periods following the treatment of patients with various drugs (3,4,9). An in vivo study by Stetka et al. (10), measured the SCEs induced in rabbit lymphocytes by weekly injections of mitomycin C. Each injection was followed by a rapid rise in SCE numbers and a return to control levels within a few days, until the fourth injection when SCE levels failed to decline after the treatment. These patterns suggested that lymphocytes could accumulate SCE-inducing lesions in vivo and that they might be utilized as an indicator of chronic exposure to mutagens/carcinogens.

Such dosimetry would obviously be influenced by the lesion-inducing potency of the chemical, its absorption, accumulation, and metabolism by the animal, and by the rate at which lesions are lost from or repaired by the lymphocyte system. We selected DMBA as a representative test mutagen. It is rather poorly absorbed by the digestive system and requires enzymatic activation to become mutagenic. Although its metabolic half-life is expected to be approximately 24 hr (1), it is retained somewhat longer by fatty tissues (2). Daily feeding would therefore be expected to result in relatively constant tissue levels of the chemical in the animal after a few days.

Since the gastrointestinal system of the pig is quite similar to that of man, this species constitutes a good model from which to predict the human response. We have previously reported that swine lymphocytes show increasing numbers of SCEs with continued ingestion of DMBA (7). Similar studies have now been conducted with sheep to obtain an indication of species variability and to evaluate the contrasting ruminant digestive system in animals of comparable size.

MATERIALS AND METHODS

Mature females of each species were individually housed and given daily doses of DMBA. The chemical was dissolved in corn oil and mixed with a small initial portion of the pig's ration; it was administered to sheep in a gelatin capsule just prior to feeding. Diets were otherwise standard maintenance rations for the species. Each animal began receiving DMBA on the day after she was bred and treatment continued through, and for a period following, the pregnancy, except where toxicity problems arose. Two groups of 3 pigs each were fed DMBA at 2.50 or 1.25 mg/kg body wt/da while the 3 sheep in each group received 1.25 or 0.625 mg/kg/da. Equal numbers of control animals were similarly housed in each case.

Blood samples were taken from pigs every 3 wk and from sheep every 2 wk during their treatment period. All animals within a species were bled on the same day so that culture conditions would be held constant. This meant, however, that length of treatment varied among animals at each sampling. A cell-rich plasma/buffy coat suspension was obtained from each sample and analyzed for white cell concentration and lymphocyte proportion. Cultures were inoculated with a quantity of each suspension to yield 0.75×10^6 lymphocytes per ml of media. Swine cells were incubated at 38°C in TC-199 plus 20% inactivated fetal bovine serum, phytohemagglutinin, and penicillin-streptomycin, while McCoy's 5a and horse serum were used for sheep lymphocytes. Bromodeoxyuridine (BrdUrd) was added to each culture (5 mg/ml) 18 hr after initiation. Total culture time was 46 hr for swine and 54 hr for sheep cells, with the last 2 and 3 hr being in the presence of 10^{-6} M colchicine. At harvest, cells were subjected to hypotonic treatment and fixed in methanol-acetic acid; air-dried preparations were stained by the fluorescence-plus-Giemsa method (8). Using coded slides, 25 randomly selected, second-division metaphases from each culture were scored for SCEs. Lymphocyte proliferation rates were evaluated by scoring the first 100 metaphases from each sample as being in their first, second, or third in vitro mitosis, based on the differential staining pattern of the chromosomes.

RESULTS

Figures 1 and 2 illustrate the measurable increase in SCEs among the lymphocytes of pigs which had consumed DMBA even for relatively short periods of time. The number of SCEs per cell increased linearly with time for approximately 80 da and then showed a tendency to plateau. This pattern is substantiated by the fact that the total data for each group showed significant ($p < 0.05$) linear correlations between SCEs and length of treatment, but an even better fit ($p < 0.01$) to a quadratic function. One animal in the lower dose group consistently had the highest level of SCEs although her values were among the lowest in 2 pretreatment samples which were scored from each pig.

Although pigs showed no indication of a physiological effect of DMBA consumption, the chemical was relatively toxic to sheep. Among the 1.25 mg/kg/da sheep, 1 died at 55 da and another was gravely ill after 50 da; treatment of this group was therefore ended. Sheep

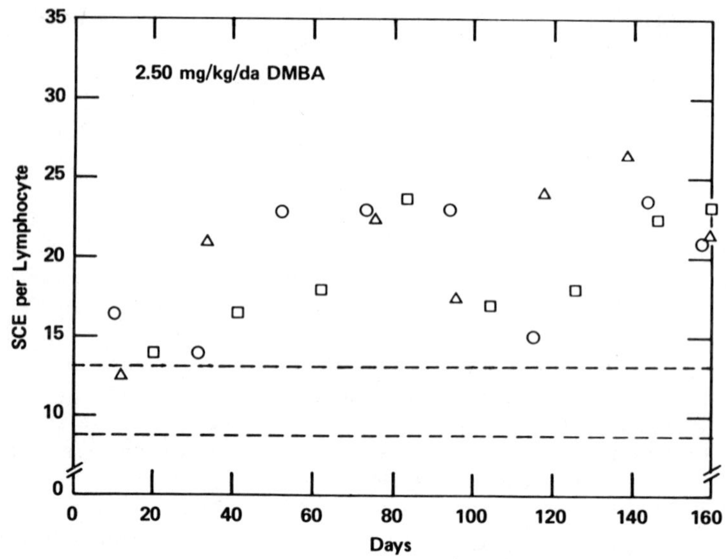

Fig. 1. Sister chromatid exchanges in lymphocytes of pigs consuming 2.5 mg/kg per da of DMBA. Three animals are represented by different symbols. Dashed lines define the 100% range of control animals.

also showed a greater sensitivity to SCE induction. Figures 3 and 4 document the rapid linear increase in SCEs among their lymphocytes during the first 40 da of treatment. No difference between the 2 dose groups was evident during this period, as their linear regression lines were almost identical. Between 60 and 120 da, the sheep maintained a relatively constant number of SCEs. In both sheep and pigs, when a tendency for SCE numbers to plateau after extended treatment was evident, it was in the vicinity of 2.5 times the control level.

A summary of the cell proliferation data is presented in Tab. 1. These values were arrived at by the formula:

$$\frac{\text{hr in culture}}{\text{I} + 2(\text{II}) + 3(\text{III})} \times 100$$

where I, II, and III are the respective percentages of first, second, and third metaphases. The resulting values are obviously not true measures of cell cycle length since they include the response time of cells to mitogenic stimulation. They are, however, valid for the comparison of total effects of treatments on both response time and cycle length. These values were usually higher among low-

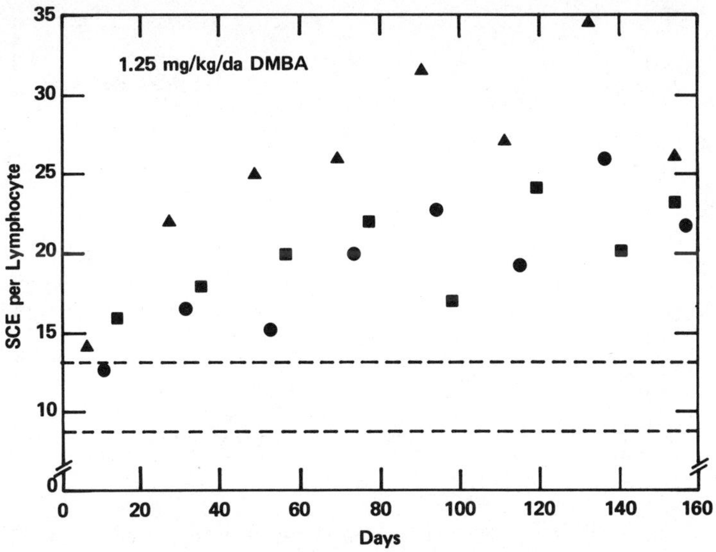

Fig. 2. Sister chromatid exchanges among the lymphocytes of pigs consuming 1.25 mg/kg body weight/da of DMBA. Note the consistently higher values in 1 animal.

dose animals than in their controls and always higher among those receiving the higher doses. The t-test comparison between treated and control groups was not significant for the low-dose group of either species; however, a highly significant ($p < .01$) increase in "cycle" length was evident in the 2.5 mg/kg pigs. In spite of a limited number of observations, the 1.25 mg/kg sheep also showed a significant ($p < .05$) increase over control values.

In Fig. 5 we present the distribution of metaphases with various numbers of SCE for sheep in the 3 groups. These values are averages for the culture obtained from each animal after approximately 40 da of exposure. An extremely wide distribution of SCE is evident and seemed to be wider among cells sampled from the highdose animals.

DISCUSSION

A great many factors are almost certain to influence SCE levels in animals under a chronic feeding regimen such as was employed for this study. The absorption of the chemical from the gastrointestinal tract is likely to vary among species. The simple-stomach pig is considered an excellent model for the extrapolation of absorption data to man. In the ruminant sheep, however, materials are

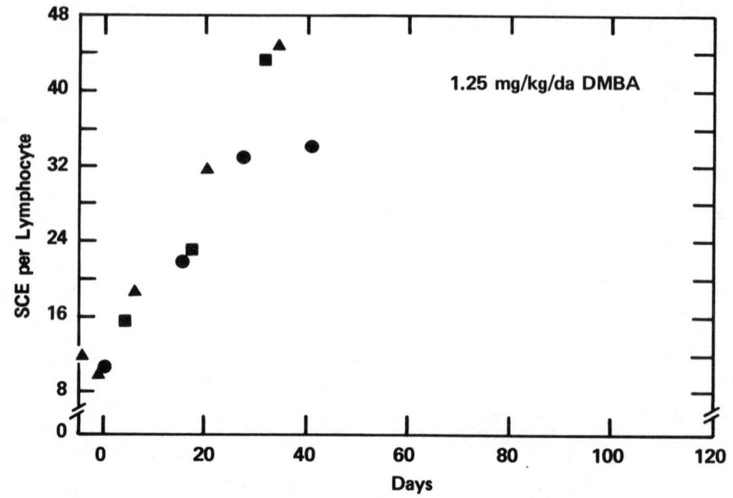

Fig. 3. Lymphocytic SCEs in sheep receiving daily oral doses at 1.25 mg/kg body weight of DMBA. Treatment was terminated at 55 da due to toxic effects of the chemical.

subjected to bacterial and fermentation actions which may alter their later absorption rate as well as influence their mutagenic potency.

If DMBA has a metabolic half-life of about 24 hr in large animal species, similar to its half-life in the mouse (1), a relatively steady-state level should be attained in the animal after a few days. Lymphocytes which are treated in vitro accumulate SCE-inducing lesions as a function of exposure time as well as dose (5); if this is also true in vivo, then a steady-state level of chemical should induce increasing levels of SCEs for a period comparable to the mean lifespan of the lymphocytes. Based on the disappearance of radiation-induced aberrations, the mean lymphocyte lifespan in pigs is a matter of only a few days (6), much too short to account for the observed increases over an 80-da period. The relatively long-term increases in SCE per cell followed by a period of little further increase would suggest that the chemical concentration continues to increase in the animal system for rather long periods.

A distinct difference in response of the 2 species to DMBA was evident for both mutagenic and toxicological endpoints. Pigs showed no evidence of a physiological effect at either 2.50 or 1.25 mg/kg/da whereas the 1.25 level was highly toxic to sheep after some 40 da. This ultimate toxicity as well as the rapid build-up of SCEs is commensurate with a continuing increase in body burden of the chemical. We cannot distinguish, of course, whether these

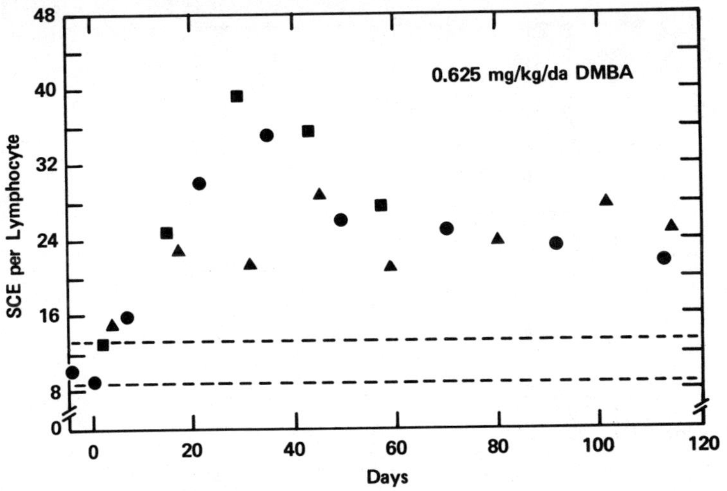

Fig. 4. Accumulation of SCEs in lymphocytes of sheep dosed daily with 0.625 mg/kg DMBA. Dashed lines include 100% of control animal values over the same sampling period.

species differences are due to differences in uptake of the chemical, its metabolic processing, or (in the case of SCEs) lesion repair. The repair of SCE-inducing chemical lesions is known to vary with the inducing chemical (5) and very likely varies also with species.

Tab. 1. "Average cell cycle" length among PHA-stimulated lymphocytes of animals ingesting various daily amounts of DMBA.

Sampling period	Pigs			Sheep		
	Control	1.25 mg/kg	2.50 mg/kg	Control	0.625 mg/kg	1.25 mg/kg
1	27.6	28.0	30.4	33.7	31.5	36.1
2	26.6	27.4	29.8	33.4	36.1	35.7
3	30.0	29.0	32.5	33.6	36.6	39.0
4	27.6	29.5	33.4	35.3	37.0	36.3
5	32.5	32.5	33.3	34.6	35.9	
6	26.9	29.4	30.5	33.1	32.2	
7	27.9	28.6	31.2	37.2	38.1	
8	29.7	30.1	31.6	33.8	35.2	

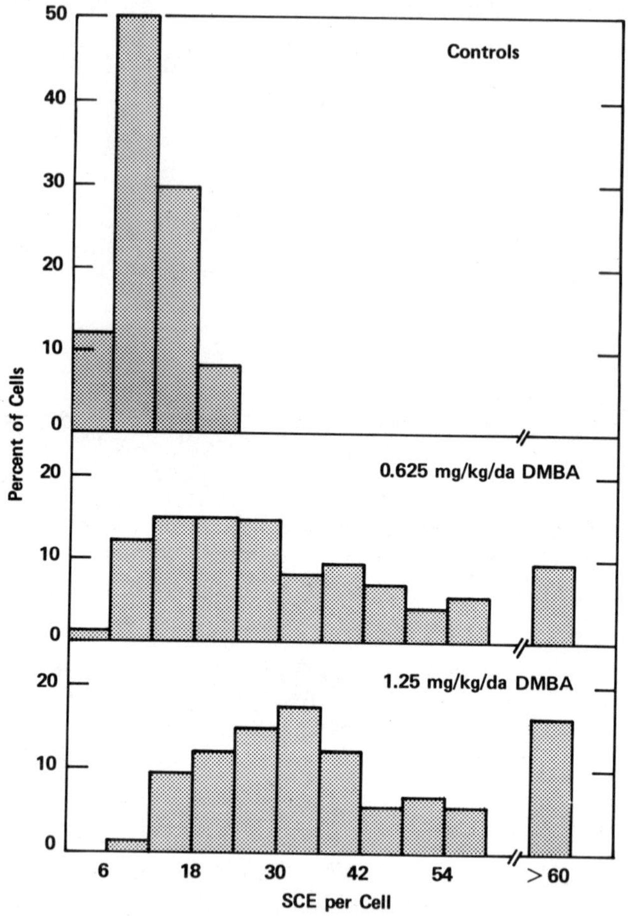

Fig. 5. Distribution of lymphocytes with various numbers of SCEs in sheep after approximately 40 da of treatment with DMBA. Each graph expresses the average for 25 metaphases from each of 3 animals.

We have also examined SCEs in mouse bone marrow following acute, intraperitoneal injections of DMBA (unpublished data). Single injections of 10 mg/kg increased bone marrow SCEs by a factor of 3, while doses of 400 mg/kg induced a 5-fold increase. The maximum oral dose to sheep was the equivalent of only 50 mg/kg spread over the first 40 da, yet this dose produced almost a 4-fold increase in SCEs. While such a comparison contains a multitude of variables, it does suggest that chronic exposure to a chemical may induce SCEs more "efficiently" than acute exposure.

The testing of other chemicals, especially those which are handled differently by the metabolic system, will be essential before the validity of evaluating chronic exposure via SCEs can be established; however, we feel that the feasibility of the approach is now defined.

ACKNOWLEDGEMENTS

This work was supported by the U.S. Environmental Protection Agency under interagency agreement 81-D-X0533 and by the U.S. Department of Energy under contract DE-AC05-760R00033 with Oak Ridge Associated Universities.

REFERENCES

1. Domsky, I.I., W. Lijinsky, K. Spencer, and P. Shubik (1963) Rate of metabolism of 9,10-dimethyl-1,2-benzanthracene in newborn and adult mice. Proc. Soc. Exp. Biol. Med. 113:110-112.
2. King, M.M., and P.B. McCay (1981) Possibilities for dietary fat and antioxidants as modulators of mammary carcinogenesis. J. Amer. College Toxicol. 2(no. 3):285-306.
3. Lambert, B., U. Ringborg, and A. Lindblad (1979) Prolonged increase of sister-chromatid exchanges in lymphocytes of melanoma patients after CCNU treatment. Mutat. Res. 59:295-300.
4. Littlefield, L.G., S.P. Colyer, and R.J. DuFrain (1980) Comparison of sister-chromatid exchanges in human lymphocytes after G_0 exposure to mitomycin in vivo vs in vitro. Mutat. Res. 69:191-197.
5. Littlefield, L.G., S.P. Colyer, A.M. Sayer, and R.J. DuFrain (1984) Persistence of SCE-inducing lesions after G_0 exposure of human lymphocytes to differing classes of DNA damaging chemicals. In Sister Chromatid Exchange: 25 Years of Experimental Research, R.R. Tice and A. Hollaender, eds. Plenum Press, New York.
6. McFee, A.F., M.W. Banner, M.N. Sherrill, and J.B. Mailhes (1972) Disappearance rates of radiation-induced chromosome aberrations from swine leukocytes. Mutat. Res. 15:325-330.
7. McFee, A.F., and M.N. Sherrill (1983) Sister-chromatid exchanges induced in swine lymphocytes by chronic oral doses of dimethylbenzanthracene. Mutat. Res. 116:349-359.
8. Perry, P., and S. Wolff (1974) New Giemsa method for the differential staining of sister chromatids. Nature (Lond.) 251:156-158.
9. Raposa, T. (1978) Sister-chromatid exchange studies for monitoring DNA damage and repair capacity after cytostatics in vitro and in lymphocytes of leukaemic patients under cytostatic therapy. Mutat. Res. 57:241:251.
10. Stetka, D.G., J. Minkler, and A.V. Carrano (1978) Induction of long-lived chromosome damage, as manifested by sister-chromatid

exchange, in lymphocytes of animals exposed to mitomycin-C. Mutat. Res. 51:383-396.

11. Stetka, D.G., and S. Wolff (1976) Sister chromatid exchange as an assay for genetic damage induced by mutagen-carcinogens, I. In vivo test for compounds requiring metabolic activation. Mutat. Res. 41:333-342.

ELIMINATION OF CYCLOPHOSPHAMIDE-INDUCED SCE IN LYMPHOCYTES OF RATS WITH TIME POST-TREATMENT

Armin Basler

Max von Pettenkofer-Institut
Bundesgesundheitsamt, C V 2
Postfach 33 00 13
1000 Berlin 33
Federal Republic of Germany

SUMMARY

Rats were treated intraperitoneally (ip) with 13.3 mg cyclophosphamide (CP)/kg. Blood was obtained by cardiac puncture 6,24,48,72 hr, 5 da and 14 da after treatment. The spontaneous sister chromatid exchange (SCE) frequency (9.94 SCEs/cell) was 4-fold increased (40.49 SCE/cell) in lymphocyte cultures set up 24 hr, after the CP treatment. Within the next 24 hr the SCE frequency began to decline and by 2 wk post-treatment the SCE values were within the control range. The results indicate that the elimination of damage to DNA by existing repair mechanisms is responsible for the reduction of SCEs with time, in addition to lymphocyte turnover.

INTRODUCTION

Lymphocyte turnover is a slow process, lasting from weeks to years. Long-term effects of induced genetic damage should therefore be demonstrable in lymphocytes of in vivo-exposed mammals. However, the results on SCE frequencies with time after in vivo exposure of humans with cytostatics are conflicting. Increased SCE values for up to several months after chronic therapy are reported (2,4), as well as a return to control values within weeks (6) and even days (5). It is unknown whether the decline in SCE values with time post-treatment is a function of lymphocyte turnover or repair mechanisms. To gain more information about the time-independent frequency of SCEs in lymphocytes, the following experiments were performed with mammals.

MATERIALS AND METHODS

Animals

The experiments were performed using female rats of the strain Wistar II, 7 wk old and weighing between 250-300 gm.

Test Compound

CP was dissolved in physiological NaCl-solution and administered ip in volumes of 0.5 ml per animal. The applied dose was 13.3 mg CP/kg. Animals of the control group were treated only with the solvent.

Lymphocyte Culture

Blood was obtained by cardiac puncture at 6, 24, 48, and 72 hr, 5 or 14 da after the CP treatment. Lymphocytes were cultured for 72 hr. Bromodeoxyuridine (BrdUrd) was added for the last 48 hr to allow determination of the SCE frequency (1).

Analysis

20 to 50 metaphases with harlequin-stained chromosomes were analyzed per animal. Statistical analysis was performed using the Student's t-test. Values of $p < 0.01$ were considered to be significant.

RESULTS

Washing of blood in a buffer (1) was conducted before lymphocyte cultures were initiated. This washing prevents a possible carry-over of the applied mutagen or its metabolites from the plasma to the cultures. Nevertheless, the frequencies of SCEs in lymphocyte cultures initiated 6 hr after the treatment of rats with 13.3 mg CP/kg increased 4-fold compared to unexposed ones (Tab. 1). Similar results were obtained when blood samples were taken 24 hr after the CP treatment. Within the next 24 hr the SCE frequency started to decline and, by 5 da post-treatment, the SCE frequency was only elevated 2-fold above the control values. Nine da later the SCE values were within the control range.

DISCUSSION

In lymphocytes of some patients undergoing cytostatic therapy, the increase in SCEs persists for up to several months after discontinuation of the chronic therapy (2,4). The plasma half-life of the drugs is much shorter than this time (4) and, therefore, cannot

Tab. 1. SCEs in lymphocytes of CP-treated rats. The rats were killed at different times after an ip injection of 13.3 mg CP/kg.

Time between injection and killing of rats	No. of rats	No. of analysed metaphases	SCEs per metaphase ± SD	range of SCE in single cells
Controls	25	1041	9.94 ± 1.73	2 - 21
6 hr	6	300	38.57^{+}± 4.03	17 - 59
24 hr	7	200	40.49^{+}± 3.51	10 - 54
48 hr	4	200	29.05^{+}± 7.67	12 - 48
72 hr	4	88	25.55^{+}± 5.22	8 - 29
120 hr	4	102	18.35^{+}± 1.91	7 - 28
14 da	5	200	12.27 ± 2.50	5 - 24

+ $p \ll 0.01$

explain the persistent increase in SCE frequency. The only explanation for the prolonged effects has to be the long in vivo half-life of these non dividing cells, which has been estimated to be as long as 3 yr for 90% of the lymphocytes in human blood (11,12).

In lymphocytes of cancer patients receiving mitomycin C (5) or CP (6) treatment, these alterations of the genetic material were not as long-lived as presumed. A distinct increase of in vivo-induced SCEs could be demonstrated only within days and weeks. The different results may be influenced by the different mode of therapy [monotherapy vs. drugs in combination; single application vs. cytostatic interval therapy (10)] and the state of health. To gain more information about the time-dependent effects of in vivo-induced SCEs, experiments were performed with mammals.

In the present experiments with rats, similar to the findings in rabbit lymphocytes (3,9), the number of induced SCEs decreased within 2 wk after a single CP injection. Huff et al. (3) suggested that the decline in SCE levels in lymphocytes of rabbits observed following increasing post-treatment time is a function of lymphocyte turnover. Lymphocyte turnover may also play a role in the decline of induced SCE in lymphocytes of rats, as the median survival of rat lymphocytes is only 4 wk (7,8). On the other hand, 5% to 8% of rat lymphocytes have a lifespan of more than 9 mo. Therefore, if only the lymphocyte turnover is responsible for the decline in SCE frequency, then the number of SCEs in some cells should be the same as the controls (9.94 SCE), while in other cells a high number of induced SCE (40.49 SCE) should still be present. However, this situation was never observed. In all cells there was a continuous decrease in SCEs with increasing post-treatment time. Similar to the controls, the range of SCE in single cells was 5 to 24, 2 wk post-treatment. Therefore, as pointed out by Raposa (6), the elimination of damage to DNA by existing repair mechanisms is responsible for the reduction of SCE with time, in addition to lymphocyte turnover.

REFERENCES

1. Basler, A. (1982) SCEs in lymphocytes of rats after exposure in vivo to cigarette smoke or to cyclosphosphamide. Mutat. Res. 102:137-143.
2. Gebhart, E., B. Windolph, and F. Wopfner (1980) Chromosome studies on lymphocytes of patients under cytostatic therapy. Human Genet. 56:157-167.
3. Huff, V., R.J. Du Frain, and L.G. Littlefield (1982) SCE frequencies in rabbit lymphocytes as a function of time after an acute dose of cyclophosphamide. Mutat. Res. 94:349-357.
4. Musilova, J., K. Michalová, and J. Urban (1979) Sister-Chromatid exchanges and chromosomal breakage in patients treated with cytostatics. Mutat. Res. 67:289-294.
5. Ohtsuru, M., Y. Ishii, S. Takai, H. Higashi, and G. Kosaki (1980) Sister chromatid exchanges in lymphocytes of cancer patients receiving mitomycin C treatment. Cancer Res. 40:477-480.
6. Raposa, T. (1978) Sister chromatid exchange studies for monitoring DNA damage and repair capacity after cytostatics in vitro and in lymphocytes of leukaemic patients under cytostatic therapy. Mutat. Res. 57:241-251.
7. Robinson, S.H., G. Brecher, I.S. Lourie, and J.E. Haley (1965) Leukocyte labeling in rats during and after continuous infusion of tritiated thymidine: Implications for lymphocyte longevity and DNA reutilization. Blood 26:281-295.
8. Röpke, C., and N.B. Everett (1973) Small lymphocyte populations in the mouse bone marrow. Cell Tissue Kinet. 6:499-507.
9. Stetka, D.G., and S. Wolff (1976) Sister-chromatid exchange as an assay for genetic damage induced by mutagen-clastogens. Mutat. Res. 41:333-342.
10. Stetka, D.G., J. Minkler, and A.V. Carrano (1978) Induction of long-lived chromosome damage, as manifested by sister-chromatid exchange, in lymphocytes of animals exposed to Mitomycin C. Mutat. Res. 51:383-396.
11. Trepel, F. (1975) Kinetik lymphatischer Zellen. In Lymphozyt und klinische Immunologie, Theml. H., and H. Begemann, eds. Springer, Berlin/Heidelberg/New York, pp. 16-26.
12. Trepel, F. (1976) Das lymphatische Zellsytem. In Hdb. Inn. Med. Bd 2, Teil 3, Begemann, H., ed. Springer, Berlin/Heidelberg/New York, pp. 1-191.

DETECTION OF SCE IN RODENT CELLS USING THE ACTIVATED CHARCOAL BROMODEOXYURIDINE SYSTEM

Pedro Morales-Ramírez, Teresita Vallarino-Kelly, and Regina Rodríguez-Reyes

Departamento de Radiobiología
Instituto Nacional de Investigaciones Nucleares
Avenida Benjamín Franklin 161
C.P. 06140 Mexico, D.F.

ABSTRACT

An in vivo method of sister chromatid differentiation based on the intraperitoneal (ip) injection of bromodeoxyuridine (BrdUrd) previously adsorbed to activated charcoal is described, as well as the protocols for the use ot this method in spermatogonial,salivary gland and bone marrow cells, will be presented.

Results on sister chromatid exchange (SCE) induction obtained in bone marrow cells using well-known mutagens, especially cyclophosphamide, are reported, and a comparison is made with the results obtained by other authors using different methods of BrdUrd administration.

The usefulness of this in vivo system for chemical mutagen detection and for chemical mutagen action on SCE induction is discussed.

INTRODUCTION

It has been considered desirable to develop systems for measuring in vivo the genetic damage caused by the increased levels of mutagenic agents now found in the environment. There is evidence that the measurement of SCE frequency is a very sensitive method for detecting mutagens (5).

Differential staining of sister chromatids involves the prior substitution of BrdUrd for thymidine in DNA during at least the first of 2 cycles of division. This BrdUrd requirement has been resolved in various ways for in vivo systems: multiple ip (1,14) continuous subcutaneous (sc) (9), or intravenous (iv) (12) infusions; and subcutaneous implantation of a BrdUrd tablet (2). An alternative method is the use of an ip injection of BrdUrd previously adsorbed to activated charcoal. This method has been used to achieve sister chromatid differentiation in spermatogonial (4), bone marrow, and salivary gland (6) cells.

The purposes of this paper are: a) to describe the protocols for the use of the BrdUrd-charcoal method for SCE analysis in spermatogonial, bone marrow, and salivary gland cells; b) to compare our results on SCE induction by well-known mutagens with those obtained by other authors using different methods of BrdUrd supply; and c) to report the results obtained with respect to the possibility of mutagen adsorption by charcoal in the peritoneal cavity.

MATERIALS AND METHODS

BrdUrd Incorporation and Chromosome Preparations

Two to 3 mo-old male Balb-C mice reproduced in our laboratory, maintained with Purina Laboratory Chow and water ad libitum, and weighing about 30-35 gm, were used.

For SCE analysis in bone marrow cells, the animals were given a single ip injection, using a 20 G needle, of BrdUrd (Sigma) (1 mg/gm bd wt) previously adsorbed to activated charcoal. Colchicine (Sigma) (15 μg/gm bd wt) was ip injected to the mice 19 hr after BrdUrd injection and 2 hr later they were killed by cervical dislocation. The femurs were removed, sectioned on either end and bone marrow cells were obtained injecting phosphate buffered saline (PBS) or Eagle's medium into one end with a 25 G needle. After centrifugation at approximately 1,000 g for 5 min, cells were incubated 15 min at 37°C in KCl 0.075 M, centrifuged again and fixed 2 times, once for 15 min and a second time for 10 min with methanol-acetic acid (3:1). Finally, the cells were dropped onto chilled slides.

For SCE analysis in salivary gland cells, cell division was induced by 2 injections of isoproterenol-HCl (Sigma) (1 μmole/gm bd wt) separated by an interval of 48 hr. BrdUrd (0.5 mg/gm bd wt) adsorbed previously to activated charcoal was ip injected 14 hr after each isoproterenol injection was administered. Colchicine (15 μg/gm bd wt) was ip injected 27 hr after the second isoproterenol injection and 4 hr before the animals were killed by ether narcosis. The appropriate times were obtained by autoradiographic analysis of tritiated thymidine incorporation and of the mitotic frequencies obtained in cuts of salivary gland tissue induced to cell division by isoproterenol in vivo (Figs. 1 and 2).

The parotid glands were removed, dissected, and cut away from the ganglia and adipose tissue, minced with scissors and processed for cell isolation using cholagenase type V (Sigma) and Hyaluronidase type V (Sigma) according to a modification of a previously reported method (3).

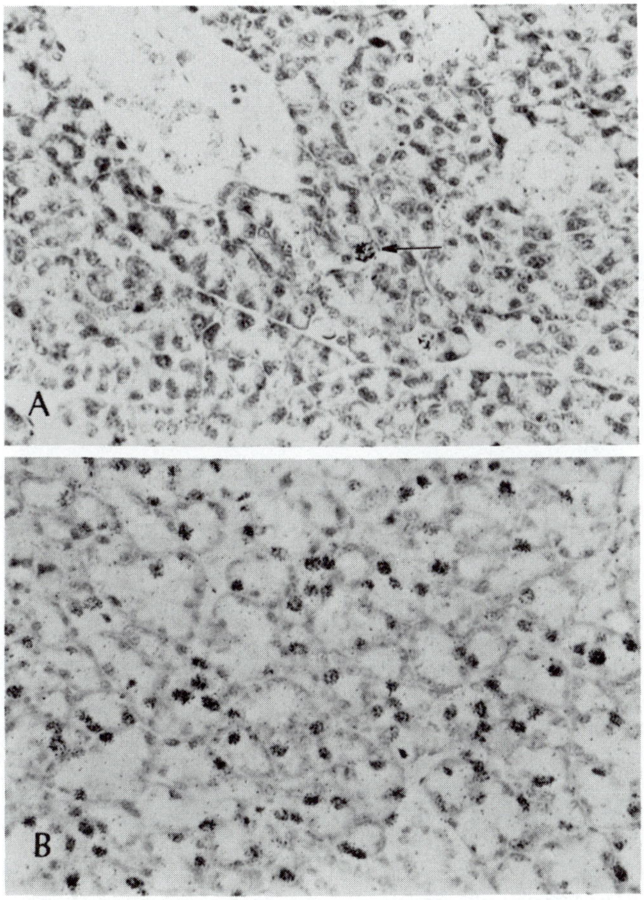

Fig. 1. Autoradiograph of salivary gland tissue unstimulated (A) and stimulated (B) to cell proliferation by a single injection of isoproterenol (1 μmole/gm bd wt). In the stimulated mouse, a single pulse of 24 μCi of ^{3}H-thymidine was given 24 hr after isoproterenol injection and 1 hr before killing.

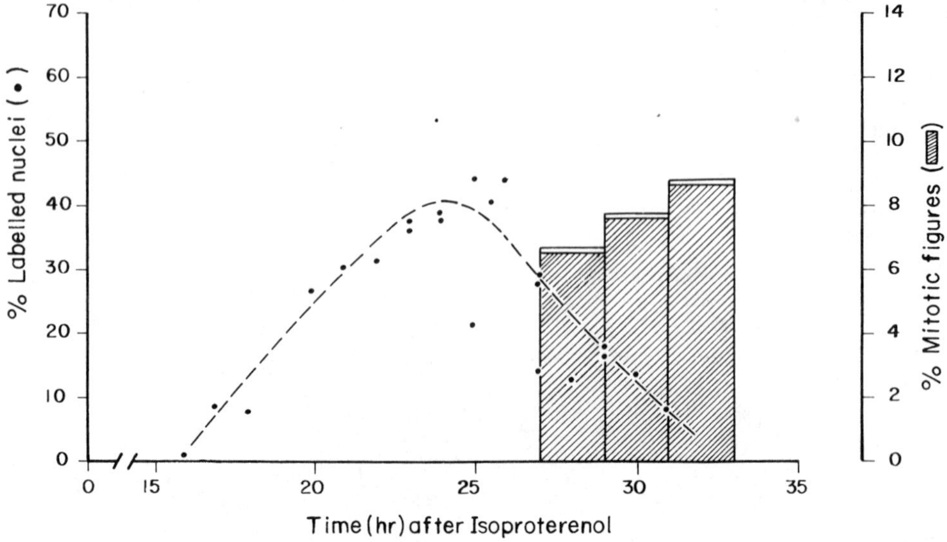

Fig. 2. Effect of isoproterenol on the frequency of labeled nuclei (%) and mitotic figures (%) at different times after a single ip injection of isoproterenol (1 μmole/gm bd wt).

The 2 parotid glands were minced with scissors in PBS, centrifuged 5 min at 1,000 g and resuspended in 2.5 ml of the first enzyme solution containing: cholagenase Sigma type V (specific activity 1,100 USP/mg 0.5 mg/ml. The mixture in a flask was incubated for 15 min at 37°C with a 130 rpm shaking in a water bath. The tissue was washed with PBS, incubated for 2 periods of 5 min at 37°C in 2 mM EDTA (Merck) and 30 min at 37°C in 2.5 ml of the second enzyme solution containing colagenase 0.66 mg/ml and Hyaluronidase 0.7 mg/ml. The tissue was repeatedly pipetted with a Pasteur pipet, incubated 15 min in the same enzyme solution at 37°C and finally drawn and expelled again using a 1 ml syringe with a 20 G needle. The isolated cells were treated with 0.075 M KCl, fixed with methanol- acetic acid and dropped on the slides in the same manner described for bone marrow cells.

For SCE analysis in spermatogonia, animals were ip injected once with BrdUrd (1 mg/gm bd wt) previously adsorbed to activated charcoal. Colchicine 15 μg/gm bd wt was injected 52 hr after BrUrd injection and 2 hr before the animals were killed. Both testes were removed and minced with scissors after the tunica was discarded. The tissue was washed 3 times with PBS and then incubated for 20 min at 37°C with skaking (130 rpm) in 2.5 ml of the first enzyme solution (the same used for salivary gland cells isolation). After the first 10 min and at the end of the incubation with enzymes, the tissue was passed through a 1.0 ml syringe with 20 G needle. The isolated cells were processed as bone marrow and salivary gland cells to obtain preparations of metaphase figures.

BrdUrd Adsorption to Charcoal

BrdUrd adsorption to activated charcoal was carried out according to Russev and Tsanev (11). One-hundred gm of activated charcoal (Merck) was washed 2 times with 500 ml 1N NaOH, 2 times with 500 ml 1N HCl, and finally with deionized water until the water pH was obtained. Deionized water was added to 2.0 liters in a graduated cylinder, mixed, and the charcoal was left to sediment. The smaller particles were removed from the surface to obtain 1.0 liter. The remaining charcoal solution was left to stand for 1 hr and then decanted. Finally, the pellet was dried overnight at 60°C and later at 105°C for 2 hr. The charcoal was enclosed in a flask.

BrdUrd solution (20 mg/ml) was mixed with charcoal (100 mg/ml of BrdUrd solution). The charcoal was desiccated and sterilized (on the flame) before it was weighed. The BrdUrd solution was previously sterilized by filtration with a 0.02 μm pore Millipore filter. The mixture was magnetically stirred in the dark for 2 hr before use. Usually, the percentage of BrdUrd not adsorbed was estimated by absorption at 280 nm in the supernatant.

BrdUrd-Cholesterol Tablets

Tablets of BrdUrd:cholesterol in the proportion of 1:2 by weight was obtained with a 0.7 cm die and pressured with 23 kg. Tablets were implanted under the skin on the back of the mice. The BrdUrd dose was approximately 1 mg/gm bd wt.

Differential Staining

The fluorescense-plus-Giemsa method (10) with slight modifications was used. Slides were stained for 10 min with an aqueous solution of 10^{-4} M Hoechst 33258; a coverslide was mounted in 0.16 M sodium-phosphate-0.04 M sodium citrate, pH 7.0, and the slides were placed beneath (1 cm) a black light (General Electric 25 W) for 40-90 min. The coverslides were removed and the slides washed with water and incubated for 20 min at 60-70°C in 2 x SSC (0.3 M NaCl-0.03 M trisodium citrate) in a Coplin jar. The slides were then washed with distilled water and stained in 6% Giemsa solution for 5 min.

Scoring

At least 30 metaphases with differentially stained sister chromatids from each mouse were scored for SCE frequency determination.

Only cells with a normal chromosome complement were considered. For the salivary gland cells, 2n and 4n complement were considered.

Mutagens

Mitomycin C (MMC) (Sigma Chemical), cyclophosphamide (CP) (Pharma Werke), and methylmethanesulphonate (MMS) (Eastman Kodak) were ip injected at 8-9 hr after the BrdUrd injection at the dose described in Tab. 1. In the experiments with CP in bone marrow cells, the mutagen was injected either 1 or 3 hr before or 8 hr after BrdUrd injection.

RESULTS AND DISCUSSION

Figure 3 shows spermatogonial (Fig. 3A), bone marrow (Fig. 3B) and salivary gland (Fig. 3C) cells with their sister chromatids differentially stained and obtained according to the previously described protocols. From the comparison of the sponteneous SCE frequencies in the different cell types (Tab. 1), we observed that the lowest frequency was obtained for spermatogonia; bone marrow had the medium frequency, and the highest frequency was obtained for salivary gland cells. The higher SCE levels in salivary gland cells could be explained on the basis that isoproterenol, in addition to stimulating cell divisions, could induce SCE. Another possibility could be that salivary gland cells accumulate unrepairable damage in their DNA which could induce SCE when they divide or that they are more sensitive to BrdUrd. Spermatogonial cells paradoxically exhibit the lowest SCE baseline frequency based on an expected gradient for BrdUrd levels (see Ref. 4).

A characteristic of the salivary gland tissue is that it was comprised of 70% of the cells with a 2n and 30% of the cells with a 4n complement. SCE basal values were nearly proportional to genomic

Tab. 1. Basal SCE frequencies in spermatogonial, bone marrow, and salivary gland cells using the BrdUrd-charcoal method.

SCE/CELL/MOUSE $\bar{X} \pm SD$			
SPERMATOGONIAL CELLS	BONE MARROW CELLS	SALIVARY GLAND CELLS	
		2n (70%)	4n (30%)
1.7 ± 0.56	3.5 ± 0.5	5.6 ± 0.9	4.8 ± 2.4
n=3	n=44	n=5	

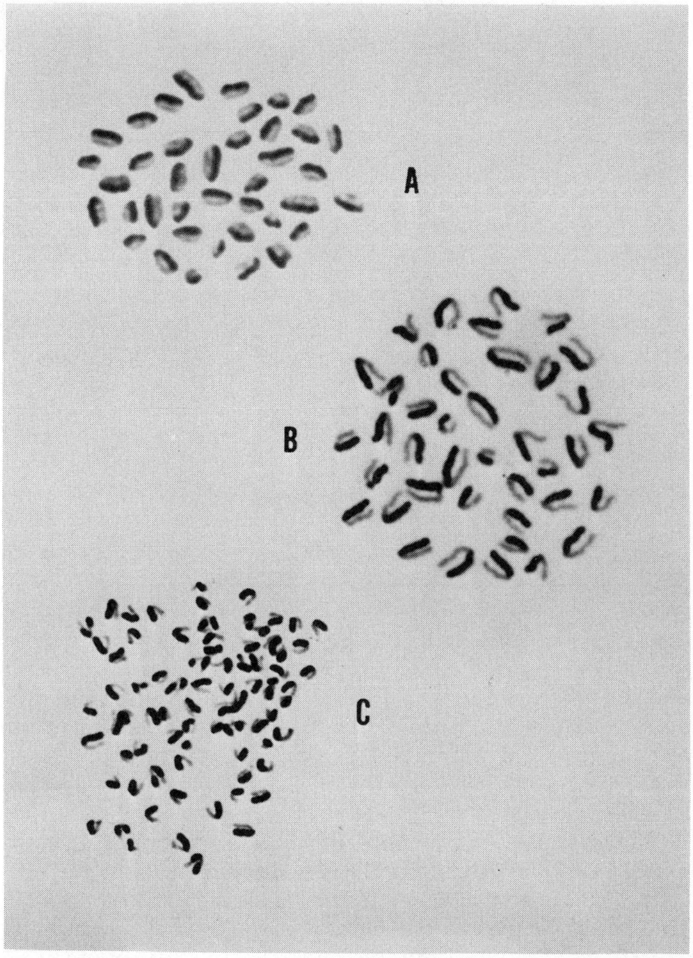

Fig. 3. Spermatogonial (A), bone marrow (B) and salivary gland (C) cells with their sister chromatids differentially stained. The BrdUrd-charcoal method was used for BrdUrd in vivo supply.

complement. This genomic size variability would allow for an analysis of SCE induction related to the size of the chromosomal complement.

A plot of percent cumulative frequency of SCE/bone marrow cell/ mouse for the SCE basal values obtained in 44 control mice (Fig. 4), shows that the distribution of basal SCE frequencies approaches a normal distribution and also shows that by using the charcoal method, 95% of the cells examined had SCE basal value lower than 4.3.

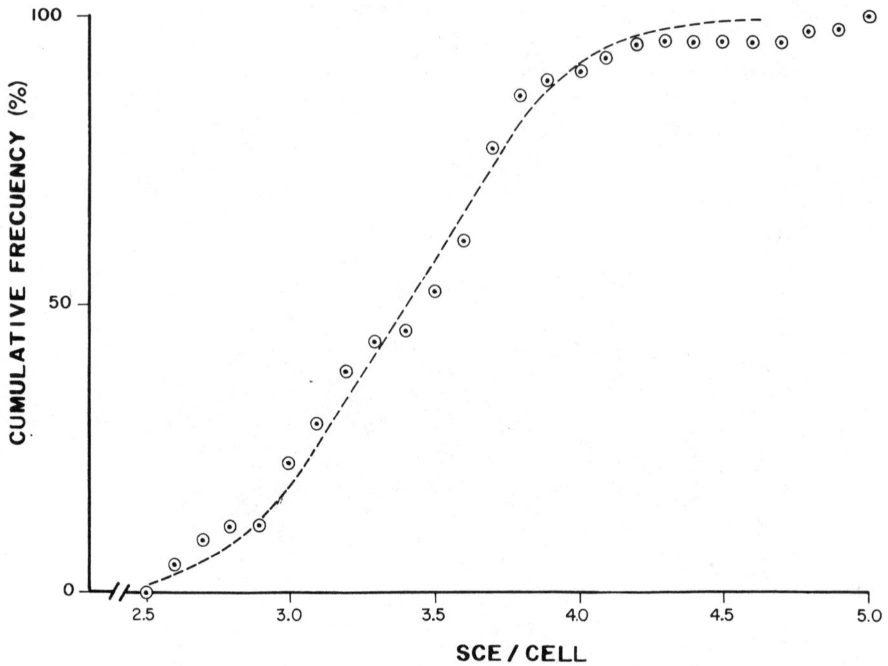

Fig. 4. Cumulative frequency (%) of SCE/cell obtained for the SCE basal values in untreated control mice. The dotted line represents the expected normal distribution (n = 44).

Table 2 shows the induction of SCEs in bone marrow cells in vivo by some well-known mutagens. The decreasing order of efficiency was MMC, CP, MMS, and EMS, a finding similar to that previously reported using other in vivo methods (5). Table 3 shows the efficiency of the SCE response induced by different mutagens using the different methods of BrdUrd supply. Efficiency is defined as the increase in SCE over the basal value per microgram of mutagen per gm bd wt. Tablet and continuous infusion (2,7) are more efficient than multiple injections (8) or the charcoal method. The fact that the results obtained with multiple injections and the charcoal method were similar suggests that the charcoal did not affect, to a great extent, SCE induction by mutagens.

The tablet and the continuous infusion methods induce higher basal SCE frequencies. These methods could allow greater incorporation of BrdUrd into DNA, thus increasing the basal SCE level and, perhaps, the efficiency of mutagens to induce SCE. The high response with the continuous infusion method could also be due to the injection of mutagens by the iv route which might be more direct and efficient than ip or sc routes.

Tab. 2. SCE induction by some mutagens in bone marrow cells in vivo using the BrdUrd-charcoal method.

	µg/g bd wt	SCE/Cell/mouse $\overline{X} \pm$ SD	animals
Mitomycin C	1.6	13.5 ± 2.0	3
Cyclophosphamide	3.5	8.3 ± 1.5	6
M M S	40.0	22.0 ± 5.0	2
E M S	380.0	8.6 ± 1.5	2
Control	---	3.6 ± 3.7	10

MMS Methylmethane sulphonate

EMS Ethylmethane sulphonate

To determine whether the possible synergy between BrdUrd incorporation and mutagen-induced SCE levels could be mutagen-dependent, the relative efficiency of the mutagens was determined for each method of BrdUrd supply (Tab. 4). From this comparison, it seems that there is not a great alteration in efficiency for MMC or CP over the range of methods considered, but there is for MMS using the continuous infusion method.

In order to eliminate the possibility of a decrease in SCE induction due to charcoal adsorption of the mutagen, and/or its active

Tab. 3. Efficiency of chemical mutagens on SCE induction in bone marrow cells using different in vivo methods of BrdUrd supply.

	TABLET (Allen et al 1977)	CONTINUOUS INFUSION (Yoshifume & Schneider 1979)	MULTIPLE INJECTIONS (Palitti et al 1982) (injections)	BrdUrd-CHARCOAL
Mitomycin C	---	16.2	6.1	6.2
Cyclophosphamide	2.5	2.3	0.7	1.2
M M S	---	---	0.32	0.46
E M S	---	0.07	---	0.013

$$\text{Efficiency} = \frac{SCE_{treated} - SCE_{control}}{\mu g/gr \text{ bd wt}}$$

Tab. 4. Relative efficiency of different mutagens on SCE induction in bone marrow cells in vivo for different methods of BrdUrd supply.

	CONTINUOUS INFUSION (Yoshifume & Schneider 1979)	MULTIPLE INJECTIONS (Palitti et al 1982)	BrdUrd-CHARCOAL
Mit C/C.P.	7.0	8.7	5.2
Mit C/MMS	---	19.1	13.5
Mit C/EMS	231.4	---	476.9
CP/MMS	---	2.2	2.6
CP/EMS	32.9	---	92.0

Mit. C Mitomycin C
C.P. Cyclophosphamide
MMS Methymethane sulphonate
EMS Ethylmethane sulphonate

metabolities, or the possibility of a synergy of SCE induction by BrdUrd (13), groups of mice were treated with CP either before (1 hr or 3 hr) or after (8 hr) BrdUrd-charcoal injection.

The results for 3 mutagen doses suggest that there was neither adsorption nor synergy (Tab. 5). Another possible explanation for such results is that these 2 events masked each other. To increase the possible effect of BrdUrd and/or of charcoal, animals were injected with an amount 50% greater than usual. No important alterations were observed (Tab. 6). To further eliminate the possibility of CP adsorption by charcoal, CP was sc injected 1 hr before or 8 hr

Tab. 5. Effect of BrdUrd incorporation on SCE induction by CP.

Cyclophosphamide µg/gbd wt	BEFORE BrdUrd: 3 hr	BEFORE BrdUrd: 1 hr	AFTER BrdUrd: 8 hr
3.5	------	7.0 ± 0.8 n = 3	8.3 ± 1.5 n = 6
7.0	15.3 ± 1.5 n = 3	17.1 ± 1.6 n = 3	14.8 ± 2.8 n = 4
14.0	------	19.2 ± 3.6 n = 4	21.0 ± 3.6 n = 4
	Control	3.6 ± 0.5	n = 16

Tab. 6. SCE induction by 14 μg/gm bd wt CP 1 hr before or 8 hr after BrdUrd supply.

	SCE/CELL/MOUSE			
	SUSPENSION BrdUrd-charcoal			TABLET BrdUrd-Chollesterol
	A	B	C	D
Cyclophosphamide	Intraperitoneal (ip)	ip	Subcutaneous (sc)	ip
BrdUrd mg/g bd wt	1.0 (ip)	1.5 (ip)	1.0 (sc)	1.0 (sc)
Charcoal mg/g bd wt	5.0 (ip)	7.5 (ip)	5.0 (sc)	---
1 hr BEFORE	19.2 ± 3.6 n=4	22.0 ± 5.1 n=4	22.4 ± 3.8* n=5	20.0 ± 3.5* n=2
8 hr AFTER	21.0 ± 3.6 n=4	24.5 ± 6.2 n=3	32.9 ± 5.8* n=5	32.4 ± 3.5* n=3
CONTROL	3.3 ± 0.36 n=4	3.5 ± 0.2 n=4	3.8 ± 0.9 n=5	2.6 ± 0.0 n=2

* $p < 0.01$ 1 hr_b vs 8 hr_a, controls did not raise the significance between them. Student's t test.

after the BrdUrd-charcoal injection. In Tab. 6 (A vs C) it is clear that the sc injection of CP resulted in an increased SCE induction 8 hr after the injection of the BrdUrd supply. This result implies that CP is adsorbed by charcoal and also that BrdUrd incorporated into DNA for 1 cycle may enhance the ability of CP to induce SCEs.

To further discount the possibility of charcoal adsorption of CP, mice were ip injected with CP 1 hr before or 8 hr after the sc implantation of a BrdUrd-cholesterol tablet. The results (Tab. 1) were similar to those obtained with 4 sc injection of CP and ip injection of BrdUrd-charcoal (Tab. 6C vs. 6D). These results imply that there is no adsorption of CP active metabolites by charcoal in the peritoneal cavity under these conditions. They also confirm the fact that BrdUrd incorporated into DNA may increase the sensitivity for SCE induction by CP and that charcoal could adsorb the mutagen if it were injected into the peritoneal cavity.

CONCLUSIONS

The system of chromatid differentiation using BrdUrd-charcoal has not been used extensively because there are some doubts with respect to its usefulness. The inconvenient and potentially deleterious effects on mice, and the possibility that charcoal could adsorb the mutagen and/or metabolites, have limited its use. However, from a comparison of the results obtained with this and other systems of

BrdUrd supply, in addition to the experimental evidence reported here, it was concluded that this method can be used if the mutagen is administered prior to BrdUrd-charcoal injection or by other than the peritoneal route. Also, in our experience, there is no apparent damage to animals resulting from exposure to charcoal for 24 hr.

From a practical point of view the charcoal method is simpler than the more favored tablet method because the tablet preparation (the weighing and the pressing) is time consuming, and the single injection is simpler than tablet implantation.

The SCE basal values in both methods are similar. The use of cholesterol, together with variations on pressure during tablet preparation, reduces the basal SCE frequencies with respect to those reported for the solid BrdUrd tablets (2).

Some comments on salivary gland cells are necessary. These cells have many interesting characteristics, the principal one being the possibility of controlling in vivo cell proliferation. There are other important ones--the presence of populations of cells with a 2n and 4n complement; that isoproterenol induces only 1 cell cycle; that these cells divide parasynchronically and that there is apparently only 1 type of cell which is induced to proliferation. These characteristics make the salivary gland cells system a very good in vivo system for the study of the phenomena involved and related to SCE formation and induction.

REFERENCES

1. Allen, J.W., and S.A. Latt (1976) Analysis of sister chromatid exchange formation in vivo in mouse spermatogonia as a new test system for environmental mutagens. Nature 260: 449-451.
2. Allen, J.W., C.F. Schuler, R.W. Mendes, and S.A. Latt (1977) A simplified technique for in vivo analysis of sister chromatid exchanges using 5-bromodeoxyuridine tablets. Cytogen. Cell Genet. 18:231-237.
3. Barka, T., and H. Van der Noen (1975) Dissociation of rat parotid gland. Lab. Invest. 32:373-375.
4. Kanda, N., and H. Kato (1979) A simple technique for in vivo observation of SCE in mouse ascites tumor and spermatogonial cells. Exp. Cell Res. 118:431-434.
5. Latt, A.S., J. Allen, S.E. Bloom, A. Carrano, E. Talke, D. Kram, E. Schneider, R. Schreck, R. Tice, B. Whitfield, S. Wolff (1981) Sister chromatid exchanges: A report of the gene-tox program. Mutat. Res. 87:17-62.
6. Morales-Ramírez, P. (1980) Analysis in vivo of sister chromatid exchange in mouse bone marrow and salivary gland cells. Mutat. Res. 74:61-69.

7. Nakanishi, Y., and E.L. Schenider (1979) In vivo sister-chromatid exchange: A sensitive measure of DNA damage. Mutat. Res. 60:329-337.
8. Palitti, F., C. Tanzarella, R. Cozzi, R. Ricordy, E. Vitagliano, and M. Fiore (1982) Comparison of the frequencies of SCE induced by chemical mutagens in bone-marrow, spleen and spermatogonial cells of mice. Mutat. Res. 103:191-195.
9. Pera, F., and P. Mattias (1976) Labeling of DNA and differential sister chromatid staining after BrdU treatment in vivo. Chromosoma 57:13-18.
10. Perry, P., and S. Wolff (1974) New Giemsa method for the differential staining of sister chromatids. Nature 251:156-158.
11. Russer, G.C., and R.G. Tsanev (1973) Continuous labeling of mammalian DNA in vivo. Analyt. Biochem. 54:115-119.
12. Schneider, E.L., J.R. Chaillet, and R.R. Tice (1976) In vivo analysis of cellular replication. Proc. Natl. Acad. Sci., USA 74:2041-2044.
13. Tice R.R., Personal communication.
14. Vogel W., and T. Bauknecht (1976) Differential chromatid staining by in vivo treatment as a mutagenicity test system. Nature 260:448-449.

ASSESSMENT OF SISTER CHROMATID EXCHANGE IN SPERMATOGONIA AND INTESTINAL EPITHELIUM IN CHINESE HAMSTERS

Steven B. Neal and Gregory S. Probst

Toxicology Division
Lilly Research Laboratories
Division of Eli Lilly and Company
Greenfield, Indiana 46140

ABSTRACT

The induction of sister chromatid exchange (SCE) has been proposed as a predictive test for the identification of mutagens/carcinogens. The in vivo application of this test was investigated by examining the chemical induction of SCE in spermatogonia, intestinal epithelium and bone marrow cells from Chinese hamsters.

Sister chromatid differentiation (SCD) was achieved in differentiating spermatogonial cells of male Chinese hamsters by the abdominal subcutaneous (sc) implantation of an agar-coated bromodeoxyuridine (BrdUrd) tablet. A number of genotoxins were administered intraperitoneally (ip) and the induction of SCE in spermatogonia and bone marrow was compared. A significant increase in SCE frequency in spermatogonia occurred following treatment with mitomycin C (MMC), cyclophosphamide (CP), or N,N',N"-triethylenethiophosphoramide (ThioTEPA). Treatment with busulfan, hycanthone (HC), or triethylenemelamine (TEM) failed to induce SCE in vivo in spermatogonia, but these compounds did induce SCE in bone marrow. Differences in cell cycle kinetics were considered to be the major factor involved in the differential induction of SCE in spermatogonia and bone marrow.

The induction of SCE in intestinal epithelium was investigated as a system for the identification of genotoxins that may result from the metabolism of xenobiotics by the gastrointestinal flora. Nitro-aromatic compounds were administered orally to Chinese hamsters. Nitro-aromatic compounds were chosen for this study since the mutagenic activity of these compounds is thought to result from their metabolism by bacterial nitroreductase. Metronidazole (MN) and 2-nitro-p-phenylenediamine (2NPPD) induced a dose-related increase in SCE formation in intestinal epithelium but not in bone marrow. Treatment with 3-nitro-o-phenylenediamine (3NOPD) or 4-nitro-o-phenylenediamine (4NOPD) did not induce the formation of SCE in either intestinal epithelium or bone marrow. These findings indicate that studies in axenic animals will be required to elucidate the contribution of the enteric flora to the metabolic activation of some genotoxins.

INTRODUCTION

Sister chromatid exchanges represent reciprocal interchange of DNA between homologous chromatids observable in metaphase chromosomes. Although neither the molecular mechanism nor biological significance of SCE formation is thoroughly understood, a variety of independent studies have established that most mutagens/carcinogens are inducers of SCEs (1,2,5,6,12).

The advent of the BrdUrd tablet method for in vivo SCD has provided a convenient approach for the identification of SCE induced in vivo (8,9). Additional modification of the BrdUrd tablet methodology by using agar-coated BrdUrd tablets (18) now allows for a more controlled, sustained release of the BrdUrd and the ability to use smaller amounts. Owing to the simplicity of this technique, the readily scored chromosomal endpoint, and excellent sensitivity to genotoxins (1,2,33), this system has emerged as an especially attractive test for chemically induced chromosomal damage.

This test has been applied both in vitro (5,6) and in vivo (1, 2,7,8); however, the in vivo application appears to offer the greatest utility as a component of a genetic toxicology test battery. The in vivo application of this test provides: 1) the advantage of multiple sites and pathways of metabolic activation, as well as the opportunity for metabolic detoxication; 2) multiple routes of administration to reflect relevant routes of exposure; and 3) systemic distribution of genotoxins to multiple target tissues. Consequently, this system may reflect the behavior of potential genotoxins in the intact animal.

In the present study, we describe the assessment of SCEs in multiple tissues of the Chinese hamster. Chinese hamsters are desirable for studies of this type due to their small size, low chromosome number, chromosome size and morphology, and low background SCE frequency.

Sister chromatid exchange was investigated in differentiating spermatogonia of the Chinese hamster as a model for detecting genotoxins that were capable of interacting with germinal tissue. To evaluate the sensitivity of the system, several known genotoxins which were identified in other systems as interacting with germinal tissue were administered and the induction of SCE in spermatogonia and bone marrow were compared.

The induction of SCE in intestinal epithelium was also examined. This tissue was investigated as a proximal target for compounds which, when administered orally, could be metabolized or detoxified by the enteric microflora. Enteric microflora can participate in the metabolism of genotoxic compounds (23-25,44-46) to form active metabolites which may be absorbed and systemically distributed. Therefore, several genotoxins were administered orally to Chinese hamsters and the induction of SCE in intestinal epithelium and bone marrow was examined.

MATERIALS AND METHODS

Animals

Inbred male and female Chinese hamsters (*Cricetulus griseus*) weighing 30-35 gm were selected from a breeding colony established in the Toxicology Division, Eli Lilly and Company. Each animal, after reaching adulthood, was allowed to acclimate to our animal facilities for at least 2 wk prior to use. Animals were housed individually in shoe-box cages containing hardwood chip bedding and excelsior nesting material. The animal quarters were maintained at 24 ± 3°C with a 40% minimum relative humidity and a reverse lighting cycle (dark: 1300-2400 hr). Purina laboratory rodent chow and tap water were provided ad libitum.

Chemicals

Bromodeoxyuridine was obtained from CalBiochem, and vinblastine sulfate (Velban[R]) from Eli Lilly and Company, Indianapolis, Indiana. Bacto-Agar[R] was purchased from Difco Laboratories, Detroit, Michigan.

All test chemicals were obtained from the chemical archives of Eli Lilly and Company or from commercial sources as indicated in the table footnotes. All chemicals were used as supplied without further purification or attempt to identify impurities. The test chemicals were administered ip as solutions in either sterile water or corn oil, or by the oral route as suspensions in 10% aqueous acacia. In one case, a compound was initially solubilized in dimethylsulfoxide (DMSO) (Fisher Scientific) and then diluted in corn oil. The final DMSO concentration in vehicles did not exceed 3%. Upper dose levels of test chemicals were chosen to produce bone marrow cytotoxicity as shown by a suppression of the mitotic index or a suppression of the frequency of second-division cells.

Preparation of BrdUrd Tablets

Tablets of pure BrdUrd were prepared using a Stokes-Eureka tablet press (Model 511-1) fitted with a 4 mm concave punch.

Agar-Coating Technique

Pure tablets of BrdUrd were coated with agar according to a modified procedure of King et al. (18). A 4.5% Difco Bacto-Agar[R] solution was prepared in Milli-Q[R] purified distilled water, autoclaved, and placed in a 55°C water bath. Aliquots of 1.5 ml of the 4.5% Bacto-Agar were dispensed into each well of a Linbro[R] multiwell culture plate (24-well multiplate, well capacity 3.5 mm, area per well 2.0 cm^2).

The agar solution in each well was allowed to cool until it was semisolidified, then a BrdUrd tablet was placed into each well. Immediately, an additional 1.5 ml of 55°C agar was overlayed onto the tablet. The agar was allowed to solidify at room temperature (cooling at 4°C resulted in condensation between the two agar layers).

Agar-coated BrdUrd (AC-BrdUrd) tablets were removed by punching each cylinder with a 8 mm glass tube. Each AC-BrdUrd tablet was trimmed to a thickness of 4 mm, placed in a petri plate lined with Saran Wrap[R], and dehydrated in a desiccator.

In Vivo Sister Chromatid Exchange Assay

Differentiating spermatogonia. Sister chromatid differentiation was obtained in vivo in differentiating spermatogonia by the sc implantation of a 40-60 mg AC-BrdUrd tablet into a 30-35 gm (3-4 mo old) male Chinese hamster.

Hamsters were anesthetized with ether and a 1 cm incision was made through the shaved abdomen. Using cuticle scissors, the skin was slightly undermined and one AC-BrdUrd tablet was placed sc into each animal. The incision edges were rejoined and the implantation site was closed using a single wound clip (Autoclip[R]). Three animals were used for each treatment group.

Forty-eight hr post-tablet implantation, the test compound was administered ip in a volume not exceeding 10 ml/kg. Forty-eight hr thereafter mitotic arrest was established by administering 10 mg/kg Velban[R] ip. A 4-hr period was used for accumulating mitotic figures and the spermatogonial metaphases were harvested using a modification of the method of Hoo and Bowles (30).

Animals were sacrificed by ether narcosis and both testicles were excised and placed into a 60 mm petri plate. The tunica albuginea was removed and the seminiferous tubules were minced using a pair of cuticle scissors. To the minced tubules was added 2 ml of 0.8% sodium citrate (NC) at 37°C. The tubule suspension was transferred, using a large bore Pasteur pipette, to a 12 ml conical centrifuge tube. An additional 8 ml of 0.8% NC was pipetted vigorously into the centrifuge tube to suspend the tubules. The tubules were allowed to sediment by gravity for 2 min after which the supernatant was discarded. An additional 10 ml of 0.8% NC was vigorously added to the tubules. After 10 min the supernatant was discarded and the tubules were fixed by the addition of methanol-acetic acid (3:1, v/v). Fixed tubules were transferred to capped vials and stored at 4°C overnight.

To obtain spermatogonial metaphases, a few tubules were removed and placed into centrifuge tubes containing freshly prepared 3:1 methanol-acetic acid. Following centrifugation for 10 min at 176 x g, the supernatant was discarded and a sufficient volume of 60% glacial acetic acid (GAA) was added to obtain a cell suspension. Spermatogonia metaphases were liberated from the seminiferous tubules by a 4-min treatment with 60% GAA. A 4-min treatment with 60% GAA was usually sufficient to obtain a cell suspension. The cell suspension was flooded onto a 65°C prewarmed slide and then was immediately withdrawn. A siliconized Pasteur pipette was used for this procedure.

Slides were dried overnight at room temperature and stained by a modification of the fluorescence-plus-Giemsa (FPG) technique of Perry and Wolff (13). Briefly, slides were stained for 12 min in an aqueous solution (1 µg/ml) of Hoechst 33258, then rinsed for 4 min in distilled water. The slides were coverslipped with phosphate buffered saline and the coverslip edges were sealed with lacquer. Following exposure to ultraviolet (UV) light (λ = 366 nm, General Electric No. F40BL) for 1 hr at a distance of 3 cm, the coverslips were removed. Final processing included a 20 min incubation in 10XSSC at 60°C, a distilled water rinse, and staining in a 3% aqueous solution of Giemsa.

Bone marrow. To obtain SCD in bone marrow cells, a 20-30 mg AC-BrdUrd tablet was sc implanted into each animal. Compounds were administered by either the ip or oral routes to animals 5 hr post-tablet implantation. Eighteen hr after compound administration, mitotic arrest was established by administering Velban[R] (ip), and 2 hr thereafter the animals were sacrificed. Bone marrow was flushed from both femurs with 75 mM KCl (37°C), maintained in this solution at 37°C for 20-25 min, then centrifuged for 5 min at 176 x g. The supernatant was discarded and the cell pellet was resuspended by hand agitation in 0.1 ml of residual supernatant. The cells were fixed by adding 5 ml of methanol-acetic acid (3:1, v/v) and subsequently, washed with 3 changes of this fixative by centrifugation for 5 min at 176 x g. Metaphase preparations were made by standard air-drying techniques. Slides containing metaphase chromosomes were stained by a modification of the FPG technique as described above.

Intestinal epithelium. Sister chromatid differentiation was achieved in intestinal epithelial cells of the ileum by the sc implantation of a 20-30 mg AC-BrdUrd tablet into a 30-35 gm female Chinese hamster. Five hr post-tablet implantation the compound was administered by the oral route in a volume not exceeding 10 ml/kg. Mitotic arrest was established by administering 1 mg/kg Velban^{R} (ip) 23 hr after tablet implantation.

Intestinal epithelial cells were isolated from the ileum by dissecting the ileum free from the animal. Gastrointestinal contents were flushed with a syringe containing 1% NC (37°C). The intestine was split longitudinally, cut into small pieces and placed in a petri plate. Small sections of intestine were placed into a conical centrifuge tube containing 10 ml of 1% NC (37°C). Total hypotonic time was 30 min at room temperature, followed by centrifugation for 2 min at 176 x g. The supernatant was discarded, and the packed tissue was resuspended by hand agitation. Tissue was fixed by the addition of cold methanol-glacial acetic acid (3:1, v/v), transferred to capped vials, and stored at 4°C.

Metaphase figures were liberated from the intestine by softening the tissue in 60% GAA for 4-5 min. The cell suspension was placed on slides and stained in a similar manner as that described for spermatogonial preparations.

Criteria for a positive response. For each animal, 25 well-spread, differentially stained metaphase figures, each containing the diploid number, were scored for SCE. Due to the number of polyploid appearing metaphases in spermatogonial preparations, SCEs were expressed as SCE/diploid genome. A dose-related increase in SCE frequency in which at least 2 successive doses were statistically different from controls, as determined by Dunnett's t-test at a probability level less than 0.05, constituted a positive response (10).

RESULTS

Influence of BrdUrd Tablet Weights on SCE Frequency

Since the BrdUrd required for SCD can influence the frequency of SCEs (4,11,19), several BrdUrd tablet weight ranges were evaluated for effects on baseline SCE response. For spermatogonia, AC-BrdUrd tablets of 40-50 mg and 50-60 mg weight ranges were compared (Tab. 1). Both tablet weights produced acceptable SCD. It was noted, however, that tablet weights of less than 40 mg did not provide sufficient levels of BrdUrd for complete SCD in spermatogonia.

Tab. 1. Baseline response for SCE in spermatogonia. (AC-BudR ≅ AC-BrdUrd in the text.)

SCE/METAPHASE[a]	
AC-BudR 40-50 mg	AC-BudR 50-60 mg
1.1 ± 0.7	1.0 ± 0.9
1.4 ± 1.0	1.4 ± 1.1
1.0 ± 0.8	1.5 ± 0.9
1.3 ± 0.9	1.4 ± 1.2
1.0 ± 0.7	1.3 ± 0.9
1.4 ± 1.3	1.3 ± 0.8

[a]Counts represent mean ± S.D. of 25 second division metaphases from each of two animals per experiment.

Acceptable SCD was obtained in bone marrow and intestinal epithelium with a 20-30 mg AC-BrdUrd tablet, and the baseline responses for SCE in these tissues were similar (Tab. 2).

Chemically Induced SCE

Spermatogonia. A variety of genotoxins were evaluated for the induction of SCE in differentiating spermatogonia and bone marrow. Compounds chosen were: CP, MMC, ThioTEPA, busulfan, TEM, and HC. Due to the dissimilar cell cycles of these two tissues, the tissues were obtained from separate groups of similarly treated animals.

A dose-related, positive response for chemically induced SCEs was found with CP, MMC, and ThioTEPA (Tab. 3). Treatment with CP, which requires metabolic activation, resulted in an increase in SCE in spermatogonia over a dose range of 10-100 mg/kg. Cytotoxicity was observed in the 100, 50, and 25 mg/kg treatment groups as evidenced by a reduction in number of metaphase figures. A dose-related induction in SCEs was noted in bone marrow cells with all doses of CP tested.

The direct-acting genotoxin MMC exhibited a cell kill effect or cell cycle delay at doses of 8 and 4 mg/kg, and a similar response has been reported for treated male mice (7). Only one treatment of MMC, 2 mg/kg, resulted in a positive SCE response in spermatogonia, whereas all doses of MMC were positive in bone marrow.

Tab. 2. Baseline response for SCE in bone marrow and intestinal epithelium. (AC-BudR ≅ AC-BrdUrd in the text.)

SCE/METAPHASE[a]	
Bone Marrow	Intestine
AC-BudR 20-30 mg	AC-BudR 20-30 mg
2.9 ± 1.3	3.9 ± 1.0
3.2 ± 1.1	4.2 ± 2.1
3.2 ± 1.3	3.4 ± 1.4
2.9 ± 1.4	3.2 ± 1.0
3.6 ± 1.9	3.5 ± 1.3
3.0 ± 1.5	3.3 ± 1.7

[a]Counts represent mean ± S.D. of 25 second division metaphases from each of two animals per experiment. Bone marrow and intestinal epithelium were obtained from each animal.

ThioTEPA, also a direct-acting genotoxin, rendered either a cell cycle delay or cell kill effect at doses of 4, 2, and 1 mg/kg, while the 1 and 0.5 mg/kg treatments produced a 2-fold increase in SCE. Although the 4 mg/kg treatment exhibited cytotoxicity in bone marrow, the remaining doses of ThioTEPA were positive for SCE induction in bone marrow.

Hycanthone, busulfan, and TEM (Tab. 3) failed to induce SCE in spermatogonia at the doses tested. All doses of these compounds induced SCE in bone marrow of similarly treated animals.

Intestinal epithelium. The induction of SCE by chemicals was also investigated in intestinal epithelium.

Oral administration of 2NPPD, 3NOPD, and 4NOPD failed to induce SCE in bone marrow (Tab. 4). However, 2NPPD induced a dose-related increase in SCE in intestinal epithelium. This response was not evident in animals treated with 3NOPD or 4NOPD.

Treatment with MN induced a statistically significant, dose-related increase in SCE in intestinal epithelial cells but not in bone marrow (Tab. 4).

Busulfan and CP were tested to evaluate the induction of SCE in intestinal epithelium by genotoxins not requiring metabolism by the enteric microflora. An equivalent induction of SCE was noted in

Tab. 3. In vivo induction of SCE in spermatogonia.

Compound	Dose mg/kg (ip)	Metaphases Scored[a]	SCE/Metaphase: Spermatogonia[b]	SCE/Metaphase: Bone Marrow[c]
Cyclophosphamide[h]	100	21	4.3 ± 1.6[d,f]	N.T.[g]
	50	31	3.5 ± 1.8[d,f]	19.1 ± 1.8[d]
	25	17	3.0 ± 1.0[d,f]	14.0 ± 1.5[d]
	10	75	2.9 ± 1.5[d]	12.0 ± 1.8[d]
	5	75	1.4 ± 0.8	9.0 ± 1.5[d]
	0	75	1.2 ± 1.1	3.4 ± 1.3
Mitomycin C[h]	8	0	--[e]	36.8 ± 9.5[d]
	4	0	--[e]	19.7 ± 7.0[d]
	2	75	3.8 ± 1.7[d]	11.2 ± 5.2[d]
	1	75	1.3 ± 1.0	8.3 ± 3.7[d]
	0.5	75	1.1 ± 0.9	6.8 ± 3.3[d]
	0	75	1.1 ± 1.1	3.5 ± 1.7
N,N',N"-Triethylenethiophosphoramide[i]	4	0	--[e]	--[e]
	2	0	--[e]	21.1 ± 6.0[d,f]
	1	64	2.7 ± 1.6[d,f]	12.7 ± 5.2[d]
	0.5	75	2.2 ± 1.9[d]	7.2 ± 2.6[d]
	0.25	75	1.2 ± 1.1	5.0 ± 2.2[d]
	0	75	1.1 ± 0.9	3.2 ± 1.3
Hycanthone[j]	200	0	--[e]	9.3 ± 3.8[d]
	100	75	1.3 ± 0.8	8.1 ± 3.5[d]
	50	75	1.4 ± 1.0	4.5 ± 2.0[d]
	25	75	1.4 ± 1.0	3.4 ± 1.9
	0	75	1.0 ± 0.7	3.2 ± 1.2
Busulfan[k]	100	0	--[e]	15.6 ± 2.1[d]
	50	35	1.5 ± 1.2[f]	8.0 ± 2.2[d]
	25	46	1.2 ± 1.0[f]	5.9 ± 2.0[d]
	12.5	35	1.5 ± 1.0[f]	4.0 ± 1.9
	0	75	1.3 ± 1.0	3.3 ± 1.6
TEM[i] Triethylenemelamine	0.062	0	--[e]	--[e]
	0.032	0	--[e]	50.0 ± 9.9[d,f]
	0.016	0	--[e]	30.7 ±10.3[d,f]
	0.008	49	1.5 ± 1.3[f]	19.9 ± 7.6[d]
	0	75	1.2 ± 0.8	3.5 ± 1.4

[a]Represents total number of spermatogonial metaphases scored from three animals per dose group.

[b]Mean ± S.D. of second division metaphases from each of three animals per dose group. Counts adjusted for polyploid metaphases. Genome of 22 chromosomes was used in calculations.

[c]Mean ± S.D. of 25 second division metaphases from each of three animals per dose group.

[d]Judged to be positive response for induction of SCE as determined by Dunnett's t-test.

[e]Toxicity noted, no observable metaphase figures.

[f]Toxicity noted. Counts for surviving cells.

[g]Not tested.

[h]Sigma Chemical Company, Rochester, NY.

[i]Lederele Laboratories Division, Pearl River, NY.

[j]Sterling Winthrop, Renssalaer, NY.

[k]ICN Pharmaceuticals Life Science Group, Cleveland, OH.

intestinal epithelium and bone marrow following the administration of busulfan. Cyclophosphamide also induced SCE in bone marrow and intestinal epithelium, but a greater response was noted in bone marrow.

Tab. 4. In vivo induction of SCE in intestinal epithelium.

Compound	Dose mg/kg (po)	Metaphases[a] Scored	SCE/Metaphase[b] Intestine	Bone Marrow
2-Nitro-p-phenylenediamine[d]	500	75	5.5 ± 1.9[c]	3.5 ± 1.6
	250	75	5.8 ± 2.3[c]	3.3 ± 1.2
	125	75	4.8 ± 2.1[c]	3.0 ± 1.7
	62.5	75	3.4 ± 1.5	3.4 ± 1.3
	0	75	3.2 ± 1.0	3.1 ± 1.5
3-Nitro-o-phenylenediamine[d]	500	75	3.3 ± 1.3	3.3 ± 1.7
	250	75	3.1 ± 1.3	3.2 ± 1.6
	125	75	3.4 ± 1.5	3.4 ± 1.6
	62.5	75	3.8 ± 1.7	3.0 ± 1.6
	0	75	3.3 ± 1.6	3.2 ± 1.4
4-Nitro-o-phenylenediamine[d]	500	75	3.5 ± 1.9	3.3 ± 1.0
	250	75	3.5 ± 1.5	3.4 ± 1.1
	125	75	3.4 ± 1.7	3.8 ± 1.3
	62.5	75	3.0 ± 1.5	3.4 ± 1.0
	0	75	3.2 ± 1.6	3.4 ± 0.9
Metronidazole[e]	500	75	7.3 ± 2.7[c]	3.6 ± 1.5
	250	75	5.7 ± 2.2[c]	3.3 ± 1.3
	125	75	5.1 ± 2.1[c]	3.5 ± 1.4
	62.5	75	4.1 ± 1.6	3.3 ± 1.6
	0	75	3.9 ± 2.0	3.2 ± 1.3
Busulfan[f]	50	75	8.0 ± 2.9[c]	7.1 ± 2.6[c]
	25	75	6.4 ± 1.9[c]	5.0 ± 2.1[c]
	12.5	75	3.9 ± 1.8	4.2 ± 2.3[c]
	6.25	75	3.7 ± 1.9	3.3 ± 1.7
	0	75	4.2 ± 2.1	3.3 ± 1.5
Cyclophosphamide[f]	100	75	8.2 ± 3.3[c]	26.4 ±12.5[c]
	50	75	8.6 ± 2.6[c]	23.8 ± 5.9[c]
	25	75	5.4 ± 1.9[c]	17.9 ± 3.5[c]
	12.5	75	3.5 ± 1.4	7.6 ± 2.4[c]
	0	75	3.3 ± 1.2	3.4 ± 1.6

[a]Represents total number of intestinal metaphases scored from three animals per dose group.
[b]Mean ± S.D. of 25 second division metaphases from each of three animals per dose group, per tissue.
[c]Judged to be positive response for induction of SCE as determined by Dunnett's t-test.
[d]Aldrich Chemical Company, Milwaukee, WI.
[e]Eli Lilly and Company (chemical archives), Indianapolis, IN.
[f]Sigma Chemical Company, Rochester, NY.

DISCUSSION

The induction of SCE was examined in vivo in somatic and germinal tissues of Chinese hamsters. A comparison of the responses in bone marrow with those in spermatogonia and intestinal epithelium revealed that routes of administration, time of dosing, cell cycle kinetics, and site of metabolism were important factors affecting the frequency of SCE in vivo.

SCE in Spermatogonia

In any battery of tests designed to detect genotoxins, it is important to include a system for the assessment of effects in germinal tissue. Tests proposed to evaluate germinal tissue include: SCE analysis in spermatogonia of mice (7), chromosomal aberration analysis in spermatocytes of rodents (3,40), dominant lethal assay in rodents (28), the sex-linked recessive lethal test in Drosophila (38), and the heritable translocation test in rodents (39). Among these, the examination of SCE in spermatogonia provides an endpoint which is the most readily evaluated. Although SCEs have been studied in mouse spermatogonia (7), similar studies have not been reported for the Chinese hamster.

The spermatogonial cell cycle of the Chinese hamster is approximately twice as long as that observed in mice (20). In mice, 52-54 hr are required for 2 periods of cell replication in differentiating spermatogonia (7). Studies presented here indicate the time required for 2 cell cycles in Chinese hamster spermatogonia is 96-100 hr. This is consistent with the studies of DeRooij et al. (17,20) which indicated that the mean cell cycle for type B differentiating spermatogonia in Chinese hamsters was approximately 60 hr. Accordingly, the development of the SCE methodology in Chinese hamsters spermatogonia required a BrdUrd delivery system other than the original tablet methodology of Allen et al. (8). Other methods for administering BrdUrd such as multiple ip injections (41), tail vein infusion (42) and sc infusion (43) were judged inappropriate for application in the Chinese hamster. Therefore, the AC-BrdUrd tablet method of King et al. (18) was adopted since this procedure represented a nontraumatic means of delivering a sufficient amount of BrdUrd for at least 1 period of DNA synthesis. Experiments with multiple tablet weights revealed that a minimum of 40 mg of AC-BrdUrd was required to provide sufficient levels of BrdUrd for SCD in differentiating spermatogonia. Tablet weights of 40-60 mg did not affect the average baseline SCE of approximately 1.0-1.6 per diploid genome in spermatogonia. This value is contrasted to a baseline response of 3.0-3.5 noted in bone marrow, which was consistent with observations for these tissues in other species (19).

The sensitivity of the Chinese hamster spermatogonia system to genotoxins was examined by testing CP, MMC, ThioTEPA, TEM, busulfan, and HC. These compounds had produced a positive response in the dominant lethal assay and, therefore, the sensitivity of spermatogonia to these compounds had been established (28,29). Additionally, some of these compounds were known to induce SCE in other tissues, including spermatogonia from mice (7).

Only 3 compounds, CP, MMC, and ThioTEPA induced SCE in both spermatogonia and bone marrow. The magnitude of SCE induction in bone marrow was always greater than that observed in spermatogonia. Similar findings have been reported for other species (7,32).

Three additional compounds, HC, busulfan, and TEM failed to induce SCE in spermatogonia but were active in bone marrow. It is interesting to note that TEM, commonly used as a positive control for the dominant lethal assay, failed to induce SCE in spermatogonia. Due to the clastogenicity of this compound very low doses were tested; however, additional doses of TEM should be evaluated before concluding that TEM is not an inducer of SCE in spermatogonia.

In comparison to bone marrow, spermatogonia represent a less sensitive tissue for the induction of SCE. This difference was most likely a reflection of the lengthy cell cycle in spermatogonia and variability in the systemic distribution of the chemical tested (31,32).

Sister Chromatid Exchange in Intestinal Epithelium

Epithelium of the small intestine represents a mitotically active tissue in which SCE may be evaluated. The cell cycle of the intestinal epithelium is similar to that of bone marrow; therefore, both tissues can be obtained from the same animal. Consequently, epithelium of the small intestine can be examined as a proximal target tissue at the site of absorption for compounds administered orally, while bone marrow may be evaluated as a secondary target tissue for compounds that are absorbed, metabolized, and systemically distributed.

Although the liver is considered the primary site of metabolism and detoxication, it is important also to consider the role of the enteric microflora in the metabolism or detoxification of compounds which are administered orally (35,36). In addition, metabolites present in bile may undergo further metabolism by the enteric microflora.

Due to the anaerobic condition of the gastrointestinal tract, the reductive xenobiotic transformations mediated by intestinal microflora were of special interest. Aromatic-nitro compounds have been shown to be mutagenic in bacteria without S-9 activation (14) and this activity is thought to result from endogenous anaerobic nitroreductase in bacteria (15,16,37). Thus, it was questioned whether the SCE frequency in intestinal epithelium was affected by the oral administration of nitro compounds which would be subject to metabolism by enteric microflora. In addition, bone marrow was evaluated as a secondary target tissue for compounds absorbed and systemically distributed.

To evaluate the SCE intestinal epithelium system, the nitrophenylenediamines were chosen because of their mutagenic activity in bacteria without exogenous metabolic activation (14). 2-Nitro-p-phenylenediamine, 3NOPD, or 4NOPD were administered orally and SCE frequencies were compared in small intestine epithelium and bone marrow. Only 2NPPD induced a dose-related increase in SCE in intestinal epithelial cells. No SCE activity was noted in bone marrow. Administrations of 3NOPD and 4NOPD failed to induce SCE in intestinal epithelium and bone marrow. It is interesting to note that 2NPPD is the only nitrophenylenediamine that has been shown to have any carcinogenic activity (22). The negative findings with 3NOPD and 4NOPD do not rule out the possibility that these compounds may undergo detoxication in vivo.

Metronidazole, a nitroimidazole, is mutagenic in bacteria without S-9 activation (14,15), and this compound is metabolized by enteric microflora in rodents and humans (21,26,34,44). Consequently, it was not unexpected that the oral administration of MN would result in a dose-related, statistically significant increase in SCE in intestinal epithelium but not in bone marrow.

Additional studies in axenic animals or in animals treated with antibiotics will be necessary to define the role of enteric microflora for the genotoxicity of orally administered compounds. Furthermore, other tissues from the alimentary canal, such as colon epithelium, should be examined as well.

CONCLUSION

The utility of SCE as a measure of genotoxicity in vivo has been defined in a few rodent species using bone marrow. New methodologies for evaluating multiple tissues permit comparison of SCE activity which allows for a more complete assessment of genotoxicity in vivo.

Sister chromatid exchange analysis in spermatogonial tissue provides a means for examining a germinal response. Preliminary studies presented here indicate that SCE can be observed in Chinese hamster spermatogonia; however, additional studies to define other genotoxic responses will be necessary. Particularly, time of dosing is a critical factor in this assay due to the extremely long cell cycle.

Sister chromatid exchange analysis in intestinal epithelium permits the investigation of additional factors that may contribute to genotoxicity. These include: 1) enzymatic or chemical alteration due to conditions found in the digestive tract, 2) metabolism/detoxication by enteric microflora, and 3) absorption and systemic distribution. Studies are underway to investigate further the participation of enteric microflora in the expression of the genotoxicity of certain compounds.

REFERENCES

1. Latt, S., R. Schreck, K. Lovejoy, and C. Shuler (1982) Sister

chromatid exchange analysis: Methodology applications, and interpretation. In Cytogenetic Assays of Environmental Mutagens, T. Hsu, ed. Allanheld, Osmum and Company, Totowa, New Jersey, pp. 29-80.

2. Neal, S., and G. Probst (1983) Chemically-induced sister chromatid exchange in vivo in bone marrow of Chinese hamsters. Mutat. Res. 113:33-43.
3. Adler, I. (1982) Male germ cell cytogenetics. In Cytogenetic Assays of Environmental Mutagens, T. Hsu, ed. Allanheld, Osmum and Company, Totowa, New Jersey, pp. 249-276.
4. Wilmer, J., and E. Soaves (1980) Sister chromatid exchange in vivo in mice: I. The influence of increasing doses of bromodeoxyuridine. Environ. Mutagen. 2:35-42.
5. Perry, P. (1980) Chemical mutagens and sister chromatid exchange. In Chemical Mutagens, Principles and Methods for Their Detection, Vol. 6, A. Hollaender and F. de Serres, eds. Plenum, New York, pp. 1-39.
6. Latt, S., R. Schreck, K. Lovejoy, and C. Shuler (1979) In vitro and in vivo analysis of sister chromatid exchange. Pharmacol. Rev. 30:501-535.
7. Allen, J., and S. Latt (1976) Analysis of sister chromatid exchange formation in vivo in mouse spermatogonia as a new test system for environmental mutagens. Nature (Lond.) 260:449-451.
8. Allen, J., C. Shuler, R. Mendes, and S. Latt (1977) A simplified technique for in vivo analysis of sister chromatid exchanges using 5-bromodeoxyuridine tablets. Cytogenet. Cell Genet. 18:231-237.
9. Allen, J., C. Shuler, and S. Latt (1978) Bromodeoxyuridine tablet methodology for in vivo studies of DNA synthesis. Somatic Cell Genet. 4:393-405.
10. Dunnett, C. (1964) New tables for multiple comparison with a control. Biometrics 20:482.
11. Mazrimas, J., and D. Stetka (1978) Direct evidence for the role of incorporated BudR in the induction of sister chromatid exchange. Exp. Cell Res. 117:23-30.
12. Perry, P., and H. Evans (1975) Cytologic detection of mutagen-carcinogen exposure by sister chromatid exchange. Nature (Lond.) 258:121-125.
13. Perry, P., and S. Wolff (1974) New Giemsa method for the differential staining of sister chromatids. Nature (Lond.) 251: 156-158.
14. Probst, G., R. McMahon, C. Thompson, J. Epp, L. Hill, and S. Neal (1981) A comparison of DNA repair in cultured rat hepatocytes with bacterial mutagenesis using 216 compounds. Environ. Mutagen. 3:11-32.
15. Rinkus, S., and W. Speck (1975) Mutagenicity of metronidazole: Activation by mammalian liver microsomes. Biochem. Biophys. Res. Commun. 66:520-525.
16. Rosenkranz, H., and W. Speck (1976) Activation of nitrofurantoin to a mutagen by rat liver nitroreductase. Biochem. Pharmacol. 25:1555-1556.

17. DeRooij, D., D. Lok, and D. Weenk (1981) Proliferation pattern of undifferentiated spermatogonia and stem cell renewal in the Chinese hamster and ram. Cell Tiss. Kinet. 6:683-684.
18. King, M., D. Wild, E. Goche, and K. Eckhardt (1982) 5-Bromodeoxyuridine tablets with improved depot effect for analysis in vivo of sister chromatid exchanges in bone marrow and spermatogonial cells. Mutat. Res. 97:117-129.
19. Kanda, N., and H. Kato (1979) In vivo sister chromatid exchange in cells of various organs of the mouse. Chromosoma 74:229-305.
20. Oud, J., and D. DeRooij (1976) Spermatogenesis in the Chinese hamster. Anat. Rec. 187:113-124.
21. Koch, R., and D. Goldman (1978) The anaerobic metabolism of metronidazole forms N-(2-hydroxyethyl)-oxamic acid. J. Pharm. Exptl. Therapeutics 208:406-410.
22. Soderman, J., ed. (1983) CRC Handbook of Identified Carcinogens and Noncarcinogens: Carcinogenicity-Mutagenicity Database. Vol.1: Chemical Class File. CRC Press, Boca Raton, p. 139.
23. Williams, R. (1972) Toxicologic implications of biotransformation by intestinal microflora. Toxicol. Appl. Pharmacol. 23:769-781.
24. Scheline, R. (1973) Metabolism of foreign compounds by gastrointestinal microorganisms. Pharmacol. Rev. 25:451-497.
25. Doolittle, D., J.M. Sherrill, and B. Butterworth (1983) Influence of intestinal bacteria, sex of the animal, and position of the nitro group on the hepatic genotoxicity of nitrotoluene isomers in vivo. Cancer Res. 43:2836-2842.
26. Koch, R., B. Beaulieu, E. Chrystal, and P. Goldman (1981) A metronidazole metabolite in human urine and its risk. Science 211:398-399.
27. Machemer, L., and D. Lorke (1975) Method for testing mutagenic effects of chemicals on spermatogonia of the Chinese hamster. Arzneim-Forsch. (Drug Res.) 25:1889-1896.
28. Epstein, S., E. Arnold, J. Andrea, W. Bass, and Y. Bishop (1972) Detection of chemical mutagens by the dominant lethal assay in the mouse. Toxicol. Appl. Pharmacol. 23:288-325.
29. Collins, T. (1972) Effect of captan and triethylenemelamine (TEM) on reproductive fitness of DBA/2J mice. Toxicol. Appl. Pharmacol. 23:277-287.
30. Hoo, S., and C. Bowles (1971) An air-drying method for preparing metaphase chromosomes from the spermatogonial cells of rates and mice. Mutat. Res. 13:85-88.
31. Dym, M., and D. Fawcett (1970) The blood-testis barrier in the rat and the physiological compartmentation of the seminiferous epithelium. Biol. Reprod. 3:308-326.
32. Allen, J. (1982) SCE and meiotic crossover exchange in germ cells. In Progress and Topics in Cytogenetics, Vol. 2: Sister Chromatid Exchange, A.A. Sandberg, ed. Alan R. Liss, New York, pp. 297-315.
33. Wolff, S. (1979) Sister chromatid exchange: The most sensitive

mammalian system for determining the effects of mutagenic compounds. In Genetic Damage in Man Caused by Environmental Agents, K. Berg, ed. Academic Press, New York, pp. 229-258.
34. Searle, A., and R. Willson (1976) Metronidazole (Flagyl): Degradation by the intestinal flora. Xenobiotica 6:457-464.
35. Zachariah, P., and M. Juchau (1974) The role of gut flora in the reduction of aromatic nitro-groups. Drug Metabol. Disposition 2:74-78.
36. McCoy, E., W. Speck, and H. Rosenkranz (1977) Activation of a procarcinogen to a mutagen by cell-free extracts of anaerobic bacteria. Mutat. Res. 46:261-264.
37. Rinkus, S., and M. Legator (1979) Chemical characterization of 465 known or suspected carcinogens and their correlation with mutagenic activity in the Salmonella typhimurium system. Cancer Res. 39:3289-3318.
38. Vogel, E., and F. Sobels (1976) The function of Drosophila in genetic toxicology testing. In Chemical Mutagens: Principles and Methods for their Detection, Vol. 4, A. Hollaender, ed. Plenum Press, New York, pp. 93-142.
39. Wiemann, H., and R. Lang (1978) Strategies for detecting heritable translocations in male mice by fertility testing. Mutat. Res. 53:317-326.
40. Snell, G. (1946) Analysis of translocation in the mouse. Genetics 31:157-180.
41. Vogel, W., and T. Bauknecht (1976) Differential chromatid staining by in vivo treatment as a mutagenicity test system. Nature (Lond.) 260:448-449.
42. Schneider, E., J. Chaillet, and R. Tice (1976) The in vivo labeling of mammalian chromosomes. Exp. Cell Res. 100:396-399.
43. Pera, F., and P. Mattias (1976) Labelling of DNA and differential sister chromatid staining after BrdU treatment in vivo. Chromosoma 57:13-18.
44. Batzinger, R., E. Bueding, B. Reddy, and J. Weisburger (1978) Formation of a mutagenic drug metabolite by intestinal microorganisms. Cancer Res. 38:608-612.
45. Spatz, M., D. Smith, E. McDaniel, and G. Laqueur (1966) Role of intestinal microorganisms in determining cycasin toxicity. Proc. Soc. Expt. Biol. Med. 124:691-699.
46. Callen, D. (1982) Microbial metabolism of environmental chemicals to mutagens and carcinogens. In Chemical Mutagens, Vol. 7, F. de Serres, ed. Plenum Press, New York, pp. 163-188.

SISTER CHROMATID EXCHANGES IN MAMMALIAN MEIOTIC CHROMOSOMES

J. W. Allen and C. W. Gwaltney

Genetic Toxicology Division
Health Effects Research Laboratory
U. S. Environmental Protection Agency
Research Triangle Park, North Carolina 27711

INTRODUCTION

Very little is known about sister chromatid exchanges (SCEs) in meiotic cells--only that they occur (1) and reveal frequency and distribution patterns apparently unaffected by cross-over (CO) exchange conditions in those cells (2). Unfortunately, the number of studies from which to draw these general conclusions is small due to a variety of methodological difficulties. Only a few organisms have been analyzed for baseline SCEs in meiotic tissue. No information pertaining to the extent or significance of SCE induction by mutagens during meiosis is available.

Interest in SCEs first arose in studies concerned with meiosis. The occurrence and frequency of SCEs in meiocytes were addressed in early cytological work with Drosophila and maize (3, 4). Sister chromatid exchanges have been likened to meiotic COs between homologs in that both exchanges are forms of chromatid recombination. Although these exchange-types are fundamentally different in many ways (i.e., in their cell-cycle times of occurrence, frequencies and distributions, inducibilities, and apparent genetic significance), their concurrence in the same chromosomes is unique and poses special analytical problems. Strictly morphological assessments have relied upon chiasmata to signal COs (5), and have been of very limited use for detecting SCEs. Opportunities to resolve SCEs and COs directly came with the autoradiographic methodology of Taylor and colleagues (6,7). Grasshopper spermatocytes were characterized for patterns of semiconservative replication and crossing-over and incidentally for SCEs as a confounding factor leading to deviations in expected label distribution patterns (7,8). Sister chromatid exchange frequencies in meiotic chromosomes were reported in subsequent autoradiographic studies of insects and plants (see Ref. 2). One such study in grasshoppers provided evidence that the formation of these SCEs was not restricted to the penultimate (mitotic) DNA synthesis period prior to their detection at meiosis (1). A binucleate spermatocyte revealed a specific chromosomal SCE in only 1 nucleus rather than in both of the nuclei, thus indicating that it occurred during the ultimate (meiotic) DNA synthesis period.

Ten years ago, Latt developed an in vitro 5-bromodeoxyuridine (BrdUrd)-fluorescent dye methodology for high resolution analyses of SCEs, chromosome structure, and replication kinetics in mammalian cells (9,10). With the significant improvement of permanent staining [i.e., the fluorescence-plus-Giemsa (FPG) step added by Perry and Wolff (11)] BrdUrd-dye methodology became widely used for studies of SCE induction by mutagens (12,13). From toxicological as well as purely genetic standpoints, the possible mechanistic and interactive relationships of SCEs with meiotic COs (homologous nonsister chromatid exchanges) were of interest. However, meiotic cells which earlier comprised a focus for SCE analyses were among the least favorable dividing tissues for BrdUrd differential staining assessments. Germ cells are difficult to grow in culture and maintain through differentiative stages of meiosis. Although in vivo trials seemed promising, initial attempts to effect differential staining in meiotic cells of locusts (14) and rodents failed--apparently because of rather exacting requirements in regard to BrdUrd dose to dividing cells. Excessive analog availability to replicating germ cell DNA results in cytotoxicity (reduced numbers of meiotic metaphase cells) whereas too little does not accomplish the levels of labeling necessary for clear differential staining and elucidation of SCEs. The unusual sensitivities of germ cells may be related to their unusually long DNA synthesis and cell-cycling times [i.e., approx. 14 hr and 30 hr, respectively, in mouse late spermatogonial cells (15)].

Eventually, suitable BrdUrd-dye methods for studying SCEs and COs in insects and rodents were developed. Although only a few studies with these systems have been conducted, several reviews concerned with the implementation of the new techniques and with the findings to date have been reported (16-18). Systems utilizing intact mammals such as mice or hamsters would be useful due to their particular relevance for genotoxicity studies; however, such studies have proven relatively difficult to perform technically. Cytotoxicity has continued to be a problem in at least some of the systems, and SCE analyses have been restricted to certain bivalents. In the present work, an Armenian hamster (Cricetulus migratorius) spermatocyte system is described in terms of its methodological features and its applications for baseline SCE analyses. Initial procedures and applications for characterization of COs have been reported elswhere (19). This system appears to have considerable promise for comparative studies of mechanisms involved in SCE and CO, and for development as an assay to detect and determine the significance of induced SCEs in meiotic cells.

METHODOLOGICAL DEVELOPMENT

Historical

Autoradiography. Taylor and associates determined that radiolabeled DNA is semiconservatively distributed in germ cells (7). Grasshopper spermatocyte chromosomes which incorporated ^{3}H-thymidine 2 cell cycles earlier during spermatogonial DNA synthesis revealed differentially labeled sister chromatids at metaphase. As progression from primary to secondary spermatocytes involves reduction division to the haploid state without further DNA synthesis, identical chromosome labeling characterizes the 2 cell types. Autoradiographic analyses revealed label switches between sister chromatids (SCEs) and between nonsister chromatids of homologous chromosomes (COs).

As noted by Taylor, random chromatid involvement in crossing-over dictates that half of the recombinants should be between like-labeled nonsisters and therein be "hidden" from detection by altered label patterns. The other half of CO recombinants should involve unlike-labeled chromatids, and lead to "visible" disruptions to the homologous chromosomal labeling patterns. Hidden COs would be detectable only by the presence of a chiasma whereas visible COs would be indicated by corresponding isolabeling in the homologs, as well as by chiasmata. However, analyses of these CO exchanges and their distinctions from SCEs is less than straightforward for several reasons. Exchanges in the synapsed regions of primary meiocyte bivalents are difficult to resolve and confusing to interpret with the limited resolution of autoradiography. Thus, assessments have frequently been carried out in secondary meiocyte (metaphase II) cells. Yet, these cells are more removed from the cross-over event and chiasmata are no longer present to signal COs. As discussed below, SCEs and COs can mimic one another's patterns in metaphase II chromosomes and pose significant problems to their individual identification and analysis.

BrdUrd differential staining. BrdUrd differential staining techniques follow the same principles of label kinetics and exchange in meiosis as those forming the basis of autoradiographic methods. Meiotic sister chromatids are differentiated through asymmetry in substitution of BrdUrd for thymidine and staining with a BrdUrd-sensitive dye. Exchange points then appear as abrupt discontinuities in chromatid staining patterns. In recent BrdUrd-dye studies of locust meiotic chromosomes, a considerable amount of information bearing on SCEs, as well as classical concepts of crossing-over, was reported (2,17). Meiotic SCEs were concluded to be similar in nature to mitotic SCEs. Comparable baseline levels are encountered and both reveal distributions which appear to be random and frequencies which depend upon chromosomal size.

A particular obstacle in carrying out in vivo studies in mammals with the BrdUrd-dye technique involved the rapidity of liver degradation of BrdUrd (20) which effectively prevented adequate levels of the analog from being incorporated into DNA. This problem was overcome by various means of continually replenishing BrdUrd to the animal. The first spermatogonial and meiotic cell SCE studies in mammals were carried out in male mice given a series of injections of BrdUrd (21). In the latter cell type, unsatisfactory chromatid differentiation, in all but the late-replicating sex bivalent, greatly restricted analyses. Extensive BrdUrd incorporation into earlier replicating autosomes appeared to be somewhat cytotoxic. Subsequently, more successful mammalian meiotic cell studies of COs and SCEs have been reported in male (22) and female (23) mice, and in Armenian hamsters (19). BrdUrd labeling techniques employed in these studies involved multiple injections, infusion, or slow release of analog from internal charcoal-BrdUrd deposition or implanted BrdUrd pellets. Mouse oocytes were labeled in vitro as part of a novel approach developed by Polani and colleagues (23). Embryonic ovaries were incorporated with BrdUrd under culture conditions and then transplanted to adults for in vivo maturation.

The first staining evidence of COs in a mammal was obtained in Armenian hamster spermatocytes (24). A subsequent study (19) provided more detail and a characterization of replication kinetics and SCEs in the meiotic cells as well. This animal was originally selected for meiosis studies because it was known to have a clear chiasma in the sex bivalent (25,26). Earlier mouse studies had revealed superior differential staining results in the sex bivalent where unfortunately COs were not discernible (21). Recent technical modifications have significantly lessened cytotoxicity problems and improved chromosome labeling results in this system. General methodology and initial results from SCE analyses using Armenian hamsters are described below.

General Methods: Armenian Hamsters

An overview of in vivo methodology for analyzing SCEs and COs in Armenian hamsters is given in Fig. 1. Young (3-5 mo) male hamster (approximately 35 gm) germ cells are labeled using the BrdUrd pellet methodology (24,27). Fifty mg pure BrdUrd pellets are pressed, coated with paraffin over most of the surface area (27) and implanted into lightly anesthetized (Metofane) animals. After 2 1/2 da, the paraffin and any remaining BrdUrd are removed.

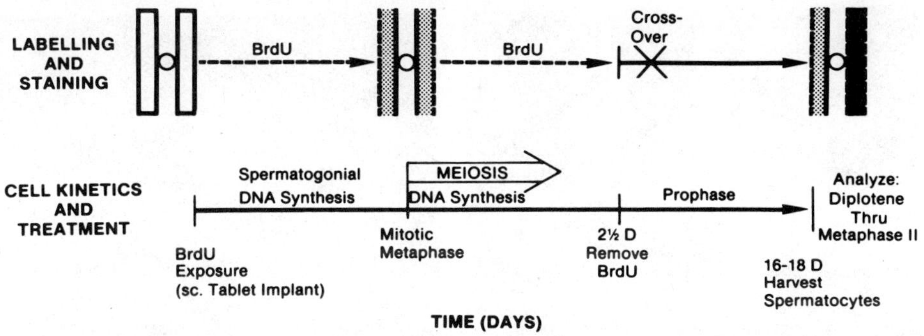

Fig. 1. Diagrammatic representation of in vivo protocol for analyzing chromatid exchanges in hamster meiotic chromosomes. BrdUrd exposure over 2 1/2 da effects analog labeling over the penultimate and ultimate germ cell DNA synthesis periods. BrdUrd substitution (broken lines) in sister chromatids is symmetrically unifilar after one round of replication, and unifilar vs. bifilar after 2 rounds of replication. The latter asymmetry of labeling characterizes diplotene through metaphase II chromosomes harvested on da 16-18. Staining of these chromosomes with 33258 Hoechst fluorescent dye results in dull fluorescence (black shading) over the more heavily labeled chromatid and relatively brighter fluorescence (grey shading) over the less heavily labeled chromatid. Various FPG procedures are also effective. Differential staining of sister chromatids enables the detection of SCEs, as well as COs between unlike-stained nonsister chromatids.

Second-division spermatogonial metaphases may be harvested at this time (after a 4-hr colchicine treatment) or harvests may be delayed until days 16-18, at which time second-division (post-labeling) spermatocyte diplotene through metaphase II cells are obtainable. Standard cytogenetic harvest, slide preparation, and fluorescent orFPG differential staining procedures (13) may be used for analyzing exchanges in meiotic chromosomes.

It is notable that the paraffin-coating step in the preparation of BrdUrd pellets has led to significantly improved results as compared with those obtained using injections and uncoated pellet methods. It would appear that the slower, steadier release of analog attained after coating the pellet reduces the heavy pulse effect which is likely to be harsh on these cells. As a result, fully labeled primary and secondary spermatocytes (Fig. 2) are common, rather than only occasionally obtainable.

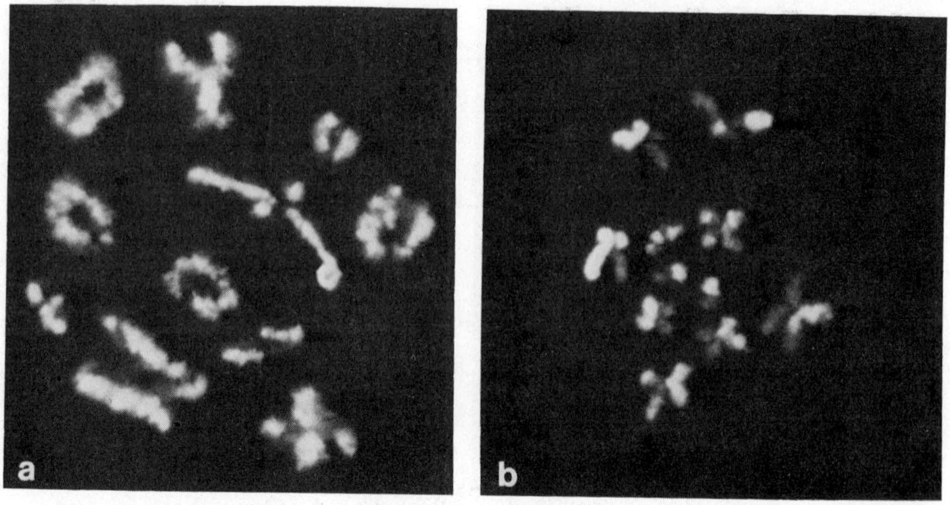

Fig. 2. Representative primary spermatocyte a and secondary spermatocyte b revealing chromatid differential staining in most or all chromosomes. Each chromosome is individually identifiable on the basis of morphology and CO recombination criteria (19,25).

HAMSTER GERM CELL ANALYSES

SCEs In Primary Spermatocytes

It is difficult to assess SCEs in all chromosomes of primary spermatocytes due to the rarity of such cells with adequate morphological clarity in all chromosomal bivalents. However, the monochiasmate sex bivalent and several of the larger autosomal bivalents (Fig. 3) which appear as mono- or multichiasmate are relatively easy to analyze (SCE frequencies per bivalent are noted below). The sex bivalent has also been particularly well studied for hidden and visible types of COs (19), and for simultaneous CO and SCE staining patterns (Fig. 4). On either side of the chiasma, X and Y short arms reveal chromatid differential staining (i.e., Fig. 4a) or isolabeling for bright or dull fluorescence (i.e., Fig. 4b). These hidden and visible COs between the sex chromosomes have been noted to occur in approximately 1:1 ratios in young Armenian hamsters (19) which agrees with similar CO-type ratios reported in locusts (28) and mice (16).

A variety of staining patterns reflecting combinations of SCEs and COs are evident in the sex bivalents of Fig. 4c-g. For example, in Fig. 4e, an SCE is apparent between the synapsed short arms in the chiasma. This observation is in keeping with the concept that pre-CO sister chromatids manifest a pairing attraction across the chiasma (5). Another exchange in this bivalent is

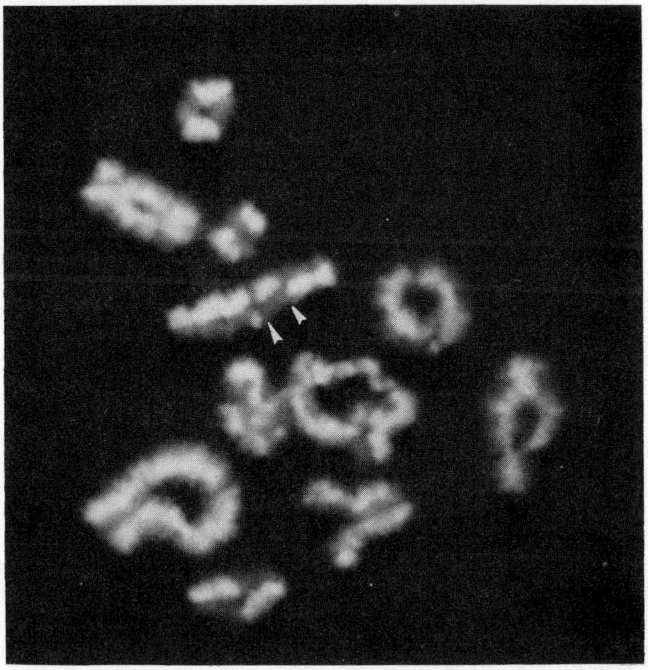

Fig. 3. Sister chromatid exchanges evident in the number 6 autosomal bivalent which, in this instance, appears monochiasmate. Two SCEs are indicated by arrows.

indicated by the mismatching of staining diagonally across the chiasma (or between long and short arms in 2 of the opposing chromatids). This pattern has been noted in locusts (28), mice (22), and hamsters (19), and it is uncertain as to whether it should be attributed to an SCE or to additional crossing-over (28).

One additional source of confusion stemming from SCEs is the production of isolabeling similar to that resulting from visible COs where, in fact, no COs occurred. This problem was addressed by Moens in an early tritium-labeling study (29). The bivalent examples in Fig. 4f and 4g are at the third division and reveal isolabeled or uniformly stained regions in the Y chromosome long arms. These patterns were likely generated by SCEs in the first and/or second of 3 replication periods. Since the sex chromosomes are generally quite late replicating, it is not uncommon to find cells in which X and Y chromosomes have replicated 3 times, and most autosomes only twice, after the start of the BrdUrd labeling.

SCEs In Secondary Spermatocytes

Meiotic metaphase II chromosomes are morphologically easy to analyze; however, potential misinterpretations of differential

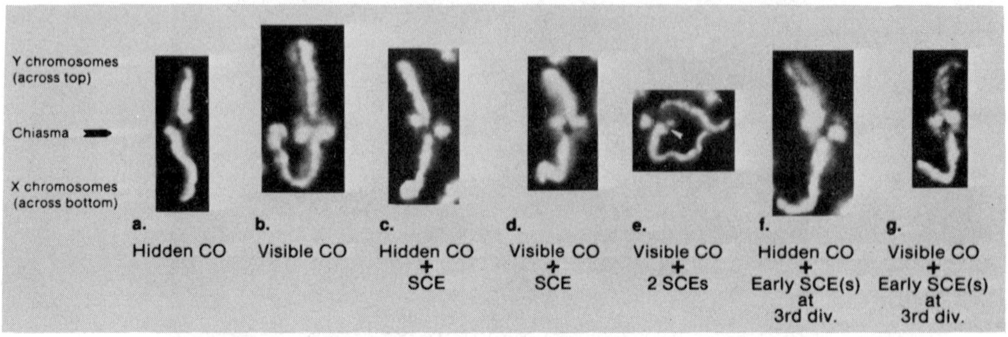

Fig. 4. Various chromatid-exchange staining patterns in monochiasmatic sex bivalents. Bivalent a exemplifies hidden crossing-over. Recombined short-arm segments have not altered the continuity of staining into the chiasma. Like-stained chromatids are in diagonal apposition across the chiasma, which is consistent with pairing attraction between regions previously in sister orientation. Bivalent b shows visible crossing-over in which a bright short-arm segment was exchanged with a dull short-arm segment to result in isolabeling for bright or dull on either side of the chiasma. Bivalents c–e reveal hidden or visible COs, and an additional exchange (SCE?, CO?) in the CO region as evidenced by the mismatched staining diagonally across the chiasma. Bivalent e also reveals an exchange between the paired X-Y short arms (indicated by arrows), which is interpreted as a pre-CO SCE, and thus is further evidence for sister chromatid pairing attraction in the chiasma. Bivalents f and g reveal dull isolabeling (long arms) consistent with that expected from visible crossing-over. However, chiasmata involving long arms of the sex chromosomes have never been observed (homology appears restricted to short arms). It is presumed that the mimicry of CO staining patterns observed here actually arose from SCEs occurring in prefinal replication periods of these third-division bivalents.

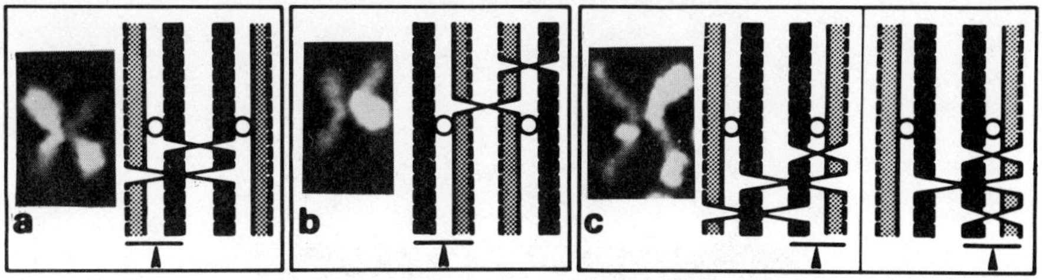

Fig. 5. Mimicry of staining appearances by SCEs and COs in representative secondary spermatocyte chromosomes (no. 1). In a, a cytological example revealing the typical appearance of an SCE is depicted with a possible recombination mechanism involving only COs (proximal hidden, distal visible). Similar, in b, a cytological example revealing the typical appearance of a visible CO (dull isolabeling) might actually have been derived via the accompanying mechanism involving a proximal hidden CO and a distal SCE. In c, a more complicated staining pattern involving the "appearance" of CO and SCE images is followed by examples of 2 possible mechanisms involving either 2 COs and 1 SCE, or 1 CO and 2 SCEs.

staining patterns present a unique problem in this cell type. Since chiasmata have disappeared, it becomes impossible to distinguish between hidden crossing-over and failure to cross-over. Further, mimicry between CO and SCE staining patterns in secondary spermatocytes prevents any precise determinations of SCE frequencies and distributions. Jones and coworkers (2,30-32) have thoroughly addressed this problem in studies with grasshoppers and locusts, and have developed means by which to indirectly assess SCE levels in such cells (see below). Figure 5 illustrates how typical SCE and CO stain images can be misleading in instances where they may not be what they appear to be. Selected hamster metaphase II chromosomes with accompanying theoretical models of recombination show how apparent switches between differentially stained sister chromatids might actually have arisen from CO events only, and how SCEs might be obscured by COs and therefore overlooked. Clearly, SCEs might be under- or overestimated in attempts to directly determine their frequencies in secondary spermatocyte cells.

Replication Kinetics In Spermatogonia

A final source of confusion in differential staining analyses of COs and SCEs in the meiocytes is that which can arise from significant intrachromosomal variations in the timing of DNA

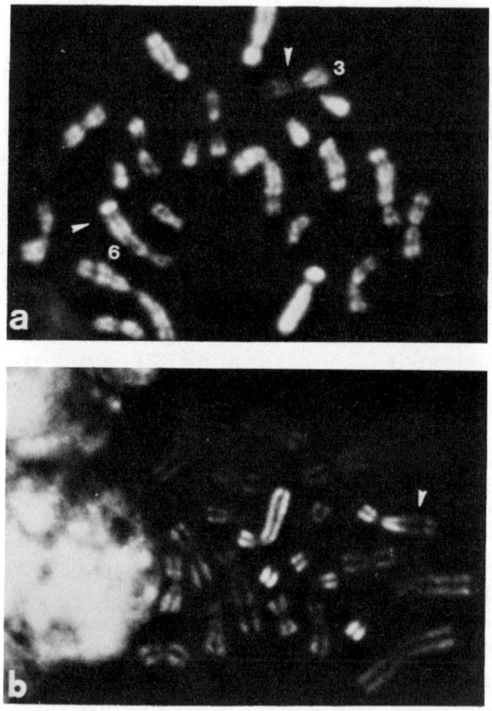

Fig. 6. Replication kinetics in Armenian hamster spermatogonial chromosomes partially labeled (early mid-S) with BrdUrd and stained at the next metaphase with 33258 Hoechst dye. Examples of chromosome nos. 3 and 6, a, and the X chromosome, b, (from Ref. 19) reveal intrachromosomal regions with markedly different times of replication, and uneven labeling with BrdUrd, to result in different staining patterns (arrows). When these regions are large or pronounced in meiotic chromosomes, they may be confused with isolabeling resulting from visible-type COs.

replication. The uneven labeling of spermatogonial DNA with trdUrd can result in an appearance of isolabeling (resembling visible COs) in meiotic metaphase I and metaphase II cells. Figure 6 reveals several Armenian hamster chromosomes which should be considered in this light. When boundaries of differentially replicating segments occur at or near frequent CO sites, as is the case in the nos. 3 and 6 chromosomes, it is difficult to assess the derivation of isolabeling detected at meiosis. By contrast, isolabeling due to differential replication in the X long arms is not problematic in its interpretation because crossing-over does not occur in that region (19). While misinterpretations of isolabel patterns have a direct effect on CO analyses, they can also indirectly affect SCE analyses in secondary spermatocytes (see below).

SCE Frequencies

SCE frequencies in Armenian hamster meiotic chromosomes can be determined by direct analysis of selected primary spermatocyte bivalents, and by indirect analysis of secondary spermatocyte chromosomal complements. As noted above, the large chromosomes no. 1, no. 6, and X-Y bivalents are especially amenable to SCE analysis. The chromosomes may also be identified in spermatogonial cells on the basis of size and centromere positions (25). Table 1 depicts comparative analyses of SCE frequencies in these chromosome pairs observed in spermatogonial and primary spermatocyte cells. No significant difference in SCE/chromosome levels

Tab. 1. SCE frequencies in specific chromosome-types analyzed at spermatogonial and primary spermatocyte metaphase stages.

Chromosome Pair	Spermatogonia[a]	1° Spermatocytes[b] (Diakinesis-Metaphase I)
No. 1	0.38	0.32
No. 6	0.37	0.36
X-Y	0.28	0.23

[a] SCE frequencies calculated from 40 cells; 2 young (3-5 mos.) Armenian hamsters
[b] SCE frequencies calculated from 25 cells for each bivalent type; 5 young (3-5 mos.) Armenian hamsters

between the 2 cell types were found. It is likely that the slightly lower values observed in the sex bivalent (similar in size to the autosomes) may be due to the differential replication kinetics in the X chromosome long arms (Fig. 6), which can obscure the uniformity of clear chromatid differentiation along the entire length of the chromosome.

Although SCEs are not always reliably identifiable in secondary spermatocyte chromosomes, frequencies may be deduced after accounting for the proportion of detectable exchange sites attributable to COs. Preliminary results following the basic approach developed by Tease and Jones (2) for analyzing SCEs in secondary spermatocyte cells are presented in Tab. 2. In 4 individual hamsters, the mean chiasma frequencies were determined in primary spermatocytes (diakinesis) and the total number of exchange boundaries (reciprocal and nonreciprocal) per cell were recorded for secondary spermatocytes. The expected number of detectable exchange boundaries due to visible crossing-over would equal one-half of the chiasma frequencies; thus, the exchanges in excess of this number are interpreted as arising from SCEs (2 excess exchanges/SCE). Values calculated for haploid secondary spermatocytes are doubled to obtain SCE frequencies for diploid cells. Average frequencies determined in this manner for the different animals ranged from 0.2 to 1.4 SCEs/meiotic cell. These frequencies are approximately those which would be predicted from the individual bivalent analyses (Tab. 1), taking into account that the bivalents analyzed were among the largest of the chromosomal complement. They also closely match the average spermatogonial cell frequencies of 1.4 ± 0.8 and 1.7 ± 0.9 observed in 2 separate hamsters (20 cells each).

It is notable that the germ cell SCE frequencies obtained for hamsters in this study are consistent with those reported by BrdUrd-labeled mice and locusts despite some differences in labeling techniques. Values in mice of approximately $\leq$ 2 SCEs/spermatogonial cell (21), 0.07 SCEs/spermatocyte bivalent (22), and 0.10 SCEs/oocyte bivalent (16) have been determined. In locusts (2), values of 0.14 to 0.37 were obtained for different chromosomes in primary spermatocytes. Slightly higher frequencies of 0.28 and 0.89 (per bivalent) were calculated in anaphase I and metaphase cells for the same chromosomes, respectively. The mammalian meiotic SCE frequencies also tend to be a little lower than those reported for other insects and plants studied with autoradiography (see Ref. 2). This is not surprising in view of the technical/procedural differences in DNA labeling.

As a final point, it remains uncertain as to whether there is a technical or biological explanation for why rodents generally reveal lower SCE frequencies in their germ cells as compared with somatic cells (18) while insects appear to show similar frequencies in somatic and germ tissues (17).

Tab. 2. Meiotic cell SCE frequencies indirectly assessed in secondary spermatocytes from Armenian hamsters.

Armemian Hamster (4-6 mos.)	No. of 2° Spermatocytes	Observed Exchange[a] Boundaries/Cell	Chiasma Frequencies[b] In 1° Spermatocytes	Expected Visible[c] Cross-Overs/Cell	SCE Frequencies[d] (2° Spermatocytes)	SCE Frequencies[e] (1° Spermatocytes)
A	23	9.7	17.4	8.7	0.5	1.0
B	19	10.0	17.2	8.6	0.7	1.4
C	14	10.0	17.4	8.7	0.7	1.4
D	36	8.8	17.3	8.7	0.1	0.2

[a] Mean number of stain-contrast junctions (reciprocal + non-reciprocal) per cell

[b] Mean number of chiasmata calculated from 30 1° spermatocytes (diplotene-diakinesis) for each animal

[c] Visible cross-over frequencies = 0.5 x chiasma frequencies

[d] SCE frequencies = 1/2 (since 2 exchange boundaries are observed/SCE) of (observed exchange boundaries - expected visible cross-overs)

[e] Double the SCE frequencies calculated for haploid secondary spermatocytes

SUMMARY

Meiotic cells have been, and remain, a relatively difficult tissue in which to study SCEs. Experimental evaluations have had to contend with various problems unique to germ cells, i.e., poor in vitro growth, unusual cytotoxicities (from BrdUrd), trying chromosome morphologies, and confusion between chromosome label patterns generated by SCEs, COs, and replication kinetics. Nevertheless, BrdUrd differential staining techniques in insects and rodents have progressed to the point where SCE frequencies can be reliably determined.

In the described Armenian hamster system, SCEs may be directly resolved in selected primary spermatocyte bivalents and indirectly assessed in secondary spermatocyte cells. SCE frequencies determined from spermatogonial, primary spermatocyte, and secondary spermatocyte cells are in good agreement. The system should be applicable for studies of possible mechanistic similarities between SCE and CO exchange, and for evaluating genotoxic effects. Experimental evidence in Drosophila (33) argues against a very close molecular relationship between these forms of breakage and recombination. However, it is likely that new information can be gained from cells in which normal SCE and CO frequencies and distributions can be characterized, and from studies of the consequences to these events after various mutagenic perturbations to premeiotic and meiotic DNA.

ACKNOWLEDGEMENTS

The authors are grateful to George Yerganian for supplying Armenian hamsters. We also wish to thank Ann Black and Everett Crosson for their able technical assistance, and Carol Kuehnhoff for her dedication in typing the manuscript.

REFERENCES

1. John, B., and K.R. Lewis (1965) The meiotic system. In Protoplasmatalogia VI/F/I, Springer-Verlag, Wien.
2. Tease, C., and G.H. Jones (1979) Analysis of exchanges in differentially stained meiotic chromosomes of Locusta migratoria after BrdU-substitution and FPG staining. II. Sister chromatid exchanges. Chromosoma 73:75-84.
3. Beadle, G.W., and S. Emerson (1935) Further studies of crossing-over in attached-X-chromosomes of Drosophila melanogaster. Genetics 20:192-206.
4. Schwartz, D. (1953) Evidence for sister-strand crossing over in maize. Genetics 38:251-260.

5. Darlington, C.D. (1937) Meiosis in diploids and polyploids. In Recent Advances In Cytology, The Blakiston Company, Philadelphia, pp. 85-133.
6. Taylor, J.H., P.S. Woods, and W.L. Hughes (1957) The organization and duplication of chromosomes as revealed by autoradiographic studies using tritium-labeled thymidine. Proc. Natl. Acad. Sci., USA 43:122-128.
7. Taylor, J.H. (1965) Distribution of tritium-labeled DNA among chromosomes during meiosis. J. Cell Biol. 25:57-67.
8. Taylor, J.H. (1984) A brief history of the discovery of sister chromatid exchanges. In Sister Chromatid Exchanges, 25 Years of Experimental Research, R.R. Tice and A. Hollaender, eds., Plenum Press, New York.
9. Latt, S.A. (1973) Microfluorometric detection of DNA replication in human metaphase chromosomes. Proc. Natl. Acad. Sci., USA 70:3395-3399.
10. Latt, S.A. (1974) Sister chromatid exchanges, indices of human chromosome damage and repair: Detection by fluorescence and induction by mitomycin C. Proc. Natl. Acad. Sci., USA 71:3162-3166.
11. Perry, P., and S. Wolff (1974) New Giemsa method for the differential staining of sister chromatids. Nature (Lond.) 251:156-158.
12. Perry, P.E. (1980) Chemical mutagens and sister chromatid exchange. In Chemical Mutagens-Principles and Methods for Their Detection, Vol. 6, F. de Serres and A. Hollaender, eds. Plenum Press, New York, pp. 1-39.
13. Latt, S.A., J. Allen, S.E. Bloom, A. Carrano, E. Falke, D. Kram, E. Schneider, R. Schreck, R. Tice, B. Whitfield, and S. Wolff (1981) Sister chromatid exchanges: A report of the Gene-Tox program. Mutat. Res. 87:17-62.
14. Jones, G.H. (1977). A test for early terminalization of chiasmata in diplotene spermatocytes of Schistocerca gregaria. Chromosoma 63:287-294.
15. Monesi, V. (1962) Autoradiographic study of DNA synthesis and the cell cycle in spermatogonia and spermatocytes of mouse testis using tritiated thymidine. J. Cell Biol. 14:1-18.
16. Polani, P.E., J.A. Crolla, and M.J. Seller (1981) An experimental approach to female mammalian meiosis: Differential chromosome labeling and an analysis of chiasmata in the female mouse. In Bioregulators of Reproduction, Vogel and Jagiello, eds. Academic Press, New York, pp. 59-87.
17. Jones, G.G., and C. Tease (1981) Meiotic exchange and analysis by molecular labelling. In Chromosomes Today, Vol. 7, M.D. Bennett, M. Bobrow, and G. Hewitt, eds. George Allen and Unwin, Ltd., London, pp. 114-125.
18. Allen, J.W. (1982) SCE and meiotic crossover exchange in germ cells. In Progress and Topics in Cytogenetics, Vol. 2, Sister Chromatid Exchange, A.A. Sandberg, ed. Alan R. Liss, Inc., New York, pp. 297-311.

19. Allen, J.W. (1979) BrdU-dye characterization of late replication and meiotic recombination in Armenian hamster germ cells. Chromosoma 74:189-207.
20. Kriss, J.P., Y. Maruyama, L.A. Tung, S.B. Bond, and L. Revesz (1963) The fate of 5-bromodeoxyuridine, 5-bromodeoxycytidine, and 5-iododeoxycytidine in man. Cancer Res. 23:260-268.
21. Allen, J.W., and S.A. Latt (1976) In vivo BrdU-33258 Hoechst analysis of DNA replication kinetics and sister chromatid exchange formation in mouse somatic and meiotic cells. Chromosoma 58:325-340.
22. Kanda, N., and H. Kato (1980) Analysis of crossing over in mouse meiotic cells by BrdU labelling technique. Chromosoma 78:113-121.
23. Polani, P.E., J.A. Crolla, M.J. Seller, and F. Moir (1979) Meiotic crossing over exchange in the female mouse visualized by BudR substitution. Nature (Lond.) 278:348-349.
24. Allen, J.W., C.F. Shuler, and S.A. Latt (1978) Bromodeoxyuridine tablet methodology for in vivo studies of DNA synthesis. Somatic Cell Genet. 4:393-405.
25. Lavappa, K.S., and G. Yerganian (1970) Spermatogonial and meiotic chromosomes of the Armenian hamster, Cricetulus migratorius. Exp. Cell Res. 61:159-172.
26. Solari, A.J. (1974) The relationship between chromosomes and axes in the chiasmatic XY pair of the Armenian hamster (Cricetulus migratorius). Chromosoma 48:89-106.
27. McFee, A., K.W. Lowe, and San Sebastian, Jr. (1983) Improved sister chromatid differentiation using paraffin-coated bromodeoxyuridine tablets in mice. Mutat. Res. Letters 119/1:83-88.
28. Tease, C., and G.H. Jones (1978). Analysis of exchanges in differentially stained meiotic chromosomes of Locusta migratoria after BrdU-substitution and FPG staining. I. Crossover exchanges in monochiasmate bivalents. Chromosoma 69:-163-178.
29. Moens, P.B. (1966) Segregation of tritium-labeled DNA at meiosis in Chorthippus. Chromosoma 19:277-285.
30. Jones, G.H., and T. Craig-Cameron (1969) Analysis of meiotic exchange by tritium autoradiography. Nature (Lond.) 223:946-947.
31. Craig-Cameron, T.A., and G.G. Jones (1969) The analysis of exchanges in tritium-labelled meiotic chromosomes. I. Schistocerca gregaria. Heredity 25:223-232.
32. Jones, G.H. (1971) The analysis of exchanges in tritium-labelled meiotic chromosomes. II. Stethophyma grossum. Chromosoma 34:367-382.

33. Gatti, M., S. Pimpinnelli, and B.S. Baker (1980) Relationships among chromatid interchanges, sister chromatid exchanges, and meiotic recombination in Drosophila melanogaster. Proc. Natl. Acad. Sci., USA 77:1575-1579.

REMOVAL AND PERSISTENCE OF SCE-INDUCING DAMAGE IN HUMAN LYMPHOCYTES IN VITRO

Bo Lambert, Margareta Sten, and Dennis Hellgren

Department of Clinical Genetics
Karolinska Hospital
104 01 Stockholm, Sweden

INTRODUCTION

The ability to induce sister chromatid exchanges (SCEs) is a well-known property of many DNA-damaging agents (for review see Refs. 20,32). There is no conclusive evidence that any particular type of DNA lesion is more frequently involved in SCE induction than other lesions (e.g., Refs. 3,11,31), although, in general, agents inducing DNA interstrand cross-links appear to be very potent SCE inducers (1,8,13,24).

Sister chromatid exchanges are formed during the S phase of the cell cycle (35). Thus, in theory, SCE induction could be influenced by any process leading to removal or modification of DNA damage before cells start to replicate DNA. The induced SCE frequency would then be dependent on the rate of removal and the time available for repair and/or modification of SCE-inducing DNA damage. However, excision repair seems to have little influence on the SCE frequency (for discussion see Ref. 28), and some SCE-inducing DNA lesions appear to be long-lived and partially resistant to repair (12,18,23-25,33).

To obtain more information about the possible removal vs. apparent persistence of SCE-inducing DNA damage, we have studied the SCE frequency in human lymphocytes exposed to different DNA-damaging agents in various phases of the cell cycle. These cells are normally in a resting (G_0) state when circulating in the body, and can be held at this state for several days in culture. When stimulated in vitro by phytohemagglutinin (PHA) or other mitogens, the cells initiate blast transformation and enter S phase after about 24-26 hr.

We have studied the effect of different post-treatment recovery times on SCE frequency in the second mitosis (M2) after PHA stimulation by exposing the cells to DNA-damaging agents during G_0 as well as at various intervals during the extended G_1 phase (Fig. 1). We have also obtained some information about the persistence of SCE-inducing damage in proliferating cells by the analysis of SCE in third-generation metaphases (M3).

REDUCTION OF SCE-INDUCING DAMAGE DURING G_0

Human lymphocytes were exposed to nitrogen mustard (HN2) or melphalan in vitro. Phytohemagglutinin stimulation was initiated either immediately after the exposure, or 24 or 48 hr later, in order to study the effect of prolonged post-treatment incubation on the induced SCE frequency. Cells exposed to HN2 2 hr before PHA stimulation were found to have a significantly higher SCE frequency than cells exposed 48 hr before PHA. Melphalan-exposed cells showed a similar response although not as pronounced as that after HN2 (Tab. 1).

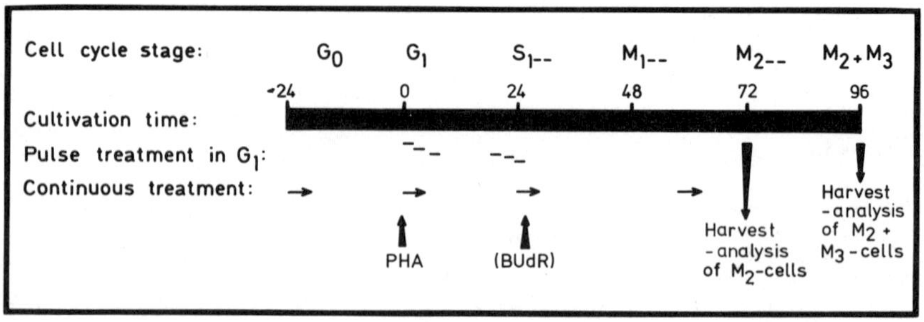

Fig. 1. Scheme of cultivation and treatment condition. Conventional lymphocyte cultures were prepared from whole blood or buffy coat of healthy donors as previously described (15,17). Treatment with the different compounds was for 2 hr (unless stated otherwise), followed by repeated washing of the cells and continued cultivation until harvest at indicated times. Phytohemagglutinin was added at time 0, and BrdUrd (100 μmoles/l) 24 hr later (unless stated otherwise). Harvest, fixation, and staining of the cells were as described (15). Sister chromatid exchanges were scored in 20 cells from each culture, and the rate of cell proliferation was estimated by the analysis of 100 cells for the proportion of first-, second-, and third-generation metaphases on the basis of their differential staining pattern.

In previous experiments (22) we found that 1-(2-chloroetyl)-3-cyclohexyl-1-nitrosurea (CCNU) (Lomustin) induces about the same increase of SCEs in cells exposed 48 hr before PHA stimulation as in cells exposed immediately before the addition of PHA (Tab. 1). A possible interpretation of these observations is that the SCE-inducing damage caused by HN2 and, to some extent, also melphalan is removed (or changed into non-SCE-inducing lesions) during prolonged incubation of G_0 lymphocytes, whereas damage caused by CCNU is not reduced to the same extent.

In similar experiments, storage of human G_0 lymphocytes for 6-8 da after exposure to mitomycin C (MMC) was found to enhance the induced SCE frequency (6). Hedner and coworkers (10) studied the SCE frequency in human lymphocytes exposed in vitro to N-acetoxy-2-acetylaminofluorene (NA-AAF) or ethylene oxide (EO), either 1 or 18 hr before PHA stimulation. Their results showed no difference in the NA-AAF-induced SCE frequency between the 2 exposure times. However, the induced SCE frequency in cells exposed to EO 18 hr before PHA was significantly lower than in cells exposed 1 hr before PHA. The authors' interpretation was that EO-induced DNA damage is more rapidly repaired than NA-AAF-induced damage (10).

Hence, it seems likely that different types of SCE-inducing damage are either persistent or can be modified in different ways in human G_0 lymphocytes. Such modifications can obviously result in either a decrease or an increase of the SCE frequency. Already from these observation it appears unlikely that nucleotide excision repair has a major role in the removal of SCE-inducing damage, since

Tab. 1. The SCE frequency in human lymphocytes after different post-treatment incubation times in G_0.

Exposure time (hours before PHA stimulation	Concentrations (moles/l) and SCE frequency [a]		
	HN2 ($2x10^{-6}$)	Melphalan (10^{-6})	CCNU ($5x10^{-5}$)
0	27.2	22.1	22.6
24	17.6*	19.0	-
48	9.1*	14.2*	17.6

[a] The data for HN2 and melphalan are compiled from 2 separate but identical experiments. Data for CCNU are from Ref. 22. The mean level of SCE/cell in exposed cells after subtraction of the SCE frequency in untreated control cells (range 11.2-15.0 SCEs/cell) are shown. At least 20 cells were analyzed from each culture. Two-tailed t-test was used for statistical analysis. The asterisk indicates a difference at the level of $p < 0.01$ from the top line data.

all of the compounds mentioned above are known to induce DNA repair replication.

Further information about post-treatment reductions in the SCE frequency induced by alkylating agents in human lymphocytes have been obtained in studies of cancer patients receiving cytostatic drugs. Immediately after administration of melphalan, CCNU and other DNA-damaging compounds, the SCE frequency is usually increased to a level 2- to 5-fold higher than that observed before any treatment (9,16). The SCE frequency then decreases, and after about 4 wk (when usually the next administration is given) it reaches a level that is slightly higher than before treatment (Tab. 2). This type of response is usually repeated after successive administrations and eventually the SCE frequency remains elevated at a constant high level for several months. A similar persistence of an elevated SCE frequency has been demonstrated in peripheral lymphocytes of rabbits after repeated exposure to MMC (30).

Hence, the removal of SCE-inducing damage caused by alkylating agents seems to be very slow in G_0 lymphocytes in vivo. It may then be inferred that repeated or continuous exposure may lead to accumulation of SCE-inducing damage, giving rise to a persistent increase of the SCE frequency.

Tab. 2. The SCE frequencies in peripheral lymphocytes from 2 melanoma patients treated with melphalan or CCNU.*

Patient and treatment schedule	Course no.	SCE/cell: mean + S.D. (no of cells) Before treatment	After treatment
A. 76 y old male patient with	1	17.0±6.2(20)	61.2±10.2(8)
melanoma treated with melphalan	3	25.4±7.4(40)	77.5±13.1(13)
(1 mg/kg i.v.) every 4 weeks	5**	28.4±8.5(20)	49.5±22.3(20)
	6	21.0±10.5(20)	71.8±22.3(20)
B.*** 48 y old female melanoma patient	1	11.3±5.6(13)	25.6±7.7(19)
treated with CCNU ($130\ mg/m^2$, p.o.)	3	24.6±17.8(20)	50.4±15.8(20)
every 4 weeks	4	59.0±21.3(20)	60.6±22.6(20)

* Extended data from Ref. 16.
** The melphalan dose was reduced to 0.5 mg/kg because of hematotoxic reaction.
*** This patient had previously been treated with actinomycin D without any effect on the SCE frequency.

THE SCE FREQUENCY IN CELLS EXPOSED AT DIFFERENT TIMES IN G_1

Incubation with PHA in vitro stimulates human lymphocytes to enter the G_1 phase of the cell cycle, which lasts until DNA replication begins after about 24-26 hr. During this extended G_1 phase, as well as later in the cell cycle, DNA repair activity may change, as has been demonstrated after exposure to ultraviolet (UV) light (2, 19), methylmethanesulfonate (MMS), and NA-AAF (29). To study the possible removal of SCE-inducing damage during G_1, PHA-stimulated human lymphocytes were exposed to different SCE-inducing agents, either immediately after initiation of the cultivation (early G_1) or 24 hr later (late G_1). The treatment lasted for 1 or 2 hr, and after washing of the cells, the cultivation continued in fresh medium. The results shown in Fig. 2 indicate that, with the exception of CCNU and adriamycin, all the agents studied induced more SCEs in late G_1 than in early G_1. The difference was most pronounced for HN2 and melphalan, but was repeatedly demonstrated also at several dose levels of MMS (17) and UV light (26).

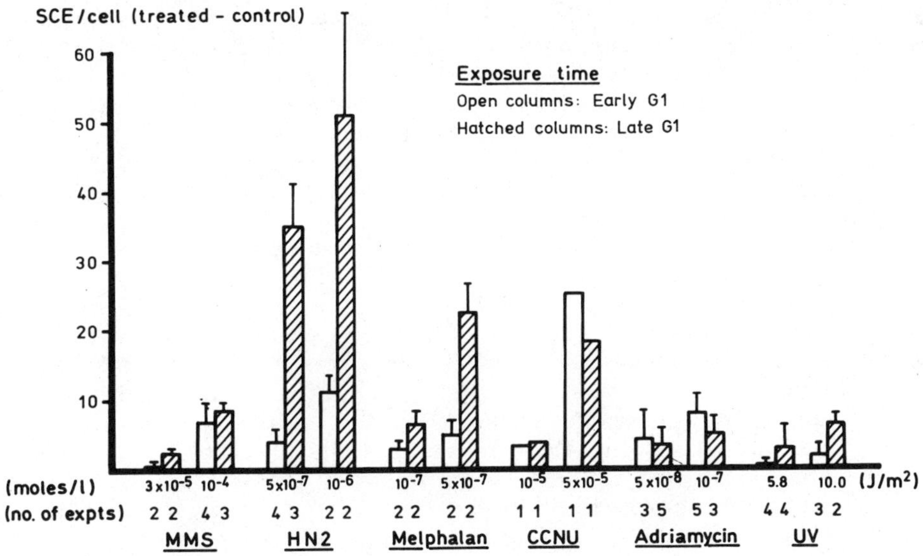

Fig. 2. Sister chromatid exchange frequencies in cells exposed to different SCE-inducing agents in early G_1 or late G_1. The exposure time was 0-2 hr (early G_1) and 24-26 hr (late G_1) for MMS, HN2, melphalan, and adriamycin, and 5-6 hr (early G_1) and 24-25 hr (late G_1) for CCNU. Ultraviolet irradiation was either at 6 hr or at 22 hr, respectively. The SCE frequency in concurrent, untreated control cultures has been subtracted. Bars indicate one S.D. of the means of indicated number of experiments. Data include results from Refs. 17, 18, 22, and 26.

As shown previously (22), treatment with low and moderate doses of CCNU induced about the same increase of SCEs in early and late G_1, but a high dose induced significantly more SCEs after treatment in early G_1 than in late G_1 (Fig. 2). This result may be explained by assuming that the cross-links caused by CCNU are more efficient SCE-inducing lesions than the corresponding monoadducts. The cross-links are formed by conversion of the chloroethyl monoadducts in a slow reaction during 6-12 hr after removal of the unbound drug (4,7, 14). Hence, after treatment in early G_1, relatively more DNA cross-links would be expected to be present during the following S phase than after treatment in late G_1. On the other hand, the delay in cross-link formation allows time for DNA repair removing the monoadducts, thus preventing their conversion to cross-links (5). This tentative balance between the removal of monoadducts and their slow conversion into cross-links may explain why the difference in the SCE frequency between early and late G_1 treatment was detected only at a high dose of CCNU.

Adriamycin was found to be a relatively poor inducer of SCEs in human lymphocytes (18). A doubling of the SCE frequency was obtained with doses up to 10^{-7} moles/l, whereas higher doses caused severe inhibition of cell proliferation. No difference in the SCE frequency was observed between cells exposed in early and late G_1 (Fig. 2). In addition, adriamycin did not induce a detectable level of DNA repair replication in human G_0 lymphocytes (18).

Ultraviolet light (254 nm) is generally considered to be an efficient SCE-inducing agent (see Ref. 20). In our own experiments with human fibroblasts (which were irradiated at confluency and then immediately reseeded at low density and cultivated for 62.5 hr in 10 μmoles/l of BrdUrd) we observed a 4-fold increase of the SCE frequency after a dose of 10 J/m^2 (26). In contrast, we found a remarkably poor response in human lymphocytes. A small but reproducible increase of SCE was observed in cells exposed to 10 J/m^2 of UV light in late G_1, whereas the response in cells irradiated in early G_1 was not different from controls (Fig. 2) (26).

The results shown in Fig. 2 clearly indicate that the DNA cross-linking agents HN2, melphalan, and CCNU were the most potent SCE inducers, and showed the greatest difference in SCE induction between early and late G_1 treatment. However, although HN2 and melphalan contain the same alkylating moiety, melphalan was found to induce more SCEs in early G_1 than HN2 and fewer in late G_1 at the same dose level (5×10^{-7} moles/l) and exposure time (2 hr) (Fig. 2). This difference between the 2 compounds may be related to the kinetics of cross-link formation. As shown by Ross et al. (27) in L1210 cells, cross-links are formed very rapidly after HN2 treatment and then decline with a half-life of about 7 hr. Melphalan, on the other hand, gives rise to an initial rapid cross-link formation followed by a slow addition of further cross-links up to about 10 hr after treatment. Moreover, the removal of cross-links after melphalan treatment seems to be slower than after HN2 treatment (27). Assuming that this difference between the 2 compounds in the formation and removal of DNA cross-links exists also in human G_1 lymphocytes, it seems likely that the smaller difference in SCE frequency between early and late G_1 treatment with melphalan than with HN2 is due to the slower formation and/or removal of DNA cross-links after melphalan treatment. The results obtained after CCNU treatment adds further support to this interpretation, as discussed above.

In order to obtain further information about the rate of removal of SCE-inducing lesions during G_1, human lymphocytes were exposed to MMS or HN2 at various times after PHA stimulation. The treatment time was 1 hr and then the cultivation continued in fresh, BrdUrd-containing medium until harvest at 75 hr. The results are shown in Fig. 3. Again, a clear difference in the response after early and late G_1 treatment was found with both compounds. However, the SCE induction by HN2 showed a different time dependence in late G_1, whereas the SCE frequency induced by MMS seemed to be linearly related to the pretreatment incubation time over the entire G_1 phase.

These results suggest that SCE-inducing MMS damage is removed at a slow and steady rate during G_1, whereas SCE-inducing HN2 damage

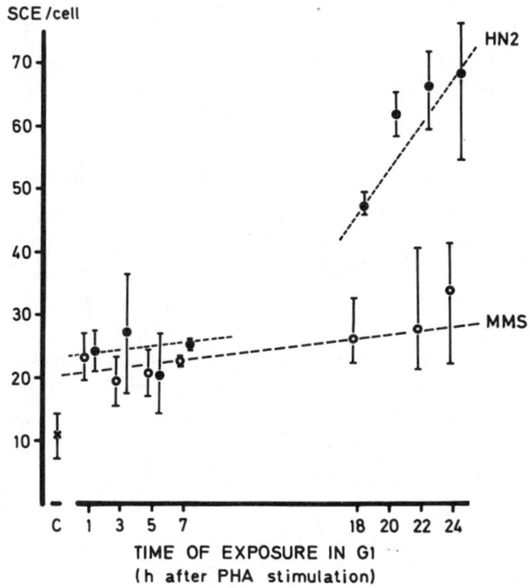

Fig. 3. The SCE frequency in cells treated with HN2 or MMS at different times in G_1. The exposure was for 1 hr ending at indicated times. C is untreated controls. The mean and range (bars) of the SCE frequency in 2 to 3 identical experiments with HN2 (filled symbols), MMS (open symbols), and in 9 concurrent, control cultures (x) are shown.

is removed at a slow rate in early G_1 and much faster in late G_1. In fact, by extrapolation, it can be calculated that the rate of removal of SCE-inducing HN2 damage is the equivalent of about 0.4 SCE/cell/hr in G_0 and early G_1, and about 4 SCE/cell/hr in late G_1.

Thus, the removal of SCE-inducing HN2 damage seems to depend on at least 2 different mechanisms operating at widely different rates. The rapid reduction of SCE-inducing HN2 damage during late G_1 in human lymphocytes appears to be due to a mechanism which becomes activated before the cells initiate DNA synthesis. It is tempting to speculate that this mechanism is involved in the first step of cross-link repair, and represents a function similar to that observed in human fibroblasts treated with MMC (8,13).

THE INDUCTION OF SCE BY MMS IN THE SECOND G_1

The SCE frequency in M2 cells is the sum of SCEs induced during the 2 cell cycles, and it cannot be deduced from the experiments presented so far if SCE-inducing damage introduced in the first G_1 phase remains and gives rise to further SCEs in the second cell cycle. To study this problem, cells were treated with SCE-inducing agents about 55 hr after PHA stimulation, because most M2 cells harvested 20 hr later would be in their second G_1 phase at this time of treatment (Fig. 1). By subtracting the induced SCE frequency in the 55 hr-treated M2 cells from that in M2 cells exposed in the first G_1 phase, it should (in theory, at least) be possible to obtain an indirect estimation of the SCE induction in each of the 2 cell cycles.

Surprisingly, it was found that MMS treatment in the second G_1 phase after PHA stimulation induced significantly more SCEs than treatment in the first G_1. This effect was observed after continuous treatment with 3 different concentrations (Fig. 4A), as well as after a 2-hr pulse treatment with MMS (Fig. 4B). Since cells treated in the second G_1 phase in these experiments contained BrdUrd-substituted DNA, it seemed important to study the effect of BrdUrd in the culture medium. However, no differences could be demonstrated in the MMS- or HN2-induced SCE frequencies between cells incubated with or without BrdUrd during G_1 (Tab. 3), and a clear depression of the MMS-induced SCE frequency was observed by increasing concentrations of BrdUrd (Tab. 4). It is not likely that the latter effect is due to the inhibition of cell proliferation observed at the high BrdUrd dose, since the same effect was found at the intermediary dose where no such inhibition was observed (Tab. 4).

Thus, the results of these experiments did not offer any explanation for the enhanced induction of SCE in cells treated with MMS in the second G_1 phase. Previous work by Ockey (25) and Natarajan and coworkers (24) have indicated that complex interactions may

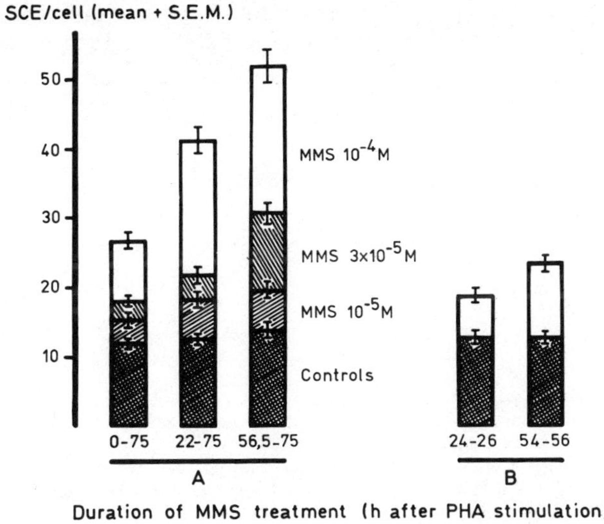

Fig. 4. The SCE frequency in cells treated with MMS in the first and second G_1. A. Three different concentrations of MMS were used, and the compound was present in the culture during the times indicated. Bars indicate S.E. of the mean frequency of SCE in 20 cells at each time and concentration. B. The concentration of MMS was 10^{-4} moles/l, and the compound was present for 2 hr as indicated. The bars indicate range of the mean SCE frequency in 2 separate experiments including the analysis of 30 and 20 cells, respectively.

occur between BrdUrd and MMS in the induction of SCE, and further studies are needed to elucidate the mechanisms behind these observations.

THE PERSISTENCE OF SCE-INDUCING DAMAGE IN CYCLING CELLS

Several different approaches have been used to estimate the persistence of SCE-inducing damage during successive cell generations in vitro as well as in vivo. The results of these studies (reviewed in Ref. 33) have indicated that some types of SCE-inducing damage are long-lived and induce new SCEs in several cell generations, whereas the SCE-inducing effect of other types of damage is restricted to the first or second S phase after exposure. We have studied the frequency of SCEs in M3 cells after exposure of human lymphocyte cultures in G_1 to 4 different SCE-inducing agents. The analysis of M3 cells permits the separation of SCEs induced during the 2 first cell cycles from those induced in the third cell cycle (33). Moreover, by the analysis of M2 cells in the same preparations it is possible to evaluate differences in the SCE frequencies

Tab. 3. The SCE frequency in MMS- or HN2-treated cells cultivated in the presence or absence of BrdUrd during G_1.

Time of BrdUrd addition (h after PHA stimulation)	The frequency of SCE/cell (mean + S.D.)[a] Control cultures	HN2 $2x10^{-6}$	MMS (10^{-4})
1	12.6 ± 0.1	29.1 ± 7.9	17.7 ± 0.1
24	13.8 ± 0.8	30.3 ± 0.4	15.8 ± 1.6

[a] Results from 2 separate experiments with blood from a single donor. Twenty cells were analyzed from each culture. Treatment with indicated concentrations (moles/l) of HN2 and MMS was for 1 hr immediately after PHA stimulation. The cells were then washed and resuspended in fresh medium. BrdUrd (100 μmoles/l) was added either immediately after treatment (1 hr) or 24 hr later.

Tab. 4. The effect of increasing concentrations of BrdUrd on the frequency of MMS-induced SCE.

Concentration (moles/l) of BrdUrd	The mean frequency of SCE/cell + S.D.[a] (replication index) A Controls	B MMS	A-B
0.5×10^{-4}	10.3 ± 1.1 (0.86)	29.5 ± 0.6 (0.79)	19.2 ± 1.7
10^{-4}	18.8 ± 3.0 (0.98)	33.5 ± 4.2 (0.85)	14.7 ± 1.3
2.0×10^{-4}	24.5 ± 4.9 (0.48)	33.1 ± 4.7 (0.34)	8.7 ± 0.2

[a] Results from 2 separate experiments with blood from different donors. Twenty cells were analyzed from each culture, except in one of the MMS-treated cultures with the highest BrdUrd concentration, where only 5 cells were suitable for analysis. The mean of these cells was 36.4 (range 31-42). BrdUrd (100 μmoles/l) and MMS (6×10^{-5} moles/l) were added simultaneously 18 hr after PHA stimulation. The figures in parentheses show the replication index based on the fraction (f) of cells in first (f1), second (f2), and third (f3) division, calculated as (1 + f3 - f1).

in slowly and rapidly proliferating cell populations in the same cultures.

In untreated control cultures, the (BrdUrd-induced) baseline frequency of SCEs in the 2 first cell cycles of M3 cells was very similar to that in M2 cells in the same cultures, and about 4 times as high as the number of SCEs induced in the third cell cycle of M3 cells (Tab. 5). This is the expected result if the same number of SCEs is induced in each cell cycle, since SCEs in the third cell cycle are visible in only half of the chromosomes (for discussion, see Ref. 33). Accordingly, depletion of BrdUrd during cultivation leading to decreasing SCE level in successive cell cycles does not seem to occur under these conditions.

Cells treated with HN2, melphalan, MMS, and adriamycin in the first G_1 phase (0-2 hr or 24-26 hr) after PHA stimulation showed a small increase in the SCE frequency in the third cell cycle of M3

Tab. 5. The SCE frequency in M2 and M3 cells in control cultures and in cultures treated with 4 SCE-inducing agents.

Concentration (moles/l) and time of treatment (h after PHA stimulation)	Increased frequency of SCE/cell[a] A: M2 cells cell cycle 1+2	B: M3 cells cell cycle 1+2	C: M3 cells cell cycle 3	A/B	B/C
Controls	14.6 ± 2.9	14.4 ± 2.3	3.5 ± 1.0	1.0	4.1
HN2 (5 x 10^{-7})					
0-2	3.2	6.6	0.3	0.5	22.0
24-26	38.4	44.8	1.4	0.9	32.0
Melphalan (5 x 10^{-7})					
0-2	11.3	10.9	1.2	1.0	9.7
24-26	25.4	29.5	1.7	0.8	17.4
MMS (10^{-4})					
0-2	5.0	4.1	1.3	1.2	3.2
24-26	7.9	9.6	1.1	0.8	8.7
Adriamycin (5 x 10^{-8})					
0-2	3.7	4.0	0.9	0.9	4.4
24-26	4.4	3.9	1.0	1.1	3.9

[a] Data for controls are the means + S.D. of 20 cells in each of 9 experiments. Other data show means of 20 cells in each of 1 or 2 separate experiments, after subtraction of the concurrent control level. The culture time was 96 hr.

cells (Tab. 5). This increase was observed in all experiments and ranged from 0.3 to 1.7 SCE/cell, which is about 30% higher than the control level (range 2.3-4.1 SCE/cell). The SCE frequency induced by all 4 agents in the 2 first cell cycles of M3 cells was very similar to that in M2 cells in the same cultures (Tab. 5,A/B). Since the M2 and M3 cells were harvested at the same time, it appears that the rate of cell proliferation did not influence the yield of induced SCEs under the experimental conditions used.

When the induced frequency of SCE in the 2 first cell cycles of M3 cells was compared with that in the third cell cycle, HN2 and melphalan were found to induce 10-30 times as many SCEs in the 2 first cell cycles as in the third cell cycle (Tab. 5,B/C). In contrast, the ratio was much lower after treatment with MMS or adriamycin. These data, therefore, suggest that most of the SCE-inducing damage caused by HN2 and melphalan in the first G_1 are removed before the S phase of the third cell cycle, whereas damage induced by MMS and adriamycin seems to be more persistent. It cannot be excluded, however, that a small fraction of SCE-inducing lesions remains for several cell generations also after exposure to HN2 and melphalan. Whether these "persisting" lesions belong to the same fraction of DNA damage as those inducing SCEs during the first cell cycle after exposure, or constitute a different type of damage or some secondary modifications of DNA is not known.

SUMMARY

The SCE frequency was studied in PHA-stimulated human lymphocytes exposed to various SCE-inducing agents in different stages of the cell cycle. Melphalan, HN2, MMS, and UV light were found to induce a higher SCE frequency in late G_1 (18-24 hr after PHA stimulation) than in early G_1 (1-6 hr after PHA) or G_0 (before PHA stimulation). In contrast, CCNU induced more SCEs in early G_1 than in late G_1, and the adriamycin-induced SCE frequency was about the same after treatment in early and late G_1. These results suggest that SCE-inducing lesions are being removed at different rates in human G_1 lymphocytes.

The removal of SCE-inducing HN2 lesions was found to be about 10 times more rapid in late G_1 than in early G_1, indicating the activation of a cross-link repair mechanism prior to DNA replication in human lymphocytes.

Cells treated with MMS in the second G_1 (after cultivation for about 55 hr in the presence of PHA and BrdUrd) showed a higher SCE frequency than cells treated with the same dose of MMS in the first G_1. This result indicates that some type of interaction occurs between MMS damage and BrdUrd lesions in the DNA during replication, which leads to an enhanced induction of SCE.

Analysis of SCEs induced during the 2 first vs. the third cell cycle in third-generation metaphases showed that most of the SCE-inducing damage caused by treatment with HN2 and melphalan in G_1 of the first cell cycle are removed before the S phase of the third cell cycle, whereas damage caused by MMS and adriamycin seem to be more persistent.

These observations suggest that the rate by which different types of SCE-inducing damage are removed or modified in resting (G_0) or PHA-stimulated human lymphocytes can have a great influence on the SCE frequency. This is of practical importance in studies using SCE analysis to evaluate human exposure to suspected genotoxic agents in the environment.

ACKNOWLEDGEMENTS

This work was supported by The Swedish Medical Research Council, The Swedish Work Environmental Fund, and The Swedish Council for Planning and Coordination of Research. We are grateful to Sigrid Sahlén and Barbro Werelius for skillful technical assistance.

REFERENCES

1. Bredberg, A., and B. Lambert (1983) Induction of SCE by DNA cross-links in human fibroblasts exposed to 8-MOP and UVA irradiation. Mutat. Res. 118:191-204.
2. Darzynkiewicz, Z. (1971) Radiation-induced DNA synthesis in normal and stimulated human lymphocytes. Exp. Cell Res. 69: 356-360.
3. Duncan, A.M.V., and H.J. Evans (1982) Molecular lesions involved in the induction of sister-chromatid exchange. Mutat. Res. 105:423-427.
4. Ericksson, L.C., M.O. Bradley, J.M. Ducore, R.A.G. Ewig, and K.W. Kohn (1980) DNA crosslinking and cytotoxicity in normal and transformed human cells treated with antitumor nitrosoureas. Proc. Natl. Acad. Sci., USA 77:467-471.
5. Erickson, L.C., G. Laurent, N.A. Sharkey, and K.W. Kohn (1980) DNA cross-linking and monoadduct repair in nitrosourea-treated human tumor cells. Nature (Lond.) 288:727-729.
6. Evans, H.J., and Vijayalaxmi (1982) Storage enhances chromosome damage after exposure of human leukocytes to mitomycin C. Nature (Lond.) 284:370-372.
7. Ewig, R.A.G., and K.W. Kohn (1977) DNA damage and repair in mouse leukemia L1210 cells treated with nitrogen mustard, 1,3-Bis(2-chloroethyl)-1-nitrosourea, and other nitrosoureas. Cancer Res. 37:2114-2122.
8. Fujiwara, Y. (1982) Defective repair of mitomycin C crosslinks in Fanconi's anemia and loss in confluent normal human and xer-

oderma pigmentosum cells. Biochim. Biophys. Acta 699:217-225.
9. Gebhart, E., B. Windolph, and F. Wopfner (1980) Chromosome studies on lymphocytes of patients under cytostatic therapy. II. Studies using the BUDR-labelling technique in cytostatic interval therapy. Human Genet. 56:157-167.
10. Hedner, K., F. Mitelman, and R.W. Pero (1983) Sister chromatid exchanges in human lymphocytes after a non-S-phase incubation period to allow excision DNA repair - In vitro exposure to N-acetoxy-2-acetylaminofluorene and ethylene oxide. In Aspects on Sister Chromatid Exchange in Human Lymphocytes, K. Hedner, Thesis. University of Lund, Sweden.
11. Heflich, R.H., D.T. Beranek, R.L. Kodell, and S.M. Morris (1982) Induction of mutations and sister chromatid exchanges in Chinese hamster ovary cells by ethylating agents. Mutat. Res. 106:147-161.
12. Ishii, Y., and M.A. Bender (1978) Factors influencing the frequency of mitomycin C-induced sister chromatid exchanges in 5-bromodeoxyuridine-substituted human lymphocytes in culture. Mutat. Res. 51:411-418.
13. Kano, Y., and Y. Fujiwara (1981) Role of DNA interstrand cross-linking: Its repair in the induction of sister chromatid exchange and a higher induction in Fanconi's anemia cells. Mutat. Res. 81:365-375.
14. Kohn, K.W. (1977) Interstrand cross-linking of DNA by 1,3-Bis-(2-chloroethyl)-1-nitrosourea and other 1-(2-haloethyl)-1-nitrosoureas. Cancer Res. 37:1450-1454.
15. Lambert, B., A. Lindblad, K. Holmberg, and D. Francesconi (1982) The use of sister chromatid exchange to monitor human populations for exposure to toxicologically harmful agents. In Sister Chromatid Exchange, S. Wolff, ed. John Wiley & Sons, New York, pp. 149-182.
16. Lambert, B., U. Ringborg, A. Lindblad, and M. Sten (1979) The effects of DTIC, melphalan, actinomycin D and CCNU on the frequency of sister chromatid exchanges in peripheral lymphocytes of melanoma patients. In Adjuvant Therapy of Cancer II, S.E. Jones and S.E. Salmon, eds. Grune & Stratton, New York, pp. 55-62.
17. Lambert, B., M. Sten, D. Hellgren, and D. Francesconi (1984) Different SCE-inducing effects of HN2 and MMS in early and late G_1 in human lymphocytes. Mutat. Res. 139:71-77.
18. Lambert, B., M. Sten, S. Söderhäll, U. Ringborg, and R. Lewensohn (1983) DNA repair replication DNA breaks and sister chromatid exchange in human cells treated with adriamycin in vitro. Mutat. Res. 111:171-184.
19. Lewensohn, R., D. Killander, U. Ringborg, and B. Lambert (1979) Increase of UV-induced DNA repair synthesis during blast transformation of human lymphocytes. Exp. Cell Res. 123:107-110.
20. Littlefield, L.G. (1982) Effects of DNA-damaging agents on SCE. In Sister Chromatid Exchange, A.A. Sandberg, ed. Alan R. Liss, New York, pp. 355-394.

21. Littlefield, L.G., S.P. Colyer, and R.J. DuFrain (1983) SCE evaluations in human lymphocytes after G_0 exposure to mitomycin C. Lack of expression of MMC-induced SCEs in cells that have undergone greater than two in vitro divisions. Mutat. Res. 107:119-130.
22. McKenzie. W., and B. Lambert (1983) Induction and reduction of sister chromatid exchange by CCNU in human lymphocytes in vitro. Cancer Genet. Cytogenet. 9:261-271.
23. Muscarella, D.E., and S.E. Bloom (1982) The longevity of chemically induced sister chromatid exchanges in Chinese hamster ovary cells. Environ. Mutagen. 4:647-655.
24. Natarajan, A.T., A.D. Tates, M. Meijers, J. Neuteboom, and N. de Vogel (1983) Induction of sister chromatid exchanges (SCEs) and chromosomal aberrations by mitomycin C and methyl methanesulfonate in Chinese hamster ovary cells. Mutat. Res. 121:211-223.
25. Ockey, C.H. (1981) Methyl methane-sulphonate (MMS) induced SCEs are reduced by the BrdU used to visualise them. Chromosoma 84: 243-256.
26. Perticone, P., and B. Lambert (1984) The SCE-frequency in human lymphocytes irradiated with UV-light at different times in G_1 (submitted for publication).
27. Ross, W.E., R.A.G. Ewig, and K.W. Kohn (1978) Differences between melphalan and nitrogen mustard in the formation and removal of DNA cross-links. Cancer Res. 38:1502-1506.
28. Sasaki, M.S. (1982) Sister chromatid exchange as a reflection of cellular DNA repair. In Sister Chromatid Exchange, A.A. Sandberg, ed. Alan R. Liss, New York, pp. 135-161.
29. Scudiero, D., A. Norin, P. Karran, and B. Strauss (1976) DNA excision-repair deficiency of human peripheral blood lymphocytes treated with chemical carcinogens. Cancer Res. 36:1397-1403.
30. Stetka, D.G., J. Minkler, and A.V. Carrano (1978) Induction of longlived chromosome damage as manifested by sister chromatid exchange in lymphocytes of animals exposed to mitomycin C. Mutat. Res. 51:383-396.
31. Swenson, D.H., P.R. Harbach, and R.J. Trzos (1980) The relationship between alkylation of specific DNA bases and induction of sister chromatid exchange. Carcinogenesis 1:931-936.
32. Takehisa, S. (1982) Induction of sister chromatid exchanges by chemical agents. In Sister Chromatid Exchange, S. Wolff, ed. John Wiley & Sons, New York, pp. 87-147.
33. Tice, R.R., and J.B. Schwartzman (1982) Sister chromatid exchange: A measure of DNA lesion persistence. In Sister Chromatid Exchange, A.A. Sandberg, ed. Alan R. Liss, New York, pp. 33-45.
34. Wolff, S. (1978) Chromosomal effects of mutagenic carcinogens and the nature of the lesions leading to sister chromatid exchange. In Mutation Induced Chromosome Damage in Man, H.J. Evans and D.C. Lloyd, eds. Edinburgh University Press, pp. 208-215.

35. Wolff, S., J. Bodycote, and R.B. Painter (1974) Sister chromatid exchanges induced in Chinese hamster cells by UV-irradiation of different stages of the cell cycle: The necessity for cells to pass through S. Mutat. Res. 25:73-81.

PERSISTENCE OF SCE-INDUCING LESIONS AFTER G_0 EXPOSURE OF HUMAN LYMPHOCYTES TO DIFFERING CLASSES OF DNA-DAMAGING CHEMICALS

L. Gayle Littlefield, Shirley P. Colyer, Anne M. Sayer, and Russell J. DuFrain

Radiation Emergency Assistance Center/Training Site
Medical and Health Sciences Division
Oak Ridge Associated Universities
P.O. Box 117
Oak Ridge, Tennessee 37831

ABSTRACT

We conducted studies to determine whether cycling human lymphocytes are equally efficient in repairing sister chromatid exchange (SCE)-producing lesions induced by differing classes of DNA-damaging chemicals. Lymphocytes were pulse-treated during G_0 with mitomycin C (MMC), N,N',N"-triethylenethiophosphoramide (ThioTEPA), ethylmethanesulfonate (EMS), or cis-diamminedichloroplatinum (cis-DDP). Bromodeoxyuridine (BrdUrd) was added to the 72 hr cultures at 0 hr or at 48 hr after phytohemmagglutinin stimulation. The concentrations of chemicals employed induced a greater than 2-fold increase in SCEs in second-division metaphases from lymphocytes cultured in the presence of BrdUrd for the entire 72 hr. The analysis of SCEs in uniformly harlequinized metaphases from G_0-treated lymphocytes cultured in BrdUrd for the terminal 24 hr showed no increase above baseline after exposure to MMC, and intermediate increases above baseline after exposures to ThioTEPA and cis-DDP. However, after G_0 treatment with EMS, the observed SCE frequency was consistent with that expected had all DNA lesions persisted and continued to give rise to SCEs during 3 cell cycles. These findings suggest that cycling human lymphocytes are not equally efficient in eliminating SCE-producing lesions after exposure to differing classes of DNA-damaging chemicals.

INTRODUCTION

The utility of SCE analysis as a method for monitoring genotoxic exposures in human populations depends to a large extent on our understanding of the basic mechanisms involved in the induction, expression, and persistence of SCE-producing lesions in human lymphocytes. Overviews of current research efforts aimed at characterizing those factors affecting SCE induction and modulation in a variety of cell systems are presented by several authors in other chapters of this volume. In this chapter I will review some of our recent findings regarding the persistence of lesions that lead to SCEs after G_0 exposure of human lymphocytes to chemicals having differing modes of action on DNA.

During the last several years, a number of studies have been conducted in efforts to determine how long lesions that can subsequently be expressed as SCEs persist in DNA of cycling and noncycling cell populations. In an early experiment, Wolff et al. (1) evaluated induced SCEs in CHO cells exposed to ultraviolet radiation during various stages of the cell cycle. Their results demonstrated that lesions induced during G_2 of one cell cycle were not expressed as SCEs until the cells had progressed through the next period of DNA replication. These early findings demonstrated that lesions capable of producing SCE can be "long-lived" in that they persisted at least from the G_2 stage of one cell cycle to the next S period. Subsequent evaluations in experimental animals (2,3) and in cancer patients (4-8) demonstrated that high frequencies of induced SCEs could be observed in cultured lymphocytes for periods ranging from several days up to several months after in vivo exposures to various alkylating chemicals. These findings provided evidence that lesions capable of producing SCEs at the time of mitogen-induced DNA replication can persist in the DNA of "resting" or G_0 lymphocytes for exceedingly long periods of time. In contrast to the pronounced longevity of SCE-producing lesions in mitotically quiescent lymphocytes, recent studies in mice have shown that mitomycin C (MMC)-induced lesions in proliferating marrow cells are quite short-lived, in that they are expressed as SCEs during the first S period after in vivo exposure, and do not persist and give rise to SCEs during subsequent cell cycles (9).

The relative lifespan of SCE-producing lesions in proliferating cells has been further characterized in several in vitro studies which have suggested that lesion persistence may be dependent on both the repair capabilities of the specific cell strain or type, and on the mode of action of the DNA-damaging agent to which the cells are exposed. In early autoradiographic studies, Kato (10) evaluated induced SCEs in Chinese hamster DON cells that had been pulse-labeled with tritiated thymidine at various time intervals after brief exposures to several DNA-damaging chemicals. After exposure of cells to 4-nitroquinoline-1-oxide (4NQO) and MMC, SCE levels did not return to control levels until the cells had passed through 6-8 post-treatment cell divisions. In contrast, after treatment with the intercalating agent proflavin, SCE levels fell rapidly, and reached control levels after only 48 hr (3 cell doublings). Using a similar approach of adding bromodeoxyuridine

(BrdUrd) to the culture medium at various time intervals after acute exposure of proliferating Chinese hamster ovary (CHO) cells to various DNA-damaging chemicals, Muscarella and Bloom (11) observed that SCE-inducing lesions were rapidly eliminated from cells exposed to quinacrine mustard and MMC, but persisted for up to 10 cell cycles after exposure to the monofunctional alkylating agent, methylmethanesulfonate (MMS). As an alternative approach to assessing lesion persistence, Wolff (12), and Linnainmaa and Wolff (13) evaluated the ratios of twin vs. single exchanges (14) in tetraploid CHO cells following pulse treatment with various classes of chemicals during the G_1 stage of the cell cycle. These studies demonstrated that the bifunctional alkylating agents MMC and 8-methoxypsoralen, and the monofunctional derivative of MMC, decarbamoyl MMC, induce short-lived lesions in CHO cells that lead to SCEs in the first period of DNA synthesis, while angelicin, MMS, EMS, and N-acetoxyacetylaminofluorene produced lesions that lasted more than one cell cycle.

As previously discussed, studies in experimental animals and man have suggested that alkylator-induced SCE-producing lesions may persist for long periods of time in the DNA of resting (G_0) lymphocytes. Less information is available regarding the relative longevity of these lesions in cycling lymphocytes during successive rounds of mitogen-induced DNA synthesis. Based on the analysis of reciprocal exchanges in M3 metaphases (15) of human lymphocytes treated during G_0 with MMC, Ishii and Bender (16) concluded that some MMC-induced lesions persisted and continued to give rise to SCEs in the third cell cycle. To further examine lesion persistence in mitogen-stimulated lymphocytes, we employed a design similar to that used by Kato (10) in his evaluations of lesion persistence in DON cells. In these studies human G_0 lymphocytes were pulse-treated with MMC, and stimulated to undergo DNA replication with phytohemagglutinin. MMC-treated and control lymphocytes were subsequently exposed to BrdUrd for the entire culture period or for the terminal 24 hr of 72 hr cultures. We observed a 3- to 4-fold increase in SCEs in second-division metaphases from lymphocytes treated with MMC and cultured in BrdUrd for the entire culture period. In contrast, in replicate cultures of MMC-treated lymphocytes that were exposed to BrdUrd for the terminal 24 hr only, SCE frequencies in uniformly harlequinized metaphases were not significantly different from those observed in control cultures. In contrast to the earlier findings of Ishii and Bender, the results of our study suggested that MMC-induced lesions in the DNA of human G_0 lymphocytes are probably expressed as SCEs during the first period of mitogen-induced DNA synthesis and do not persist and give rise to SCEs in subsequent cell divisions (17).

These findings with MMC prompted us to undertake additional studies using this methodology to determine whether other classes of DNA-damaging agents would yield similar results in the human lymphocyte system. We therefore evaluated several chemicals having differing modes of action on DNA to determine whether induced lesions in G_0 lymphocytes persist and give rise to SCEs in cells that have undergone greater than 2 in vitro cell divisions.

MATERIALS AND METHODS

Chemicals Tested

The chemicals evaluated in this series of experiments included: EMS and MMC, purchased from Sigma Chemicals; ThioTEPA purchased from Lederle; and *cis*-DDP purchased from Bristol Laboratories. These compounds are effective SCE inducing-agents in several cell systems (see Ref. 18 for review), and while each is known to induce a variety of molecular lesions, EMS is generally classified as a monofunctional alkylating agent, while *cis*-DDP, ThioTEPA, and MMC are bi- or polyfunctional alkylating agents. In addition to other types of molecular lesions, these 3 latter compounds all produce DNA-DNA and/or DNA-protein cross-links.

Prior to undertaking the studies described in this report, human lymphocytes were treated during G_0 with these chemicals to determine the concentrations required to yield a > 2-fold increase in SCEs in second-division metaphases of 72 hr cultures. Subsequently, we conducted the experiments described below.

G_0 Treatment Protocol

On two different treatment dates, heparinized whole blood from a total of 4 donors (i.e., A,B,C,D) was allowed to gravity-sediment at room temperature for sufficient time to collect several milliliters of cell-rich plasma (CRP). After thorough mixing, 1.0 ml aliquots of CRP from each donor were inoculated into a series of flasks containing 9.0 ml TC 199 culture medium (Gibco). In each experiment 2 flasks containing cells from each donor were used as controls. To replicate flasks we added a volume of 0.1 ml distilled water containing appropriate dilutions of the test chemicals to yield final concentrations in the culture medium of 1.0×10^{-2}, 1.5×10^{-2}, or 2.0×10^{-2} M EMS; 0.5 μg/ml MMC, 2.5 μg/ml ThioTEPA, or 10 μg/ml *cis*-DDP. All exposures were set up in duplicate. Treated and control flasks from each donor were incubated for 2 hr at 37°C, after which the cells were collected by centrifugation and washed 3 times in 10 ml medium. Afterwards, the control and G_0-treated lymphocytes were inoculated into 8.0 ml TC 199 culture medium containing 2.0 ml cell-free autologous plasma from each respective leukocyte donor, 0.1 ml antibiotic solution (Gibco; penicillin and streptomycin), and 0.1 ml reconstituted phytohemagglutinin (Gibco; PHA-M). All cultures were incubated at 37°C, in the dark in closed vessels.

BrdUrd Exposure Schedule

To 1 flask containing untreated lymphocytes, and to 1 flask containing lymphocytes that had been exposed during G_0 to each concentration of chemical, 30 μM BrdUrd was added at the time of culture initiation and the cells were harvested at 72 hr (i.e., harvest time/time in BrdUrd, 72 hr/72 hr). To replicate flasks of control and G_0-treated lymphocytes from each donor, BrdUrd was added 48 hr after culture initiation, and the cells were harvested at 72 hr, after being cultured for the terminal 24 hr in the presence of BrdUrd (i.e., 72 hr/24 hr).

All cultures were exposed to colchicine (10 μg/ml) for 3 hr prior to harvest. Cells were exposed to a hypotonic solution of 1 part 0.9% sodium citrate and 1 part 0.075 M KCl and fixed in aceto-methanol (1:3). Slides were prepared and stained using a previously described modification (19) of the fluorescence-plus-Giemsa technique (20).

SCE Analysis

From cultures of control and G_0-treated lymphocytes that had been incubated in the presence of BrdUrd for 72 hr, we scored 20 well-differentiated second-division metaphases to obtain an estimate of baseline, and of baseline + induced SCEs that occurred during the first and second in vitro cell divisions.

From replicate cultures of control and G_0-treated lymphocytes that had been exposed to BrdUrd for the terminal 24 hr of the 72 hr culture period, we assessed SCEs in 20 uniformly harlequinized metaphases from each culture. Earlier studies in our laboratory have provided evidence that these metaphases are derived from populations of lymphocytes that completed their first mitogen-induced in vitro cell division prior to the addition of BrdUrd to the culture medium and their second and third cell cycles during the terminal 24 hr exposure to BrdUrd (17).

RESULTS

SCE Analysis in 72 hr/72 hr Cultures

The results of our analyses of baseline and of baseline + induced SCEs in replicate cultures of control and of G_0-treated lymphocytes that had been cultured in the presence of BrdUrd for the entire culture period are shown in Tab. 1. Baseline SCE frequencies in second-division metaphases in these 72 hr/72 hr cultures ranged from 9.4 to 14.1 SCE/metaphase. In preparations from all 4 subjects, a greater than 2-fold increase in SCEs was observed in M2

Tab. 1. SCEs in second-division metaphases 72 hr/72 hr cultures. (Estimate of baseline and of baseline + induced SCEs that occurred during the first and second cell cycles after G_0 exposure.)

Subject	G_o Treatment*	Time in BrdU (hr)	# Meta Scored	SCE/meta ± SEM	Range
A	None	72	20	12.6 ± 1.0	4-20
A	EMS ($2x10^{-2}$M)	72	11	37.2 ± 3.3	18-51
A	ThioTEPA	72	20	45.2 ± 1.8	20-61
A	cis-PDD	72	20	36.4 ± 2.3	20-70
A	MMC	72	20	34.0 ± 2.5	13-58
B	None	72	20	14.1 ± 1.2	8-26
B	EMS ($2x10^{-2}$M)	72	No data		
B	ThioTEPA	72	20	41.0 ± 2.3	20-68
B	cis-PDD	72	20	45.4 ± 4.1	15-109
B	MMC	72	20	33.1 ± 1.6	19-49
C	None	72	20	10.8 ± 0.9	4-18
C	EMS ($1x10^{-2}$M)	72	20	20.0 ± 1.2	10-28
C	EMS (1.5×10^{-2}M)	72	20	27.4 ± 1.4	15-41
C	ThioTEPA	72	20	34.0 ± 1.6	25-53
D	None	72	20	9.4 ± 0.7	3-18
D	EMS ($1x10^{-2}$M)	72	20	25.2 ± 1.4	15-39
D	EMS (1.5×10^{-2}M)	72	20	29.8 ± 1.7	15-45
D	ThioTEPA	72	20	36.7 ± 1.7	25-51

* 2hr G_o exposure to EMS, 2.5 μg/ml ThioTEPA, 10 μg/ml cis-PDD, or 0.5 μg/ml MMC.

metaphases of lymphocytes that had been exposed during G_0 to EMS, ThioTEPA, cis-DDP, or MMC. Because BrdUrd was added to the medium at the time of culture initiation, the SCEs in these M2 metaphases are estimates of the sum of baseline and of baseline + induced SCEs that were expressed during the first and second cell cycles after G_0 exposure to the various chemicals.

SCE Analysis in 72 hr/24 hr Cultures

In Tab. 2 are shown the results of our analyses of SCEs in uniformly harlequinized metaphases from replicate cultures of control and G_0-treated lymphocytes that had been cultured in the presence of BrdUrd for the teriminal 24 hr of the 72 hr culture period. In preparations from 3 of the 4 subjects, baseline SCEs were lower than those observed in M2 metaphases of the corresponding 72 hr/72 hr cultures. Relative to findings in the 72 hr/24 hr control cultures, SCE analyses in uniformly harlequinized metaphases in replicate

Tab. 2. SCEs in uniformly harlequinized metaphases 72 hr/24 hr cultures. (Estimate of baselines and of baseline + induced SCEs that occurred during the second and third cell cycles after G_0 exposure.)

Subject	G_o Treatment*	Time in BrdU (hr)	# Meta Scored	SCE/meta ± SEM	Range
A	None	24	20	12.3 ± 1.0	4-21
A	EMS ($2x10^{-2}$M)	24	20	22.6 ± 1.4	13-24
A	ThioTEPA	24	20	14.8 ± 1.0	8-28
A	cis-PDD	24	20	14.8 ± 1.1	5-24
A	MMC	24	20	12.9 ± 0.8	8-21
B	None	24	20	10.0 ± 0.6	5-15
B	EMS ($2x10^{-2}$M)	24	No data		
B	ThioTEPA	24	20	13.4 ± 2.3	11-18
B	cis-PDD	24	20	12.8 ± 0.9	6-24
B	MMC	24	20	11.8 ± 0.9	6-21
C	None	24	20	7.1 ± 0.9	3-12
C	EMS ($1x10^{-2}$M)	24	20	11.9 ± 1.2	5-23
C	EMS (1.5×10^{-2}M)	24	20	12.0 ± 1.6	5-36
C	ThioTEPA	24	20	11.3 ± 1.5	7-17
D	None	24	20	7.4 ± 0.6	4-12
D	EMS ($1x10^{-2}$M)	24	20	14.9 ± 1.3	7-28
D	EMS (1.5×10^{-2}M)	24	20	18.1 ± 2.0	8-41
D	ThioTEPA	24	20	10.3 ± 1.0	4-17

* 2hr G_o exposure to EMS, 2.5 μg/ml ThioTEPA, 10 μg/ml cis-PDD, or 0.5 μg/ml MMC.

72 hr/24 hr cultures of G_0-treated lymphocytes showed: 1) no apparent increase above baseline in lymphocytes that had been exposed during G_0 to MMC, 2) intermediate increases above baseline after G_0 exposure to ThioTEPA and cis-DDP, and 3) large, apparently dose-related, increases above baseline in cultures of lymphocytes that had been exposed during G_0 to EMS.

To determine whether the observed SCE frequencies in the 72 hr/24 hr cultures of G_0-treated lymphocytes were significantly elevated relative to baseline SCEs in the 72 hr/24 hr control cultures, we compared sets of data using a 2-way ANOVA with $\sqrt{x}$ transformation of the SCE/metaphase data (Tab. 3). These comparisons demonstrated highly significant differences in SCEs between control vs. treated 72 hr/24 hr cultures after G_0 exposure to EMS, ThioTEPA, and cis-PDD, but no significant differences after G_0 exposure to MMC.

Comparison of Observed vs. Expected SCEs Occurring During the Second and Third Cell Cycle after G_0 Exposure to DNA-Damaging Chemicals

For comparisons with the number of SCEs observed in metaphases

Tab. 3. Summary of statistical analyses.

Treatments Compared*	Subjects	SCE/metaphase	F	Probability
72/24 control	C,D	7.2		
vs.			45.926	<.001
72/24 EMS (1.0×10^{-2}M)	C,D	13.4		
72/24 control	C,D	7.2		
vs.			43.425	<.001
72/24 EMS (1.5×10^{-2}M)	C,D	15.0		
72/24 control	A,B,C,D	9.2		
vs.			40.241	<.001
72/24 ThioTEPA	A,B,C,D	12.4		
72/24 control	A,B	11.2		
vs.			7.670	<.01
72/24 *cis*-PDD	A,B	13.8		
72/24 control	A,B	11.2		
vs.			2.189	N.S.
72/24 MMC	A,B	12.3		

* Two-way analysis of variance with a $\sqrt{x}$ transformation of SCE/metaphase data.

from the 72 hr/24 hr cultures, we calculated for each chemical the sum of baseline + induced SCEs that would have been expected in metaphases that completed their second and third post-treatment cell division in the presence of BrdUrd. These calculations assumed that all lesions induced in the DNA of the lymphocytes during the G_0 exposure period would persist and have equal probability of being expressed as SCEs during the first, second, and third cell cycles. To obtain an estimate of the number of induced SCEs that occurred during the first and second cell cycles after G_0 exposure, we subtracted the observed SCE frequency in M2 metaphases of the 72 hr/72 hr control cultures from the SCE frequency observed in M2 metaphases of the respective 72 hr/72 hr G_0-treated cultures. We next assumed that lesion-bearing DNA is segregated in a ratio of 4:2:1 during the first, second, and third cell cycles, and calculated the number of induced SCEs that would have been expected in lymphocyte metaphases that completed their second and third post-treatment cell divisions in the presence of BrdUrd. To this calculated number we added the baseline SCEs observed in the 72 hr/24 hr control culture from the respective blood donor.

Figure 1 shows our comparisons of observed baseline + induced SCEs in uniformly harlequinized metaphases of 72 hr/24 hr cultures with the expected frequencies of baseline + induced SCEs in metaphases of G_0-treated lymphocytes that completed their second and

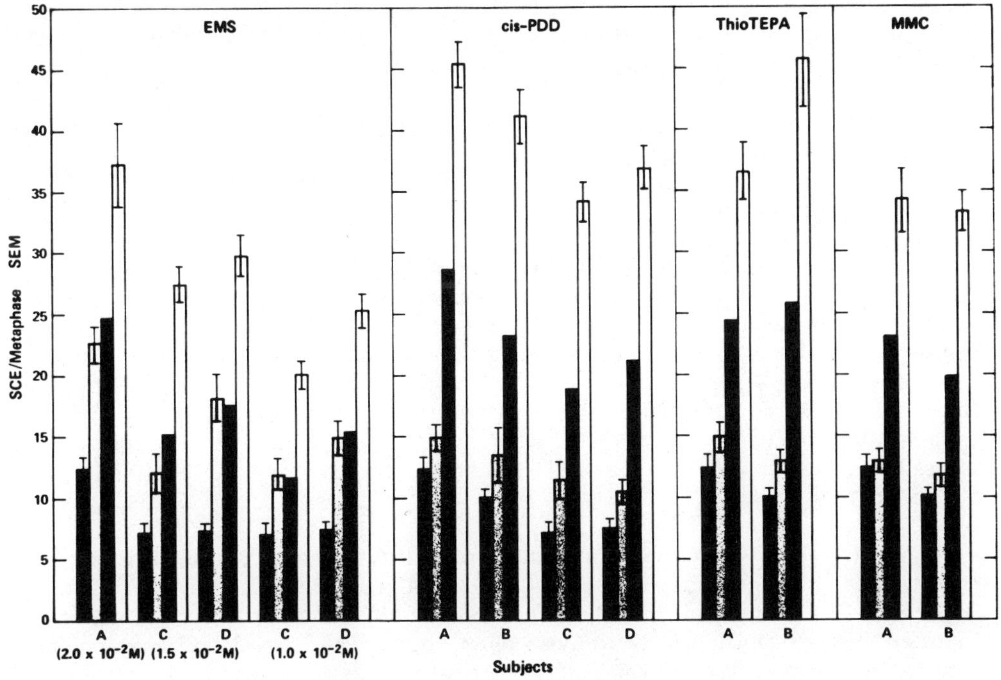

Fig. 1. Comparisons of observed SCE frequencies in 72 hr/24 hr metaphases of G_0-treated lymphocytes with those expected in cells that completed their second and third post-treatment cell divisions in the presence of BrdUrd (see text for detail). Reading left to right for each subject, the vertical bars represent: data from uniformly harlequinized metaphases in 72 hr/24 hr control cultures; data from uniformly harlequinized metaphases in 72 hr/24 hr treated cultures; calculated estimate of baseline + induced SCEs that would be expected if all SCE-inducing lesions persisted and were expressed as SCEs during second and third cell cycle; data from M2 metaphases in 72 hr/72 hr treated cultures.

third post-treatment cell divisions in the presence of BrdUrd. These comparisons suggested: 1) for EMS, the observed SCE frequency was consistent with that expected if all DNA lesions induced during the G_0 exposure had persisted and had continued to produce SCEs during the 3 cell cycles; 2) for ThioTEPA and cis-DDP, the observed SCE frequency was consistent with that expected if a small proportion of DNA lesions had persisted and continued to give rise to SCEs during the second and third cell cycles; 3) for MMC, the observed SCE frequency was consistent with that expected had all induced SCEs been expressed during the first period of mitogen-induced DNA synthesis, with no subsequent expression of induced SCEs in the second or third cell cycle.

DISCUSSION

In earlier experiments we employed an identical experimental protocol in assessing lesion persistence in cycling human lymphocytes that had been exposed to varying concentrations of MMC during the G_0 stage of the cell cycle (17). In the present study we included MMC as a reference for comparing data obtained after G_0 exposures of human lymphocytes to 3 other DNA-damaging chemicals. As in our previous study, we did not observe any increase above baseline SCEs in metaphases of MMC-treated G_0 lymphocytes that completed 2 rounds of DNA synthesis in the presence of BrdUrd during the terminal 24 hr of the 72 hr culture period. In contrast, highly significant increases above baseline SCE were observed in 72 hr/24 hr cultures after G_0 exposure of human lymphocytes to EMS, cis-DDP, and ThioTEPA. These findings suggest that some proportion of lesions induced by these 3 chemicals during the G_0 exposure interval continued to be expressed as SCEs during the second and third post-treatment cycles.

As a means of comparing the relative degree of lesion persistence after G_0 exposure to the 4 chemicals, we used the frequency of induced SCEs observed in M2 metaphases for calculating the number of induced SCEs that would have been expected in lymphocytes that completed their second and third post-treatment cell divisions in the presence of BrdUrd. We assumed that all lesions induced during the G_0 treatment interval would have equal probability of being expressed as SCEs during the first, second, and third cell cycle, and that lesion-bearing DNA would be segregated in a ratio of 4:2:1 during the first 3 cell cycles. These comparisons suggested that SCE-producing lesions induced by EMS during G_0 exposure are not eliminated in cycling human lymphocytes during the first or second cell cycles after mitogen-induced lymphocyte transformation. On the other hand, although increased frequencies of SCEs were observed in 72 hr/24 hr cultures of lymphocytes treated during G_0 with cis-DDP and ThioTEPA, the observed frequencies were substantially lower than those expected if all SCE-producing lesions had persisted for 3 rounds of DNA replication. Our findings with these 4 chemicals provide evidence that cycling human lymphocytes are capable of eliminating or repairing the vast majority of the SCE-producing lesions induced by MMC, a sizable proportion of those lesions induced by ThioTEPA and cis-DDP, and virtually none of the DNA lesions induced by the monofunctional alkylating agent EMS.

Our observation that MMC-induced lesions in the DNA of G_0 lymphocytes are expressed as SCEs only during the first period of DNA synthesis is in agreement with earlier findings in proliferating marrow cells of mice treated in vivo with MMC (9), and with data obtained from studies of lesion persistence in cycling CHO cells (11,13). The findings in these 3 cell systems are contrary to the observation of long-term persistence of MMC-induced SCE-producing lesions in the Chinese hamster cell line, DON D-6. As previously suggested by Wolff (21), such observed differences in lesion persistence may reflect differences in the repair capabilities of various cell types.

Data from the present study also suggest that cycling human lymphocytes are similar to CHO cells in their lack of ability to remove SCE-producing lesions that are induced after G_0 exposure to the monofunctioning alkylating agent EMS. Based on the analysis of the ratio of twin and single exchanges in tetraploid cells, Wolff concluded that no repair of DNA damage occurred between the 2 cell cycles after exposure of CHO cells to either EMS or MMS (12). Similar long-term persistence of MMS lesions in cycling CHO cells has also been reported by Muscarella and Bloom (11).

We are not aware of earlier studies of lesion persistence in proliferating mammalian cells after exposure to ThioTEPA or *cis*-DDP. Previous evaluations in our laboratory demonstrated that each induces SCE-producing lesions in human lymphocytes after G_0 exposure (22,23), but that neither is as effective as MMC. We included evaluations of these 2 chemicals in the present study to determine whether lesion persistence in cycling human lymphocytes would be similar after G_0 exposures to different bifunctional alkylating agents. Our experimental observations suggest that while all MMC-induced lesions are expressed as SCEs during the first period of mitogen-induced DNA synthesis, a small, but significant, proportion of lesions induced by *cis*-DDP and ThioTEPA persist and give rise to SCEs during the second and third cell cycle. These findings provide evidence that cycling human lymphocytes are not equally efficient in eliminating SCE-producing lesions after exposure to different bifunctional alkylating chemicals. This conclusion is not unexpected, since human lymphocytes exhibit widely different sensitivities to SCE induction after exposure to these 3 chemicals. Likewise, although all 3 chemicals have in common their ability to induce DNA cross-links, each also induces a variety of other molecular lesions that might be involved in SCE formation, and it is highly probable that the longevity of different types of lesions would not be identical in human cells. Because of the extreme potency of bifunctional alkylating agents in inducing SCEs, DNA cross-links have been implicated as causative lesions in the formation of SCEs. However, experimental studies with monofunctional derivatives of MMC (13,24) and of 8-methoxypsoralen (13) have indicated that monoadducts rather than cross-links are responsible for SCE formation. Certainly, definitive explanations for the observed differences in lesion persistence in cycling cells will not be forthcoming until more information is available on the mechanisms of lesion induction after exposure to specific DNA-damaging chemicals.

In summary, data from the present evaluations in cycling human lymphocytes suggest that the induction and persistence of SCE-producing lesions is dependent on as-yet-undefined molecular mechanisms of action of various DNA-damaging chemicals, as has previously been observed in other mammalian cell systems (11,12). Attempts to derive definitive conclusions from published findings on the persistence of SCE-producing lesions after exposure of mammalian cells to various classes of DNA-damaging chemicals are immediately fraught with difficulty. As previously noted, it appears that cells of differing species origin, or of differing strains from the same species, may vary in their initial sensitivity to lesion induction, as well as in their abilities to remove or repair specific DNA lesions. Furthermore, although DNA-damaging chemicals are frequently classified according to their primary mode of action, virtually all induce a variety of lesions at the molecular level. Keeping in mind these complexities, some generalities may be drawn from studies of lesion persistence in mammalian cells. Although exceptions have been noted, studies in several cell types have shown that a number of different bifunctional alkylating agents, and the DNA intercalating agents proflavin and quinacrine mustard, induce short-lived lesions that are rapidly removed from cycling cells. In contrast, studies of lesion persistence after exposure to several monofunctional alkylating agents, including EMS, MMS, MNNG, and angelicin, have generally shown that SCE-producing lesions are relatively longlived in that they persist through several successive cell cycles. Additional work is obviously needed to define more precisely the biochemical and cellular processes involved in those molecular events leading to SCE formation, and in the removal or repair of lesions that are expressed as SCEs.

ACKNOWLEDGEMENTS

This chapter is based on work performed under Contract No. DE-AC05-76OR00033 between the Department of Energy, Office of Energy Research, and Oak Ridge Associated Universities.

REFERENCES

1. Wolff, S., J. Bodycote, and R.B. Painter (1974) Sister chromatid exchanges induced in Chinese hamster cells by UV irradiation of different stages of the cell cycle: The necessity for cells to pass through S. Mutat. Res. 25:73-81.
2. Stetka, D.G., and S. Wolff (1976) Sister chromatid exchange as an assay for genetic damage induced by mutagen-carcinogens. I. In vivo test for compounds requiring metabolic activation. Mutat. Res. 41:333-342.
3. Stetka, D.G., J. Minkler, and A.V. Carrano (1978) Induction of long-lived chromosome damage as manifested by sister-chromatid exchange, in lymphocytes of animals exposed to mitomycin C. Mutat. Res. 51:383-396.
4. Raposa, T. (1978) Sister chromatid exchange studies for monitoring DNA damage and repair capacity after cytostatics in

vitro and in lymphocytes of leukaemic patients under cytostatic therapy. Mutat. Res. 57:241-251.
5. Lambert, B., R. Ulrik and A. Lindblad (1979) Prolonged increase of sister-chromatid exchanges in lymphocytes of melanoma patients after CCNU treatment. Mutat. Res. 59:295-500.
6. Lambert, B., U. Ringborg, E. Harper, and A. Lindblad (1978) Sister chromatic exchanges in lymphocyte cultures of patients receiving chemotherapy for malignant disorders. Cancer Treatment Reports 62:1413-1419.
7. Ohtsuru, M., Y. Ishii, S. Taki, H. Higashi, and G. Kosaki (1980) Sister chromatid exchanges in lymphocytes of cancer patients receiving mitomycin C treatment. Cancer Res. 40:477-480.
8. Littlefield, L.G., S.P. Colyer, and R.J. DuFrain (1980) Comparison of sister-chromatid exchanges in human lymphocytes after G_0 exposure to mitomycin C in vivo vs. in vitro. Mutat. Res. 69:191-197.
9. Kram, D., G.D. Bynum, R. Dean, E.L. Schneider, W.H. Farland, and J.R. Williams (1981) Effects of acute and chronic administration of MMC on the induction of sister chromatid exchanges in vivo. Environ. Mutagen. 3:489-495.
10. Kato, H. (1974) Induction of sister chromatid exchanges by chemical mutagens and its possible relevance to DNA repair. Exp. Cell Res. 85:239-247.
11. Muscarella, D.E., and S.E. Bloom (1982) The longevity of chemically induced sister chromatid exchanges in Chinese hamster ovary cells. Environ. Mutagen. 4:467-475.
12. Wolff, S. (1978) Chromosomal effects of mutagenic carcinogens and the nature of lesions leading to sister chromatid exchange. In Mutagen-Induced Chromosome Damage in Man, H.J. Evans and D.C. Lloyd eds. Yale University Press, New Haven, pp. 208-215.
13. Linnainmaa, K., and S. Wolff (1982) Sister chromatid exchange induced by short-lived monoadducts produced by the bifunctional agents mitomycin C and 8-Methoxypsoralen. Environ. Mutagen. 4: 239-247.
14. Taylor, J.H. (1958) Sister chromatid exchanges in tritium-labeled chromosomes. Genetics 43:515-529.
15. Tice, R., J. Chaillet, and E.L. Schneider (1975) Evidence derived from sister chromatid exchanges of restricted rejoining of chromatid subunits. Nature (Lond.) 256:642-644.
16. Ishii, Y., and M.A. Bender (1978) Factors affecting the frequency of mitomycin C-induced sister-chromatid exchanges in 5-bromodeoxyuridine-substituted human lymphocytes in culture. Mutat. Res. 51:411-418.
17. Littlefield, L.G., S.P. Colyer, and R.J. DuFrain (1983) SCE evaluations in human lymphocytes after G_0 exposure to mitomycin C. Lack of expression of MMC-induced SCEs in cells that have undergone greater than two in vitro divisions. Mutat. Res. 107: 119-130.
18. Littlefield, L.G. (1982) Effects of DNA-damaging agents on SCE.

In Sister Chromatid Exchange, A.A. Sandberg, ed. Alan R. Liss, Inc., New York, pp. 355-394.

19. Littlefield, L.G., S.P. Colyer, E.E. Joiner, and R.J. DuFrain (1979) Sister chromatid exchanges in human lymphocytes exposed to ionizing radiation during G_0. Rad. Res. 78:514-521.
20. Perry, P., and S. Wolff (1974) New Giemsa method for differential staining of sister chromatids. Nature (Lond.) 251:156-158.
21. Wolff, S. (1981) Induced chromosome variation. Chromosomes Today 7:226-241.
22. Littlefield, L.G., S.P. Colyer, A.M. Sayer, and R.J. DuFrain (1979) Sister-chromatid exchanges in human lymphocytes exposed during G_0 to four classes of DNA-damaging chemicals. Mutat. Res. 67: 259-269.
23. Morrison, W.D., V. Huff, S.P. Colyer, R.J. DuFrain, and L.G. Littlefield (1981) Cytogenetic effects of cis-Platinum(II) Diamminedichloride in vivo. Environ. Mutagen. 3:265-274.
24. Carrano, A.V., L.H. Thompson, D.G. Stetka, J.L. Minkler, J. A. Mazrimas, and S. Fong (1979). DNA crosslinking, sister-chromatid exchange and specific mutations. Mutat. Res. 63:175-188.

PROLIFERATIVE KINETICS AND CHEMICAL-INDUCED SISTER CHROMATID EXCHANGES IN HUMAN LYMPHOCYTE CULTURES

Kanehisa Morimoto

Department of Public Health
Faculty of Medicine
University of Tokyo
Tokyo 113, Japan

INTRODUCTION

Short-term cultures of phytohemagglutinin (PHA)-stimulated human lymphocytes are widely used to detect chromosome-damaging agents and possible human exposure to mutagenic carcinogens (1), and to study the immune response of blood (2,3). After stimulation of blood lymphocytes with PHA, the cultures soon contain different generations of cells, i.e., cells that have divided different numbers of times (4,5). This heterogeneity of cell division has been ascribed variously to a difference in cell cycle times, or to a difference in the times when the cells start blastogenesis in response to PHA (6-10).

It is necessary for cells to pass through the S phase before DNA damage results in the formation of sister chromatid exchanges (SCEs) (11). If induced lesions are completely repaired before cells enter S, then those cells show no increase in SCE frequency (12). On the contrary, if there is no repair of induced lesions within the consecutive 2 cell cycles before sampling, those lesions can contribute to the formation of SCEs in both the first and second cycles. Therefore, the induction of SCEs by DNA-damaging agents depends on the cell-cycle stage in which cells are treated, and on the nature of repair of induced lesions. Treatments resulting in SCE formation also induce cell-cycle delays (12-14), which might deform the true dose-response relationship in the SCE induction.

It has recently been shown that repeated pretreatments of Chinese hamster and human cell lines, as well as bacteria (15-17), with alkylating agents can render them resistant to a following challenge treatment with a higher dose of the alkylating agent (18). Liver cells from rats chronically pre-exposed to alkylating agents have also been found to show an increased removal of induced DNA alkylation damage (19,20). The possible mechanism for this adaptation-like phenomenon is the induction of the repair enzymes, methyltransferase and glycosylase (21-23). Human lymphocytes are known to contain these enzymes, and PHA stimulation results in a many-fold increase in enzyme activities (24,25). Thus, it should be examined whether repeated pretreatments with alkylating agents can also render human lymphocytes resistant to the formation of alkylation damage by a following chemical challenge.

In the present studies on human lymphocytes, I examine (i) the cell proliferation kinetics in human lymphocyte cultures and their delay induced by chemical treatments measured by a simple combination of autoradiography and sister chromatid differential staining, (ii) the cell-stage dependence of the SCE induction by alkylating agents, and (iii) the induction of "adaptation-like" resistance in human lymphocytes by repeated pretreatments with alkylating agents.

MATERIALS AND METHODS

Lymphocyte Culture

Human whole blood cultures were initiated using medium RPMI 1640, fetal bovine serum (15%), and antibiotics (100 µg/ml penicillin and 100 units/ml streptomycin). The culture medium also contained 5-bromodeoxyuridine (BrdUrd) 20 µM) when necessary for sister chromatid differential staining by the fluorescence-plus-Giesma (FPG) method.

Chemical Treatment

For chemical treatment, the chemicals to be tested were first dissolved in serum-free culture medium, Dulbecco's phosphate buffered saline (PBS), or dimethylsulfoxide (DMSO). Aliquots of this freshly made solution were added to the medium to give the appropriate final concentration. The highest concentrations of these solvents (PBS or DMSO) in the culture medium were less than 0.5%, and control cultures in each experiment contained PBS or DMSO at the highest concentration added when these solvents were employed. To terminate the pulse treatments cells were washed at least 3 times with pre-warmed PBS or serum-free culture medium. The cells were then resuspended in complete culture medium containing BrdUrd for further incubation. Each chemical treatment was conducted at 37°C because I have found that treatment with mitomycin C (MMC) or ethylmethanesulfonate (EMS) induces considerably more SCEs at 37°C than at 25°C (data not shown, manuscript in preparation).

Sister Chromatid Differential Staining

Slides were stained by a modification of the FPG technique (26) to obtain differentially stained chromosomes as previously described (5,12). In these preparations, cells dividing for the first (X1), second (X2), or third or more (X3+) time in cultures containing BrdUrd were determined by the differential staining pattern of sister chromatids. First-division (X1) cells contain chromosomes with both sister chromatids stained uniformly darkly. Second division (X2) cells contain only differentially stained chromosomes, with one chromatid darkly stained and its sister chromatid lightly stained, whereas 3rd or subsequent-division (X3+) cells contain some differentially stained chromosomes and chromosomes with both sister chromatids stained uniformly lightly. When necessary, fourth or subsequent-division (X4+) cells were distinguished from X3 cells by counting the number of differentially stained chromosomes in a cell (4).

For SCEs, 30-50 consecutive X2 cells were scored blindly on coded slides per point per person.

Labeling of Cells with ^{3}H-Thymidine (^{3}H-dThd)

Cells were treated with ^{3}H-dThd (sp. act., 52 Ci/mmole) at a final concentration of 0.1 μCi/ml. Because the medium also contained 20 μM BrdUrd, the effective specific activity of ^{3}H-dThd in the medium was estimated to be 5.0 mCi/mmole, provided there is no difference in the efficiency of uptake of dThd and BrdUrd. In a previous study, I showed that labeling with 0.1 μCi/ml ^{3}H-dThd did not disturb cell proliferative kinetics (27).

Autoradiography

For autoradiography of FPG-stained cells, slides were dipped in a 1% Formvar solution (28) to prevent the formation of chemical grains by direct contact between the autoradiographic nuclear emulsion and the Giemsa stain. The slides were then dipped in Ilford L-4 emulsion (1:1 emulsion:distilled water) and developed in Kodak D19 developer for 6 min at 20°C after a 45-da exposure at 4°C.

Scoring of Labeled Metaphases

For cells that have incorporated ^{3}H-dThd in their first S phase, an X1 cell has label on both chromatids of a chromosome, whereas an X2 cell has label only on lightly stained chromatids. An X3 cell has label on lightly stained chromatids only on those portions of chromosomes with both chromatids stained uniformly lightly. For cells that have incorporated ^{3}H-dThd in the second S phase, an X1 cell has no label, an X2 cell has label on both chromatids, and an X3 cell has label on the lightly stained chromatids of both differentially stained and uniformly lightly stained chromosomes.

Such combination patterns of labeling and sister chromatid differential staining are the same in whichever division SCEs take place.

The degree of labeling was determined by the number of chromosomes labeled in a cell as previously reported (29). An X1 cell was designated as labeled if more than one-third of the chromosomes were labeled on both chromatids. If more than two-thirds of the chromosomes were so labeled the cell was considered to be heavily labeled. Similarly, an X2 cell was considered to be labeled if more than one-3rd of the chromosomes were labeled on the light chromatid. If more than two-thirds of the chromosomes were so labeled, the cell was designated as being heavily labeled. An X3 cell was considered labeled if labeling occurred in the lightly stained chromatids of more than one-sixth of those chromosomes with both chromatids lightly stained in the centromere regions. If such labeling in more than one-third of such chromosomes, the cell was designated as being heavily labeled. An X3 cell has approximately 23 such chromosomes (4). These criteria were applied for labeled metaphase cells that had incorporated ^{3}H-dThd in their first S phase after initiation.

For metaphase cells whose labeling pattern indicated ^{3}H-dThd incorporation in the second S phase, an X2 cell was considered labeled if more than one-third of the chromosomes were labeled on both chromatids; when more than two-thirds of them were so labeled, the metaphase was considered heavily labeled. Again, an X3 cell was designated as labeled (or heavily labeled) if more than one-third (or two-thirds) of the chromosomes were labeled on the lightly stained chromatid.

Two hundred metaphases were scored blindly on coded slides for each point.

RESULTS AND DISCUSSION

Proliferative Kinetics in PHA-Stimulated Human Lymphocyte Cultures

Lymphocyte proliferation measured by a simple combination of autoradiography and sister chromatid differential staining. After stimulation of blood lymphocytes with PHA, the cultures soon contain different generations of cells, i.e., cells that have divided different numbers of times (4-5). In a previous study (27), a simple combination of autoradiography, to determine when a cell synthesized DNA, and sister chromatid differential staining, to determine how many times a cell has divided, was used to evaluate the effect of ^{3}H-dThd labeling on human lymphocyte proliferation, and to investigate the time when cells begin cycling after initiation of cultures. The results showed that labeling with more than 0.1 μCi/ml ^{3}H-dThd slows cell cycling, and that the older generation metaphase cells in culture commence their first DNA synthesis at earlier times after the start of incubation (27). The present study (29) further shows that the cycling lymphyocytes, whether they have begun cycling at earlier or later times after stimulation, have about the same generation times of 12 to 14 hr.

When cells from the first set of cultures, which were treated with ^{3}H-dThd between 24 and 32 hr, were examined at 4-hr intervals from 36 to 72 hr after initiation (Fig. 1), labeled metaphases were seen only in the fast-growing cells in each generation. Up to 40 hr after initiation, all dividing cells were in their first metaphase and all of them were labeled. At subsequent fixation times, the frequency of labeled X1 cells decreased quickly, the frequency of heavily labeled cells decreased even more quickly, and after 56 hr, no X1 cells were labeled. X2 cells first appeared in cultures at 44 hr; at that time all of them were labeled. The relative frequency of total X2 cells peaked at 60-64 hr, whereas the frequency of labeled X2 cells peaked at 52 hr. At 72 hr no X2 cells were labeled. The frequency of X_{3+} cells increased continuously with increasing culture times. The relative frequencies of both labeled and heavily labeled X_{3+} cells, however, peaked at 68 hr. At later sampling times, X4+ cells were occasionally observed: at 72 hr about 20% of metaphase cells were in their fourth division, almost all of them heavily labeled (data not shown).

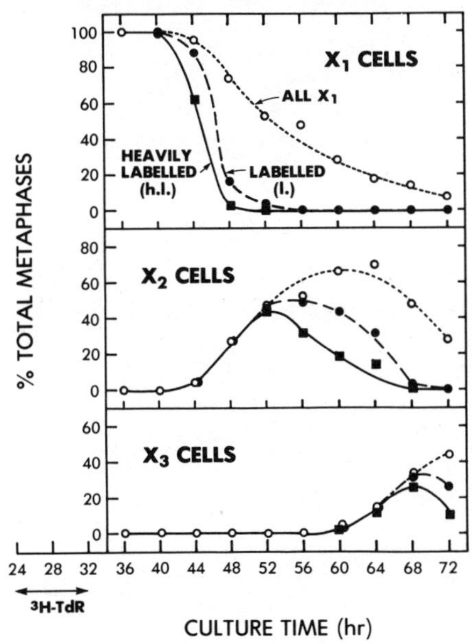

Fig. 1. Fate of proliferating lymphocytes after PHA stimulation. Cells were incubated continuously in medium containing BrdUrd and labeled with 0.1 μCi/ml ^{3}H-dThd at 24-32 hr. Relative percentages of total, labeled (l.), and heavily-labeled (h.l.) metaphases in X1, X2, and X3 were plotted as a function of culture time. (^{3}H-TdR ≅ ^{3}H-dThd in the test in this and subsequent figures.)

With these data, the mean generation times can be estimated by measuring the time intervals between the points when 50% of cells are heavily labeled in each generation of metaphases. 50% labeling was observed at 45 hr in X1 cells, at 57 hr in X2 cells, and at 70 hr in X3 cells (Fig. 1). The mean generation time was thus estimated to be approximately 12 hr for the first to second generation cell cycle and 13 hr for the second to third generation cycle.

When the labeling pattern in each generation of metaphases from the second set of cultures was examined (Fig. 2), it was noted that those cells that had entered their first S phase between 32 and 40 hr after initiation had basically the same cell division kinetics as did the cells that had entered their first S phase between 24 and 32 hr, except that the times were shifted by 8 hr. In the same way, the mean generation time in this subpopulation of cells was estimated to be 14 hr for both first to second, and second to third generation cycles.

Mitomycin C-induced delay in lymphocyte proliferation. To substantiate the above-stated findings on unaffected lymphocyte cultures, we examined the distribution of X1, X2, and X_{3+} cells from 72 hr cultures exposed to BrdUrd for various times before fixation

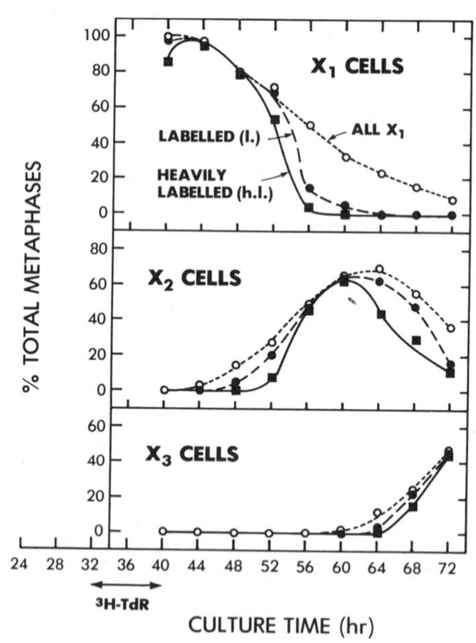

Fig. 2. Fate of proliferating lymphocytes after PHA stimulation. Cells were incubated continuously in medium containing BrdUrd and labeled with 0.1 μCi/ml ^{3}H-dThd at 32-40 hr. Relative percentages of total, labeled (l.), and heavily-labeled (h.l.) metaphases in X1, X2, and X3 were plotted as a function of culture time.

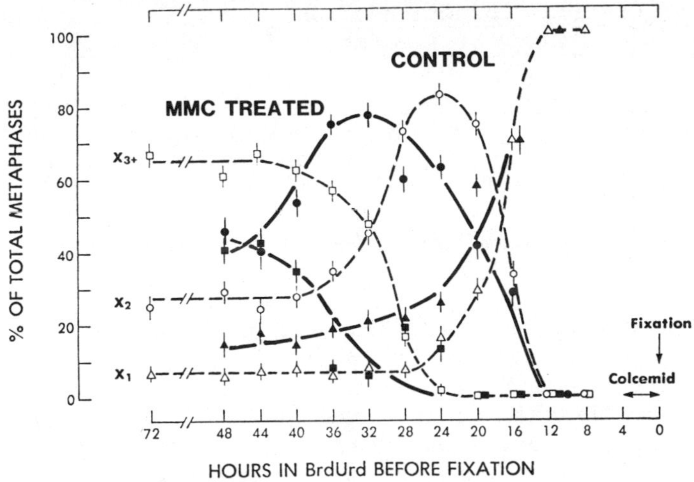

Fig. 3. Proportions of X1 (triangles), X2 circles), and X_{3+} (squares) metaphases in MMC-treated (solid symbols) and control (open symbols) cultures exposed to BrdUrd for different times before fixation. Cells were treated with MMC for 1 hr at 24 hr. All cultures were fixed at 72 hr after PHA stimulation.

(Fig. 3). It was found that the proportions of cells in different divisions changed with the duration of BrdUrd treatment. With 12 hr or shorter treatments the cultures contained exclusively X1 cells. With longer treatments, the relative frequency of X1 cells decreased continuously, whereas that of X2 cells increased, peaking at 24 hr, and then decreased. With treatments longer than 24 hr, X_{3+} cells appeared and their frequency increased continuously. The proportions of X1, X2, and X_{3+} cells, however, stayed constant with 40 hr or longer treatments with BrdUrd. These results confirm the finding (Figs. 1 and 2) that cycling lympocytes have about the same mean cell-cycle times of 12-14 hr whether they began their first DNA synthesis at earlier or later times after PHA stimulation.

In cultures exposed to 3 x 10^{-6} M MMC for 1 hr at G0/G1, however, it was found that the distributions of X1, X2, and X3+ cells changed in a basically similar way, but more slowly with the duration of BrdUrd treatment (Fig. 3, closed symbols), as compared to those in untreated control cultures (Fig. 3, open symbols): for instance, the relative frequency of X2 cells in MMC-treated cultures peaked at 32 hr whereas that in untreated cultures peaked at 24 hr. Because these second-division cells are mixed subpopulations of cells that have been in their first G_1 or the first half of their first S phase when BrdUrd was added, and because Colcemid was added for the last 4 hr of the cultures, the data indicate that the cells exposed to 3 x 10^{-6} M MMC for 1 hr at G_0 had elongated mean cell-cycle times of about 16-20 hr.

Dependency of Chemical Induction of SCEs on the Cell-Cycle Stages at Treatment

Induction of SCEs by MMC treatment in cells at the transient stages from G_0 to G_1. Although circulating blood lymphocytes are dormant, or in a resting G_0 stage, soon after PHA stimulation they undergo various biochemical and morphological changes (2,3), and show elevated concentrations of repair enzymes (24,25) and increased susceptibilities to chromosomal aberration formation by irradiation (30-31).

It is necessary for cells to pass through the S phase before the cellular lesions can result in the formation of SCEs (11). If induced cellular lesions are repaired completely before cells enter S, then those cells show no increase in SCE frequencies. In a previous study (12), I have shown that, the closer to the beginning of the first S phase cells had been pulse treated with benzene plus S-9 mix, which produce SCE-forming cellular lesions, the larger the increase in SCE frequency the cells showed.

In the present experiment, cells were pulse-treated with 3×10^{-6} M MMC for 1 hr at various times up to 32 hr after PHA stimulation of cultures. The results showed that the cells had almost linearly increasing frequencies of SCEs with increasing times of treatment after stimulation: treatment with 3×10^{-6} M MMC immediately before stimulation resulted in a net induction of about 25 SCEs per cell whereas the identical treatment at the 32nd hr led to an 80% larger induction of SCEs (data not shown, manuscript in preparation).

Induction of SCEs by chemical treatment at different times during the consecutive 2 cell cycles before sampling. When cells that had been pulse-treated with an equitoxic concentration of either MMC, EMS, or 4-nitroquinoline-1-oxide (4NQO) at various times through the 2 cell cycles before sampling were examined, it was found that the induction of SCEs was highly dependent on the cell cycle stage at treatment. In general, treatments at around the beginning of the S phase led to larger productions of SCEs than a treatment at the end of the S (Tab. 1; also see Fig. 3 for the cell kinetics in the exposed cultures). Especially, the SCE frequency decreased sharply as the time of treatment approached the end of the 2nd S phase. It also seems that treatment of cells with EMS or 4NQO, but not with MMC, at the first cell cycle resulted in a relatively higher induction of SCEs than treatment at the 2nd cycle, indicating that some portions of the lesions produced by EMS or 4NQO in G_1 or S of a cell cycle can persist and form SCEs in the next cell cycle.

In a previous study, Wolff (32) showed by scoring the ratio of single to twin SCEs in tetraploid Chinese hamster ovary cells (CHO) that no repair of EMS-induced lesions occurred but that partial repair of 4NQO-induced lesions could take place between the 2 cell

Tab. 1. Frequencies of SCEs in lymphocytes treated with MMC, EMS, or 4NQO at different times after stimulation of cultures.

Time of treatment (h after stimulation)	Chemical treatment (M)			
	0	MMC 3×10^{-6} M	EMS 1×10^{-2} M	4NQO 2×10^{-6} M
0	6.5 (0.6)[a]	24.1 (1.4)	10.7 (0.5)	7.9 (0.7)
12	7.6 (0.5)	30.6 (1.3)	12.9 (0.7)	9.0 (0.6)
24	7.0 (0.7)	29.9 (1.2)	17.0 (1.0)	7.9 (0.8)
36	7.6 (0.5)	46.8 (1.8)	22.5 (1.3)	12.0 (1.0)
39	---	38.1 (2.6)	18.5 (1.0)	9.5 (0.7)
42	---	37.2 (2.6)	17.9 (1.0)	11.2 (1.0)
45	---	31.0 (1.4)	16.9 (1.2)	9.9 (0.6)
48	7.9 (0.7)	38.3 (2.5)	10.5 (0.7)	10.2 (0.6)
51	---	32.3 (1.6)	13.7 (0.6)	8.9 (0.6)
54	---	37.8 (1.9)	15.1 (0.7)	9.9 (0.8)
57	---	25.1 (1.7)	14.5 (0.7)	10.7 (1.1)
60	7.0 (0.7)	18.9 (1.4)	14.9 (0.9)	10.8 (0.8)
63	---	16.6 (0.8)	13.3 (1.1)	12.1 (0.8)
66	---	7.8 (0.6)	11.3 (0.8)	9.4 (0.7)
69	6.9 (0.6)	6.8 (0.4)	8.3 (0.6)	6.8 (0.6)

[a]SEs in parentheses.

cycles. Natarajan (33) has also found that treatment with methylmethanesulfonate (MMS), another monoalkylating agent, at G_1 in the first cell cycle led to about twice the induction of SCEs than did the identical treatment in the second cycle. They interpreted these data as evidence for the complete absence of repair of MMS-induced lesions that can form SCEs. Our present data also showed that EMS treatment around the beginning of the S phase in the first cycle produced several times more SCEs as compared to the similar treatment in the second S, which agree well with the above-stated findings.

The number of SCE-forming lesions induced in a cell, however, decreases by 50% through each mitosis. Thus, even if there is no repair of induced lesions in the 2 cell cycles, the ratio of single to twin SCEs in tetraploid cells should be 1.5, and cells treated at the G_1/S boundary in the first cell cycle should show 1.5 times higher frequencies of net-induced SCEs than those identically treated in the second cycle before sampling. The previous and present results (Refs. 32,33; also see Tab. 1), however, show that the ratio of single to twin SCEs in MMS-treated tetraploid cells was more than 1.5, and that the EMS- or MMS-induced SCE frequencies were more than 1.5 times higher in cells treated in the first cell cycle than those in the second cycle. These data suggest several possibilities: (i) The combination of EMS- or MMS-induced lesions and incorporated BrdUrd as a template synergistically enhances the induction of SCEs; (ii) there is a selective loss against more severely damaged cells, or a differential cell-cycle delay between more and less severely damaged cells as the cells go through the first cell cycle and the following G_1 phase; (iii) new lesions that cause SCEs in the subsequent S phase are formed probably when cells synthesize DNA using EMS- or MMS-damaged DNA as template, and/or when cells repair EMS- or MMS-induced initial lesions. Because Natarajan et al. (33,34) have recently found that treatment of cells with MMS, but not MMC, resulted in larger inductions of SCEs when BrdUrd-incorporated DNA was used as template than when dThd-incorporated DNA was used. These data provide evidence favorable for the first possibility, but the latter two might also be likely at present.

Contrary to the effect of the monofunctional alkylating agents, EMS and MMS, the bifunctional alkylating agent, MMC, produces a quite different spectrum of lesions. It has been found that treatment of cells with MMC at the G_1/S in the first cell cycle induced about the same number of SCEs as did that in the second cell cycle (Ref. 33, see also Tab. 1). Littlefield et al. (35) have also shown the lack of expression of MMC-induced lesions as SCEs in cells that have undergone greater than 2 in vitro divisions after treatment. These data indicate that MMC-induced SCE-forming lesions are almost completely repaired before the cells enter the second post-treatment S phase, that there is a strong selection at mitosis against the cells that could not remove the lesions causing SCEs, and/or that there is almost no repair of MMC-induced SCE-forming lesions, and SCEs formed in the second cell cycle cancel out the SCEs formed in the first cell cycle.

Effect of MMC- and 4NQO-induced cell cycle delays on the SCE dose-response relationship. Treatments that produce larger numbers of SCEs also induce longer cell-cycle delays (5,12). Even in cultures exposed to the same dose of MMC, cells that had longer cell cycles showed higher SCE frequencies (Tab. 2). Because there was no significant difference between the baseline SCE frequencies in cells that had relatively longer cell cycles and those that had shorter cell cycles (Tab. 1), and because cells treated with 3×10^{-6} M MMC had, on average, 5 hr longer cell-cycle time than untreated cells (Fig. 3), these data indicate that the cells having longer cell cycle delays have higher frequencies of SCEs.

When cells pulse-treated with increasing concentrations of either MMC or 4NQO at the 0th, 24th, or 48th hr after PHA stimulation were examined, it was found that treatments at the different

Tab. 2. Effect of BrdUrd treatment time on the SCE frequency observed in human lymphocytes.

h in BrdUrd before fixation	Treatment	
	MMC[a]	Control
20	12.1 ± 0.5[b]	- - -
24	15.8 ± 0.7	9.3 ± 0.4
28	21.1 ± 1.1	- - -
32	27.0 ± 2.1	- - -
36	43.2 ± 3.7	9.2 ± 0.7
48	49.1 ± 2.5	- - -

[a] Cells were treated with 3×10^{-6} M MMC for 1 h at the 24th h after PHA stimulation.

[b] Mean ± S.E.

times gave different dose-response curves; the most efficient treatment time was 24 hr for MMC whereas it was 48 hr for 4NQO (Fig. 4a,b). It was also noted that the dose curves tend to decrease in slope at very high doses. This effect might be caused by differential sampling of less severely damaged cells; modestly or severely damaged cells entered their second mitosis later than 72 hr after stimulation because of very long cell-cycle delays induced.

To test this possibility, cells were treated with 3×10^{-5} M MMC, or 3×10^{-4} M 4NQO at 24 hr and then harvested at 72, 84, and 96 hr. I found that cells sampled at 84 hr had higher SCE frequencies than those at 72 hr, but that cells harvested 96 hr had lower frequencies than those at 84 hr. The latter finding of the decrease in SCE frequencies can be explained if later harvesting leads to a differential sampling of cells that have entered cycling at later times, and thus had a longer time for the repair of SCE-forming lesions. To further test this possibility, cells were pulse-treated with MMC or EMS at 24 hr and then fixed at intervals from 66 hr to 144 hr after stimulation. The results clearly showed that cells sampled at later times had lower frequencies of SCEs (Fig. 5).

Adaptation-Like Response of Human Lymphocytes to Alkylating Agents Measured by the SCE Frequency

The induction of resistance to MNNG-induced alkylation damage. It is now widely known that chronic exposure of *E. coli* to very low nontoxic doses of the monoalkylating agent, N-methyl-N'-nitro-N-nitrosoguanidine (MNNG), leads to the de novo synthesis of an adap-

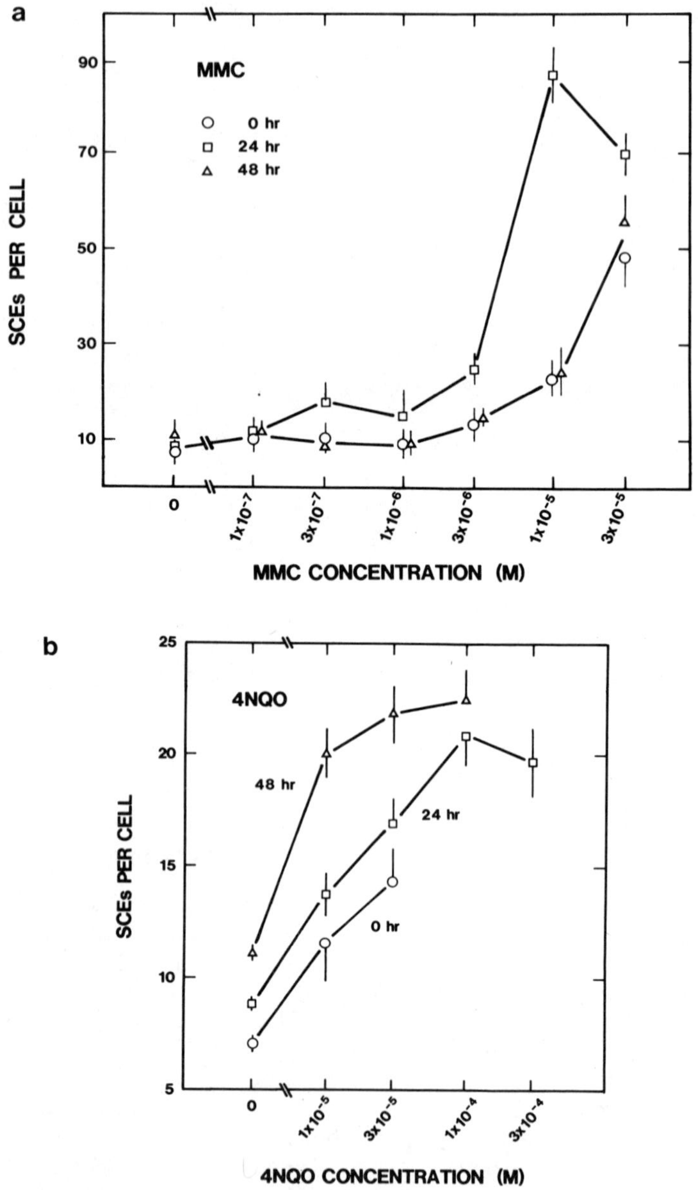

Fig. 4. Frequencies of SCEs in cells exposed to MMC (a), or 4NQO (b) at increasing concentrations for 1 hr at 0 hr (circles), 24 hr (squares), or 48 hr (triangles) after stimulation. All cultures were fixed at 72 hr after initiation of cultures.

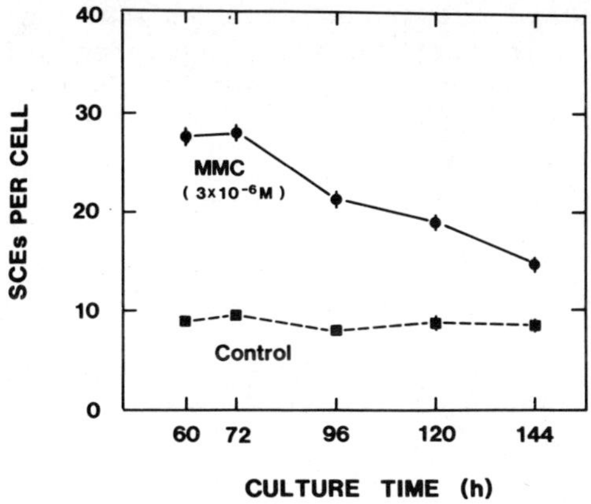

Fig. 5. MMC-induced SCE frequencies in cells fixed at different times after PHA stimulation. Cells were treated with MMC (3×10^{-6} M) for 1 hr immediately before stimulation of cultures, and exposed to BrdUrd for 44 hr before fixation.

tive repair pathway (15-17). There is also evidence for the existence of an adaptation-like repair pathway in Chinese hamster and human cell lines (18,36). To examine whether there is also an inducible resistance of human lymphocytes to the induction of SCEs by alkylating agents, cells were repeatedly treated with very low concentrations of MNNG once every 6 hr for 72 hr starting from 24 hr after stimulation, and subsequently challenged with a high concentration of MNNG. The results demonstrate that repeated pretreatments of cells with 10 ng/ml MNNG provided the strongest resistance to the induction of SCEs by a following challenge treatment with a high MNNG dose (Tab. 3).

Tab. 3. Effect of repeated pretreatment with MNNG of human lymphocytes on the SCE induction by the following challenge with MNNG (4 μg/ml).

MNNG	SCEs/cell		
dose (ng/ml)	Challenged MNNG(4μg/ml)	Control	Net induction
0	51.4 ± 2.2[a]	8.6 ± 1.8	42.8
5	42.8 ± 2.3	10.0 ± 0.5	32.8
10	32.7 ± 2.2	7.9 ± 0.9	24.8
25	45.8 ± 2.9	14.6 ± 0.9	31.2
50	50.6 ± 2.6	18.2 ± 1.1	32.4

[a] Mean ± S.E.

Effect of repeated pretreatments with MNNG on SCE induction by MMC, MNU, or 4NQO. To test whether "adapting" pretreatments with MNNG also make the lymphocytes resistant to the induction of SCEs by a challenge treatment with other aklylating agents, cells were repeatedly exposed to 10 ng/ml MNNG for 72 hr, and then pulse-challenged with MMC, ENU, or 4NQO. The data showed that the frequency of SCEs in the pretreated cells that had been challenged with ENU was less than half that in the nontreated cells whereas there was no difference between the SCE frequencies in the pretreated and nontreated cells which had been challenged with MMC. When challenged with 4NQO, the pretreated cells also seemed to have a decreased requency of SCEs compared to the nontreated cells, but the data are rathr limited at present (data not shown, manuscript in preparation).

Samson and Schwartz (18) have shown that repeated pretreatments with MNNG at a very low concentration rendered CHO and SV-40 transformed human fibroblasts resistant to the induction of SCEs by the subsequent challenge treatment with MNNG, MNU, or ENU, but not with ultraviolet light. The probable mechanism for the "adaptive" response is the induction of the repair enzymes, because Samson and Schwartz (36) have recently excluded the other three possibilities: (i) cells were becoming impermeable to, or somehow detoxifying, alkylating agents and therefore not actually incurring damage; (ii) adapting pretreatment might cause a redistribution of cells in the cell cycle, generating a population containing a larger than normal proportion of cells in a resistant phase of the cell cycle; (iii) pretreated cells somehow contrived to remain in the G_1 phase relatively longer than nontreated cells and gained extra time to repair damage before the commencement of DNA replication in the S phase. Several enzymes have been suggested strongly as adaptation repair enzymes, i.e., methyl- (ethyl-) transferase and glycosylase (16-18). Waldstein et al. (23) also showed an adaptive increase of O^6-methylguanine-acceptor protein, i.e., methyltransferase, in HeLa cells following MNNG treatment. Both methyltransferase and glycosylase activities have also been found in human lymphocytes (24,25). Thus, measurement of methyltransferase and glycosylase activities in pretreated cells might shed light on the precise mechanism(s) for the observed induction of resistance to SCE production in human lymphocytes.

SUMMARY

1. Although human lymphocyte cultures contain cells that have divided for different numbers of times after PHA stimulation, this heterogeneity of different cell divisions can be explained by a difference in the times when cells start their first DNA synthesis in responding to PHA. Cycling lymphocytes, whether they entered cycling earlier or later after stimulation, have about the same mean cell cycle

times of 12-14 hr. Treatment with 3 x 10^{-6} M MMC was found to induce an approximately 5-hr delay in the cell cycle.

2. The induction of SCEs by chemical treatment depends on the stage in the cell cycle at treatment and on the persistence of the induced SCE-forming lesions. The most efficient time of treatment is the G_1/S boundary in the first cell cycle of the 2 consecutive cycles before sampling. Among 3 alkylating agents tested here, EMS and 4NQO induce quite long-lived lesions that lead to SCE formation, whereas MMC-induced lesions seem to be completely removed within a cell cycle.

3. Treatments with increasing concentrations of the chemical induce a larger increase in SCE frequency and longer delays in the cell cycle. However, with a fixed experimental regimen, treatments with relatively higher doses cause a deformity of the dose-response relationship. The data also show that longer BrdUrd treatment before fixation results in a sampling of cells that have higher SCE frequencies.

4. Repeated pretreatments of human lymphocytes with a very low concentration of MNNG render them resistant to a following challenge with MNNG, or ENU, but not with MMC. The data at present indicate that there is a quite large variability among blood donors in this induction of resistance in lymphocytes.

REFERENCES

1. Evans, H.J., and D.C. Lyoid (1978) Mutagen-Induced Chromosome Damage in Man. Yale University Press, New Haven.
2. Wedner, H.J., and C.W. Parker (1976) Lymphocyte activation. Prog. Allergy 20:195-300.
3. Oppenheim, J.J., and D.L. Rosenstreich (1976) Signals regulating in vitro activation of lymphocytes. Prog. Allergy 20:65-194.
4. Tice, R.R., E.L. Schneider, and E.L. Rary (1976) The utilization of bromodeoxyuridine incorporation into DNA for the analysis of cellular kinetics. Exp. Cell Res. 102:232-236.
5. Morimoto, K., and S. Wolff (1980) Increase of sister chromatid exchanges and perturbations of cell division kinetics in human lymphocytes by benzene metabolites. Cancer Res. 40:1189-1193.
6. Tice, R., P. Thorne, and E.L. Schneider (1979) Bisack analysis of the phytohemagglutinin-induced proliferation of human peripheral lymphocytes. Cell Tissue Kinet. 12:1-9.
7. Crossen, P.E., and W.F. Morgan (1979) Proliferation of PHA-stimulated lymphocytes measured by combined antoradiography and

sister chromatid differential staining. Exp. Cell Res. 118: 423-427.

8. Craig-Holmes, A.P., and M.W. Shaw (1976) Cell cycle analysis in asynchronous cultures using the BUdR-Hoechst technique. Exp. Cell Res. 99:79-87.

9. Soren, L. (1973) Variability of the time at which PHA-stimulated lymphocytes initiate DNA synthesis. Exp. Cell Res. 78: 201-208.

10. Auf der Maur, P., and K. Berlincourt-Bohni (1979) Human lymphocyte cell cycle: Studies with the use of BrdUrd. Human Genet. 49:209-215.

11. Wolff, S., J. Bodycote, and R.B. Painter (1974) Sister chromatid exchanges induced in Chinese hamster cells by UV irradiation at different stages of the cell cycle: The necessity for cells to pass through S. Mutat. Res. 25:73-81.

12. Morimoto, K. (1983) Induction of sister chromatid exchanges and cell division delays in human lymphocytes by microsomal activation of benzene. Cancer Res. 43:1330-1334.

13. Miura, K., K. Morimoto, and A. Koizumi (1983) Proliferative kinetics and mitomycin C-induced chromosome damage in Fanconi's anemia lymphocytes. Human Genet. 63:19-23.

14. Craig-Holmes, A.P., and M.W. Shaw (1977) Effects of six carcinogens on SCE frequency and cell kinetics in cultured human lymphocytes. Mutat. Res. 46:375-384.

15. Samson, L., and J. Cairns (1977) A new pathway for DNA repair in *Escherichia coli*. Nature (Lond.) 267:281-283.

16. Robins, P., and J. Cairns (1979) Quantitation of the adaptive response to alkylating agents. Nature (Lond.) 280:74-76.

17. Karran, P., and T. Lindahl (1979) Adaptive response to alkylating agents involves alteration *in situ* of O^6-methylguanine residues in DNA. Nature (Lond.) 280:76-77.

18. Samson, L., and J.L. Schwartz (1980) Evidence for an adaptive DNA repair pathway in Chinese hamster ovary and human skin fibroblast cell lines. Nature (Lond.) 287:861-863.

19. Montesano, R., H. Bresil, and G.P. Margison (1979) Increased excision of O^6-methylguanine from rat liver DNA after chronic administration of dimethylnitrosoamine. Cancer Res. 39:1798-1802.

20. Montesano, R., H. Bresil, G.P. Plnch-Martel, and A.E. Pegg (1980) Effects of chronic treatment of rats with dimethylnitrosoamine on the removal of O^6-methylguanine from DNA. Cancer Res. 40:452-458.

21. Bridges, B., and A. Lehmann (1982) Inducible responses to DNA damage. Nature (Lond.) 298:118.

22. Evensen, G., and E. Seeberg (1982) Adaptation of alkylation resistance involves the induction of a DNA glycosylase. Nature (Lond.) 296:773:775.

23. Waldstein, E.A., E-H. Cao, and R.B. Setlow (1982) Adaptive increase of O^6-methylguanine-acceptor protein in HeLa cells following N-methyl-N'-nitro-N-nitrosoguanidine treatment. Nucleic Acids Res. 10:4595-4604.

24. Sirover, M.A. (1979) Induction of the DNA repair enzyme uracil-DNA glycosylase in stimulated human lymphocytes. Cancer Res. 39:2090-2095.
25. Waldstein, E.A., E-H. Cao, M.A. Bender, and R.B. Setlow (1982) Abilities of extracts of human lymphocytes to remove O^6-methylguanine from DNA. Mutat. Res. 95:405-416.
26. Goto, K., Y. Maeda, Y. Kano, and T. Sugiyama (1978) Factors involved in differential Giemsa-staining of sister chromatids. Chromosoma 66:351-359.
27. Morimoto, K., and S. Wolff (1980) Cell cycle kinetics in human lymphocyte cultures. Nature (Lond.) 288:604-606.
28. Rutledge, M.H. (1979) A simple procedure for obtaining autoradiographs of G-banded chromosomes. Chromosoma 70: 259-262.
29. Morimoto, K., M. Sato, and A. Koizumi (1983) Proliferative kinetics of human lymphocytes in culture measured by autoradiography and sister chromatid differential staining. Exp. Cell. Res. 145:349-356.
30. Sasaki, M.S. (1978) Radiation damage and its repair in the formation of chromosome aberrations in human lymphocytes. In Mutagen-Induced Chromosome Damage in Man, H.J. Evans and D.C. Lloyd, eds. Yale University Press, New Haven, pp. 62-76.
31. Wolff, S. (1972) The repair of X-ray-induced chromosome aberrations in stimulated and unstimulated human lymphocytes. Mutat. Res. 15:435-444.
32. Wolff, S. (1978) Chromosomal effects of mutagenic carcinogens and the nature of lesions leading to sister chromatid exchange. In Mutagen-Induced Chromosome Damage in Man, H.J. Evans and D.C. Lloyd, eds., Yale University Press, New Haven, pp. 208-215.
33. Natarajan, A.T., A.D. Tates, M. Meijers, I. Neuteboom, and N. de Vogel (1983) Induction of sister chromatid exchanges (SCEs) and chromosomal aberrations by mitomycin C and methyl methanesulfonate in Chinese hamster ovary cells. An evaluation of methodology for detection of SCEs and of persistent DNA lesions towards the frequencies of observed SCEs. Mutat. Res. 121:211-223.
34. Natarajan, A.T., and G. Obe (1982) Mutagenicity testing with cultured mammalian cells: Cytogenic assays. In Mutagenicity --New Horizons in Genetic Toxicology, J.A. Heddle, ed. Academic Press, New York, pp. 551-559.
35. Littlefield, L.G., S.P. Colayer, and R.J. DuFrain (1983) SCE evaluations in human lymphocytes after G_0 exposure to mitomycin C: Lack of expression MMC induced SCEs in cells that have undergone greater than two in vitro divisions. Mutat. Res. 107:119-130.
36. Samson, L., and J.L. Schwartz (1983) The induction of resistance to alkylation damage in mammalian cells. (submitted for publication)

IMPORTANCE OF THE MITOGEN IN SISTER CHROMATID EXCHANGE STUDIES

Ghislain Deknudt

Laboratory of Mammalian Genetics
Department of Radiobiology
S.C.K.-C.E.N.
B-2400 Mol, Belgium

INTRODUCTION

The occurrence of homologous exchanges between sister chromatids (SCEs) was first demonstrated by Taylor (1) in autoradiographic studies on plant chromosomes labeled with tritiated thymidine. More recently, the pioneering work of Latt (2) on the substitution of thymidine by 5-bromo-2'-deoxyuridine (BrdUrd) and the application of the bis-benzimidazole fluorochrome Hoechst 33258, as well as the simultaneous development of the fluorescence-plus-Giesma (FPG) techniques by Kim (3), Korenberg and Freedlender (4), and Perry and Wolff (5), has provided differentially stained cytological preparations with a far greater resolution for SCE visualization than that obtained with autoradiography. These more sensitive techniques have led to an intensive proliferation of research in this new area of cytogenetics.

Although neither the exact mechanism nor the biological significance of SCE formation is fully understood, most mutagens and/or carcinogens have been shown to increase the frequency of SCEs in animal cells in vivo as well as in vitro, either directly or after metabolic activation (6). These increases in SCE frequency occur at agent concentrations which are noncytotoxic and relatively nonclastogenic, i.e., at dose levels far below those required for chromosomal aberration production. Unlike chromosomal aberrations, SCEs are easy to identify and occur at much higher frequencies; therefore, fewer cells are needed to be scored for chemical testing purposes. The utility of the SCE assay for human population studies is potentially limited by the variability associated with individual baseline SCE frequencies. Whether spontaneous or BrdUrd-induced, the SCE baseline frequency can be influenced, particularly in vitro, by different factors such as BrdUrd concentration (7-11), type of tissue culture medium (12-14), amino acid composition (15), source and concentration of serum (16-18), cell density (8,19,20), temperature (21-24) and length (25) of cell incubation, duration of the cell cycle (21,26), the harvesting time (27,28), or the method of differential staining (13).

Inherent, however, to the lymphocyte culture technique is the need of a mitogenic agent, which may also affect the SCE frequency. Since the discovery by Nowell in 1960 (29) of the mitogenic capacity of phytohemagglutinin (PHA), other plant extracts such as concanavalin A (Con A), Wistaria floribunda (WFA), and Lens culinaris (LcH-A) have also been found to stimulate primarily thymus-derived (T) lymphocytes (30) in the peripheral blood to divide in culture. Stimulation of lymphocytes by these different plant mitogens have revealed statistically significant variations in mitotic index (MI), cell kinetic response, and SCE frequency, but not in differential radiosensitivity (31-33).

In contrast to the numerous reports published on the induction of SCEs by chemical mutagens in proliferating human lymphocytes stimulated by PHA (for example, Refs. 32,34-39), only a few studies have been carried out on the induction of SCEs in lymphocytes chemically exposed prior to PHA activation (35,40-44). Little information is, therefore, available on a possible relationship between the rate of induced SCEs and the stage of the cell cycle at treatment and the nature of the mitogenic agent.

To gain additional information on this subject, lymphocytes were exposed to mitomycin C (MMC) alone, or to cyclophosphamide (CP) in the presence of S-9 mix during G_0 prior to mitogenic activation and compared to a lymphocyte exposure 48 hr after initiation and stimulated by PHA, Con A, WFA, or LcH-A extracts.

MATERIALS AND METHODS

Chemicals

Both chemical mutagens, MMC per se and CP after metabolic activation, are known as potent inducers of SCEs in proliferating mammalian cells exposed in the presence of BrdUrd (45).

Mitomycin C, an antitumor antibiotic isolated from *Streptomyces caespitosus* (46), behaves chemically as a bifunctional alkylating agent, and interacts with DNA by cross-linking complementary strands (47). Although an enzymatic reduction of MMC is required for its activity, most cell types, including lymphocytes, are capable of reducing MMC (48).

Cyclophosphamide, synthesized in 1958 by Arnold and Bourseaux (49), is in itself ineffective as a cytostatic or clastogenic drug but, after in vivo metabolic activation, becomes a very active bifunctional alkylating intermediate (50).

For both compounds, the doses of 10^{-8} M for MMC (Sigma) and 10^{-4} M for CP (Endoxan, Asta) used in our experiments are based on data from the literature, reporting to produce a significant increase of SCEs in PHA-activated lymphocytes (51).

S-9 Mix

In an attempt to incorporate the activation aspect of the mammalian metabolism into an in vitro test for SCEs, microsomes from homogenates of rat liver were added to the culture system together with a reduced nicotinamide adenine dinucleotide phosphated (NADPH)-regenerating system (S-9 mix) (52). The S-9 was prepared as described by Ames (52) from Aroclor-1254-induced rats (250-300 gm) and was obtained from the supernatant after centrifugation at 9000 X g. The S-9 mix was prepared immediately before use according to the protocol of Stetka and Wolff (53) and contained 10% S-9 rat liver extract, 5 mM NADP, 5 mM glucose 6-phosphate, 8 mM $MgCl_2$, 33 mM KCl, and 0.1 M Na_2HPO_4 - NaH_2PO_4. The S-9 mix was then diluted by a factor of 20 (1/20 strength) in mitogen-free medium.

Mitogenic Agents

The 4 mitogenic agents under investigation in our studies are sugar-binding proteins of plant origin. The mitogenic properties of PHA extracted from the red kidney bean (Phaseolus vulgaris), Con A isolated from the Jack bean (Canavalia ensiformis), LcH-A from Lens culinaris and the Wistaria floribunda extract WFA, were first described by Nowell (29), Powell and Leon (54) Toyoshima et al. (55), and Barker and Farnes (56), respectively. These mitogenic lectins can be divided into 2 groups, based on their sugar-binding specificity, which enables them to bind to specific glycoproteins on the cell surface of the lymphocytes. Concanavalin A and LcH-A comprise one group which bind α-D-glucosyl or α-D-mannosyl residues, while PHA and WFA comprise a second group, which bind N-acetyl-D-galactosamine (57). These plant mitogens can also be divided into two subgroups depending upon their lectin valency. Wistaria floribunda (58) and LcH-A (59) have 2 sugar-binding sites per molecule whereas PHA (60) as well as Con A (61) possess 4 sugar-binding sites per molecule.

Cell Culture

Peripheral blood samples were taken by venipuncture from 4 hematologically normal and healthy adult male individuals with no recent history of significant illness, medication, or X-ray exposure. These subjects were also free of any drug, alcohol, or tobacco habit and had not been knowingly exposed to environmental or occupational hazards. From these donors, lymphocyte cultures were prepared, 1 hr after sampling, by adding 0.5 ml whole blood to 5 ml Ham's F10 (62) medium supplemented with 10% bovine serum (Difco) and penicillin-streptomycin (Gibco), to which either 100 μg/ml PHA (Wellcome), 50 μg/ml Con A (P-L Biochemicals), 50 μg/ml WFA (P-L Biochemicals), or 40 μg/ml LcH-A (P-L Biochemicals) extracts had been added. These optimal stimulating concentrations, used in our studies, had been determined in previous experiments (31,33). For each donor, all cultures were set up in the same medium to which the 4 different mitogens were added and these cultures were run simultaneously to avoid differences in medium or cell composition. BrdUrd (Sigma), at a concentration of 10 μg/ml, was present in all cultures for the whole culture duration of 72 hr. In order to avoid photolysis of the BrdUrd-substituted DNA, all cultures were kept in the dark (63).

Experimental Procedure

Experiment I: Treatment 48 hr after stimulation. Forty-eight hr after initiation, cells in all cultures from the first 2 donors were collected by centrifugation and resuspended in mitogen-free medium. Half of the cultures were then exposed to MMC alone or to CP in the presence of S-9 mix for 1 hr at 37°C; the remainder serving as controls. Thereafter, all cultures were washed 2 times with mitogen-free medium and reincubated for an additional 24-hr period in fresh media containing the corresponding appropriate mitogen and BrdUrd.

Experiment II: Treatment before stimulation. Whole blood (0.5 ml) from the last 2 donors was exposed to MMC alone or to CP with metabolic activation for 1 hr, just before stimulation in 5 ml mitogen-free medium at 37°C. After exposure, the cells were rinsed twice in mitogen-free medium to remove residual chemical and stimulated by the 4 different mitogens for 72 hr in the presence of BrdUrd. All corresponding control cultures were treated in the same way.

Harvesting and Staining Procedure

Metaphases were collected by adding 1 ml colchicine (Merck, 10^{-5} M) to each culture 3 hr prior fixation. The cells were then harvested by centrifugation, treated with hypotonic KCl solution (0.075 M) for 8 min and fixed with 3 changes of 3:1 methanol-acetic acid. Air-dried slides were stained with Hoechst 33258 in phosphate buffered saline (PBS) for 10 min, rinsed with tapwater, mounted in the same buffer with a coverglass and exposed for 24 hr to artificial light to allow the photochemical reaction to take place. After removal of the coverslips, the slides were washed in tapwater, treated during 2 hr with concentrated standard saline (SSC) at 65°C, rinsed with tapwater, stained with Giemsa (5% in SSC) for 15 min, and allowed to dry before examination. The whole procedure is known as the fluorescence-plus-Giesma (FPG) method of Perry and Wolff (5).

Scoring and Statistical Analysis

For SCE analysis, metaphases in their second in vitro division were selected on the basis of cell spreading, chromatid differentiation and chromosome number. For each culture, 25 well-spread and clearly differentiated metaphases with the full chromosome complement were scored. The MI, as well as the cell cycle kinetic analysis, were carried out on the same slides used for SCE evaluation. The MI was determined for all control cultures and was based on the number of dividing cells per 1,000 activated cells. For cell-cycle scoring, 100 cells in division were examined in all control cultures and in the treated cultures in the G_0 exposure experiment. For statistical evaluation of the experimental data, the chi-square (χ^2) test was used for the MI and for the cell kinetic comparisons, whereas for the SCE frequencies, the two-level nested ANOVA test was applied.

RESULTS AND DISCUSSION

Mitotic Indices

Table 1 groups the mitotic indices observed in the 72 hr control cultures from the 4 donors examined in both experiments. Lens culinaris activation resulted in the lowest MI (around 4.5%) and PHA the highest (about 8%), whereas an intermediate MI, with some variation among donors, was observed for Con A and WFA stimulation. Statistical analysis of these data revealed significant χ^2 values among the cultures activated by PHA and LcH-A (χ^2 = 12.24 ; P = 0.008), Con A and LcH-A (χ^2 = 4.66 ; P = 0.03), and WFA and LcH-A (χ^2 = 6.73; P = 0.009) but not between PHA and Con A, PHA and WFA, or Con A and WFA activated cultures. Comparable MI values were recorded in a previous study (31) in which lymphocytes from 4 other donors were stimulated from 42 to 54 hr by PHA, WFA, or LcH-A extracts without

Tab. 1. Mitotic index in 72 hr control cultures stimulated by PHA, Con A, WFA, or LcH-A (1,000 cells counted per point).

Stimulating agent	*Donor I*	*Donor II*	*Donor III*	*Donor IV*	*Mean*
PHA	*8.3*	*7.8*	*7.8*	*8.9*	*8.2*
Con A	*7.0*	*5.2*	*5.5*	*8.8*	*6.6*
WFA	*6.9*	*5.5*	*7.8*	*8.2*	*7.1*
LcH-A	*4.9*	*4.2*	*3.6*	*4.8*	*4.4*

Experiment I (Donors I and II)
Experiment II (Donors III and IV)

culture interruption or cell washing processes. In that experiment, the percentage of cells in mitosis at different fixation times never exceeded 5.5% for LcH-A-stimulated cells and generally was far below 5%; it was about 6.5% in the WFA-activated cultures and around 7% in the PHA-stimulated ones. From this, it appears that neither a culture interruption by discarding the medium 48 hr after initiation and incubation for 1 hr at 37°C in mitogen-free medium followed by 2 changes of mitogen-free medium and an additional 24 hr stimulation with the appropriate mitogen (Tab. 1, donors I and II) nor cell preincubation in mitogen-free medium for 1 hr at 37°C and washing (2 times) followed by a normal 72 hr culture procedure (Tab. 1, donors III and IV) had altered the mitotic response for a given mitogen. Our results are also comparable to those obtained by Wolff and Arutyunyan (64), who reported a MI of 10.9% in 72 hr PHA-activated control cultures, but are in sharp contrast with those recorded in untreated human lymphocytes stimulated by PHA for 48 hr (4.55 and 0.8) by Crossen and Morgan (65) and Morimoto and Wolff (39), respectively, or for 72 hr (2.95, 3.3, and 4.0%) by Crossen and Morgan (65), Morimoto and Wolff (39), and Hatcher et al. (66), respectively. Crossen and Morgan (65) ascribed that drastic drop in MI between the 48 and 72 hr cultures (4.6 versus 2.48%) to a toxic effect of BrdUrd, relying on the contention of Chaganti et al. (67) that BrdUrd treatment caused a reduction in the number of metaphases. However, a number of other factors such as the origin and concentration of colchicine, serum, and PHA, variation in the number of inoculated cells, or individual differences in cell response can also be invoked to explain these discrepancies in MI.

Cell Kinetics

The recently developed FPG technique which facilitates the detection of SCEs also allows the study of cell cycle kinetics in a far simpler and more reliable way than tritiated thymidine autoradiography. Using this staining technique, Tab. 2 reveals that, regardless of the mitogen or donor used, cells in their first, second, third, or subsequent division were simultaneously present in all untreated lymphocyte cultures harvested at 72 hr. These data indicate that the human lymphocyte population contains fast and slow cycling cells. In a previous study (31), we demonstrated that third-generation cells occurred at 54 hr, following human lymphocyte stimulation by PHA, WFA, or LcH-A extracts. Confirming our data, other investigators have observed third-generation cells in PHA-stimulated cultures at a harvest time of 48 hr (38) or 60 hr (64,68). As shown in Tab. 2, differences in the proportion of cells in division were, however, observed between the donors of both experiments. In the first study (donor I and II) most cells were found to be in their third cell cycle, while in the second experiment (donors III and IV), the majority of cells had only undergone 2 mitotic divisions. Comparable results on cell cycle kinetics were observed in 72 hr PHA-activated human control cultures by others (21,34,39,64,65,68-70) but contrast with those of Bianchi et al. (12) and Giulotto et

Tab. 2. Frequency of cells in first (1st), second (2nd), or subsequent (3rd+) division in 72 hr control cultures stimulated by PHA, Con A, WFA, or LcH-A (100 cells examined per point).

Stimulating agent	*Donor I*			*Donor II*			*Donor III*			*Donor IV*			*Mean*		
	1st	*2nd*	*3rd+*	*1st*	*2nd*	*3rd+*	*1st*	*2nd*	*3rd+*	*1st*	*2nd*	*3rd+*	*1st*	*2nd*	*3rd+*
PHA	14	27	59	14	19	67	33	53	14	7	57	36	17.00	39.00	44.00
Con A	18	41	41	19	38	43	21	44	35	28	44	28	21.50	41.75	36.75
WFA	17	31	52	14	43	43	20	31	49	21	49	30	18.00	38.50	43.50
LcH-A	24	30	46	36	27	37	41	33	26	47	32	21	37.00	30.50	32.50

Experiment I (Donors I and II)

Experiment II (Donors III and IV)

al. (28) where most of the cells were observed in their first cell cycle. Morimoto and Wolff (39) claimed that cell-division kinetics were not perturbed by washing the cells and replacing the medium; therefore, we believe that the differences we observed can probably be ascribed to minor changes in culture conditions or to variable responses of lymphocytes from different individuals to the stimulating agent.

Evidence for a considerable variation among donors in the proportions of first-, second-, and third-generation metaphases in 72 hr PHA- activated control cultures was offered by Crossen and Morgan (65). These authors, analyzing cell-cycle kinetics in 30 donors, have observed individual variations in first, second, or third cell divisions ranging from 13-97%, 3-83%, or 0-28%, respectively. Statistical differences in the frequency of first-generation cells in the pooled results from the 4 different donors occurred between PHA- and LcH-A-activated (χ^2 = 10.17; P = 0.002), Con A- and LcH-A-activated (χ^2 = 5.9; P = 0.015), and WFA- and LcH-A-activated (χ^2 = 9.08; P = 0.03) cells but not between the PHA and Con A, PHA and WFA, or Con A and WFA ones, reflecting the results of the MI analysis. The lowest frequency of cells in first division was observed in PHA-activated cells and the highest frequency in LcH-A-stimulated cultures, confirming our previous findings (31,33) in which the frequency of cells in first division at the conventional culture time of 48 hr varied around 75%, 87%, 85%, or 95% for PHA-, Con A-, WFA-, or LcH-A-activated cultures, respectively.

An exposure to MMC alone (Tab. 3) or to CP in the presence of S-9 mix (Tab. 4) for 1 hr prior to stimulation by the 4 different mitogens resulted in a significant increase ($P < 0.01$) in the proportion of first-generation metaphases and in a decrease of third-generation divisions compared to the corresponding stages in control cultures. These results indicate that a chemical exposure caused a delay in the cell cycle proliferation resulting in a significant lengthening of cell cycle duration as already observed for a variety of chemicals (10,26,68,70,71). In our experiment (Tabs. 3 and 4), most cells exposed to MMC or CP were found in their first cell cycle for Con A-, WFA-, and LcH-A-activated cultures while, in the case of PHA stimulation, most of them reached the second cell division.

SCEs

Table 5 provides the average frequency of SCEs in second division metaphases from the 72 hr untreated lymphocyte cultures of 4 donors stimulated by 4 different mitogens. Dependent on the donor, the mean SCE frequency per cell varied between 4.60 and 7.04, 5.44 and 9.04, 6.60 and 9.48, or 6.00 and 9.76 for the PHA-, Con A-, WFA-, or LcH-A-activated lymphocytes, respectively. Because no statistically significant differences ($F = 0.36$; $P > 0.25$) in these SCE frequencies were recorded between the different stimulating agents, we can conclude that neither a culture interruption 48 hr after

Tab. 3. Frequency of cells in first (1st), second (2nd), or third and subsequent (3rd+) division in 72 hr controls and in cultures treated with MMC before stimulation by PHA, Con A, WFA, or LcH-A (100 cells analyzed per point).

Stimulating agent	Controls									MMC								
	Donor I			Donor II			Mean			Donor I			Donor II			Mean		
	1st	2nd	3rd+	1st	2nd	3rd+	1st	2nd	3rd+	1st	2nd	3rd+	1st	2nd	3rd+	1st	2nd	3rd+
PHA	33	53	14	7	57	36	17.00	39.00	44.00	36	57	7	31	60	9	33.50	58.50	8.00
Con A	21	44	35	28	44	28	21.50	41.75	36.75	59	32	9	63	35	2	61.00	33.50	5.50
WFA	20	31	49	21	49	30	18.00	38.50	43.50	56	42	2	56	39	5	56.00	40.50	3.50
LcH-A	41	33	26	47	32	21	37.00	30.50	32.50	48	39	13	60	35	5	54.00	37.00	9.00

Tab. 4. Frequency of cells in first (1st), second (2nd), or third and subsequent (3rd+) division in 72 hr controls and in cultures treated with CP before stimulation by PHA, Con A, WFA, or LcH-A (100 cells analyzed per point).

Stimulating agent	Controls									CP								
	Donor I			Donor II			Mean			Donor I			Donor II			Mean		
	1st	2nd	3rd+	1st	2nd	3rd+	1st	2nd	3rd+	1st	2nd	3rd+	1st	2nd	3rd+	1st	2nd	3rd+
PHA	33	53	14	7	57	36	17.00	39.00	44.00	35	52	13	38	59	3	36.50	55.50	8.00
Con A	21	44	35	28	44	28	21.50	41.75	36.75	66	29	5	72	27	1	69.00	28.00	3.00
WFA	20	31	49	21	49	30	18.00	38.50	43.50	59	38	3	54	44	2	56.50	41.00	2.50
LcH-A	41	33	26	47	32	21	37.00	30.50	32.50	54	34	12	56	38	6	55.00	36.00	9.00

Tab. 5. Frequency of SCEs/cell (mean ± S.E.) in control cultures stimulated by PHA, Con A, WFA, or LcH-A (25 cells in second mitosis examined per point).

Stimulating agent	*Donor I*	*Donor II*	*Donor III*	*Donor IV*	*Mean*
PHA	*6.48 ± 0.65*	*6.52 ± 0.50*	*4.60 ± 0.49*	*7.04 ± 0.69*	*6.16 ± 0.58*
Con A	*9.04 ± 0.85*	*7.40 ± 0.45*	*5.44 ± 0.40*	*8.48 ± 0.72*	*7.59 ± 0.60*
WFA	*8.52 ± 0.68*	*9.48 ± 0.59*	*6.60 ± 0.62*	*8.28 ± 1.04*	*8.22 ± 0.73*
LcH-A	*9.76 ± 0.67*	*7.44 ± 0.52*	*6.00 ± 0.60*	*8.96 ± 0.69*	*8.04 ± 0.62*

Experiment I (Donors I and II)

Experiment II (Donors III and IV)

initiation and cell washing (Experiment I), or cell preincubation in mitogen-free medium and washing (Experiment II), nor the proliferation properties of lymphocytes activated in culture by these different mitogens have had any influence on the SCE baseline level. Regardless of the mitogen, our SCE background levels correspond to those normally reported for PHA-activated control cultures, ranging from 4.02 (37) to 22.2 (10). Tables 6-9 summarize the results of the chemically induced SCEs from both experiments. The outcome of cell exposure prior to initiation or 48 hr after stimulation confirmed that both alkylators are capable of inducing DNA lesions during all stages of the cell cycle (36,44,72), lesions that can subsequently be expressed as SCEs after culturing for 2 subsequent cell cycles (35,73-75). Likewise, additional evidence for the existence of long-lived DNA lesions induced by various alkylating agents in G_0 lymphocytes which can elicit an SCE response during in vitro culturing, comes from experimental in vivo studies on laboratory animals or from therapeutical treatments in man (41,53,76,77).

An in vitro exposure of human lymphocytes for 1 hr to 10^{-8} M MMC (Tab. 6) or to 10^{-4} M CP (Tab. 7) in the presence of rat-liver microsomes (S-9 mix) 48 hr after stimulation by 4 different mitogens followed by an additional 24 hr incubation in medium containing the appropriate mitogen resulted in a significant increase ($P < 0.05$) in the SCE level compared to that of the control values. It also appears from these experiments that, between cell exposure to chemicals 48 hr after initiation and SCE examination in second-division cells 24 hr later, human lymphocytes had completed, regardless of the mitogen used, 2 rounds of DNA synthesis in the presence of BrdUrd. These findings are in agreement with the statement by Morimoto and Wolff (39) that in PHA-responsive cells the mean generation time

Tab. 6. Frequency of SCEs/cell (mean ± S.E.) in controls and in cultures treated with MMC 48 hr after initiation and stimulated by PHA, Con A, WFA, or LcH-A (25 cells in second mitosis examined per point).

Stimulating agent	Controls Donor I	Controls Donor II	Controls Mean	MMC Donor I	MMC Donor II	MMC Mean
PHA	6.48 ± 0.65	6.52 ± 0.50	6.50 ± 0.57	22.04 ± 1.46	26.27 ± 0.92	24.16 ± 1.19
Con A	9.04 ± 0.85	7.40 ± 0.45	8.22 ± 0.65	37.00 ± 2.01	32.60 ± 1.68	34.80 ± 1.84
WFA	8.52 ± 0.68	9.48 ± 0.59	9.00 ± 0.63	38.16 ± 1.78	40.28 ± 1.77	39.22 ± 1.77
LcH-A	9.76 ± 0.67	7.44 ± 0.52	8.60 ± 0.60	35.56 ± 2.22	37.68 ± 1.60	36.62 ± 1.91

Tab. 7. Frequency of SCEs/cell (mean ± S.E.) in controls and in cultures treated with CP 48 hr after initiation and stimulated by PHA, Con A, WFA, or LcH-A (25 cells in second mitosis examined per point).

Stimulating agent	Controls Donor I	Controls Donor II	Controls Mean	CP Donor I	CP Donor II	CP Mean
PHA	6.48 ± 0.65	6.52 ± 0.50	6.50 ± 0.57	17.12 ± 2.18	32.04 ± 3.17	24.58 ± 2.67
Con A	9.04 ± 0.85	7.40 ± 0.45	8.22 ± 0.65	41.52 ± 3.96	47.48 ± 3.16	44.50 ± 3.56
WFA	8.52 ± 0.68	9.48 ± 0.59	9.00 ± 0.63	42.04 ± 3.20	51.32 ± 2.80	46.68 ± 3.00
LcH-A	9.76 ± 0.67	7.44 ± 0.52	8.60 ± 0.60	58.16 ± 5.24	58.20 ± 3.76	58.18 ± 4.50

Tab. 8. Frequency of SCEs/cell (mean ± S.E.) in controls and cultures treated with MMC before stimulation by PHA, Con A, WFA, and LcH-A (25 cells in second mitosis examined per point).

Stimulating agent	Controls			MMC		
	Donor I	Donor II	Mean	Donor I	Donor II	Mean
PHA	4.60 ± 0.49	7.04 ± 0.69	5.82 ± 0.59	17.44 ± 0.88	19.04 ± 1.40	18.24 ± 1.14
Con A	5.44 ± 0.40	8.48 ± 0.72	6.96 ± 0.56	19.40 ± 0.80	18.84 ± 1.10	19.12 ± 0.95
WFA	6.60 ± 0.62	8.28 ± 1.04	7.44 ± 0.83	16.60 ± 0.96	18.44 ± 1.09	17.52 ± 1.02
LcH-A	6.00 ± 0.60	8.96 ± 0.69	7.48 ± 0.64	17.56 ± 0.78	20.76 ± 1.28	19.16 ± 1.03

Tab. 9. Frequency of SCEs/cell (mean ± S.E.) in controls and cultures treated with CP before stimulation by PHA, Con A, WFA, or LcH-A (25 cells in second mitosis examined per point).

Stimulating agent	Controls			CP		
	Donor I	Donor II	Mean	Donor I	Donor II	Mean
PHA	4.60 ± 0.49	7.04 ± 0.69	5.82 ± 0.59	18.84 ± 0.95	19.76 ± 1.06	19.30 ± 1.00
Con A	5.44 ± 0.40	8.48 ± 0.72	6.96 ± 0.56	17.40 ± 0.97	24.20 ± 1.10	20.80 ± 1.03
WFA	6.60 ± 0.62	8.28 ± 1.04	7.44 ± 0.83	17.28 ± 0.96	22.92 ± 0.99	20.10 ± 0.97
LcH-A	6.00 ± 0.60	8.96 ± 0.69	7.48 ± 0.64	16.08 ± 0.70	23.84 ± 1.30	19.96 ± 1.00

corresponds roughly to a 12-hr period interval. Nevertheless, various responses in SCE levels were observed in cells exposed to both chemicals 48 hr after stimulation by different mitogens. In general, MMC exposure caused a 4-fold increase in SCEs in PHA- and Con A-activated cells and a 4- to 5-fold augmentation in WFA- and LcH-A-stimulated cells whereas exposure to CP after metabolic activation induced a 4-fold, 5- to 6-fold or even 7-fold increase in the SCE frequencies for PHA-, Con A-, and WFA- or LcH-A-activated lymphocytes, respectively. On the contrary, chemical exposure to both agents at the same concentrations for 1 hr in the G_0 resting stage prior to stimulation by the 4 different mitogens resulted in an approximately 3-fold increase in SCE response as compared to controls, irrespective of donor, mutagen, or mitogen used. An interesting comparable effect had been observed by White and Hesketh (40), in which human lymphocytes exposed to CP and S-9 mix at the beginning of the cell culture (0 hr) or after 48 hr resulted in different SCE responses, with the later treatment being more effective. An exposure for 1 hr to 10^{-5} M CP in the presence of S-9 mix 48 hr after stimulation by PHA resulted in a doubling of the SCE yield as compared to a 1 hr exposure at initiation. These discrepancies in the SCE rate observed between both exposures (G_0 and 48 hr after initiation) could be due to the heterogeneity of the peripheral blood lymphocyte population which contains "fast" and "slow" cycling lymphocytes (31,33,65,68,78-80). Fast-responding cells are primarily T cells while slow responders contain a higher proportion of B cells (81). Probably, the lymphocytes with a very long G_1 phase of several days represent such B cells (82-84) which become stimulated probably via a helper function of the previously stimulated T cells (85,86).

However, our findings on DNA damage induced in G_0 lymphocytes by chemical mutagens and expressed as SCEs during the second post-treatment cell cycles, 72 hr after initiation, indicates identical SCE rates regardless of the mitogen used. Littlefield et al. (43) have also treated cells in G_0 with MMC and then stimulated them with PHA for 48 or 72 hr in the presence of BrdUrd for the entire culture period. The outcome of their experiments resulted in similar frequencies of MMC-induced lesions expressed as SCEs in "early dividing" as "late dividing" lymphocyte populations. These findings are also in agreement with those of Beek and Obe (27) in which the frequency of Trenimon-induced SCEs in human lymphocytes harvested from 66 to 102 hr after stimulation by PHA remained constant. Our previous investigations on cell-cycle kinetics (31,33) revealed large variations in cell proliferation at 48 hr after stimulation by the 4 different mitogens. At that time the frequency of cells in first division varied around 75%, 87%, 85%, and 95% for PHA-, Con A-, WFA-, or LcH-A-stimulated lymphocytes, respectively. If the different stages of the cell cycle were not uniformly sensitive to a mutagenic agent, different yields of SCEs would have been obtained in cells treated in the various parts of the cell cycle, depending on the stimulation capacity of different mitogens. Therefore, we believe that the higher frequencies of SCEs observed in cells exposed to chemicals 48 hr after stimulation by different mitogens as compared to those treated in the G_0 stage before stimulation by these mitogens can be ascribed to differences in the cell cycle stage at treatment and not to lymphocyte subpopulations displaying different sensitivities.

Several conclusions can be derived from our data. Stimulation of human lymphocytes by different T mitogens (PHA, Con A, WFA, or LcH-A) resulted in differences in the MI and cell-cycle kinetics. Considering the MI, PHA gave the highest and LcH-A the lowest response, while intermediate responses were observed for Con A- and WFA-activated cells. PHA also resulted in the lowest number of cells in first mitosis, whereas the opposite was true for LcH-A. Chemical SCE induction in proliferating human lymphocytes is much more pronounced than that in G_0 cells. It is obvious that lymphocytes which were stimulated by different mitogens, resulting in different mitotic indices and cell-cycle kinetics, will be in different stages of the cell cycle 48 hr after initiation. Therefore, depending on the sensitivity of the cell stage at treatment, various yields of SCEs will be observed 24 hr later, whereas cell exposure at time 0 before initiation followed by stimulation in the presence of BrdUrd for the entire culture period will yield an equivalent SCE rate regardless of the mitogen used.

ACKNOWLEDGEMENT

This work was supported by Research Contract Euratom-S.C.K. Nr:BIO- D-378-81-B.

REFERENCES

1. Taylor, J.H. (1958) Sister chromatid exchanges in tritium-labeled chromosomes. Genetics 43:515-529.
2. Latt, S.A. (1973) Microfluorometric detection of deoxyribonucleic acid replication in human metaphase chromosomes. Proc. Natl. Acad. Sci., USA 70:3395-3399.
3. Kim, M.A. (1974) Chromatidaustausch und Heterochromatinveränderungen menschlicher Chromosomen nach BUdR-Markierung: Nachweis mit Benzimidazolfluorochrom und Giemsafarbstoff. Humangenetik 25:179-188.
4. Korenberg, J.R., and E.F. Freedlender (1974) Giemsa technique for the detection of sister chromatid exchanges. Chromosoma 48:355-360.
5. Perry, P., and S. Wolff (1974) New Giemsa method for the differential staining of sister chromatids. Nature (Lond.) 251: 156-158.
6. Sandberg, A.A. (1982) Sister Chromatid Exchange. Alan R. Liss, Inc., New York.

7. Carrano, A.V., J.L. Minkler, D.G. Stetka, and D.H. Moore (1980) Variation in the baseline sister chromatid exchange frequency in human lymphocytes. Environ. Mutagen. 2:325-337.
8. Davidson, R.L, E.R. Kaufman, C.P. Dougherty, A.M. Ouellette, C.M. DiFolco, and S.A. Latt (1980) Induction of sister chromatid exchanges by BUdR is largely independent of the BUdR content of DNA. Nature (Lond.) 284:74-76.
9. Kato, H. (1974) Spontaneous sister chromatid exchanges detected by a BUdR-labeling method. Nature (Lond.) 251:70-72.
10. Lambert, B., K. Hansson, J. Lindsten, M. Sten, and B. Werelius (1976) Bromodeoxyuridine-induced sister chromatid exchanges in human lymphocytes. Hereditas 83:163-174.
11. Latt, S.A., and L.A. Juergens (1977) Determinants of sister chromatid exchange frequencies in human chromosomes. In Population Cytogenetics, E.B. Hook and I.H. Porter, eds. Academic Press, New York, pp. 217-236.
12. Bianchi, N.O., M.S. Bianchi, and M. Larramendy (1979) Kinetics of human lymphocyte division and chromosomal radiosensitivity. Mutat. Res. 63:317-324.
13. Morgan, W.F., and P.E. Crossen (1981) Factors influencing sister-chromatid exchange rate in cultured human lymphocytes. Mutat. Res. 81:395-402.
14. Sharma, T., and B.C. Das (1981) Culture media and species-related variations in the requirement of 5-bromodeoxyuridine for differential sister-chromatid staining. Mutat. Res. 81: 357-364.
15. Schempp, W., and W. Krone (1979) Deficiency of arginine and lysine causes increase in the frequency of sister chromatid exchanges. Human Genet. 51:315.
16. Ghosh, P.K., and R. Nand (1979) Reduced frequency of sister chromatid exchanges in human lymphocytes cultured in autologous serum. Human Genet. 51:167-170.
17. Kato, H., and A.A. Sandberg (1977) The effect of sera on sister chromatid exchanges in vitro. Exp. Cell Res. 109:445-448.
18. McFee, A.F., and M.N. Sherrill (1981) Mitotic response and sister chromatid exchanges in lymphocytes cultured in sera from different sources. Experientia 37:27.
19. Mazrimas, J.A., and D.G. Stetka (1978) Direct evidence for the role of incorporated BUdR in the induction of sister chromatid exchanges. Exp. Cell Res. 117:23-30.
20. Stetka, D.G., and A.V. Carrano (1977) The interaction of Hoechst 33258 and BrdU substituted DNA in the formation of sister chromatid exchanges. Chromosoma 63:21-31.
21. Abdel-Fadil, M.R., C.G. Palmer, and N. Heerema (1982) Effect of temperature variation on sister-chromatid exchange and cell-cycle duration in cultured human lymphocytes. Mutat. Res. 104: 267-273.
22. Kato, H. (1980) Temperature-dependence of sister chromatid exchange: An implication for its mechanism. Cancer Genet. Cytogenet. 2:61.

23. Livingston, G.K., and L.A. Dethlefsen (1979) Effect of hyperthermia and X-irradiation on sister chromatid exchange (SCE) frequency in Chinese hamster ovary (CHO) cells. Radiat. Res. 77:512-520.
24. Speit, G. (1980) Effect of temperature on sister chromatid exchanges. Human Genet. 55:333-336.
25. Santesson, B., K. Lindahl-Kiessling, and A. Mattsson (1979) SCE in B and T lymphocytes. Possible implications for Bloom's syndrome. Clin. Genet. 16:133-135.
26. Snope, A.J., and J.M. Rary (1979) Cell cycle duration and sister chromatid exchange frequency in cultured human lymphocytes. Mutat. Res. 63:345-349.
27. Beek, B., and G. Obe (1979) Sister chromatid exchanges in human leukocyte chromosomes: Spontaneous and induced frequencies in early- and late-proliferating cells in vitro. Human Genet. 49:51-61.
28. Giulotto, E., A. Mottura, R. Giorgi, L. de Carli, and F. Nuzzo (1980) Frequencies of sister-chromatid exchanges in relation to cell kinetics in lymphocyte cultures. Mutat. Res. 70:343-350.
29. Nowell, P.C. (1960) Phytohemagglutinin: An initiator of mitosis in cultures of normal human leukocytes. Cancer Res. 20:462-466.
30. Castellani, A. (1980) Lymphocyte Stimulation. Plenum Press, New York, pp. 1-13.
31. Deknudt, Gh. (1982) Cell kinetics and radiosensitivity of human lymphocytes stimulated by Phytohemagglutinin, Wistaria floribunda or Lentil lectin. Can. J. Genet. Cytol. 24:761-769.
32. Deknudt, Gh., and O. Kamra (1983) Influence of various mitogens on the yield of sister-chromatid exchanges, induced by chemicals, in human lymphocytes. Mutat. Res. 111:161-170.
33. Deknudt, Gh., and A. Leonard (1980) Stimulation of irradiated human lymphocytes by different mitogens. Int. J. Radiat. Biol. 38:361-364.
34. Craig-Holmes, A.P., and M.W. Shaw (1976) Cell cycle analysis in asynchronous cultures using the BUdR-Hoechst technique. Exp. Cell Res. 99:79-87.
35. Latt, S.A. (1974) Sister chromatid exchanges, indices of human chromosome damage and repair: Detection by fluorescence and induction by mitomycin C. Proc. Natl. Acad. Sci., USA 71:3162-3166.
36. Ishii, Y., and M.A. Bender (1978) Factors influencing the frequency of mitomycin C-induced sister-chromatid exchanges in 5-bromodeoxyuridine-substituted human lymphocytes in culture. Mutat. Res. 51:411-418.
37. Ristow, H., and G. Obe (1978) Acetaldehyde induces cross-links in DNA and causes sister-chromatid exchanges in human cells. Mutat. Res. 58:115-119.
38. Morimoto, K., S. Wolff, and A. Koizumi (1983) Induction of sister-chromatid exchanges in human lymphocytes by microsomal activation of benzene metabolites. Mutat. Res. 119:355-369.

39. Morimoto, K., and S. Wolff (1980) Cell cycle kinetics in human lymphocyte cultures. Nature (Lond.) 288:604-606.

40. White, A.D., and L.C. Hesketh (1980) A method utilizing human lymphocytes with in vitro metabolic activation for assessing chemical mutagenicity by sister-chromatid exchange analysis. Mutat. Res. 69:283-291.

41. Littlefield, L.G., S.P. Colyer, and R.J. DuFrain (1980) Comparison of sister-chromatid exchanges in human lymphocytes after G_0 exposure to mitomycin in vivo vs. in vitro. Mutat. Res. 69: 191-197.

42. Littlefield, L.G., S.P. Coyler, and R.J. DuFrain (1981) Physical, chemical, and biological factors affecting sister-chromatid exchange induction in human lymphocytes exposed to mitomycin C prior to culture. Mutat. Res. 81:377-386.

43. Littlefield, L.G., S.P. Colyer, and R.J. DuFrain (1983) SCE evaluations in human lymphocytes after G_0 exposure to mitomycin C. Mutat. Res. 107:119-130.

44. Littlefield, L.G., S.P. Colyer, A.M. Sayer, and R.J. DuFrain (1979) Sister-chromatid exchanges in human lymphocytes exposed during G_0 to four classes of DNA-damaging chemicals. Mutat. Res. 67:259-269.

45. DeSerres, F.J., and M.D. Shelby (1981) Comparative Chemical Mutagenesis. Plenum Press, New York, pp. 539-547.

46. Hata, T., Y. Sano, R. Sugawara, A. Matsumae, K. Kanamori, T. Shima, and T. Hoshi (1956) Mitomycin, a new antibiotic from streptomyces. I. J. Antibiotics, Ser. A9:141-146.

47. Fishbein, L., W.G. Flamm, and H.L. Falk (1970) Chemical Mutagens. Academic Press, New York.

48. Kersten, H. (1975) Mechanism of action of mitomycins. In Antineoplastic and Immunosuppressive Agents, Vol. II, A.C. Sartorelli and D.G. Johns eds. Springer-Verlag, New York, pp. 47-64.

49. Arnold, H., and F. Bourseaux (1958) Synthese und Abbau cytostatisch wirksamer cyclischer N-Phosphamidester des Bis- (β-chloräthyl)-amins. Angew. Chem. 70:539-544.

50. Connors, T.A., P.J. Cox, R.B. Farmer, A.B. Foster, and M. Jarman (1974) Some studies of the active intermediates formed in the microsomal metabolism of cyclophosphamide and isophosphamide. Biochem. Pharmac. 23:115-129.

51. Wolff, S., P. Perry, and A.T. Natarajan (1981) Mutagenicity of selected chemicals in sister chromatid exchange assays. In Comparative Chemical Mutagenesis, F.J. de Serres and M.D. Shelby, eds. Plenum Press, New York, pp. 539-547.

52. Ames, B.N., J. McCann, and E. Yamasaki (1975) Methods for detecting carcinogens and mutagens with the Salmonella/mammalian-microsome mutagenicity test. Mutat. Res. 31:347-364.

53. Stetka, D.G., and S. Wolff (1976) Sister chromatid exchange as an assay for genetic damage induced by mutagens-carcinogens, II. In vitro test for compounds requiring metabolic activation. Mutat. Res. 41:343-350.

54. Powell, A.E., and M.A. Leon (1970) Reversible interaction of human lymphocytes with the mitogen concanavalin A. Exp. Cell Res. 62:315-325.
55. Toyoshima, S., T. Osawa, and A. Tonomura (1970) Some properties of purified phytohemagglutinin from Lens culinaris seeds. Biochem. Biophys. Acta 221:514:521.
56. Barker, B.E., and P. Farnes (1967) Mitogenic property of Wistaria floribunda seeds. Nature (Lond.) 215:659-660.
57. Toyoshima, S., Y. Akiyama, K. Nakano, A. Tonomura, and T. Osawa (1971) A Phytomitogen from Wistaria floribunda seeds and its interaction with Human Peripheral Lymphocytes. Biochem. 10: 4457-4463.
58. Toyoshima, S., and T. Osawa (1975) Lectins from Wistaria floribunda seeds and their effect on membrane fluidity of human peripheral lymphocytes. J. Biol. Chem. 250:1655-1660.
59. Howard, I.K., H.J. Sage, M.D. Stein, N.M. Young, M.A. Leon, and D.F. Dyckes (1971) Studies on a phytohemagglutinin from the lentil II. Multiple forms of Lens culinaris hemagglutinin. J. Biol. Chem. 246:1590-1595.
60. Sharon, N. (1976) Lectins as Mitogens. Academic Press, London, pp. 31-41.
61. Weber, T.H., H. Aro, and C.T. Nordman (1972) Characterization of lymphocyte stimulating blood cell-agglutinating glycoproteins from red kidney beans (Phaseolus vulgaris). Biochem. Biophys. Acta 263:94-105.
62. Ham, R.G. (1963) An improved nutrient solution for diploid Chinese hamster and human cell lines. Exp. Cell Res. 29:515-526.
63. Ikushima, T., and S. Wolff (1974) Sister chromatid exchanges induced by light flashes to 5-bromodeoxyuridine and 5-iododeoxyuridine substituted Chinese hamster chromosomes. Exp. Cell Res. 87:15-19.
64. Wolff, S., and R. Arutyunyan (1979) The apparent decrease in thiotepa-induced chromosome aberrations in human lymphocytes caused by an effect of WR2721 on the cell cycle as found by the definitively determined division method. Environ. Mutagen. 1:5-13.
65. Crossen, P.E., and W.F. Morgan (1977) Analysis of human lymphocyte cell cycle time in culture measured by sister chromatid differential staining. Exp. Cell Res. 104:453-457.
66. Hatcher, N.H., P.S. Brinson, and E.B. Hook (1976) Sister chromatid exchanges in Ataxia telangiectasia. Mutat. Res. 35:333-336.
67. Chaganti, R.S.K., S. Schonberg, and J. German (1974) A manyfold increase in sister chromatid exchanges in Bloom's syndrome lymphocytes. Proc. Natl. Acad. Sci., USA 71:4508-4512.
68. Tice, R., E.L. Schneider, and J.M. Rary (1976) The utilization of bromodeoxyuridine incorporation into DNA for the analysis of cellular kinetics. Exp. Cell Res. 102:232-236.
69. Hatcher, N.H., and E.B. Hook (1976) Early in vitro division of PHA stimulated cord blood lymphocytes implications for study of

chromosome breakage. Am. J. Human Genet. 28:290-293.

70. Mutchinick, O., M.E. Gonsebatt, L. Ruz, P. Mauleón, R. Lisker, and R. Lichtenberg (1983) Sister-chromatid exchanges and cell kinetics in human and rabbit lymphocytes exposed in vivo and in vitro to 2-bromo-α-ergocryptine. Mutat. Res. 117:163-171.

71. Craig-Holmes, A.P., and M.W. Shaw (1977) Effects of six carcinogens on SCE frequency and cell kinetics in cultured human lymphocytes. Mutat. Res. 46:375-384.

72. Evans, H.J., and Vijayalaxmi (1980) Storage enhances chromosome damage after exposure of human leukocytes to mitomycin C. Nature (Lond.) 284:370-372.

73. Kato, H. (1973) Induction of sister chromatid exchanges by chemical mutagens and its possible relevance to DNA repair. Exp. Cell Res. 85:383-390.

74. Perry, P., and H.J. Evans (1974) Cytological detection of mutagen-carcinogen exposure by sister chromatid exchange. Nature (Lond.) 251:121-125.

75. Wolff, S., J. Bodycote, and R.S. Painter (1974) Sister chromatid exchanges induced in Chinese hamster cells by UV irradiation of different stages of the cell cycle: The necessity for cells to pass through S. Mutat. Res. 25:73-81.

76. Ohtsuru, M., Y. Ishii, S. Takai, H. Higashi, and G. Kosaki (1980) Sister chromatid exchanges in lymphocytes of cancer patients receiving mitomycin C treatment. Cancer Res. 40:477-480.

77. Stetka, D.G., J. Minkler, and A.V. Carrano (1978) Induction of long-lived chromosome damage as manifested by sister-chromatid exchange, in lymphocytes of animals exposed to mitomycin. Mutat. Res. 51:383-396.

78. Bender, M.A., and J.G. Brewen (1969) Factors influencing chromosome aberration yields in the human peripheral leukocyte system. Mutat. Res. 8:383-399.

79. Dutrillaux, B., and A. Fosse (1977) Utilisation du BrdU dans l'étude du cycle cellulaire de sujets normaux et anormaux. Ann. Genet. (Paris) 19:95-102.

80. Heddle, J.A., H.J. Evans, and D. Scott (1967) Sampling time and the complexity of human leukocyte culture system. In Human Radiation Cytogenetics, H.J. Evans, W.M. Court Brown, A.S. McLean, eds. Amsterdam, North Holland, pp. 6-19.

81. Riedel, L., and G. Obe (1980) Trenimon-induced SCEs and structural chromosomal aberrations in early- and late-dividing lymphocytes. Mutat. Res. 73:125-131.

82. Jasinska, J., J.A. Steffen, and M. Michalowski (1970) Studies on in vitro lymphocyte proliferation in cultures syncrhonized by the inhibition of DNA synthesis. II. Kinetics of the initiation of the proliferative response. Exp. Cell Res. 61:333-341.

83. Sören, L. (1973) Variability of the time at which PHA-stimulated lymphocytes initiate DNA synthesis. Exp. Cell Res. 78:201-208.

84. Steffen, J.A., and W.M. Stolzman (1969) Studies on in vitro

lymphocyte proliferation in cultures synchronized by the inhibition of DNA synthesis. I. Variability of S plus G2 periods of first generation cells. Exp. Cell Res. 56:453-460.

85. Phillips, B., and I.M. Roitt (1973) Evidence for transformation of human B lymphocytes by PHA. Nature (Lond.) 251:156-158.

86. Vischer, T. (1972) Mitogenic factors by lymphocyte activation: Effect on T- and B- cells. J. Immunol. 109:401-402.

SERUM- AND PLASMA-DEPENDENT VARIATIONS OF BENZO(A)PYRENE-INDUCED SISTER CHROMATID EXCHANGE IN HUMAN LYMPHOCYTES

J. K. Wiencke,[1*] K. Kelsey,[2] R. Kreiger,[1] and V. F. Garry[1]

[1]Laboratory of Environmental Pathology
Department of Laboratory Medicine and Pathology
University of Minnesota Medical School
Minneapolis, Minnesota 55414

ABSTRACT

Sister chromatid exchange (SCE) is frequently used to assess the potential mutagenicity of chemical agents to human beings. We demonstrate here that levels of SCE induced by benzo(a)pyrene (BP) in the widely used blood lymphocyte assay are influenced by serum and plasma supplements. Sister chromatid exchange induction by BP was greatest when using fetal calf serum (FCS), intermediate with newborn calf serum (NCS), and lowest with autologous human plasma (AHP). This new finding adds to a growing list of factors capable of modulating the SCE response and underscores the need for researchers to consider serum and plasma supplements in the standardization of the SCE approach in human mutagen assessments. The data also demonstrate the potential of SCE to aid in the study of serum factors which modify the mutagen sensitivity of human cells towards environmental carcinogens.

INTRODUCTION

Detecting environmental contaminants which demonstrate mutagenic activity toward human cells requires the application of highly

---

[2] Department of Occupational Medicine, Harvard School of Public Health, Boston, Massachusetts 02111.

* Current address: Laboratory of Radiobiology and Environmental Health, University of California, San Francisco, California 94143.

sensitive and reproducible tests of genotoxicity. Observations on the induction of SCEs by primary DNA-alkylating agents support the idea that SCE is a sensitive indicator of mutagenesis (1,2). Important to these considerations has been the attempt to define those parameters which can influence background and mutagen-induced SCEs in mammalian cells. While several investigators have identified factors affecting background SCE rates (3,4,5), it has only recently been appreciated that the yield of mutagen-induced SCEs in vitro can be influenced by age (6), gender and reproductive status (7), pharmacological agents (8), disease states (9), red blood cells (10), genetic conditions (11), and mutagen exposure itself (12). These and other sources of variation in SCE measurements are of paramount importance in the design of experimental studies because of the serious implications of human mutagen exposure.

Since serum and plasma supplements are necessary components of all human SCE assays, we felt that additional studies were required to indicate whether this parameter could influence the rate of mutagen-induced SCEs in the widely used blood lymphocyte system. The present study examined the effects of serum and plasma supplements on SCE induction by BP. Benzo(a)pyrene was chosen as a model mutagenic carcinogen because it is often used as a reference chemical in mutagenesis studies and as a standard for estimating carcinogenic risk in environmental assessments (13).

To determine the effect of serum and plasma supplements on BP-induced SCEs, cultures were established from 49 healthy adult volunteers utilizing 3 different conditions: lymphocytes were either grown in 20% AP, 20% NCS, or 20% FCS. Control and BP-treated cultures were established for each individual; a level of BP treatment was selected which produced a maximal SCE response (10 μg/ml).

MATERIALS AND METHODS

Cell Culture and SCE Studies

To culture cells, 0.5 ml of heparinized whole blood was added to a final volume of 5 ml media containing 0.2 ml phytohemagglutinin (PHA-M) (Gibco), 18 μg/ml bromodeoxyuridine (BrdUrd) (Sigma) and RPMI 1640. Blood was cultured in 25 cm^2 flasks (Corning) for 72 hr in complete darkness at 37.5°C, 5% CO_2 and 98% relative humidity. Benzo(a)pyrene (Eastman Kodak, Rochester, New York) was dissolved in dimethylsulfoxide (DMSO) (0.2% in culture) and added to blood cultures at an initial concentration of 10 μg/ml (39.6 μM) for each of the 3 serum/plasma conditions. Benzo(a)pyrene remained in culture for the entire culture period. Cells were harvested, slides prepared by a flame-dry method and stained for 10 min in Hoechst bisbenzamid 33258 (2.5 μg/ml). Slides were stained for SCE analysis by exposure to ultraviolet (UV) light (40 watt Blak-ray) at 2 cm for

2 hr and 37°C. Following UV, slides were counterstained with 3% Giemsa for 3 min (pH 6.8). To determine mean SCE/cell frequencies, 20 second-division cells with 40 or more clearly stained chromosomes were scored. All slides were analyzed by one cytogeneticist.

BP Solubility

In studies of BP solubility, FCS and NCS were the same as those used in previous studies. Three sources of human plasma were utilized: 1) platelet-rich plasma from heparinized whole blood), 2) platelet-poor plasma (platelet-rich centrifuged 20,000 x g for 1 hr), and 3) citrated plasma (Red Cross, courtesy of Dr. Klein). Since all plasma sources gave the same results the data presented are combined from all sources (i.e., 3 experiments).

Solubility was determined by preparing 20% serum/plasma with 80% RPMI 1640. The samples (3 ml each) were treated with a combination of 3H-BP and cold BP and incubated 1 or 14 hr at 37°C in a shaker waterbath. Radiolabeled BP was obtained (66 Ci/mM) from the NCI Chemical Carcinogen Standard Repository, courtesy of Dr. D. G. Longfellow. Following incubations, the culture media were centrifuged at 20,000 x g for 1 hr. After centrifugation a 1 ml sample of the supernatant fraction was carefully recovered and counted in a Beckman LS 7500 liquid scintillation counter. Quench was corrected by automatic externalization (H number). Counting efficiencies were routinely 50-60%. Incubations conducted for 1 hr gave identical results to those lasting 14 hr.

BP Uptake by Isolated Mononuclear Cells

For uptake studies, venous blood was drawn from volunteer donors and within 15 min of venipuncture applied to Ficoll-Hypaque gradients and centrifuged (400 x g; 30 min 22°C). Isolated mononuclear cells (MNCs) (95%) were washed 3 times with phosphate buffered saline (PBS) and resuspended in RPMI 1640 at a density of 2×10^6 viable cells/ml. Mononuclear cells were incubated in 5 ml cultures at 37°C, 5% CO_2, 98% humidity for 6 hr with either 20% AHP or 20% FCS and 2% PHA in all cases. Plasma was centrifuged prior to incubations to decrease platelet contamination. Cultures were treated with a mixture of radiolabeled and cold BP to give a final concentration of 10 µg/ml. All samples were run in duplicate. Following incubations (6 hr) cells were washed 6 times with PBS containing 10% DMSO and 20% serum which reduced residual BP to a background level of less than 20 ng. Background counts were subtracted from final uptake values. A final wash in PBS was performed to remove solvent. The cell pellet was taken up in PBS and counted in a Beckman LS 7500 liquid scintillation counter. Quench was corrected by automatic external standardization (H-number). Counting efficiencies were routinely 43%. An uptake value of 80.4 pg/10^5 cells represents 12,730 dpm.

Cell Growth Studies

To determine cell growth, lymphocyte counts were made initially (time 0) and subtracted from counts made following 72 hr of culture. Whole blood cultures were established as in SCE studies. White blood cell and differential counts were used to estimate the initial number of cells cultured. At 72 hr, red blood cells (RBCs) were lysed with 0.87% NH_4Cl and viable cell counts determined by trypan blue exclusion. Aliquots of culture suspensions were prepared by cytocentrifuge (Shandon Southern Corp.) and Wright stained for differential white cell morphology. Culture flasks were scraped with a rubber policeman and rinsed with PBS to obtain a reproducible yield of cells in culture. All cell counts were performed in triplicate using a hemocytometer and at least 500 cells scored for differential white blood cell analysis.

RESULTS

The results of our cytogenetic measurements comparing different serum and plasma supplements are presented in Fig. 1, Tabs. 1 and 2. A total of 10 different persons were studied using platelet-rich autologous plasma (AHP), 15 persons with NCS and 30 subjects analyzed using FCS. The distributions of background and BP-treated SCE levels are illustrated in Fig. 1. The control distributions (Fig. 1A, C, and E) and mean SCE/cell (approximately 7 SCEs/cell) were comparable to our previous work (14). No significant differences in untreated baseline SCEs were detected between the 3 serum/plasma conditions tested. In other studies where comparisons have been made between SCE rates in lymphocytes cultured in FCS or AP, similar results were reported (15). However, comparisons among different types of sera did reveal serum-related variations in background SCE levels (16,17).

In our study, culture-related differences in SCEs became apparent by comparing BP-treated cultures. While a significant increase ($p < 0.001$) in SCEs was detected in cases using AHP, the mean SCE/cell was only 2.3 SCEs/cell higher than control values. A significant shift toward higher SCE scores was observed when cells were cultured in NCS. Benzo(a)pyrene-induced SCEs/cell values in NCS were on the average 2-fold higher (compare Fig. 1B and D). Even more dramatic is the effect of FCS; an increase of 3.5-fold as compared to AHP (Fig. 1F). Analysis of BP-induced SCEs/cell for individuals in each group (Tab. 1) showed that SCE induction by BP was significantly higher in NCS compared to AHP ($p < 0.001$) and that induced levels in FCS were elevated above the levels observed in NCS ($p < 0.001$).

Because these comparison groups were comprised of different individuals, a subset of 5 persons cultured on the same day using both AHP and FCS were studied. An analysis-of-variance was performed to

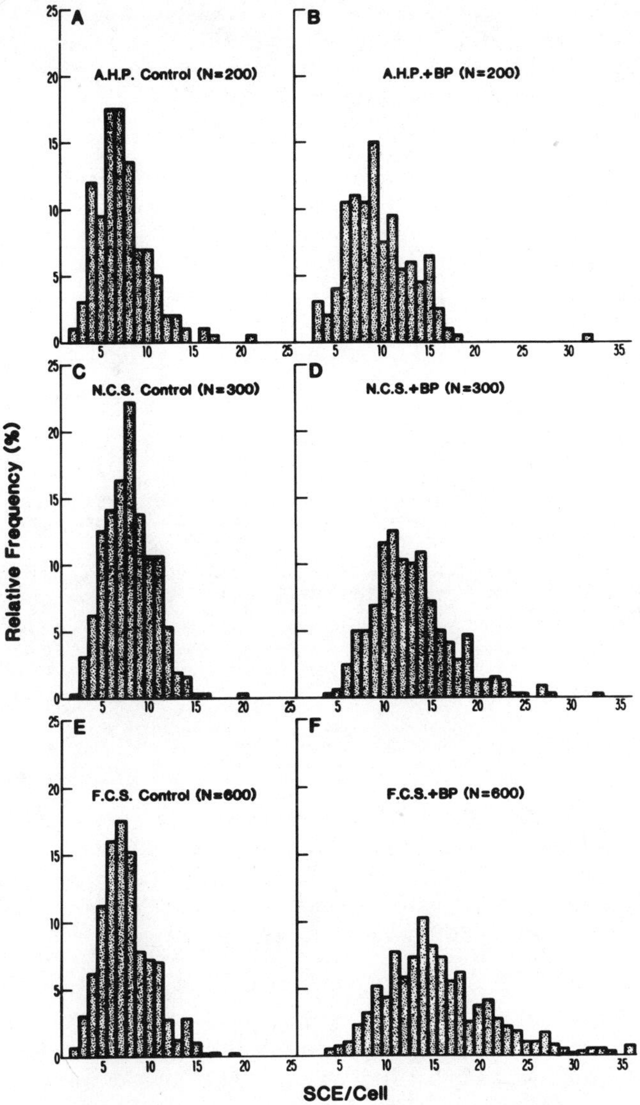

Fig. 1. The distributions of baseline (untreated) and BP-treated lymphocyte SCE rates are illustrated. In panels A and B are shown data collected from 10 control individuals whose lymphocytes were grown in 20% AHP. Panel B depicts the distribution of BP-treated cells. In panels C and D are shown data collected from 15 persons whose cells were cultured in 20% NCS (Biocell Corp., California.) Panel D shows the SCE values for BP-treated cultures. In panels E and F are illustrated SCE scores for 30 individuals whose cells were grown in 20% FCS (Reheis, Armour Co., Michigan). Panel F shows the effect of BP treatment on SCE frequencies.

Tab. 1. Benzo(a)pyrene-induced SCE frequencies in AHP, NCS, and FCS. Cells were cultured for 72 hr without and with BP for each individual in the study. BP was added at the initiation of the experiment (time 0) at a concentration of 10 μg/ml.

Culture Condition	N	SCE/Cell ± SEM	BP-Induced SCE/Cell ± SD
AHP-Control	10	7.3 ± 0.3	------
AHP- +BP	10	9.6 ± 0.4	2.3 ± 1.4[a]
NCS-Control	15	6.8 ± 0.2	------
NCS- +BP	15	11.9 ± 0.3	5.1 ± 1.1[b]
FCS-Control	30	7.6 ± 0.2	------
FCS- +BP	30	15.6 ± 0.4	8.1 ± 1.8[c]

a/b/c denotes, for each experimental group, the mean of individual BP-induced SCEs/cell. Using Student's t-test:

(a)- significantly different from b and c; $p < 0.001$ **
(b)- significantly different from a and c; $p < 0.001$
(c)- significantly different from a and b; $p < 0.001$

** Taken from Statistical Methods by G.W.Snedecor and W.G.Cochran, Iowa State University Press(1967): pp. 114-116.

identify the contributions of serum type, BP treatment, donor, and replicate observations on the variations in SCE rates observed. The results of this analysis are presented in Tab. 2. The data show that the serum or plasma supplement used had a very significant effect on the BP-induced SCE values observed.

In one set of experiments, the effects of AP on dl-1,3-butadiene diepoxide (DEB)-induced SCEs were studied. Cultures were established from a control individual using AHP and FCS, to which DEB (a direct-acting alkylating agent) was added at 0.125 μg/ml. In contrast to experiments using BP, almost identical yields of DEB-induced SCEs were observed in FCS and AHP. Baseline SCEs/cell were 7.2 and 7.6 for AHP and FCS, respectively. The DEB-induced SCE levels for AHP were 15.9 SCEs/cell (8.7 induced SCEs/cell) and 15.7 SCEs/cell (8.3 induced SCEs/cell) for cells cultured in FCS. The consistency of DEB-related SCE response under different serum conditions implies that differences in BP-induced SCEs among serum/plasma supplements are unique. In other words, the effect of serum supplementation on SCE is dependent on the inducing agent.

To explore the mechanisms underlying these observations we compared BP solubility, cellular uptake, and cell growth characteristics under differing serum and plasma conditions. Benzo(a)pyrene

Tab. 2. Factor analysis-of-variance: sister chromatid exchanges response by BP treatments, serum type, donor, and replicate observations. Each of the 20 cells scored per culture per condition was considered a replicate observation. P values are indicated to denote the main effect of each factor considered, a level of $p < 0.01$ being required for statistical significance.

Term[1]	DF	Sum Of Squares	Mean Squares
T	1	2218.40	2218.40
S	1	772.84	772.84
D	4	190.30	47.58
R	19	302.10	15.90
ERROR-1	374	6087.30	16.28
TOTAL	399	9571	
GRAND AVERAGE	1	37,249.0	

F-Test

FACTOR[2]	F-VALUE	p-VALUE
BP treatment	136.3	1×10^{-12}
Serum type	47.5	2×10^{-11}
Donor	2.9	0.021
Replicates	0.9	0.4878

[1]Terms of the model = Y(IJKL) = T(I) + S(J) + D(K) + R(I) + E1(IJKL) where T = BP treatment (10 µg/mL), (S) = serum/plasma, (D) = donor and (R) = replicate observations

[2]FCS and AHP were used, 5 different blood donors with 20 cells counted to estimate mean SCE rates

solubility was estimated by adding radiolabeled BP to 20% serum or plasma solutions containing nutrient media (RPMI 1640) and allowing the samples to incubate at 37°C in a shaker waterbath. Following centrifugation, the supernatant fraction was assayed for soluble BP by liquid scintillation counting. No significant differences were noted in the solubility of BP in FCS, NCS, or human plasma over a concentration range of 1-250 µM BP (see Fig. 2).

Since BP is noncovalently associated with the lipoprotein fraction of human plasma (18) and serum uptake is proportional to lipoprotein concentration (19), we thought it appropriate to examine BP uptake in isolated mononuclear cells cultured with different serum/plasma supplements. Of further interest, lipoprotein carriers have been shown to facilitate cellular BP uptake (20). The results of our uptake study (shown in Tab. 3) indicated that BP absorption in

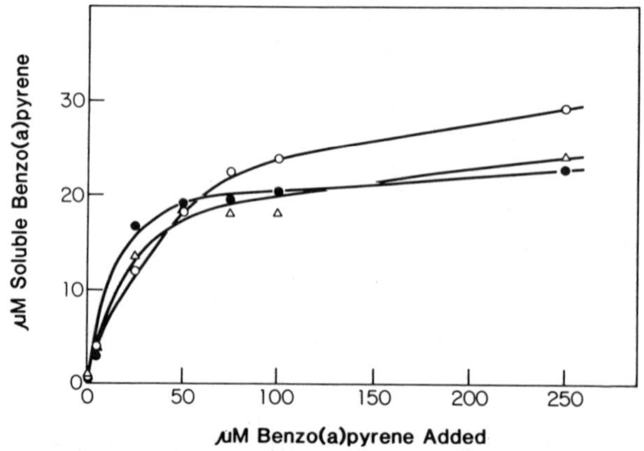

Fig. 2. The solubility of BP in tissue culture medium supplemented with either 20% FCS, NCS, or human plasma is depicted. Open circles represent plasma, solid circles FCS and open triangles NCS. Standard BP treatments for SCE studies were 10 μg/ml or 39.6 μM added BP.

AHP was 4-fold greater than that observed in cells incubated in FCS 6 hr following BP addition. These results are consistent with findings from the previously cited work. However, it is quite evident that BP-induced SCEs (see Tab. 1) are unrelated to initial cellular uptake. The BP-induced SCE levels in FCS were substantially higher than in AHP even though FCS contains lower cholesterol and lipoprotein concentrations relative to adult human plasma (21). The pro-

Tab. 3. Cell growth characteristics in serum and AP at 72 hr. Duplicate cultures for each condition from a single individual were analyzed. Fold increase in lymphocytes was calculated by dividing total lymphocyte counts at 72 hr by the initial number planted at time 0. Cell morphology on Wright-stained cytocentrifuge preparations was used to differentiate PMNs.

Experiment Number	Culture Conditions	BP Uptake ($pg/10^5$ cells)
1.	Plasma 20%	46.3
2.	Plasma 20%	46.4
3.	Plasma 20%	80.4
4.	FCS 20%	13.6
5.	FCS 20%	13.0
6.	FCS 20%	12.3

tein and lipid constituents of different lipoproteins classes in the 2 species are also widely dissimilar (22,23). Our results raise the question as to whether the presentation of BP by divergent classes of lipoprotein carriers may affect the intracellular fate and mutagenicity of this compound. Resting lymphocytes do possess surface lipoprotein receptors (24) and, indeed 192 low-density lipoproteins have been shown to facilitate the induction of unscheduled DNA synthesis following BP diol-expoxide treatments in vitro (19).

As cell cycle kinetics have been related to baseline SCEs (25,26) and the activation of BP to its reactive metabolites may be dependent upon the extent of lymphocyte proliferation, we compared the growth characteristics of cells cultured in AHP, NCS and FCS. The results of these comparisons are presented in Tab. 4. In all cases, the total number of lymphocytes increased 72 hr following PHA stimulation. There was a uniform loss of polymorphonuclear cells (PMNs), and PMNs present at 72 hr showed degranulation, as well as nuclear pyknosis and cytoplasmic vacuolization. Lymphocytes were

Tab. 4. Benzo(a)pyrene uptake by PHA-stimulated mononuclear cells at 6 hr. Experiments were conducted using 3 healthy control donors, each cultured in both AHP and FCS. Experiments 1 and 4, 2 and 5, and 3 and 6 consist of paired observations from each of the blood donors. Ficold-Hypaque density gradients were used to isolate MNCs (> 95% MNCs and 98% viability by trypan blue exclusion). Uptakes were run in duplicate, all samples were within ± 10% of each other.

Culture Condition	cells/mL	[1] %viability	[2] Fold Increase in Lymphocytes	[3] % Decrease PMNs
20%FCS-Control	1.82×10^6	90.6	5.95	57
20%FCS- +BP	1.05×10^6	86.0	2.72	60
20%NCS-Control	2.05×10^6	93.6	6.80	55
20%NCS- +BP	1.10×10^6	91.1	3.52	75
20%AHP-Control	3.88×10^6	94.6	13.88	55
20%AHP- +BP	2.29×10^6	94.4	8.00	65

[1] Viability determined by trypan blue exclusion

[2] Lymphocyte count determined by differential WBC analysis at 0 and 72 hours. Initial total WBC = 1.0×10^6/mL.

[3] Polymorphonuclear cell counts determined by differential WBC analysis at 0 and 72 hours.

almost entirely blast forms. Increases in lymphocyte counts were always lower in BP-treated cultures. No obvious differences in proliferation were seen between NCS and FCS. However, the proliferation of lymphocytes was greatly increased in AHP. Thus, serum and plasma supplements differ in their ability to support lymphocyte blastogenesis under these experimental conditions. In 2 experiments, platelets were depleted from plasma by centrifugation and BP-induced SCE levels measured. Removal of platelets reduced the proliferation of lymphocytes as evidenced by a greater proportion of first-division mitotic cells at 72 hr, but SCE induction scores were not significantly higher than in cultures employing platelet-rich plasma (data not shown).

It should be noted that although lymphocyte aryl-hydrocarbon hyroxylase (AHH) activity has been reported to vary with proliferation (27,28,29), other workers have failed to detect a relationship between BP metabolism and the extent of PHA-induced blastogenesis (30,31). Finally, comparative studies of BP metabolism and BP-induced SCEs in human lymphocytes indicated that AHH activity may not reflect the enzymatic reactions pertinent to the production of DNA lesions leading to SCE formation (32). Other enzymatic pathways may prove to be important in controlling SCE induction by BP in mammalian cells. Recent studies show that glutathione depletion is associated with increased BP-induced SCEs in vitro (33) and in vivo (34). These reports and our present study indicate the need to examine in greater detail the relationships among cellular metabolism, proliferation, and SCE induction.

DISCUSSION

The data presented here illustrate several problems encountered in attempts to optimize the sensitivity of the lymphocyte system toward chemical mutagens. Sensitivity of the assay appears greatest toward BP under conditions where cell proliferation is diminished. Thus, attempts to optimize the system in terms of cellular blastogenesis may run counter to efforts to increase its sensitivity. Further, when considering what type of serum or plasma supplement to use, it is prudent to consider the effects of serum/plasma on SCE baselines separately from effects on mutagen-induced SCEs. This adds a new dimension to the problem of standardization heretofore not appreciated. Finally, it is essential that the possible effect of serum supplementation on in vivo SCE induction be examined in light of the present in vitro results. In fact, in vivo SCE induction has been shown to be affected by cell culture parameters in studies of cigarette smokers (35). Because increases in SCEs observed in humans during chronic low-level mutagen exposures are of the same order of magnitude as those observed in the present study, the effects of serum on SCE reported here take on additional significance. We conclude that serum must be considered as a possible interfering factor in human mutagen assessments.

Prior to using the blood lymphocyte assay in mutagen detection it may be desirable to "screen" serum for the capacity to support SCE induction by BP. Until a more complete understanding of the interactions of serum components with human lymphocytes is reached, the rationale for such maneuvers will be based solely on empiric observation. In fact, culture conditions favoring SCE induction may even diminish the usefulness of an in vitro system in detecting other endpoints of genotoxicity, depending on the agent used. For example, the clastogenic activity of cadmium was greatest when Chinese hamster ovary cells were cultured in human plasma, reduced in NCS, and lowest under conditions where FCS was used (36). The present study underscores the need for further study of culture parameters which influence the sensitivity of SCE analysis in the study of environmental mutagens. The results also indicate the potential for SCEs to provide a rapid and practical tool to investigate the mechanisms by which serum and plasma factors alter the expression of chemical mutagenesis in normal human cells.

CONCLUSIONS

1. The number of SCEs induced by BP in human lymphocytes in vitro is influenced by the type of serum/plasma supplement used. Benzo(a)pyrene-related SCE is greatest when lymphocytes are cultured in FCS, intermediate in NCS, and lowest when AHP is used.

2. Serum-related effects on carcinogen-induced SCE are agent-specific. Identical yields of DEB-induced SCEs are observed in lymphocytes cultured in FCS and AHP.

3. Serum-related changes in BP-induced SCEs are independent of any effect on baseline untreated SCE levels. Thus, screening serum for effects on both background and mutagen-induced SCEs are necessary to insure optimal sensitivity of the SCE assay as a mutagen detection method.

4. Both lymphocyte proliferation and initial cellular BP uptake are increased in cultures treated with AHP as compared with FCS or NCS. Thus, carcinogen absorption and cellular proliferation are not limiting factors in the induction of SCEs by BP in human lymphocytes.

REFERENCES

1. Carrano, A.V., and L.H. Thompson (1982) Sister chromatid exchange and single gene mutation. In Sister Chromatid Exchange, S. Wolff, ed. John Wiley and Sons, New York, pp. 59-86.
2. Abe, S., and M. Sasaki (1982) SCE as an index of mutagenesis and carcinogenesis. In Sister Chromatid Exchange, A.A. Sandberg, ed. Alan R. Liss, New York, pp. 461-514.

3. Morgan, W.F., and P.E. Crossen (1981) Factors influencing sister chromatid exchange in cultured human lymphocytes. Mutat. Res. 81:395-402.
4. Carrano, A.V., and D.H. Moore (1982) The rationale and methodology for quantifying sister chromatid exchange in humans. In Mutagenicity: New Horizons in Genetic Toxicology, John Heddle, ed. Academic Press, New York, pp. 267-304.
5. Lindblad, A., K. Holmberg, and D. Francesconi (1982) The use of sister chromatid exchange to monitor human populations exposed to toxicologically harmful agents. In Sister Chromatid Exchange, S. Wolff, ed. John Wiley and Sons, New York, pp. 149-182.
6. Schneider, E.L., and B. Gilman (1979) Sister chromatid exchange and aging. III. The effect of donor age on mutagen-induced sister chromatid exchange in human diploid fibroblasts. Human Genet. 46:57-63.
7. Hill, A., and S. Wolff (1982) Increased induction of sister chromatid exchange by diethylstilbesterol in lymphocytes from pregnant and premenopausal women. Cancer Res. 42:893-896.
8. Morgan, W.F., and J.E. Cleaver (1982) 3-aminobenzamide synergistically increases sister chromatid exchanges in cells exposed to methyl methane sulfonate but not ultraviolet light. Mutat. Res. 104:361-366.
9. Rudiger, H.W., W. Harder, P. Maack, F.V. Kohl, and U.J. Schmidt-Preuss (1980) Decreased rate of benzo(a)pyrene-induced sister chromatid exchange in fibroblast cultures from patients with lung cancer. Cancer Clin. Oncol. 102:169-175.
10. Norppa, H., H. Vainio, and M. Sorsa (1983) Metabolic activation of styrene by erythrocytes detected as increased sister chromatid exchanges in cultured human lymphocytes. Cancer Res. 43:3579-3582.
11. Evans, H.J. (1982) Sister chromatid exchanges and disease states in man. In Sister Chromatid Exchange, S. Wolff, ed. John Wiley and Sons, New York, pp. 183-228.
12. Samson, L., and J.L. Schwartz (1980) Evidence for an adaptive DNA repair pathway in CHO and human skin fibroblast cell lines. Nature (Lond.) 287:861-863.
13. Holmberg, B., and U. Ahlborg (1983) Consensus Report: Mutagenicity and carcinogenicity of car exhausts and coal combustion emissions. Environ. Hlth. Perspec. 47:1-30.
14. Wiencke, J.K., J. Cervenka, and H. Paulus (1979) Mutagenic activity of anticancer agent cis-dichlorodiamine platinum II. Mutat. Res. 68:69-74.
15. Crossen, P.E. (1982) SCE in lymphocytes. In Sister Chromatid Exchange, A.A. Sandberg, ed. Alan R. Liss, New York, pp. 175-193.
16. Kato, H., and A.A. Sandberg (1977) The effect of sera on sister chromatid exchange in vitro. Exp. Cell Res. 109-445-448.
17. Ghosh, P., and R. Nand (1979) Reduced frequency of sister chromatid exchanges in human lymphocytes cultured with autologous serum. Human Genet. 51:167-170.

18. Shu, H., and A.V. Nichols (1979) Benzo(a)pyrene uptake by human plasma lipoproteins in vitro. Cancer Res. 39:1224-1230.
19. Busbee, D.L., P.W. Rankin, D.M. Payne, and D.N. Jasheway (1982) Binding of benzo(a)pyrene and intracellular transport of a bound electrophilic benzo(a)pyrene metabolite by lipoproteins. Carincogenesis 3:1107-1112.
20. Remsen, J., and R.B. Shireman (1981) Effect of low density lipoproteins on the incorporation of benzo(a)pyrene by cultured cells. Cancer Res. 41:3179-3185.
21. Price, P., and E.A. Gregory (1982) Relationship between in vitro growth promotion and biophysical and biochemical properties of the serum supplement. In Vitro 18:576-584.
22. Goodman, D.S. (1965) Cholesterol ester metabolism. Physiol. Rev. 45:747-839.
23. Evans, L., S. Patton, and R.D. McCarthy (1961) Fatty acid composition of the lipid fractions from bovine serum lipoproteins. J. Dairy Sci. 44:475.
24. Strazullo, B.P., A.M. Scanu, M.C. Ritter, A. Postiglone, and M. Mancini (1978) Binding, internalization, and degradation of human serum low density lipoproteins by human leukocytes in vitro: Effect on sterol metabolism. In International Conference on Atherosclerosis, L.A. Carson, R. Paoletti, C.R. Sitori, and G. Weber, eds. Raven Press, New York, pp. 477-483.
25. Beek, B., and G. Obe (1979) Sister chromatid exchange in human leukocyte chromosomes: Spontaneous and induced frequencies in early and late proliferating in vitro. Human Genet. 49:51-61.
26. Snope, A.J., and J.M. Rary (1979) Cell cycle duration and sister chromatid exchange frequency in cultured human lymphocytes. Mutat. Res. 63:345-349.
27. Kellerman, G., E. Cantrell, and C.R. Shaw (1973) Variations in extent of arylhydrocarbon hydroxylase in cultured human lymphocytes. Cancer Res. 33:1654-1656.
28. Hart, P., W. Cooksley, G.C. Farrel, and L.W. Powell (1977) 3-methylcholanthrene inducibility of aryl hydrocarbon hydroxylase in cultured human lymphocytes depends upon the extent of blast transformation. Biochem. Pharmacol. 26:1831-1834.
29. Kouri, R.E., R.L. Inblum, R.G. Sosnowski, D.J. Slomiany, and C.E. McKinney (1979) Parameters influencing quantitation of 3-methylcholanthrene induced aryl-hydrocarbon hydroxylase activity in cultured human lymphocytes. J. Env. Path. and Toxicol. 2:1079-1098.
30. Prasad, N., R. Prasad, J.E. Harrell, J. Thornby, and L. Fahr (1978) Relationship between mitogen response and aryl hydrocarbon hydroxylase in cultured human lymphocytes. Life Sci. 23:247-252.
31. Prasad, R., N. Prasad, J.E. Harrell, J. Thornby, J.H. Liem, P.T. Hudgins, and J. Tsuang (1979) Aryl hydrocarbon hydroxylase inducibility and lymphoblast formation in lung cancer patients. Int. J. Cancer 23:316-320.
32. Rudiger, H.W., F.V. Kohl, V. Mangels, P. Wichert, C.R. Bartram, W. Wohler, and W.E. Passarage (1976) Benzo(a)pyrene induced

sister chromatid exchanges in cultured human lymphocytes. Nature (Lond.) 262:290-292.

33. Schurer, C.C., C.R. Bartram, H.R. Glatt, F.V. Kohl, W. Mangels, F. Oesch, and H.W. Rudiger (1980) Benzo(a)pyrene-4,5 oxide discrepancy between induction of sister chromatid exchange and binding to DNA in cultured human fibroblasts. Biochim. Biophys. Acta 609:272-277.

34. Bayer, U., M. Younes, and C.P. Siegers (1981) Enhancement of benzo(a)pyrene-induced sister chromatid exchange as a consequence of glutathione depletion in vivo. Toxicol. Lett. 9:339-349.

35. Lambert, B., A. Lindblad, K.G. Holmberg, and D. Francesconi (1982) The use of sister chromatid exchange to monitor human populations for exposure to toxicologically harmful agents. In Sister Chromatid Exchange, S. Wolff, ed. John Wiley and Sons, New York, pp. 149-151.

36. Deaven, L.L., and E.W. Campbell (1980) Factors affecting the induction of chromosomal aberrations by cadmium in Chinese hamster cells. Cytogenet. Cell Genet. 26:251-260.

ANALYSIS OF SISTER CHROMATID EXCHANGES IN BLOOM SYNDROME BY USE OF ENDOMITOTIC AND THREE-WAY DIFFERENTIATION PROCEDURES

Yukimasa Shiraishi

Laboratory of Human Cytogenetics
Department of Anatomy
Kochi Medical School
Nankoku-City 781-51, Kochi, Japan

INTRODUCTION

Bloom syndrome (BS) is an autosomal recessive disorder characterized by pre- and postnatal growth retardation, a narrow face with dolicocephaly, sun-sensitive eruption of the face, an increased rate of chromosomal instability and sister chromatid exchanges (SCEs), and an increased risk of cancer (1). The most prominent cytogenetic characteristic is an increased rate of SCEs in cells labeled with bromodeoxyuridine (BrdUrd) for 2 cell cycles (2,3), although the exact mechanisms leading to, and the biological significance of, SCEs remain unknown.

Cell fusion studies have reported normalization of BS SCEs in somatic cell hybrids (4-6). In our report (6), when BS cells labeled with BrdUrd for 1 round of DNA replication were fused with nonlabeled normal cells, the hybrid cells had a normal level of SCE at the first mitosis after fusion. However, recent studies of single and twin SCEs in endoreduplicated normal cells have shown that an equal number of SCEs occur in each of the 2 cell cycles (7,8). On the basis of these data, only a 50% reduction in BS SCE frequency could theoretically be expected to be observed 1 cell cycle after fusion. However, a clearcut normalized level of SCE was observed; therefore, the theoretical half reduction was clearly inconsistent with our cell fusion data. In order to clarify this point, the exact frequency of SCEs induced in the first, second, and third cell cycles should be analyzed in BS cells labeled with BrdUrd. Endomitotic analysis is one way to analyze which SCEs occur during the first and second cell cycles. Each SCE that arises during the S_1 phase appears as a twin SCE at the second mitosis (M2 phase), and an exchange that occurs during the S_2 phase gives rise to a single SCE in only 1 of the pair of daughter chromosomes at the M2 phase. Three-way differentiation analysis is also available for analyzing SCE induction during the first, second, and third cell cycles. The present report describes the analysis of BS SCEs with the use of endomitosis and three-way differentiation in cells labeled with BrdUrd.

To induce and analyze endomitoses and three-way differentiation, a relatively longer period of culture is necessary. Endomitosis requires 4 da and three-way differentiation requires that the cell cycle proceed smoothly for 3 cell cycles in the presence of BrdUrd. Therefore, phytohemagglutinin (PHA)-stimulated short-term lymphocyte cultures are not adequate for these analyses. The establishment of B-lymphoid cell lines is particularly well-suited for these purposes. Lymphoid cell lines derived from blood cells appear to have an infinite life-span in vitro, and once established in culture, propagate rapidly with comparative ease, when compared to skin fibroblasts. Fortunately, using Epstein-Barr virus (EBV), we have established several BS B-lymphoid cell lines which retained the original high levels of SCEs. Using these BS B-lymphoid cell lines, we have analyzed the timing of BS SCE occurrence using endomitotic and three-way differentiation analyses.

Analysis of Sister Chromatid Exchanges in Endoreduplicated Normal and BS Cells

A permanent BS B-lymphoid cell line (BS_{1-2}), which retained its original increased SCE character in 100% of the cells, and a normal B-lymphoid cell line (KS-86), established from a patient with BS and from a normal subject by using EBV, were used for the endomitotic analysis. To induce endoreduplication of human cells, we used the spindle inhibitor Colcemid technique (8). The method involved an 8-hr treatment with Colcemid (0.1 μg/ml) in the middle of the BrdUrd incubation--that is, between 22 and 30 hr, which is during the late part of first mitosis. After this treatment, the Colcemid-containing medium was discarded, and the cells were reincubated in fresh medium containing BrdUrd for an additional 48 hr. Table 1 summarizes an analysis of baseline S1 phase and S2 phase exchanges in endoreduplicated normal and BS cells. In normal cells, the ratio of single (6.30 SCEs per cell) to twin (2.92 SCEs per cell) SCEs was 2:1 on the endoreduplicated-cell basis, and 1:1 on the diploid-cell basis as is usually observed in human diploid cells by the ordinary SCE method (8).

Based on the above background, we analyzed the frequencies of twin and single SCEs in more than 45 endoreduplicated BS metaphases. A many-fold increase in single SCEs was detected; 144.8 single SCEs were counted on the average in endoreduplicated BS metaphases (Tab. 1). This increase was present in all endomitoses labeled with

Tab. 1. Frequencies of twin and single SCEs in endoreduplicated chromosomes in normal and BS cells.*

	Normal	Ratio	BS	Ratio
Endoreduplicated cell scored	27		45	
Mean twin SCEs/cell (Mean S_1 SCEs)	2.92		5.9	
Mean single SCEs/cell	6.30	singles/twin = 2	144.8	singles/twin = 25.1
Mean SCEs/S_2 diploid cell	3.15	$S_2/S_1 = 1$	72.4	$S_2/S_1 = 12.3$
Total SCEs($S_1 + S_2$)/diploid cell	6.07		78.3	
Mean SCEs in diploid cells	5.84†		79.6†	

*S_1 and S_2 refer to S_1 and S_2 phases of the cell cycle.

†Number of cells counted = 100.

BrdUrd for 2 cell cycles. In contrast, twin SCEs were rare, and only 5.9 twin SCEs were counted in spite of the many-fold increase in single SCEs.

The localization of twin and single SCE break-points was carefully analyzed in BS endomitoses because the exact evaluation of single and twin SCEs depends on the resolution limits of the SCE break-points. Figure 1 arranges several sets of diplochromosomes from BS endomitoses, which show single (short arrows) and twin (long arrows) SCEs. As clearly seen, a difference in the exact site of the break-point makes a clear distinction between single and twin SCEs though both of them are closely linked. In BS cells, the ratio of single (144.8 SCEs per cell) to twin (5.9 SCEs per cell) SCEs was

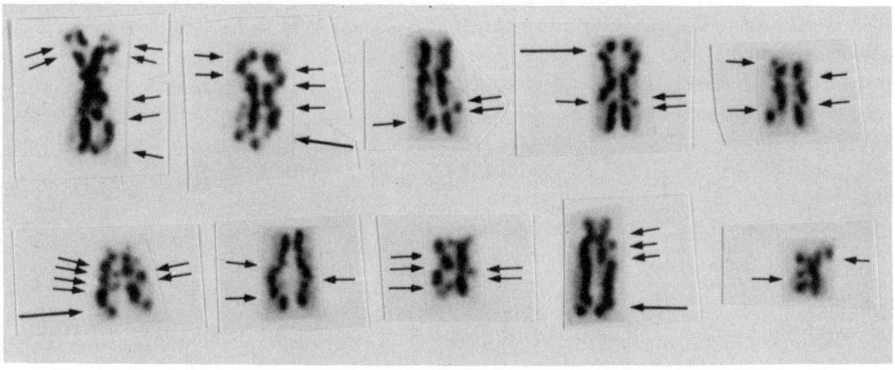

Fig. 1. Several pairs of BrdUrd-labeled BS diplochromosomes, indicating single (short arrows) and twin (long arrows) SCEs. A difference in the exact site of break-point can be seen in the single and twin SCEs, although both of them are closely linked.

25:1 on the endoreduplicated-cell basis, and 12:1 on the diploid-cell basis; the rates of S1 phase and S2 phase exchanges were 5.9 and 72.4 SCEs per cell, respectively. The total SCE number (S1 + S2 = 78.3) corresponds well to that of BrdUrd-labeled diploid BS cells (79.6). This finding strongly indicates that most of the BS SCEs occur during the second cell cycle.

Analysis by the Three-way Differentiation of BS SCE in BrdUrd-Substituted Chromosomes

Schvartzman (10) demonstrated that 3 levels of staining densities (three-way differentiation) could be achieved in the chromatids by either increasing or decreasing the amount of BrdUrd incorporated into the chromosomal DNA during each of the 3 consecutive generations.

The procedures for three-way differentiation make it possible to distinguish 3 levels of staining (dark, intermediate, and light staining) and to analyze the 4 different types of SCEs among these 3 densities of chromatid labeling. There are 2 types of three-way differentiation due to the differing amounts of BrdUrd incorporated during each of the 3 successive cell cycles (10,11) (Figs. 2 and 3). Even though both kinds of labeling patterns were observed in normal cells, only 1 of the labeling patterns proved to be usable in BS cell three-way differentiation because of the extreme sensitivity of BS cells to BrdUrd. BS cells which incorporated BrdUrd heavily during the first S phase do not survive to the third mitosis making three-way differentiation impossible with this protocol. Therefore, BS three-way differentiation analyses were performed according to the BrdUrd labeling pattern of Fig. 3 in Schvartzman (10). Figure 3 arranges several sets of three-way differentiated metaphase chromosomes from normal (Fig. 2a) and BS (Fig. 2b) cells. The exact SCE break-points were indicated by 3 sizes of arrows. As clearly seen from these figures, a difference of staining intensity at the SCE break-point makes a clear distinction between SCEs which occur during each of the 3 cell cycles. Sister chromatid exchanges that occurred during the first cell cycle (SCE_1) appear as a straight arrangement (large arrows in Fig. 2) of dark and intermediate stainings in a single chromatid with the opposite chromatid being lightly stained. Sister chromatid exchanges that occurred during the second cell cycle (SCE_2) are recognized as an exchange between dark and intermediate stainings of chromatids with the remaining part of both chromatids being lightly stained (middle arrows in Fig. 2). SCE_3 appears as 2 types of SCEs, 1 between dark and dark staining of the chromatids and the other between light and light staining of the chromatids with the remaining part of both chromatids being lightly stained (small arrows in Fig. 3).

Table 2 summarizes the analysis on SCE_1, SCE_2, and SCE_3 in three-way differentiated normal and BS cells. In KS-86 cells 45 M3 cells and 50 M2 were analyzed. In the normal M3 cells the mean frequencies of SCE_1, SCE_2, and SCE_3 were 2.44 ± 0.11, 2.88 ± 0.15, and

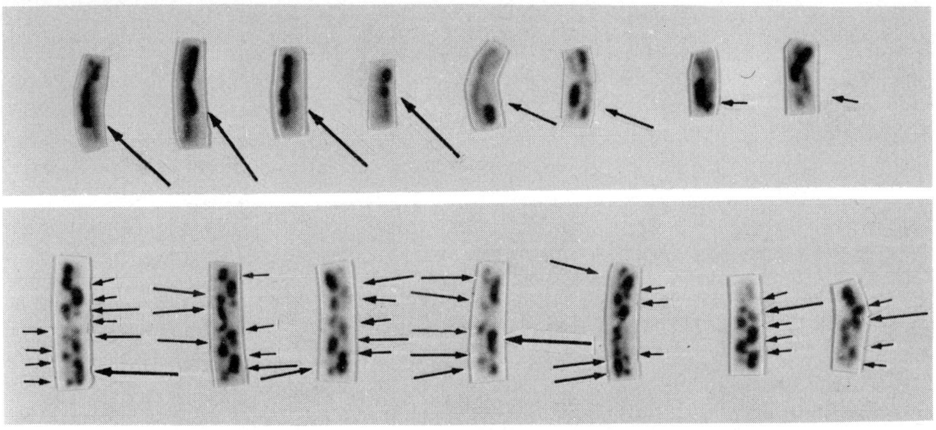

Fig. 2. Arrangements of several sets of three-way differentiated metaphase chromosomes from normal (upper row) and BS (lower row) cells. The left 2 chromosomes in the upper row (normal) show the incorporation pattern of Schvartzman's first model (10). The other arranged chromosomes in that row correspond to the other model (10). Large arrows denote SCE_1; moderately sized arrows denote SCE_2; small arrows denote SCE_3.

4.08 ± 0.14, respectively. The ratio of SCE_1 to SCE_2 was approximately 1:1, and the difference was not statistically significant ($P > 0.1$), while the value of SCE_3 was significantly higher than SCE_1 and SCE_2 ($P < 0.01$). The combined frequency of SCE_1 and SCE_2 was 5.82 ± 0.13 per M3 metaphase which corresponds almost exactly to the 5.6 ± 0.18 exchange frequency of SCEs observed in the M2 cells from the same cell line. In BS_{1-2} cells 51 M3 cells and 50 M2 cells were carefully analyzed. In the BS M3 cells the exact frequency of BS SCE_1 was rare and the mean value of SCE_1 was 3.28 ± 0.13. This value corresponds well to that of twin SCEs observed in BS endomitoses in our previous study (9). In contrast, high frequencies of SCE_2 and SCE_3 were observed in three-way differentiated BS cells, with mean frequencies of 63.32 ± 2.13 (SCE_2) and 73.08 ± 2.15 (SCE_3), respectively. Mean SCE_3 was slightly higher than that of SCE_2, although there was variation in SCE_2 and SCE_3. As seen from Tab. 1, the values of SCE_2 and SCE_3 in BS cells vary from a high of 77 and 89, respectively, to a low of 38 and 46, respectively, although SCE_1 value was fairly constant at 2-6. In BS cells, the rates of SCE_1 and SCE_2 were 3.28 ± 0.13 and 63.32 ± 2.13, and the ratio of SCE_1 to SCE_2 (1:19) corresponds well to that of single (= 144.3 SCE/cell) to twin (= 4.6 SCE/cell) SCEs, 1:25 on the endoreduplicated-cell basis, and 1:12 on the diploid cell basis. The combined frequencies of SCE_1 and SCE_2 frequencies (3.28 + 63.32 = 66.60) corresponds well to that of BrdUrd-labeled diploid BS cells (75.9).

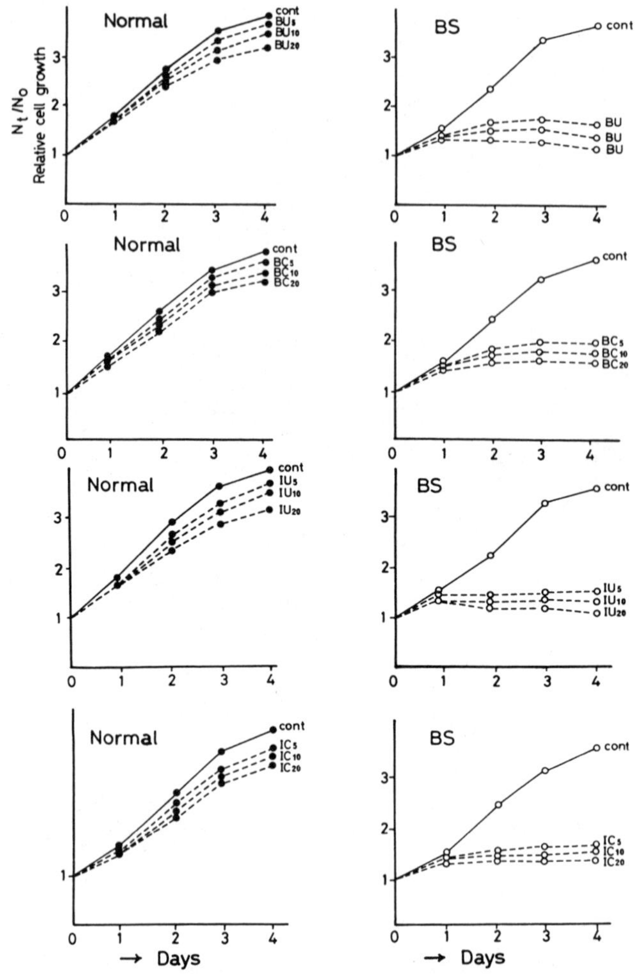

Fig. 3. Cell growth effects of BrdUrd (BU), BrdCyd (BC), IudUrd (IU), and IudCyd (IC). N_o, viable cell number at time 0 (initial); N_t, viable cell number at a given time; cont, control.

Tab. 2. Sister chromatid exchange analysis of normal and BS cells in the first (SCE_1), second (SCE_2), and third (SCE_3) cell cycles with the three-way differential staining technique.

		SCE_1	SCE_2	SCE_3
Normal cells (KS86)	Mean ± S.E.	2.44 ± 0.11	2.88 ± 0.15	4.08 ± 0.15
BS_{1-2}	Mean ± S.E.	3.28 ± 0.13	63.32 ± 2.13	73.08 ± 2.15

Effects of Bromo-Compounds on Cell Growth and Cell Killing

In order to evaluate further the mechanism of BS SCEs occurring when BrdUrd-containing DNA is used as the template for replication, we have tested the effects of bromo- and iodine-compounds on cell growth and on SCEs in BS as well as in normal cells.

Effects of BrdUrd, bromodeoxycytidine (BrdCyd), iododeoxyuridine (IudUrd), iododeoxycytidine (IudCyd), 5-bromouracil (5BU), 5-bromouridine (5BUN), deoxyuridine (dUrd), and deoxycytidine (dCyd) in various concentrations on cell growth and on the frequency of SCEs were examined in normal (KS86) and BS cells. Chemical treatment effects were evaluated on relative cell growth, which was measured with the ratio of counted cells at 1, 2, 3, and 4 da to initial cell number (5×10^5 cells) in normal as well as in BS cultures (12). All experiments were performed 3 times with qualitatively similar results.

In the control cultures, the growth patterns of these lines were very similar during exponential growth phase, with doubling times ranging from 30-36 hr (Figs. 3-4), although BS cell lines grew slightly more slowly than the normal cells. The generation times (cell cycle time) of these lines were about 24 hr. Even though BS cells grew at nearly the same rate as control cells during the first day of culture in the presence of BrdUrd, BrdCyd, IudUrd, and IudCyd, very little cell growth was observed in BS_{1-2} in the presence of these compounds after 2 da of culture. These compounds also produced differential staining of sister chromatids enabling an evaluation of SCE frequencies. However, the growth curve for treatment with dUrd, dCyd, 5BU, and 5BUN for both the normal and the high SCE BS cell lines were almost the same, and dUrd and dCyd exhibited almost no inhibition of cell growth. Interestingly, these compounds did not produce differential staining of the sister chromatids, indicating that the bromo-compounds, 5BU and 5BUN, were probably not incorporated into DNA.

Sister chromatid exchange and the differential staining characteristics of various compounds are presented in Tabs. 3 and 4. Table 3 presents the data on SCEs and differential staining using BrdUrd, IudUrd, BrdCyd, IudCyd, dUrd, dCyd, 5BU, and 5BUN in normal cells at concentrations of 5, 10, and 20 µg/ml. Differential staining and SCE were detected at concentrations of 10 and 20 µg/ml in IudUrd- and BrdCyd-treated cells and SCE increased with dosage. The mean value of SCEs in KS86 cells varied from 4.6 to 5.2 per cell. Bromodeoxyuridine and IudUrd were slightly more toxic than BrdCyd and IudCyd. Table 4 presents the data on SCEs induced in BS_{1-2} (the high SCE BS cell lines). The same concentrations of the same compounds produced differential staining as in normal cells. Mean SCE values in BS cells ranged from 68.4 to 75.3 SCE per cell. The same staining characteristics as before were observed in these cells. The difference from normal cells (KS86) in SCE rates (4.8-5.4 per

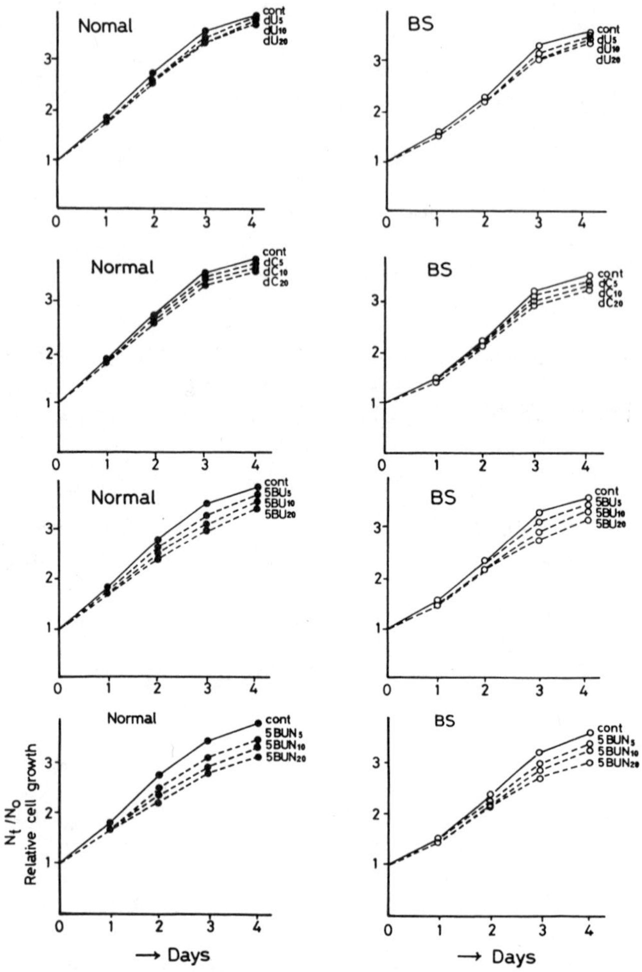

Fig. 4. Cell growth effects of dUrd (dU), dCyd (dC), 5BU, and 5BUN. N_o, viable cell number at time 0 (initial); N_t, viable cell number at a given time; cont, control.

cell) is not statistically significant. When the cell line was cultured in the presence of dUrd, dCyd, 5BU, and 5BUN at concentrations of 5, 10, and 20 µg/ml, differential staining was not obtained, indicating that 5BU and 5BUN were probably not incorporated into DNA.

DISCUSSION

The data from endomitotic analysis and three-way differentiation reveal the cell cycle timing of the occurrence of BS SCEs. Most BS SCEs in BrdUrd-labeled cultures occur during the second and

Tab. 3. Sister chromatid exchanges and differential staining for sister chromatid differentiation in normal cells (KS86) treated with bromo- and iodine-compounds, dUrd, and dCyd. (BUdR = BrdUrd; IUdR = IrdUrd; BCdR = BrdCyd; ICdR = IudCyd.)

Treatment (μg/ml)	Differential staining for SCD	Cells scored	SCE/cell Mean ± S.E.
BUdR(5)	Yes	100	4.6 ± 0.21
BUdR(10)	Yes	100	4.9 ± 0.29
BUdR(20)	Yes	85	5.1 ± 0.31
IUdR(5)	No	—	—
IUdR(10)	Yes	95	4.7 ± 0.28
IUdR(20)	Yes	85	5.2 ± 0.32
BCdR(5)	No	—	—
BCdR(10)	Yes	100	4.6 ± 0.26
BCdR(20)	Yes	95	4.8 ± 0.24
ICdR(5)	No	—	—
ICdR(10)	Yes	100	4.7 ± 0.25
ICdR(20)	Yes	100	4.9 ± 0.29
dU(5)	No	—	—
dU(10)	No	—	—
dU(20)	No	—	—
dC(5)	No	—	—
dC(10)	No	—	—
dC(20)	No	—	—
5BU(5)	No	—	—
5BU(10)	No	—	—
5BU(20)	No	—	—
5BUN(5)	No	—	—
5BUN(10)	No	—	—
5BUN(20)	No	—	—

third cell cycles when BrdUrd-containing DNA is used as the template for replication. The analysis of single and twin SCEs in BrdUrd-labeled endomitosis demonstrated that although an equal number of SCEs occurred in each of the 2 cell cycles in normal cells, most of BS SCEs appeared during the second cell cycle as single SCEs. Three-way differentiation was analyzed by the 2 models proposed by Schvartzman (10,11). In 1 of the 2 models, BrdUrd is heavily incorporated during the first cell cycle, decreasing in each successive cycle. In the other model, the BrdUrd incorporation rate is comparatively low in the first cell cycle, increasing in successive cell cycles. Even though three-way differentiation according to both models was detected in normal cells, only three-way differentiation

Tab. 4. Sister chromatid exchanges and differential staining for sister chromatid differentiation in BS cells (BS_{1-2}) treated with bromo- and iodine-compounds, dUrd, and dCyd.

Treatment (μg/ml)	Differential staining for SCD	Cells scored	SCE/cell Mean ± S.E.
BUdR(5)	Yes	100	69.8 ± 3.02
BUdR(10)	Yes	100	70.2 ± 3.14
BUdR(20)	Yes	85	71.9 ± 3.21
IUdR(5)	No	—	—
IUdR(10)	Yes	85	72.2 ± 3.09
IUdR(20)	Yes	85	75.3 ± 3.41
BCdR(5)	No	—	—
BCdR(10)	Yes	90	68.4 ± 2.84
BCdR(20)	Yes	85	69.8 ± 3.02
ICdR(5)	No	—	—
ICdR(10)	Yes	84	67.9 ± 2.98
ICdR(20)	Yes	81	70.5 ± 3.21
dU(5)	No	—	—
dU(10)	No	—	—
dU(20)	No	—	—
dC(5)	No	—	—
dC(10)	No	—	—
dC(20)	No	—	—
5BU(5)	No	—	—
5BU(10)	No	—	—
5BU(20)	No	—	—
5BUN(5)	No	—	—
5BUN(10)	No	—	—
5BUN(20)	No	—	—

according to the latter model was possible with BS cells. This finding indicates that BS cells are highly sensitive to BrdUrd and that BS cells which incorporated BrdUrd heavily during the first cell cycle can not proceed to the third cell cycle. This is supported also by other experiments in which no cell growth was seen in BS cultures after 2 da in the presence of BrdUrd at concentrations of 5, 10, and 20 μg/ml, although normal cells grew even in the presence of BrdUrd at these concentrations. Since the SCE level during the first cell cycle is nearly normal in three-way differentiated BS cells which have proceeded for 2 or 3 successive cell cycles, it seems plausible that if BS cells incorporated BrdUrd heavily and a high frequency of SCEs occurred in the first cell cycle, these cells could no survive to the third cell cycle. Therefore, all the data from endomitotic analysis and three-way differentiation point to the conclusion that most BS SCEs are caused by BrdUrd, although the exact spontaneous SCE level in vivo still remains unknown.

Moreover, it is unreasonable to suppose that the high SCE frequency is occurring in vivo in BS patients because when normal cells were treated with various alkylating agents (4-nitroquinoline-1-oxide, mitomycin C, etc.), cells with above 70 SCEs per cell cannot survive for a long term. It is unthinkable that the BS SCE levels of up to 70-80 SCE per cell could exist and the BS patients still be surviving. Further, even though anemia and other severe symptoms are expected to be associated with a situation where an elevated level of SCEs occur in vivo, anemia and other severe symptoms are not seen in BS patients.

Another indication that SCE levels in vivo may be near normal (unless stimulated by BrdUrd and some carcinogens or mutagens) is to be found in the establishment of IgM-positive BS B-lymphoid cell lines with the characteristic high level of SCEs. The cell lines used in our experiments are all B-lymphoid cell lines which are producing immunoglobulin both in the cytoplasm and at the cell membrane. Fortunately, we have established 3 IgM-positive cell lines that have high SCE values in 100% of the cells. Recently, the model that immunoglobulin differentiation (IgM--IgG) occurs with an SCE has been proposed by Honjo et al. (13). Three IgM-positive cell lines have maintained their IgM character for more than 1 yr since their establishment. Therefore, if a high rate of spontaneous SCEs in BS cells existed, it should have resulted in the differentiation of these B-cell lines from IgM to IgG, an event which did not occur. From these considerations it seems reasonable to suppose that the spontaneous SCE level in vivo and in vitro is normal. Our recent research using chemical agents (SCE-inducing agents) support the SCE model for immunoglobulin differentiation. IgM-positive BS B-lymphoid cell lines have changed to IgG or IgA following mutagen treatment. This finding is of special interest in clarifying the biological significance of SCEs and further experiments are now in progress.

Since a high level of SCEs can be induced easily in these cell lines, they also should be usable as a highly sensitive system for detecting mutagens and carcinogens in the environment.

ACKNOWLEDGEMENTS

This work was supported by grants-in-aid from the Ministry of Education, Science and Culture, from the Ministry of Health and Welfare of Japan, and from the Nissan Science Foundation. We are grateful to Michiyo Ozawa for her technical assistance.

REFERENCES

1. German, J. (1969) Bloom's syndrome. I. Genetical and clinical observation in the first twenty-seven patients. Am. J. Human Genet. 21:196-227.
2. Chaganti, R.S.K., S. Schonberg, and J. German (1974) A manyfold increase in sister chromatid exchanges in Bloom's syndrome lymphocytes. Proc. Natl. Acad. Sci., USA 71:4508-4512.
3. Shiraishi, Y., A.I. Freeman, and A.A. Sandberg (1976) Increased sister chromatid exchange in marrow and blood cells from Bloom's syndrome. Cytogenet. Cell Genet. 17:162-173.
4. Bryant, E.M., H. Hoehn, and G.M. Martin (1979) Normalisation of sister chromatid exchange frequencies in Bloom's syndrome by euploid cell hybridisation. Nature 279:795-796.
5. Alhadeff, B., M. Velivasakis, I. Pagan-Charry, W.C. Wright, and M. Siniscalco (1979) High rate of sister chromatid exchanges of Bloom's syndrome chromosomes is corrected in rodent human somatic cell hybrids. Cytogenet. Cell Genet. 27:8-23.
6. Shiraishi, Y., S.I. Matsui, and A.A. Sandberg (1981) Normalization by cell fusion of SCE in Bloom syndrome lymphocytes. Science 212:820-822.
7. Sasaki, M.S. (1977) Sister chromatid exchange and chromatid interchange as possible manifestation of different DNA repair processes. Nature 269:623-625.
8. Kano, Y., and Y. Fugiwara (1982) Higher inductions of single and twin sister chromatid exchanges by crosslinking agents in Fanconi's anemia cells. Human Genet. 60:233-238.
9. Shiraishi, Y., T.H. Yosida, and A.A. Sandberg (1982) Analysis of single and twin sister chromatid exchanges in endoreduplicated normal and Bloom syndrome B-lymphoid cells. Chromosoma (Berl.) 87:1-8.
10. Schvartzman, J.B. (1979) Three-way differentiation of sister chromatids in 5-bromodeoxyuridine-substituted chromosomes. J. Heredity 70:423-424.
11. Schvartzman, J.B., and V. Goyanes (1980) A new method for the identification of SCEs per cell cycle in BrdU-substituted chromosomes. Cell Biol. Int. Rev. 4:415-419.
12. Isida, R., and M. Buchwald (1982) Susceptibility of Fanconi's anemia lymphoblasts to DNA-cross-linking and alkylating agents. Cancer Res. 42:4000-4006.
13. Honjo, T., S. Nakai, Y. Nishida, T. Kataoka, Y. Yamawaki-Kataoka, N. Takahashi, M. Obata, A. Shimizu, Y. Yaoita, T. Nikaido, and N. Ishida (1981) Rearrangements of immunoglobulin genes during differentiation and evolution. Immunol. Rev. 59.

COMPLEMENTATION STUDIES IN MURINE/HUMAN HYBRIDS SUGGEST MULTIPLE ETIOLOGY FOR INCREASED RATE OF SISTER CHROMATID EXCHANGE IN MAMMALIAN CELLS

Becky Alhadeff and Marcello Siniscalco

Sloan-Kettering Institute for Cancer Research
New York, New York 10021

SUMMARY

Two mutational changes which occurred in culture and are associated with a high rate of sister chromatid exchange (SCE) phenotype have been identified in the L-A9 murine cell genome by means of complementation studies with somatic cell hybrids. Preliminary cytogenetical evidence suggests that the retention of human autosome 6 (namely the region comprised between Xq12 and Xqter) or human autosome 19 is required in the hybrid metaphases for complementation to occur, independently of their being derived from normal human or Bloom syndrome (BS) cells. These data and other complementation studies previously reported by our group and by other investigators suggest that mammalian cells may possess several independent systems involved in the control of SCEs during chromatid replication. Thus, the high rate of SCE can be regarded as the common phenotype resulting from a variety of qualitative or quantitative changes affecting the mammalian cell genome. Bloom syndrome is evidently an example of homozygosity for a recessive mutation occurring in nature. The high SCE mutants found among rodent cells (as those seen in unstable rodent-human hybrid cells) are more likely the result of chromosomal loss or rearrangement occurring in culture at one or more of the genetic systems hypothesized above. The occurrence of complementation within or between the species barrier, following cell hybridization or cocultivation, indicates the recessive nature of the corresponding mutations and the possible homology of the relevant genetic systems in different mammalian species.

The isolation of rodent clonal cell lines with a stable high rate of SCEs and the production of somatic cell hybrids between them and BS cells offer a promising experimental tool for studying the biology of SCEs in general and the genetics of BS in particular.

INTRODUCTION

We have briefly described in earlier reports (1,2) the finding of an unusually high rate of SCE in an established rodent cell line (L-A9) which is an 8-azaguanine-resistant mutant derivative (3) of the C3H mouse tumor L cell line NTCS 929 (4). Since both the L cells and the C3H mouse diploid fibroblasts were found to have a normal rate of SCEs, we concluded that the increased rate of SCE constantly observed in 2 subpopulations of the L-A9 cell line must have been the result of one or more irreversible changes which occurred in culture. We showed also that the aberrant rate of SCEs in the L-A9 chromosomes could be fully corrected in A9 human somatic cell hybrids and that, in general, the rate of SCEs in the hybrid cell chromosomes is inversely correlated with the number of human chromosomes retained by each hybrid metaphase. In particular, we found that in a hybrid line which had retained only the human X chromosome, the rate of SCE had reverted to the high level of the parental murine line. Additional studies clearly indicate now that, whatever its (their) nature, the lesion(s) responsible for the high SCE frequency of the A9 chromosomes must be different from that which characterizes the chromosomes of patients with BS, since somatic cell hybrids between high rate SCE L-A9 cells and BS fibroblasts show mutual complementation for SCEs. Complementation experiments with somatic cell hybrids of the kind referred to above represent a convenient experimental tool for studying the biology of SCE in general and the genetics of BS in particular. With this perspective in mind we wish to report in full the results of an extensive study carried out in our laboratory on the variation of SCE frequency in somatic cell hybrids between the high rate SCE L-A9 cells with normal and BS human cells.

MATERIAL AND METHODS

Murine Cells

The murine L-A9 cell line is a forward deficient mutant at the locus for hypoxanthine-guanine phosphoribosyl transferase (HGPRT) (E.C.: 2.4.2.8) isolated by Littlefield (3) from the NCTS clone 929 (4) of the mouse L cells originally established by Earle (5) from normal subcutaneous tissue from a C3H-An male mouse. In addition to the HGPRT deficiency, the L-A9 cells, as well as the L cells, are also deficient (6) in adenine phosphoribosyl transferase (APRT) (E.C.: 2.4.2.7).

The L-A9 cell populations used in the present study are labeled L-A9-1 (SKI), L-A9-2 (AE) and L-A9-3 (NIH) to identify their origin. The first 2 are subcultures derived from the same thawing of the original mutant cell line L-A9 of Littlefield (3), that one of us (M.S.) obtained from Dr. H. Harris in 1968 (L-A9-Oxford). However, these 2 L-A9 subcultures were stored for different lengths of time in liquid nitrogen and since 1975 were propagated separately in 2 different laboratories, i.e., the Somatic Cell Genetics section of the Sloan-Kettering Institute and the Department of Genetics of the Albert Einstein College of Medicine. The line L-A9-3 (NIH) was separately derived (7) by subcloning of a much later thawing of the original L-A9 mutant cell line (3).

The RAG/Ag cell line is a forward HGPRT-deficient mutant derived by Dr. Klebe et al. (8) from a renal adrenal carcinoma transplantable tumor of the Balb/cd mouse strain (9). Both the L-A9 and RAG/Ag cells are resistant to 8-azaguanine and unable to grow in the HAT selective medium of Szybalski and Szybalska (10). L cells (clone NCTC 929) were obtained from the American Type Culture Collection. A quasi-diploid murine fibroblastic strain was established from a primary explant of the spleen of a C3H/An mouse. The genealogies of the above-described murine cell lines and strains are summarized in Tab. 1.

Tab. 1. Genealogy of the murine cell lines and strains use.

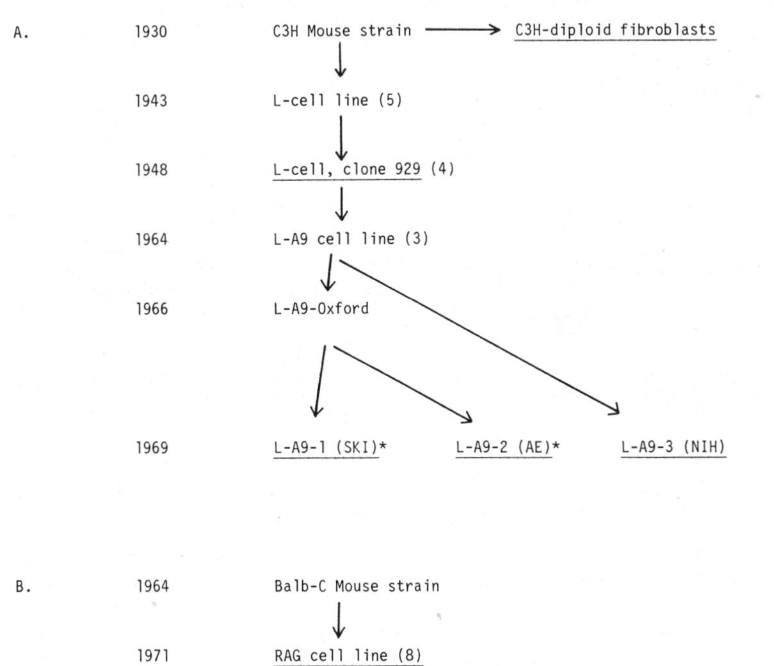

The cell types underlined are those tested for the SCE phenotype. The ones with a high rate are marked by an asterisk.

Human Cells

Fibroblastic strain ME, established from a patient with Gardner syndrome (11), was kindly made available to us by Dr. Kopelovich. Human fibroblastic strains, GM73 and GM97, were obtained from the Camden Repository. These were established (12,13) from heterozygous carriers of balanced X-autosome translocations involving respectively the X chromsome with autosome 14 [46X,X,t(X;14)(q13,q32)] and the X chromosome with autosome 1 [46X,X,t(X;1)(q26,q12)]. Human fibroblast strain IT was established by Dr. Schwinger from a carrier of a balanced translocation between autosomes 6 and 7 [46,XX,t(6;7)(q12;-q14)] and made available to us by Dr. Grzeschik (14).

Several peripheral blood short-term cultures established from lab staff members (B.A., M.P.) were used to determine the rate of SCE in normal humans with our experimental protocol. Finally, the rate of SCE from patients with BS was determined on the fibroblastic strain GM1492 obtained from the Camden Repository.

Hybrid Cells

The murine human hybrid cells analyzed in the present study for their rate of SCE in murine and human chromosomes were derived at different times from independent fusions carried out in our own laboratory (when not otherwise stated) between the L-A9 cells with a high SCE rate and human cells of different types:

(i) A9/ME-10-4 and A9/ME-10-5 are secondary hybrid clones derived from the Sendai virus-induced fusion of L-A9-1 (SKI) cells with normal rate SCE human diploid fibroblasts (ME).

(ii) A9/ME-10-5-B and the series of clones A9/ME-10-5:1, 2, 4, 7, 12, 16, 20, 25, 28 and 30 are tertiary clones derived from the secondary hybrid A9/ME-10-5.

(iii) A9/GM73-2b is a secondary hybrid clone derived from a Sendai virus-induced fusion of L-A9-2 (AE) cells with diploid fibroblastic cells GM73 (15).

(iv) A9-IT-1 and A9-IT-1-13 are respectively a primary and a secondary hybrid clone from a PEG-induced fusion between L-A9-2 (AE) cells and diploid fibroblasts from a carrier of a 6-7 reciprocal balanced translocation (14).

(v) A9/HRBC2 is a mouse-human hybrid cell line from our cell bank, which, after long propagation in HAT selective medium, has retained only the human X chromosome and a fragment of autosome 2 (16,17). This hybrid line had been obtained in 1968 by Dr. O.J. Miller through Sendai virus-induced fusion between L-A9 (Oxford) cells and nucleated peripheral blood cells from an erythroblastotic newborn male infant, with red cell precursors making up 50% of the nucleated cells (17).

The other hybrid lines referred to in the following pages were freshly prepared (for the reasons given in the Results section) by using the PEG fusion protocol described in Ref. 2. The genealogies of all hybrid lines analyzed are summarized in Tab. 2.

Media

All the cell strains and lines were grown in monolayer cultures in Falcon T flasks. The basic culture medium for both cell strains and lines was MEM supplemented with 10% FCS, 2 mM glutamine, nonessential amino acids, chlorotetracycline (25 μg/ml), penicillin (50 units/ml), and streptomycin (50 μg/ml). The hybrid lines A9/HRBC2 and A9/GM73-2b were normally propagated in the HAT selective medium of Szybalski and Szybalska (10), prepared according to Littlefield (3) and by using the MEM basic culture medium described above. However, prior to the termination of the rate of SCE, these cell lines were thoroughly washed with phosphate buffered saline (PBS) and transferred to HAT-free MEM medium to avoid the competitive effect of thymidine on the incorporation of bromodeoxyuridine (BrdUrd). Parental and hybrid lines were periodically tested for pleuropneumonialike organisms (PPLO) contamination with Hoechst 33258, which produces a characteristic fluorescence in the cytoplasm of infected cells (18). PPLO-contaminated lines were excluded from the study.

Tab. 2. Genealogy of the hybrid cell lines analyzed for their SCE phenotype. (Cell strains or lines with high SCE are marked with an asterisk).

Parental cells:		Types of Hybrid Clones:		
1. First round hybrids				
Murine	Human	Primary	Secondary	Tertiary
L-A9-1 (SKI)*	ME	A9/ME-10	A9/ME-10-4 A9/ME-10-5	Clones: 1,2,4,7,12,16,20,23, 25,28,30 and 10-5-B*
"	GM 1492*	A9/GM-1492-1		
2. Hybrids from previous studies				
Murine	Human			
L-A9-2 (AE)*	GM 73		A9/GM 73-2-b*	
"	GM 97		A9/GM 97-13-c*	
"	IT	A9/IT-1	A9/IT-1-13	
3. Second round hybrids				
Hybrid line	Human			
L-A9/ME-10-5-B*	BA	A9-ME/10-5-B/BA-1 A9-ME/10-5-B/BA-2 A9-ME/10-5 B/BA-3		
	GM 1492*	A9-ME/10-5-B/GM 1492-1		

Cytogenetic Studies

In each hybrid line, the presence and the identification of the residual human chromosomes were established by G banding (19) followed by destaining and restaining with Giemsa-11 (20). The simultaneous detection of SCE and of the human chromosomes in the man-mouse hybrid cells were performed with our fluorescence-plus-Giemsa (FPG)-Giemsa-11 protocol (21), which combines the standard FPG method of Perry and Wolff (22) with the Giemsa-11 staining procedure of Bobrow and Cross (20), as modified by Friend et al. (23). However, the concentration of BrdUrd (Sigma B5002) used for the detection of SCE in the hybrid cell was 30 µg, since this concentration was found to give sharper contrast between the different chromatids stained with the FPG-Giemsa-11 protocol.

All A9 cell lines referred to above as well as the L cells (clone 929) were found to have an average of 50 chromosomes, of which about one-third are bi-armed as a result of centromeric fusion. The latter phenomenon is of very common occurrence in established murine cell lines and it was observed also in the freshly established diploid fibroblasts from C3H mice. The RAG cells, though equally heteroploid as the L-A9 cells, have only a small number of bi-armed chromosomes.

Criteria for Assessment of SCE Rate

Throughout this study, SCE rates are expressed as the average numbers of SCEs per chromosome per metaphase. In the hybrid cells, these estimates were computed per total number of chromosomes and separately per "murine" and "human" chromosomes. The calculation of the rate of SCEs per chromosome presents a problem that has been the subject of argumentation among specialists (24). This is due to the circumstance that, on the one hand, SCEs occurring at the centromeric region of bi-armed chromosomes cannot be distinguished from the accidental twisting of one arm with respect to the other during the spreading of the metaphases on the slide. On the other hand, the occurrence of SCEs in the centromeric region of acrocentric chromosomes can be entirely overlooked. Thus, if it is true, as suggested by some studies (25), that the frequency of SCEs at the centromeric region is higher than elsewhere, the 2 kinds of biases will probably neutralize one another, since they affect the estimates of SCEs in opposite directions. In our study, the estimation of SCE rates is further complicated by the high incidence of murine bi-armed chromosomes resulting from centromeric fusion of 2 acrocentric ones. However, since this phenomenon is equally frequent among the L-A9, their murine progenitor cells L929 and the A9-human hybrid cells, it makes no difference whether the estimates of SCE are expressed per chromosome or per chromosome arm. For this reason, and for the consideration that a bi-armed chromosome--whatever its origin--replicates as a unique entity, we decided to express the SCE rate always as the ratio between the total number of SCEs and the total number of chromosomes found per metaphase. However, to attenuate the risk of overestimating the number of SCEs across the centromeric region of bi-armed chromosomes (both murine and human), we adopted the conservative criterion of considering as true SCEs only half of those amenable to this type of misclassification.

Statistical Analysis

The distributions of SCE per chromosome per metaphase have been calculated from a random sample of 25 or 50 well-spread metaphases for each cell strain or line studied. These distributions are reasonably continuous and unimodal so that the differences in their means could be evaluated in terms of the standard statistical parameters of the normal distribution. In addition to this, the different batches of data have been subjected to a nonparametric analysis with the use of the "boxplot" graphical technique of Tukey (26). Boxplots offer the advantage of displaying in a concise manner a large number of distribution data with respect to the most important features exhibited by each of them, such as the group size, the median, and the extremes of the distribution. The fiducial limits of the median at a chosen confidence level are expressed by a notched area around the median, and a rough idea of whether or not separate distributions differ significantly in their medians may be deducted from the overlap of their notched areas.

RESULTS

Rate of SCEs in L-A9 Cells and Other Murine Cell Lines and Strains

Among the 3 clones of L-A9 cells examined, 3 [L-A9-1 (SKI) and L-A9-2 (AE)] were repeatedly examined over a period of 3 yr and constantly found to exhibit a rate of SCEs per chromosome per metaphase which is significantly higher (0.79 ± 0.04 and 0.79 ± 0.2, respectively) than that found among the parental L cells (0.41 ± 0.02) and among cultured diploid fibroblasts of the C3H-An mice (0.44 ± 0.02), from which the L cells themselves were originally derived. The third clone of L-A9 cells [L-A9-3(NIH)] showed instead the same rate of SCE (0.55 ± 0.03) as the L cells and the C3H fibroblasts. The latter observation, together with the finding that the 8-azaguanine-resistant RAG cells, derived from a BalbC transplantable renal adenocarcinoma (8,9), have a normal rate of SCEs (0.35 ± 0.01), indicate that the increase in SCEs observed in the first 2 batches of L-A9 cells must be the result of a lesion which occurred in culture and that such a lesion is neither related to their HGPRT deficiency nor to their tumorogenicity. The distribution of SCEs in the above-mentioned murine line is shown in Fig. 1.

Rate of SCE in Human Cell Strains and Lines

Peripheral blood lymphocytes from a normal female (BA) and a normal male (MP) have been routinely used as control cultures. The

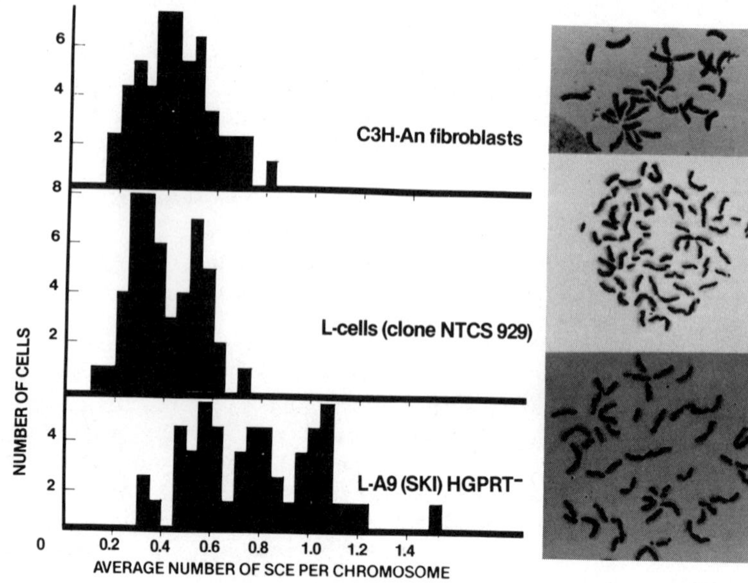

Fig. 1. Left side panels: distribution of SCEs/chromosome/metaphase from a total of 50 metaphases examined per murine cell line. The increase of SCEs is limited to line L-A9 (SKI) thus suggesting that the mutational change(s) responsible for it must have arisen in culture. Right side panels: typical appearance of normal and high SCE metaphases from the cell lines described.

rate of SCEs per chromosome in these normal short-term culture lymphocytes was found to be 0.32 ± 0.02 and 0.22 ± 0.02, respectively. BA lymphocytes have been used as parental cells in some of the hybridization experiments described below. The other fibroblastic strains ME, GM73, and GM97, used as parental human partners in several fusion experiments, were also found to have a clearly normal rate of SCE. Direct measurements of SCE rates for the human cells HRBC and IT could not be performed due to the unavailability of biological specimens at the time of the present study. However, from the clinical history of the donors it can be reasonably assumed that they had a normal rate of SCE. The rate of SCE per chromosome in the BS fibroblastic strain GM1492 was found to be 1.31 ± 0.06 when using a BrdUrd concentration of 3 μg and 1.89 ± 0.07 at a concentration of 30 μg/ml, which was the concentration regularly used for all other experiments. The increase in SCEs found with the higher concentration of BrdUrd is highly significant ($P < .001$), thus favoring the contention that BrdUrd enhances the rate of SCE in BS cells (27).

Complementation Studies for SCE in Hybrids between LA9 Cells and Human Cells of Different Types

Our previous studies (1,2) indicated that hybrid cells originating from the fusion between the high-rate SCE L-A9 cells and normal SCE human cells fall into 3 groups: (a) those with a homogeneous population of metaphases with an SCE rate as low as that of the human parental cells; (b) those with a homogeneous population of metaphases with an SCE rate as high as that of the murine parental cells; (c) those with a mixed population of metaphases with a high or low rate of SCEs. Further investigation indicated that the hybrids of the first and second groups were those which had retained the highest or the lowest number of human chromosomes, respectively (1). On the other hand, the heterogeneity in the distribution of SCEs within the same cell line was found to be at its maximum among primary hybrid clones (notoriously unstable with respect to the retention of human chromosomes) and at its minimum among secondary or tertiary clones. Within the individual hybrid metaphases, full concordance for SCE frequency was found between murine and human chromosomes (observed and expected coefficients of correlation measured on 18 hybrid cell lines: $r_{obs} = +0.59$; $r_{exp} = +0.57$, at $P = 0.01$). In particular, in 2 secondary clones derived from the fusion between the high-rate SCE 1-A9-1 (SKI) cells and normal human SCE fibroblasts (ME) we found that the SCE rate in both types of parental chromosomes was homogeneously suppressed in all metaphases (Fig. 2) inspite of the retention of only a few human chromosomes, namely 4, 6, 12, 21, and the X in A9/ME-10-4 and 6, 20, and 21 in A9/ME-10-5. Since the only human chromosomes common to both these clones are 6 and 21, we decided to further investigate the possibility that the retention of one of them was the critical factor required for the suppression of the high SCE rate of the parental murine cells.

This study was conducted with the following 3 approaches: by subcloning the secondary hybrid clone A9/ME-10-5, by testing for SCE additional hybrid lines from previous fusions between the high rate SCE L-A9-2(AE) cells and normal SCE human fibroblasts, and by producing new hybrids between the L-A9-1 (SKI) cells and the BS fibroblasts GM1492. The first approach was meant to induce the segregation of the human autosomes in the tertiary clones so that the complementing one could be identified. The second approach was required to substantiate the synteny between the suppression of SCE and the retention of a specific human chromosome with the use of a completely independent series of hybrids. The third approach would serve the double purpose of providing more data on the synteny under discussion and to confirm the previously reported (1,2) occurrence of reciprocal complementation in hybrids between L-A9-1 (SKI) murine cells and BS fibroblasts.

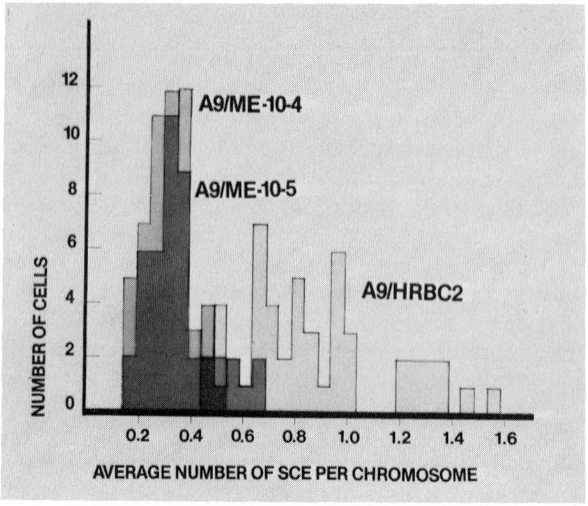

Fig. 2. Distribution of SCEs/chromosome/metaphase in 3 hybrid cell lines derived from the fusion of high SCE L-A9 cells and normal SCE human cells. Suppression of SCEs has occurred in the 2 hybrid lines A9/ME-10-4 and A9/ME-10-5 which have retained a minimum of 3 and a maximum of 6 human chromosomes among which autosomes 6, 20, and 21. The hybrid line A9/HRBC2 has retained only the human X chromosome and shows reversion to high levels of SCEs.

Suggestive Synteny between the Retention of Human Autosome 6 and Suppression of High SCE Rate of L-A9 Cells

All the 11 tertiary clones derived from the subcloning of the secondary clone A9/ME-10-5 were found to exhibit a homogeneously low rate of SCE (for both murine and human chromosomes) in each of the 25 metaphases that were analyzed per subclonal line. Cytogenetical analysis revealed the presence of autosome 6 in every metaphase. The other 2 autosomes (autosomes 20 and 21) were found to be present with a variable frequency and were completely lost in 2 clones (A9-ME-10-7 and A9-ME-10-25).

To further substantiate this suggestive synteny between the retention of autosome 6 and suppression of SCE frequency, we tested 5 additional hybrid clones derived from independent fusion experiments. Three ot these were known to have lost autosome 6, while the other 2 (derived from a human parental strain with a reciprocal translocation between autosomes 6 and 7) had retained either both translocation chromosomes (i.e., the entire autosome 6) or only the one including the region 6q12-6pter. The SCE phenotype and the present chromosomal constitution of these and of the other hybrid lines mentioned above are reported in Tab. 3a, and 3b. These data support the conclusion that an SCE-complementing human gene is probably located on the short arm of autosome 6.

Hybrids L-A9-1/BS Fibroblasts

Twenty-five metaphases were examined from a primary cell hybrid pseudoclone derived from the fusion of L-A9-1 (SKI) cells and the BS fibroblasts GM1492. The distribution of SCEs in the metaphases was found to be homogeneously low (0.32 ± 0.02 for the murine chromosomes and 0.24 ± 0.04 for the human ones) and autosome 6 was found to be regularly present in all metaphases analyzed, which had a minimum of 10 human autosomes per metaphase (Fig. 3). These results give strong support to our previous claim (1,2) that mutual complementation for SCE occurs in these types of cell hybrids and they are also compatible with the hypothesis that autosome 6 may be specifically required to suppress the high rate of SCEs typical of the L-A9 parental cells.

Tab. 3. Summary of segregation of human chromosomes versus SCE phenotype (average of 25 metaphases/cell).

Cell line or clone	Type of Clone	SCE phenotype	Human chromosomes retained in 80% of metaphases
(a) 1st round hybrids:			
A9-1 (SKI)/ME-10-4	Secondary	low	4,6,12,21,X
" /ME-10-5	"	low	6,20,21
Subclone: 2,12,16,20	Tertiary	low	6,20,21
23,28,30	"	low	6,20,21
1	"	low	6,20
7	"	low	6,21
4,25	"	low	6
10-5-B	"	high	6,20,21
A9-1 (SKI)/1492	Uncloned fusion mixture	low	1,3,4,6,7,10,11,13,14, 15,16,17,18,19,20,21, 22,X
(b) Old series hybrids			
A9-2 (AE)/GM73-2-b	Secondary	high	13,15,18,21
" /GM97-13-c	"	high	14,15,21
" /IT-1	Primary	low	4,5,t(6;7) (q12;q14), t(7;6) (q14; q12), 11, 12,14,17,18,20,21
" /IT-1-13	Secondary	low	5, t(6;7) (q12; q14), 15, 17,18,21
A9 (Oxford)/HRBC2-2	Secondary	high	X, fragment of 2
(c) 2nd round hybrids			
A9/ME-10-5-B/BA-1	Primary	low	2,3,6,11,14,16,19,20,21
" 2	"	low	6,9,13,14,15,16,19,20, 21,X
" 3	"	low	2,3,5,6,7,9,17,19,20,21
" /GM1492-1	Primary	low	1,2,3,6,16,17,19,20,21,X

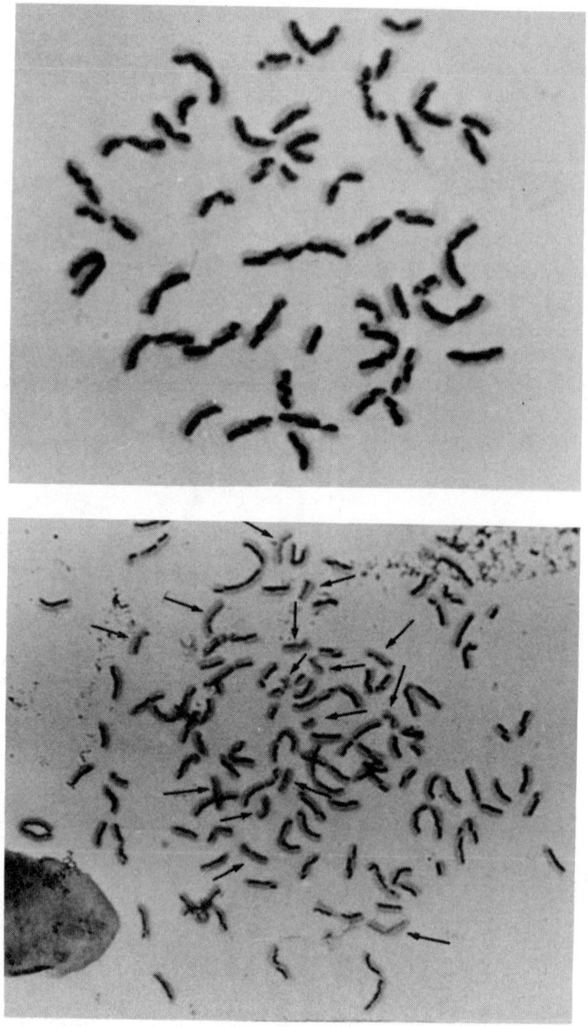

Fig. 3. Top: typical appearance of the high SCE phenotype that characterizes the L-A9 parental cells (SKI & AE). Bottom: a hybrid metaphase between the above L-A9-1 (SKI) cells and BS fibroblasts (GM1492). The suppression of SCEs in both murine and human chromosomes (indicated by an arrow) is self-evident. In the original slide preparation the murine chromosomes appear stained in red magenta whereas the human ones are stained in pale blue.

A High Rate of SCE in Cultured Murine Cells Can Be Due to the Occurrence of Different Lesions

During the last 3 yr, we have repeatedly subcultured, from small inocula, the original secondary clone A9/ME-10-5 and its derivative tertiary clones in the hope of deriving hybrid metaphases in which the loss of human autosome 6 could be directly associated with the reversal of the SCE phenotype to high levels. Unfortunately, we have not been able as yet to isolate a clone without it. However, in one occasion, the recovery of the secondary clone (L-A9/ME-10-5) from a very small inoculum yielded a tertiary pseudoclone (L-A9-ME-10-5-B) with a rate of SCE for both murine and human chromosomes even higher than that of the original L-A9 parental cells, though still retaining the same average number of murine chromosomes and the human autosomes 6, 20, and 21 in all of its cells (Tab. 3a). This hybrid pseudoclone was expanded and repeatedly checked for its rate of SCE numerous times over the last 2 yr (with intermitting periods of storage) without showing further changes either with respect to the SCE rate or to the retention of human chromosomes. These circumstances suggest that the subclonal derivative of the A9/ME-10-5 hybrid secondary clone, though remaining complemented with respect to the original lesion(s) present in the L-A9 parental cells has suffered an additional genomic loss that made it again susceptible to a high rate of BrdUrd-induced SCEs. A comprehensive description of the SCE rates observed in the entire series of parental and hybrid cell lines thus far discussed is reported in the boxplot reproduced in Fig. 4.

Second-Round Hybridization of the High SCE Tertiary Hybrid Clone ME-10-5-B with Normal Human and BS Cells

If the high rate of SCE found in the tertiary hybrid clone A9/ME-10-5-B was due to an altogether new lesion, it had to be expected that a second-round hybridization of its cells with normal cells should result in complemented metaphases only when 1 or more additional human chromosomes were retained. This was indeed found to be the case in 2 sets of second-round hybrid metaphases derived from fusions of the tertiary hybrid clone A9/ME-10-5-B with normal human lymphocytes (BA) as well as with BS fibroblasts (GM1492). All the metaphases screened were found to have retained a large number of human chromosomes (Fig. 5) and to exhibit a low SCE rate in both murine and human chromosomes, independently of the fact that these were being derived from normal or BS parental cells. This obviously suggests that also the second lesion that arose in the A9/ME-10-5 hybrid cells is different from the BS mutation which characterizes the GM1492 cells. From the cytogenetical analysis thus far completed in 4 of the above-mentioned second-round primary hybrid clones (Tab. 3c), it appears that the only human chromosome regularly present in every complemented metaphase (in addition to autosomes 6, 20, and 21) is autosome 19.

DISCUSSION AND FUTURE PERSPECTIVES

Nonspecialized readers of this volume will be impressed to find out that interest in SCE shows no sign of abating as we celebrate its 25th birthday. However, the addicts to the field will no doubt ponder about the disturbing reality that some of the basic questions about SCE remain still unanswered today or are at least very much debated: (i) Do SCEs precede or follow the incorporation of BrdUrd necessary for their identification? (ii) Is the high SCE phenotype the expression of a DNA repair defect or rather the efficient repair of chromosomes highly susceptible to DNA damages? (iii) Does the primary lesion associated with SCEs lead to the production of an abnormal gene end product or merely to the absence of the normal one? (iv) Do SCEs have a multiple etiology and, in particular, do different BS mutations exist? (v) Do SCEs have a role in normal and abnormal differentiation?

We chose to discuss the results of our studies in the framework of the above questions, not so much for the little useful information that we think they contribute, but because this gives us an opportunity to express our own bias about the situation and where and how to search for the most pressing answers.

(i) Do SCEs precede or follow exposure to BrdUrd?

This question cannot obviously have a clear-cut answer as long as BrdUrd will remain the indispensable reagent to visualize SCE. The recent observation of Shiraishi et al. (27,28) and the finding reported here of a higher rate of SCE per BS chromosome when using 30 μg of BrdUrd per ml medium, favor the contention that the drug itself is an effector of SCEs in BS cells. This, however, does not necessarily rule out the possibility that SCEs occur also in vivo and in culture independently of, and prior to, the incorporation of BrdUrd. On the other hand, until an altogether new method for SCE detection is invented, all reported effects of specific drugs and/or diseases on SCEs should be regarded as the result of their synergistic action with BrdUrd.

(ii) SCEs: a DNA repair defect or a symptom of increased chromosomal proneness to DNA damage?

The literature on this particular question is too immense to attempt a concise review. Thus, we will only briefly summarize the reasons that motivate our bias in favor of the latter possibility. An overwhelming body of experimental data referred to in this volume and in other recent reviews on SCEs (29,30) shows that the cell's exposure to a variety of clastogenic agents is followed by an intense repair activity provided that the damage is repairable. In particular, the evidence presented by Dr. Cremer (31) that the increased level of SCE is limited only to the few microirradiated chromosomes, gives strong support to the hypothesis that SCEs are a repair phenomenon following an inflicted DNA damage. Thus, it seems to us reasonable to assume that, under a critical concentration threshold of a clastogenic agent, normal and BS cells [as well as the high-rate SCE rodent cells described by this study (L-A9 cells) and other studies (CHO-EM9 cells) (32)] are all capable of repairing

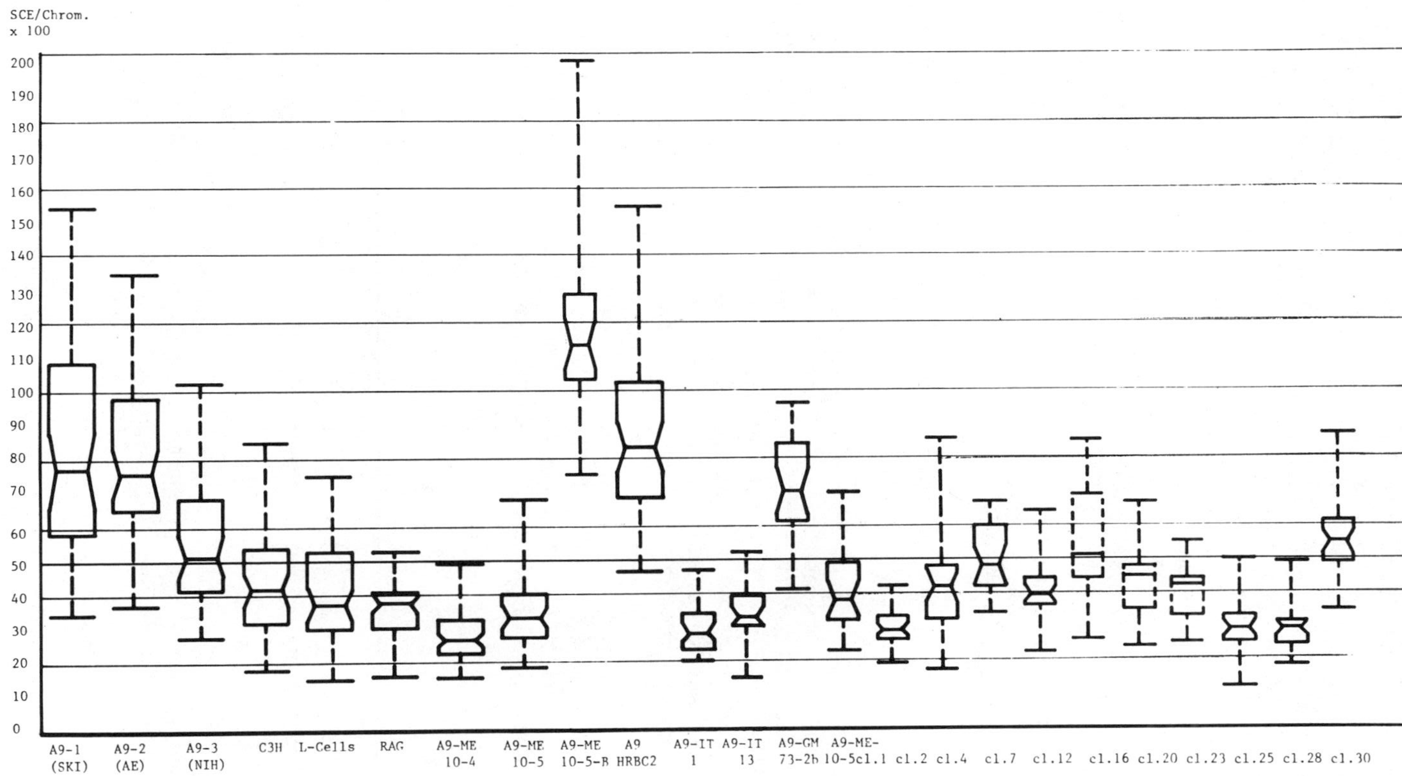

Fig. 4. Box plot of SCE in murine cells and murine-human somatic cell hybrids (cfr. J. Tuckey: Exploratory Data Analysis). Each box plot gives a nonparametrical description of the group size, the median and the extremes of the distribution. The fiducial limits of the median (at a 5% level) are expressed by the notched areas around the median. Distributions which are nonoverlapping in these notched areas differ significantly in their medians.

DNA damage but they do it with a different intensity because of the different degree of the damage inflicted. This is to say that the different rate of SCEs observed in various cell types could be regarded as the expression of the cell's sensitivity to the clastogenic agent rather than of its different ability in DNA repair. However, the homogeneous distribution of SCEs per chromosome observed in all the above mentioned high-rate SCE cell types--as opposed to the localization of SCE to a few chromosomes in microirradiated metaphases (31)--indicates that their sensitivity is the result of mutational changes affecting the cell as a whole and not the response to localized damages of the chromosomal architecture. Accordingly, the homogeneous distribution of SCEs in the chromosomes of the complemented hybrid metaphases (derived from the fusion of high- and low-rate SCE parental cells) should be regarded as the restoration of the cell's ability to neutralize the DNA-damaging effect of the clastogenic factor.

(iii) SCE mutation: an abnormal gene end product or merely the absence of a normal one?

This question emphasizes the difficulties of defining the concept of dominance and recessivity. Some experiments (33,34) have indicated the possibility that SCE-inducing factors are produced by BS cells which accumulate in spent media and in the plasma of BS patients. However, contrasting results have been obtained by other studies (2,35-38) and clear-cut evidence for the recessive behavior of the high SCE phenotype of BS chromosomes has been provided by hybridization experiments between BS and normal SCE human or rodent cells (2,39,40). Similarly, a recessive behavior is indicated for the high-rate SCE mutants of the murine genome described in the present studies as witnessed by the complementation for SCE observed in murine/human hybrid cells derived from the fusion of these cell mutants with human cells.

Fig. 5. From top to bottom: a typical complemented metaphase from the secondary clone A9/ME-10-5 showing the retention of autosomes 6, 20, and 21 (arrows) and suppression of SCEs; a metaphase from the tertiary clone A9/ME-10-5-B, which has reverted to high rate of SCEs (for both the human and murine chromosome) in spite of the retention of the same autosomes 6, 20 and 21; a second-round cell hybrid metaphase from a hybrid cell line derived from the fusion of the above tertiary clone with normal human lymphocytes (BA). Many human chromosomes have been retained and the SCE rate is again suppressed in both parental chromosomes.

A9/ME-10-5

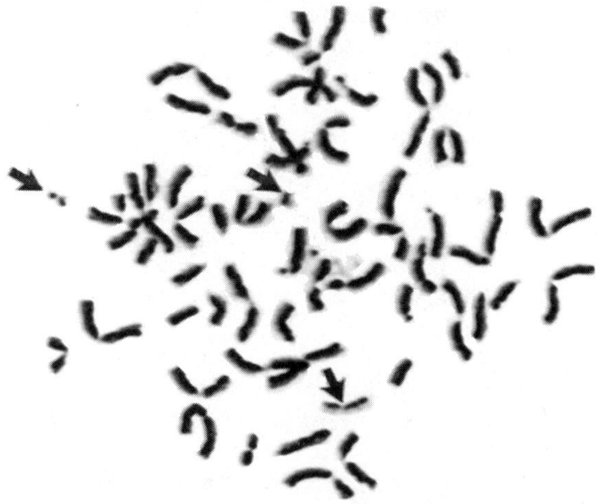

A9/ME-10-5B

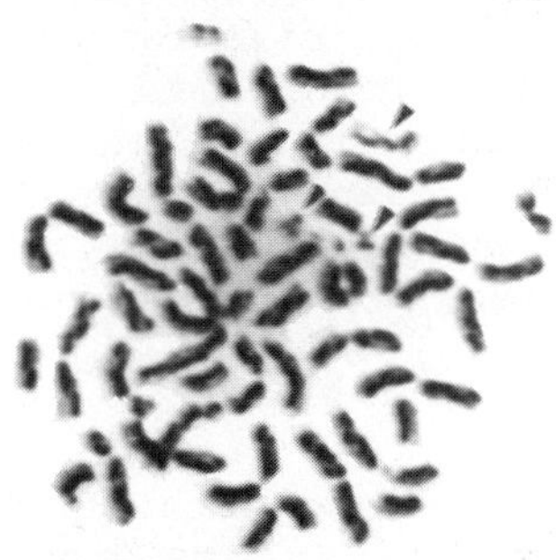

A9/ME-10-5B/BA

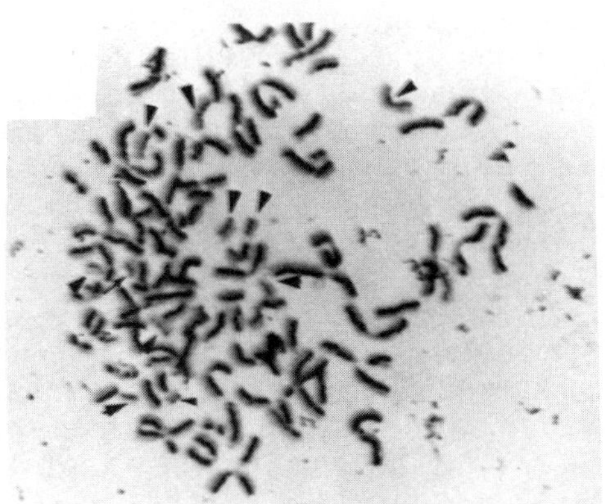

An obvious way to reconcile these contrasting observations is to assume that, as in most classical examples of inborn errors of metabolism, the genetic damage suffered by the BS cells is mediated through the accumulation of incompletely degraded metabolites, whose concentration may be enhanced by a variety of factors and, notably, by growth in BrdUrd medium. If this is the case, a variable degree of clastogenic effect could be expected from the BS spent medium or the BS plasma concentrate in view of the several factors that may be capable of influencing it, such as cell crowding, type of medium used, rate of medium changes, production of detoxicants by the cells, etc. (41,42).

(iv) Is there a heterogeneous etiology for high SCE? In particular, are there different BS mutations?

Several observations favor the existence of a heterogeneous etiology for the high SCE phenotype in mammalian cells. On the one hand, the complementation experiments reported in our study indicate that 2 different types of SCE mutation have arisen in the L-A9 cells, as witnessed by the fact that their complementation in murine human hybrids requires the retention of different human chromosomes. Secondly, the hybridization of either one of the above mutants with BS cells GM1492 generates hybrid metaphases with a normal SCE phenotype, thus suggesting that the mutational changes arisen in the murine parental line are both different from the BS mutation of this cell strain and that at least 1 of them must involve a different genetic system.

Furthermore, previous studies from our own and other groups (2,35) indicate that the high SCE phenotype of BS cells can be complemented also through cell fusion with normal SCE Chinese hamster cells (CHO-YH21). On the other hand, Thompson et al. (32) have described a clonal derivative of the same CHO cells that display a very high rate of SCE and which is in turn complemented in hybrids with normal human cells (Thompson and Carrano, personal communication). These observations suggest a multiple etiology for the high SCE phenotype in mammalian cells. The relative low frequency of the high SCE mutants among established cell lines indicates that the mutational changes responsible for them must be recessive and that a high SCE phenotype may result only when a cell line has become homozygous for the same mutational change. Both the multiplicity and recessiveness of the inborn and de novo SCE mutations are supported by the complementation studies reported. It is to be expected that sooner or later one of the mutants isolated in culture will turn out to be homologous to the BS mutation(s), thus paving the way to an experimental verification of the possible heterogeneity of BS and to the mapping of the BS locus or loci.

(v) Do SCEs have a role in normal and abnormal differentiation?

The possible role of SCEs in the production of somatic diversity has been discussed in general by German (43) and specifically invoked to explain the generation of antibody diversity in tumoral and normal immunoglobulin-producing cells (44-46). The hypothesis is a fascinating one, but has no direct experimental support as yet. One question that comes spontaneously to mind is whether there is any relationship between the chromosomal sites, where chromatid exchanges take place, and the sites of exchange between homologous and nonhomologous chromosomes during meiotic division (genetic recombination and inborn reciprocal translocations) and those found to be specifically associated with certain tumor malignancies (47). There can be little doubt that several of these issues will soon be efficiently investigated through the combined application of the experimental strategies of somatic cell genetics, restriction enzyme analysis and DNA recombinant research.

The recent report of Cavenee et al. (48), suggesting the occurrence of mitotic recombination to account for the molecular changes found between normal and tumoral tissue of retinoblastoma patients, is an excellent guide on how the harvest of data on the genetics of DNA restriction sites (49) can now be profitably applied to the study of inborn and de novo somatic variation in man.

ACKNOWLEDGMENT

The studies described here have been supported by NIH grants CA25342, AG-00541 and core grant CA-08748. We wish to thank Mike Chopan for his participation in the initial stages of the study, Mark Brown and Susan Groshen for their advice in the statistical evaluation and presentation of the data, and Iraida Pagan-Charry for her valuable technical assistance.

REFERENCES

1. Siniscalco, M. (1979) Human gene mapping and cancer biology. In Genetics and Human Biology: Possibilities and Realities, R. Porter, ed. Ciba Foundation Symposium, Lond. North Holland Publishing Co., Amsterdam/New York, pp. 283-309.
2. Alhadeff, B., M. Velivasakis, I. Pagan-Charry, W.C. Wright, and M. Siniscalco (1980) High rate of sister chromatid exchanges of Bloom's syndrome chromosomes is corrected in rodent human somatic cell hybrids. Cytogenet. Cell Genet. 27:8-23.
3. Littlefield, J.W. (1964) Selection of hybrids from matings of fibroblasts in vitro and their presumed recombinants. Science 145:709-710.
4. Sanford, K.K., W.R. Earle, and G.D. Likely (1948) The growth in vitro of single isolated tissue cells. J. Nat. Cancer Inst. 9:229-246.

5. Earle, W.R. (1943) Production of malignancy in vitro. IV. The mouse fibroblast cultures and changes seen in the living cells. J. Nat. Cancer Inst. 4:165-212.
6. Cox, R.P., M.R.Krauss, M.E. Balis, and J. Dancis (1974) Mouse fibroblasts Ag are deficient in HGPRT and APRT. Am. J. Human Genet. 26:272-273.
7. Olsen, A.S., O.W. McBride, and D.E. Moore (1981) Number and size of human X chromosome fragments transferred to mouse cells by chromosome-mediated gene transfer. Mol. & Cell Biol. 1:439-448.
8. Klebe, R.J., T. Chen, and F.H. Ruddle (1971) Mapping of a human genetic regulator element by somatic cell genetic analysis. Proc. Natl. Acad. Sci., USA 66:1220-1227.
9. Felluga, B., A. Claude, and E. Mrena (1969) Electron microscope observations on virus particles associated with a transplantable renal adenocarcinoma in Balb mice. J. Nat. Cancer Inst. 43:319-333.
10. Szybalski, W., and E. Szybalska (1962) Drug sensitivity as a genetic marker for human cell lines. Univ. Mich. Med. Bull. 28:277-293.
11. Chopan, M. and L. Kopelovich (1981) The suppression of tumorigenicity in human X mouse cell hybrids. I. Derivation of hybrid clones, chromosome analysis and turmorigenicity studies. Exp. Cell Biol. 49(2):78-89.
12. Optiz, J., P.D. Pallister, and F.H. Ruddle (1973) An (X;14) translocation, balanced, 46 chromosomes. Repository identification No. GM 73. Cytogenet. Cell Genet. 12:289-290.
13. Punnett, H.H., M.L. Kistermacher, A.E. Greene, and L.L. Coriell (1974) An (X;1) translocation, balanced, 46 chromosomes. Repository identification No. GM 97. Cytogenet. Cell Genet. 13:406-407.
14. Seravalli, E., P. de Bona, M. Velivasakis, I. Pagan-Charry, A. Hershberg, and M. Siniscalco (1976). Further data on the cytologic mapping of the human X-chromosome with man-mouse cell hybrids. In Baltimore Conference (1975): Third International Workshop on Human Gene Mapping. Birth Defects: Original Article Series. Vol. 12 No. 7, The National Foundation, New York, pp. 219-222.
15. Bauch, W., B. Hellkuhl, and K.-H. Grzeschik (1978) Regional assignment of the gene for human β-glucuronidase by the use of human-mouse cell hybrids. Cytogenet. Cell Genet. 22:434-436.
16. Di Cioccio, R.A., R. Voss, M. Krim, K.-H. Grzeschik, and M. Siniscalco (1975) Identification of human RNA transcripts among heterogeneous nuclear RNA from man-mouse somatic cell hybrids. Proc. Natl. Acad. Sci., USA 72:1868-1872.
17. Miller, O.J., P.R. Cook, P. Meera Khan, S. Shin, and M. Siniscalco (1971) Mitotic separation of two human X-linked genes in man-mouse somatic cell hybrids. Proc. Natl. Acad. Sci., USA 68:116-120.
18. Russel, W.C., C. Newman, and D.H. Williamson (1975) A simple

cytochemical technique for demonstration of DNA in cells infected with mycoplasmas and viruses. Nature (Lond.) 253:461-462.
19. Seabright, M. (1972) The use of proteolytic enzymes for the mapping of structural rearrangement in the chromosomes of man. Chromosoma 36:204-210.
20. Bobrow, M., and J. Cross (1974) Differential staining of human and mouse chromosomes in interspecific cell hybrids. Nature (Lond.) 251:77-79.
21. Alhadeff, B., M. Velvasakis, and M. Siniscalco (1977) Simultaneous identification of chromatid replication and of human chromosomes in metaphases of man-mouse somatic cell hybrids. Cytogent. Cell Genet. 19:236-239.
22. Perry, P., and S. Wolff (1974) New Giemsa method for the differential staining of sister chromatids. Nature (Lond.) 251:156-158.
23. Friend, K.K., S. Chen, and F.H. Ruddle (1976) Differential staining of interspecific chromosomes in somatic cell hybrids by alkaline Giemsa stain. Somatic Cell Genet. 2: 183-188.
24. Tice, R., J. Chaillet, and E.L. Schneider (1975) Evidence derived from sister chromatid exchanges of restricted rejoining of chromatid subunits. Nature (Lond.) 256:642-644.
25. Gibson, A.D., and D.M. Prescott (1974) Frequency and sites of sister chromatid exchanges in rat kangaroo chromosomes. Exp. Cell Res. 86:209-214.
26. McGill, R., J.W. Tukey, and W.A. Larsen (1978) Variations of box plots. The American Statistician 32:12-16.
27. Shiraishi, Y., T.H. Yosida, and A.A. Sandberg (1983) Analyses of bromodeoxyuridine-associated sister chromatid exchanges (SCEs) in Bloom's syndrome based on cell fusion: Single and twin SCEs in endoreduplication. Proc. Natl. Acad. Sci., USA 80:4369-4373.
28. Shiraishi, Y. (1984) Analysis of SCEs in Bloom syndrome by use of endomitotic and three-way differentiation to techniques. In Sister Chromatid Exchange: 25 Years of Experimental Research, R.R. Tice and A. Hollaender, eds. Plenum Press, New York.
29. S. Wolff, ed. (1982) Sister Chromatid Exchange. Wiley-Interscience, New York, pp. 1-306.
30. A.A. Sandberg, ed. (1982) Progress and topics in Cytogenetics, Vol. 2, Sister Chromatid Exchange Alan R. Liss, Inc., New York, pp. 1-706.
31. Cremer, T., M. Raith, and C. Cremer (1984) Induction of sister chromatid exchanges (SCE) by micro-irradiation of photolesions and the distribution of SCEs. In Sister Chromatic Exchange: 25 Years of Experimental Research, R.R. Tice and A. Hollaender, eds. Plenum Press, New York.
32. Thompson, L.H., K.W. Brookman, L.E. Dillehay, A.V. Carrano, J.A. Mazrimas, C.L. Mooney, and J.L. Minkler (1982) A CHO-cell strain having hypersensitivity to mutagens, a defect in DNA strand-break repair, and an extraordinary baseline frequency of sister chromatid exchange. Mutat. Res. 95:427-440.

33. Tice, R., G. Windler, and J.M. Rary (1978) Effect of co-cultivation on sister chromatid exchange frequencies in Bloom's syndrome and normal fibroblast cells. Nature (Lond.) 273:538-540.
34. Barnabei, V.M., and T.E. Kelly (1982) Bloom syndrome fibroblasts secrete a metabolite which enhances SCE rate in normal fibroblasts. Am. J. Med. Genet. 12:245.
35. Van Buul, P.P.W., A.T. Natarajan, and E.A.H. Verdegaal Immerzeel (1978) Suppression of the frequencies of sister chromatid exchanges in Bloom's syndrome fibroblasts by co-cultivation with Chinese hamster cells. Human Genet. 44:187-189.
36. Bartram, C.R., H.W. Rüdiger, and E. Passarge (1979) Frequency of sister chromatid exchanges in Bloom syndrome fibroblasts reduced by co-cultivation with normal cells. Human Genet. 46:331-334.
37. Rüdiger, H.W., C.R. Bartram, W. Harder, and E. Passarge (1980) Rate of sister chromatid exchanges in Bloom syndrome fibroblasts reduced by co-cultivation with normal fibroblasts. Human Genet. 32:150-157.
38. Schonberg, S., and J. German (1980) Sister chromatid exchange in cells metabolically coupled to Bloom's syndrome cells. Nature (Lond.) 284:72-74.
39. Bryant, E.M., H. Hoehn, and G.M. Martin (1979) Normalization of sister chromatid exchange frequencies in Bloom's syndrome by euploid cell hybridization. Nature (Lond.) 279:795-796.
40. Shiraishi, Y., S.I. Matsui, and A.A. Sandberg (1981) Normalization by cell fusion of sister chromatid exchange in Bloom syndrome lymphocytes. Science 212:820-822.
41. Emerit, I., and P. Cerutti (1981) Clastogenic activity from Bloom syndrome fibroblast cultures. Proc. Natl. Acad. Sci., USA 78:1868-1872.
42. Emerit, I., P.A. Cerutti, A. Levy, and P. Jalbert (1982) Chromosome breakage factor in the plasma of two Bloom's syndrome patients. Human Genet. 61:65-67.
43. German, J. (1982) Biological role of chromatid exchange. In Gene Amplification, R. Shimcke, ed. Cold Spring Harbor Laboratory, New York, pp. 307-312.
44. Obata, M., T. Kataoka, S. Nakai, H. Yamagishi, N. Takahashi, Y. Yamawaki-Kataoka, T. Nikaido, A. Shimizu, and T. Honjo (1981) Structure of an rearranged γ1 gene and its implication to immunoglobulin class-switch mechanism. Proc. Natl. Acad. Sci., USA 78:2437-2441.
45. Van Ness, B.G., C. Coleclough, R.P. Perry, and M. Weigert (1982) DNA between variable and joining gene segments of immunoglobulin κ light chain is frequently retained in cells that rearrange the κ locus. Proc. Natl. Acad. Sci., USA 79:262-266.
46. Höchtl, J., C.R. Müller, and H.G. Zachau (1982) Recombined flanks of the variable and joining segments of immunoglobulin genes. Proc. Natl. Acad. Sci., USA 79:1383-1387.
47. Klein, G. (1981) The role of gene dosage and genetic transpositions in carcinogenesis. Nature (Lond.) 294:313-318.
48. Cavenee, W.K., T.P. Dryja, R.A. Phillips, W.F. Benedict, R.

Godbout, B.L. Gallie, A.L. Murphree, L.C. Strong, and R.L. White (1983) Expression of recessive alleles by chromosomal mechanisms in retinoblastoma. Nature (Lond.) 305:779-784.
49. C.T. Caskey, and R. White, eds. (1983) Banbury Report 14. Recombinant DNA Applications to Human Disease. Cold Spring Harbor Laboratory, New York, pp. 1-371.

DIFFERENT PROPERTIES IN LYMPHOBLASTOID CELL LINES FROM PATIENTS WITH BLOOM SYNDROME

Tomoko Hashimoto,[1] Takahiko Sukenaga,[2] Patricia Lopetegui,[1] and Jun-ichi Furuyama[1]

[1]Department of Genetics
[2]Department of Radiology
Hyogo College of Medicine
Mukogawa-cho, 1-1
Nishinomiya 663, Japan

ABSTRACT

In 3 lymphoblastoid cell lines derived from 3 patients with Bloom syndrome (BS), the baseline frequency of sister chromatid exchanges (SCEs), chromosomal breakage, sensitivity to ethylmethanesulfonate (EMS), and bromodeoxyuridine (BrdUrd) were examined. The SCE frequency of 2 BS lines (EB-BS-NoKi-2 and EB-BS-AkSak) was about 2 to 2.5 times those of the normal cell lines, while that of another BS line (EB-BS-2KA) was about 10 to 11 times. The net increase in the number of SCEs in EB-BS-NoKi-2 and EB-BS-AkSak lines induced by EMS was similar to that of normal cell lines, but it was high in the BS-2KA line. Sensitivity to BrdUrd was examined using SCE induction at different concentrations of BrdUrd and by cell cycle analysis. EB-BS-NoKi-2 and EB-BS-AkSak lines were no more sensitive than normal cell lines, while the EB-BS-2KA line was more sensitive than controls. High frequency of chromosomal breakage was found only in the EB-BS-2KA line. These results suggest that 2 types of cells exist in the B-lymphoblastoid cells of BS.

INTRODUCTION

The remarkable high frequency of SCEs in peripheral blood lymphocytes (PBLs) has been described as an important indicator of BS (1). However, in BS-lymphoblastoid cell lines established by transformation with Epstein-Barr virus (EBV), 2 kinds of cell populations have been found. One cell population has a high frequency of SCEs, while in a few cases, normal levels of SCEs have been observed (2,3,4). In the cell lines with a low frequency of SCEs, this frequency was about 10% of that in the PBLs (3,4). Krepinsky et al. (5) noted that the number of EMS-induced SCEs in BS PBLs was several times the EMS-induced SCEs in normal cells. The authors also mentioned that lymphocytes with a low frequency of SCEs might not be affected by EMS. We have demonstrated that the net increase in SCE frequency induced by EMS in the BS cell lines with a low frequency of SCEs was the same as that of normal cell lines (3).

However, Shiraishi et al. (6), using the endoreduplication technique, reported that the incidence of SCEs in a BS line with a high frequency of SCEs was due to the incorporation of BrdUrd into the chromosomes. These authors also reported that this BS line exhibited hypersensitivity to BrdUrd.

Another important characteristic of BS is chromosomal instability (7). A high frequency of chromosomal aberrations (CAs) is found in the PBLs and fibroblasts of BS individuals (8). However, there has been no report on CA analysis in lymphoblastoid cells.

In spite of the available data, there are still many questions concerning SCEs and BS to be answered. In this publication, the data concerning SCEs, EMS-induced SCEs, sensitivity to BrdUrd, and CA analysis obtained from 3 BS-lymphoblastoid cell lines is reported. The origin of the 2 different kinds of cell populations is discussed.

MATERIALS AND METHODS

Lymphoblastoid Cell Lines

From 2 patients with BS (BS-NoKi and BS-AkSak) and normal adult controls (NL-Mu, NL-Ha, NL-Ch, and NL-Sh), lymphoblastoid cell lines (EB-BS-NoKi, EB-BS-AkSak, EB-NL-Mu, EB-NL-Ha, EB-NL-Ch, and EB-NL-Sh) were established by EBV (3). The line from BS-2KA (EB-BS-2KA) was supplied by Dr. H. Tohda (Tohoku University, Sendai). These cell lines were maintained in RPMI 1640 supplemented with 10% fetal bovine serum (FBS), 100 μg/ml of kanamycin, and 0.3 μg/ml of fungizone. For chromosomal preparation, lymphoblastoid cell lines were cultured in RPMI 1640 supplemented with 20% FBS. Chromosomal preparation of all lymphoblastoid cell lines were obtained from cultures 72 hr after subculturing (3×10^5 cells/ml) by adding colchicine (0.1 μg/ml) for the last 60 min.

SCEs

The baseline frequency of SCEs was determined for cells cultured in the presence of BrdUrd at a concentration of 5 μg/ml. For determination of EMS-induced SCEs, EMS was added to obtain a final concentration of 10^{-3} M and the cells were cultivated for 72 hr. The net increase in the number of SCE frequencies was calculated by subtracting control SCEs from EMS-induced SCEs.

For an assessment of the influence of BrdUrd on SCEs, concentrations of BrdUrd varying from 2.5 μg/ml to 20 μg/ml were employed. The Hoechst-Giemsa technique was used for differential staining of sister chromatids. Normal cell lines cultured under the same conditions were used as controls.

Cell Growth Inhibition Experiments

Lymphoblastoid cells (1 x 10^5/ml/well) were seeded in 12-well plates (Falcon). The next day (da 0), the number of viable cells in 1 of the wells was counted (the initial cell number) by the Trypan blue dye exclusion method. The cells in the other wells were treated with various concentrations of BrdUrd. The cell number was counted every other day and after da 5, when cells in the untreated wells had increased from 3- to 5-fold in number, the cell number in the BrdUrd-treated wells was counted. The growth percentage defined by the method of Ishida and Buchwald (9) was calculated as follows:

$$100 \times \frac{\text{number of BrdUrd-treated cells on the 5th day} - \text{initial cell number}}{\text{number of cells in the control well on the 5th day} - \text{initial cell number}}$$

RESULTS

Baseline SCE Frequency in Lymphoblastoid Cell Lines

The average SCE frequency in the PBLs of the individuals affected with BS were as many as 20 to 25 times that observed in controls. Two cell lines were established 9 mo apart from the PBLs of BS-NoKi (EB-BS-NoKi-1 and -2) (3). In each cell line, the frequency of SCEs was about 10% of that observed in the PBLs. In this experiment, the EB-BS-NoKi-2 cell line was used. From the patient, BS-AkSak, one cell line was established and the frequency of SCEs was 9% that observed in the PBLs (3). The frequency of SCEs of the EB-BS-2KA cell line was 39.8 per cell and this value is 42% of that observed in the PBLs (94.7 per cell). Normal control lymphoblastoid cell lines exhibited the same SCE frequency as that observed in the PBLs.

SCEs Induced by EMS

As shown in Tab. 1, the increase in the number of SCEs per metaphase of EB-BS-NoKi-2 and EB-BS-AkSak induced by EMS (10^{-3} M) was the same value as that observed in the normal cell lines. In the EB-BS-2KA cell line, the number of SCEs induced by EMS was 3 times higher than that observed in the normal cell lines and in the other 2 BS lines.

SCEs Induced by BrdUrd

To study the effect of BrdUrd concentration on the induction of SCEs in the BS-lymphoblastoid cells, 2.5 μg/ml to 20 μg/ml of BrdUrd was added to the culture medium. In EB-BS-2KA cells, the mitotic index in the culture with 20 μg/ml of BrdUrd decreased to 40% of that in the culture without BrdUrd (data not shown). No metaphase in the second or third cycle could be detected in 72-hr cultures containing 20 μg/ml of BrdUrd. However, the other BS lines exhibited the same mitotic index and the same cell cycle as control cell lines under the same conditions. Figure 1 shows the percentage of cells in the first, second, and third or greater cell cycle. In the EB-BS-2KA cell line, delay in the cell cycle was BrdUrd concentration-dependent.

Figure 2 shows the number of SCEs observed in the presence of various concentrations of BrdUrd. In the EB-BS-2KA cell line, the SCE frequency at 5 μg/ml of BrdUrd was not significantly different from the SCE frequency at 2.5 μg/ml BrdUrd (at $p = 0.02$). However, the SCE frequency at 10 μg/ml BrdUrd was significantly different from the frequency observed at 5 μg/ml BrdUrd (at $p = 0.02$). No significant increase in the SCE frequency could be observed in the 4 other cell lines cultured with 10 μg/ml of BrdUrd when compared with that of cells cultured with 5 μg/ml of BrdUrd (at $p = 0.02$).

Tab. 1. Increase in SCE frequency induced by EMS in lymphoblastoid cell lines.

Cell lines	SCEs/metaphse			SCEs/chromosome
	Base line	EMS ($\times 10^{-3}$M)	Net increase[c]	Net increase[d]
EB-BS-NoKi-2[a]	8.29	20.48	12.19	0.260
EB-BS-AkSak[a]	8.44	22.08	13.64	0.315
EB-BS2KA[b]	39.80	78.90	39.10	1.048
EB-NL-Mu[a]	3.72	19.66	15.94	0.343
EB-NL-Ha[a]	3.57	17.06	13.49	0.293

a Average number of SCEs from two experiments. In one experiment, 30 cells were scored.

b 30 cells were scored.

c Net increase of SCEs/metaphase above the spontaneous SCE frequency.

d Net increase of SCEs/chromosome above the spontaneous SCE frequency.

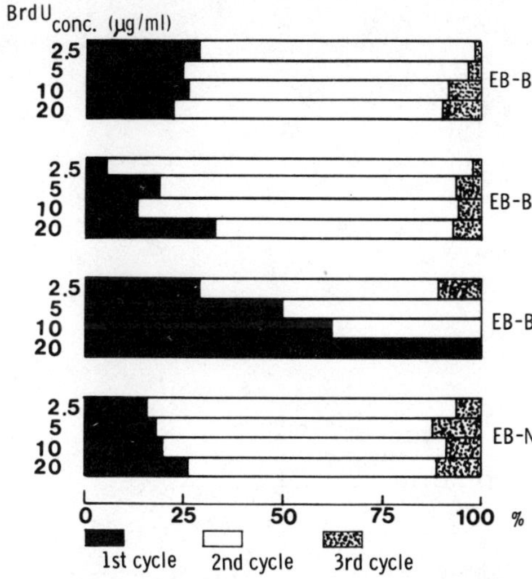

Fig. 1. Percentage of metaphases in the first, second, or third (and greater) cell cycle after 72 hr of culture in the presence of different concentrations of BrdUrd (2.5, 5, 10, and 20 μg/ml). In the EB-BS-2KA line, a delay in the cell cycle was observed in relation with increase of the BrdUrd concentration.

Cell Growth Inhibition by BrdUrd

Lymphoblastoid cells were cultured in medium supplemented with BrdUrd at final concentrations of 3, 5, 10, 30, and 50 μg/ml. EB-BS-2KA cells grew slowly, with a doubling time of more than 72 hr in medium without BrdUrd. At a final concentration of 30 μg/ml BrdUrd, the number of EB-BS-2KA cells did not vary. The other BS cell lines, as well as the normal cell lines, however, exhibited an increase in cell number under the same conditions. Figure 3 shows the rate of growth inhibition after a 5-da culture. Since EB-BS-2KA cells grew slowly, as mentioned above, the growth percentage could not be calculated. EB-BS-NoKi-2 and EB-BS-AkSak cell lines showed the same percentage inhibition curve as that of controls.

Chromosomal Aberrations

From each culture, at least 100 metaphases were examined. Breaks were classified as either chromatid (sb) or chromosome (db) breaks. Acentric fragments (Frag) were also scored. Structural rearrangements, including dicentric (Dic), chromosomal interchanges (Int), and rings (Ring), were also considered. All chromosomal breaks were scored as single events, while the structural rearrangements were scored as 2 breaks. Cytogenetic damage is presented as the number of total breaks per cell (10). The results are shown in

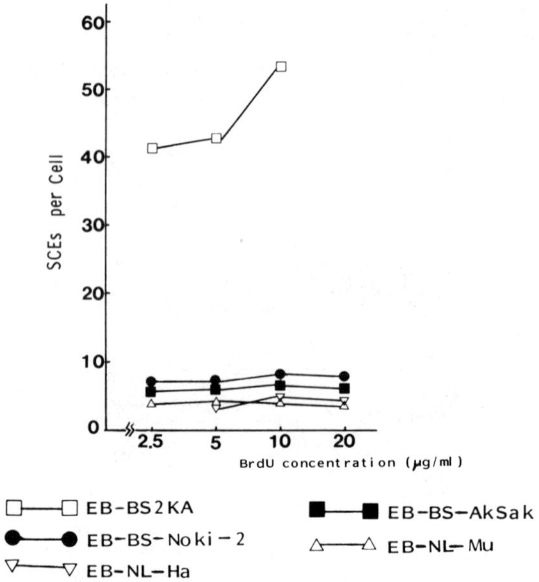

Fig. 2. Effect of various concentrations of BrdUrd (2.5, 5, 10, and 20 μg/ml) on the induction of SCEs. In the EB-BS-2KA line, no metaphase in the second cell cycle could be observed. The SCE frequency in the presence of 10 μg/ml of BrdUrd was significantly different from that in the presence of 5 μg/ml.

Tab. 2. In the EB-BS-NoKi-2 cell line, the frequency of breaks was the same as that observed in the normal cell lines. In the EB-BS-AkSak cell line, 9 dicentric chromosomes were observed in the 110 metaphases scored. The total breakage per cell was about 6 times higher in this line than in the normal cell lines. In the EB-BS-2KA cell line, however, 60% of the metaphases had CAs. In these cells, 1.59 breaks per metaphase were observed. Characteristic chromosomal interchanges between 2 isochromosomes were found. The average number of breaks per cell was 1.09; this frequency is 24 times the average value observed in normal cell lines.

DISCUSSION

According to the data presented above, dimorphism of B-lymphoblastoid cells in BS was found. This dimorphism is not only in terms of SCE frequency (11), but also in terms of sensitivity to EMS and BrdUrd, and in CAs.

Shiraishi et al. (6) noted that an EB-BS cell line with a high frequency of SCEs showed normal cell growth accompanied by hypersensitivity to BrdUrd, as determined by growth inhibition. On the

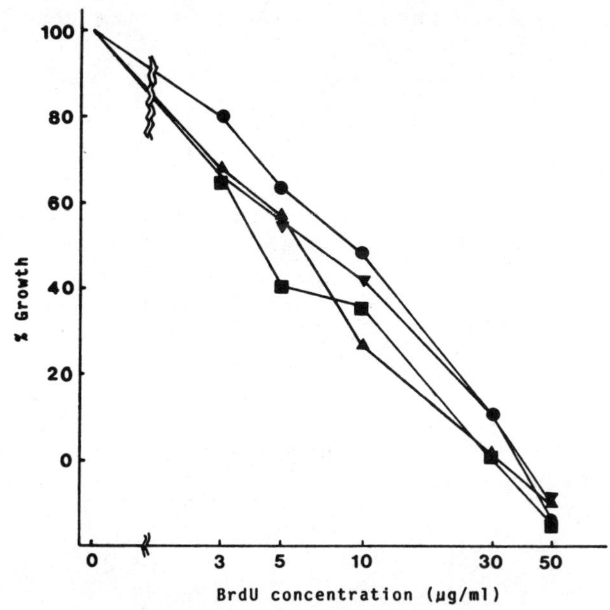

Fig. 3. Growth inhibition test of lymphoblastoid cell lines in the presence of BrdUrd. EB-BS-NoKi-2 and EB-BS-AkSak lines showed the same growth inhibition curve as normal cell lines. Since the EB-BS-BS-2KA line grew slowly, the growth percentage could not be calculated. The growth percentage was the average of 3 experiments.

Tab. 2. Chromosomal aberrations in lymphoblastoid cell lines.

Cell line	Cells with aberrations (%)	Aberrations / cell[a] sb	db	Frag	Dic	Int	Ring	Total breaks/cell[b]
EB-BS-NoKi-2	6.0	0.02	0.02	0.01	0.01	0	0	0.07
EB-BS-AkSak	14.5	0.04	0.03	0.01	0.08	0	0.01	0.26
EB-BS2KA	58.9	0.43	0.05	0.20	0.09	0.03	0	0.92
EB-NL-Mu	5.0	0.03	0.02	0	0	0	0	0.05
EB-NL-Ha	3.7	0.02	0.02	0	0	0	0	0.04

sb = chromatid break, **db** = chromosomal break, **Frag** = fragment, **Dic** = dicentric, **Int** = chromosomal interchange.

a 100-160 cells were scored.

b All chromosomal breaks were scored as single events. The structural rearrangement were scored as two breaks.

other hand, we found in the EB-BS-2KA cell line that the doubling time was over 72 hr. Therefore, it seems as though there was no correlation between the doubling time of the cells and the frequency of SCEs. Shiraishi et al. (4) also reported in BS cell lines established with EBV or adult T-cell leukemia virus (ATLV) that there were 2 populations of cells--one with a high frequency of SCEs and another with a low frequency of SCEs. In the PBLs of BS individuals they did not find cells with a low frequency of SCEs. We also reported (3) that we had established, 9 mo apart, cell lines with a low frequency of SCEs from peripheral blood lymphoycytes of a BS patient. In the PBLs stimulated with phytohemagglutinin (PHA) or concanavalin A (Con A), less than 1% of the cells were found to exhibit a low frequency of SCEs. In our studies, selection of cells with a low frequency of SCEs might have occurred in vitro. The lines established by Shiraishi et al. (4), however, were probably caused by a change of the phenotype rather than a selection in vitro. The dimorphism in terms of SCE frequency in the PBLs and lymphoblastoid cell lines has been widely discussed (3,4,11,12).

Warren et al. (13) reported that the rate of spontaneous mutation to a 6-thioguanine-resistant phenotype in fibroblasts was 16 times higher in BS than in control cells. Vijayalaxmi et al. (14), using the 6-thioguanine-resistant cell assay, found that BS peripheral blood T cells stimulated by PHA showed a high mutation rate. They reported a mutation rate in BS cells 8 times that of normal cells, both rates being about 100 times the mutation rate reported by Warren et al. (13). These data led Vijayalaxmi et al. to suggest that the high frequency of chromosomal breakage might be the origin of the high frequency of mutations at the HGPRT locus in BS cells.

Ray and German (11), on the other hand, suggested that one possible explanation for the B-lymphoblastoid cell dimorphism was a high frequency of reverse mutation. These authors also reported that they had observed a normal frequency of CAs in BS PBLs that had low (normal) baseline SCE frequency. Our data confirm that the dimorphism exists not only in terms of SCE frequency but also in terms of CAs and sensitivity to EMS and BrdUrd.

Shiraishi et al. (6) reported that after fusion of B-lymphoblastoid cells with a high frequency of SCEs and normal control lymphoblastoid cells, normalization in the SCE frequency took place by the first mitosis in hybrid cells. Rüdiger et al. (15) observed that the SCE frequency in BS fibroblasts was reduced after cocultivation with normal fibroblasts. These authors also reported that they have observed that chromosomal instability disappeared in the same BS cells after cocultivation with control cells. Attention has been drawn to the possible existence of a "Bloom corrective factor" produced by normal cells that are capable of reducing the increased rate of SCE and chromosomal breaks in BS cells. Accordingly, the "corrective factor" would not be produced in BS cells (16). In this respect, the "Bloom corrective factor" gene might be reactivated in

BS-lymphoblastoid revertant cell lines with low to normal SCE frequency and breakage frequency and normal sensitivity to EMS and BrdUrd by reverse mutation or other unknown mechanisms.

However, in ataxia telangiectasia, it was found that all lymphoblastoid cell lines lose their high frequency of chromosomal breakage after transformation with EBV (Ref. 8 and our unpublished data). Besides, in the lymphoblastoid cell lines established from the peripheral blood of patients with Fanconi anemia, all cell lines showed a high frequency of breakage (Ref. 17 and our unpublished data). These data suggest that different mechanisms exist for the production of CAs in EBV established cell lines derived from these 3 syndromes.

Our observation on 3 BS cell lines revealed that hypersensitivity to EMS and BrdUrd, besides a high frequency of CAs, are reduced in cells with reduced SCE frequency. These results show that the cell lines with low or normal SCEs have cytogenetic properties similar to those of normal cells. If this phenomenon were derived from reverse mutation of the BS gene (bl), these lines could be very helpful in studies to elucidate the phenomena involved in the pathogenesis of Bloom syndrome.

ACKNOWLEDGEMENTS

We are grateful to Dr. H. Tohda, Department of Pharmacology, Research Institute for Tuberculosis and Cancer, Tohoku University, who kindly supplied the EB-BS-2KA cells, and to Dr. K. Tatsumi, Radiation Biology Center, Kyoto University, for valuable suggestions.

This work was supported by grant "Special Coordination Funds for Promoting Science and Technology" from The Science and Technology Agency of Japan.

REFERENCES

1. Chaganti, R.S.K., S. Schonberg, and J. German (1974) A manyfold increase in sister chromatid exchanges in Bloom's syndrome lymphocytes. Proc. Natl. Acad. Sci., USA 71:4508-4512.
2. Henderson, E., and J. German (1978) Development and characterization of lymphoblastoid cell lines (LCLs) from "chromosome breakage syndrome" and related genetic disorders. J. Supramol. Struct. (Suppl.) 2:83.
3. Hashimoto, T., S. Gamo, J. Furuyama, and H. Chiyo (1983) Loss of high frequency of sister chromatid exchanges in Epstein-Barr virus-established lymphoblastoid cell lines from two patients with Bloom's syndrome. Human Genet. 63:75-76.
4. Shiraishi, Y., S. Yoshimoto, I. Miyoshi, N. Kondo, T. Orii, and A.A. Sandberg (1983) Dimorphism of sister chromatid exchange in

Bloom's syndrome B- and T-cell lines transformed with Epstein-Barr and adult T-cell leukemia viruses. Cancer Res. 43:3836-3840.

5. Krepinsky, A.B., J.A. Heddle, and J. German (1979) Sensitivity of Bloom's syndrome lymphocytes to ethyl methanesulfonate. Human Genet. 50:151-156.
6. Shiraishi, Y., T.H. Yoshida, and A.A. Sandberg (1983) Analysis of bromodeoxyuridine-associated sister chromatid exchanges (SCEs) in Bloom syndrome based on cell fusion: Single and twin SCEs in endoreduplication. Proc. Natl. Acad. Sci., USA. 80: 4369-4373.
7. German, J., R. Archibald, and D. Bloom (1965) Chromosomal breakage in a rare and probably genetically determined syndrome of man. Science 148:506-507.
8. Cohen, M.M., M. Sagi, Z. Ben-Zur, T. Schaap, R. Voss, G. Kohn, and H. Ben-Bassat (1979) Ataxia telangiectasia: Chromosomal stability in continuous lymphoblastoid cell lines. Cytogenet. Cell Genet. 23:44-52.
9. Ishida, R., and M. Buchwald (1982) Susceptibility of Fanconi's anemia lymphoblasts to DNA-cross-linking and alkylating agents. Cancer Res. 42:4000-4006.
10. Cohen, M.M., K. Hirschhorn, and W.A. Frosch (1967) In vivo and in vitro chromosomal damage induced by LSD-25. New Engl. J. Med. 277:1043-1049.
11. Ray, J.H., and J. German (1983) The cytogenetics of the "chromosome-breakage syndromes." In Chromosome Mutation and Neoplasia, J. German, ed. Alan R. Liss, Inc., New York, pp. 135-167.
12. German, J., S. Schonberg, E. Louie, and R.S.K. Chaganti (1977) Sister-chromatid exchanges in lymphocytes. Am. J. Human Genet., 29:248-255.
13. Warren, S.T., R.A. Schultz, C.-C. Chang, M.H. Wade, and J.E. Trosko (1981) Elevated spontaneous mutation rate in Bloom syndrome fibroblasts. Proc. Natl. Acad. Sci., USA 78:3133-3137.
14. Vijayalaxmi, H.J. Evans, J.H. Ray, and J. German (1983) Bloom's syndrome: Evidence for an increased mutation frequency in vivo. Science 221:851-853.
15. Rüdiger, H.W., C.R. Bartram, W. Harder, and E. Passarge (1980) Rate of sister chromatid exchanges in Bloom syndrome fibroblasts reduced by co-cultivation with normal fibroblasts. Am. J. Human Genet. 32:150-157.
16. Passarge, E. (1983) Bloom's syndrome. In Chromosome Mutation and Neoplasia, J. German, ed. Alan R. Liss, Inc., New York, pp. 11-21.
17. Cohen, M.M., C.E. Fruchtman, S.J. Simpson, and A.O. Martin (1982) The cytogenetic response of Fanconi's anemia lymphoblastoid cell lines to various clastogens. Cytogenet. Cell Genet. 34:230-240.

STUDY OF BASAL CELL NEVUS SYNDROME FIBROBLASTS AFTER TREATMENT WITH DNA-DAMAGING AGENTS

H. Nagasawa, F. F. Little, M. J. Burke, E. F. McCone,
H. S. Targovnik, G. L. Chan, and J. B. Little

Department of Cancer Biology
Harvard University School of Public Health
665 Huntington Avenue
Boston, Massachusetts 02115

INTRODUCTION

Basal cell nevus syndrome (BCNS) is a rare autosomal dominant inherited disorder (1). About 20% of gene carriers develop brain tumors in early childhood. Basal cell nevus syndrome patients are abnormally susceptible to radiation-induced cancer; several patients treated with radio-therapeutic doses have developed large numbers of basal cell tumors in the irradiated field within 6 mo to 3 yr of exposure (2-5). Featherstone et al. (6) reported that fibroblasts from BCNS patients showed no increased susceptibility to X-ray-induced cell killing; however, G_0-irradiated lymphocytes from BCNS patients were found to have a significantly higher level of X-ray-induced chromosomal aberrations (CAs) compared with normal cells. On the other hand, Chan and Little (7) reported that fibroblasts from BCNS patients were slightly hypersensitive to X-ray-induced lethality.

It has been suggested that cancer proneness might be associated with a systemic defect in the ability of somatic cells to repair some types of DNA damage (8). In the present study, a number of different techniques, including measurements of cell survival, biochemical and cytogenetic assays, have been used to examine the response of BCNS cells to exposure to the DNA-damaging agents: X-irradiation, ultraviolet light (UV), and mitomycin C (MMC) treatments.

MATERIALS AND METHODS

Cell Culture

Human diploid fibroblasts were obtained from the Genetic Cell Repository at the Institute for Medical Research, Camden, New Jersey (Tab. 1). The cells were seeded into either 100 or 30 mm plastic petri dishes, or T-25 plastic flasks, and cultured at 37°C until nearly confluent. The culture medium (Eagle's minimum essential medium supplemented with 15% fetal calf serum) was then changed twice at daily intervals, and the experiments began 48 hr after the second medium change. At this time, only 0-3% of the cells in these confluent, density-inhibited cultures following a 15-min incubation with medium containing 10 μci/ml of ^{3}H-thymidine (27 Ci/mmole) incorporated label, indicating that they were in S phase.

X-Irradiation

The confluent cultures were X-irradiated at a dose rate of 80 rads/min with a G.E. Maximar X-ray Generator (220 KeV, 15 mA).

UV Exposure

The confluent cultures were rinsed with Eagle's balanced salt solution (EBSS) and exposed at room temperature to UV light (predominately 254 nm) in a specially constructed sterile chamber, as previously described (9).

Tab. 1. Characteristics of human cell strains.

STRAIN	CLINICAL CLASS	SEX	AGE	SOURCE
AG 1522	normal	M	3 days	IMR
GM 730	normal	F	45 years	IMR
AT5BI	ataxia telangectasia	M	18 years	Dr. C.F. Arlett
GM 2990	xeroderma pigmentosum	F	8 years	IMR
GM 2995	xeroderma pigmentosum	M	10 years	IMR
GM 1309	Fanconi anemia	M	12 years	IMR
GM 2098	Basal Cell Nevus Syndrome	M	31 years	IMR
GM 1575	Basal Cell Nevus Syndrome	F	31 years	IMR
14	Basal Cell Nevus Syndrome			Dr. L.C. Strong

Mitomycin C

Mitomycin C was obtained from Sigma and prepared in Dulbecco's phosphate buffered saline (pH 7.4). The cultures were treated with MMC for 1 hr at 37°C, in the dark.

Cell Survival

Cell survival was measured by a standard colony-formation assay. Cell numbers were adjusted to yield approximately 50-100 viable colony-forming cells per dish. Colonies were scored under a dissecting microscope 10-15 da after treatment.

Alkaline Elution

The procedure used is a modification of that previously described by Kohn et al. (10). Cellular DNA was bulk-labeled by growing cells for 24 hr (37°C, 5% CO_2) in medium containing 0.02 μCi/ml (2-^{14}C)thymidine (New England Nuclear, 50 Ci/mmole). For single-strand break analysis, cells were layered on polycarbonate filters (Nucleopore, 2 μm pore size), and lysed with 2% SDS/0.2 M Na_4-EDTA, then washed with 0.04 M Na_4-EDTA (pH 10.0). Cellular DNA was then eluted off the filter with 0.1 M tetrapropylammonium hydroxide/0.1% SDS/0.02 M EDTA (free acid form), pH 12.1. For cross-link analysis, cells were layered on polyvinyl chloride filters (Millipore, 2 μM pore size) and lysed with 0.2% Sarkosyl (NL-30, Ciba-Geigy)/2 M NaCl/0.02 M Na_4-EDT (pH 10.0).] Samples which received proteinase K were treated as described by Fornace and Little (11) except that the final concentration of proteinase K in the lysis was 40 μg/ml. Cellular DNA was then eluted off the filter with 0.01 M tetrapropylammonium hydroxide/0.02 M EDTA (free acid form), pH 12.1. In both single-strand break and cross-link analysis the pump speed was 0.035 ml/min. Elution fractions were collected directly into 6.0 ml liquid scintillation counting vials at 90-min intervals. Data were analyzed as described by Kohn and coworkers (10).

Cross-link factors were calculated as follows: log (relative retention, Controls) - log (relative retention, X-rays)/log (relative retention, treated, no X-rays) - log (relative retention, treated, X-rays). Relative retention was taken as the fraction of DNA retained on the filter after 12 hr of elution. With no cross-links this parameter is 1.0 by definition.

Unscheduled DNA Synthesis (UDS) Measurements

The cells were grown in culture on glass cover slips in P-30 plastic petri dishes. When the cells reached confluence, they were washed with EBSS and exposed to UV. The cells were cultured in medium containing 10 μCi/ml of ^{3}H-thymidine (specific activity 27 Ci/mmole) for 1 hr at 37°C. The medium was then washed off, and the cells fixed on the cover slips. The cells were treated with cold 5% trichloroacetic acid (TCA) for 5 min and stained with aceto-orcein. Cells were covered with Kodak nuclear emulsion NTB2 to obtain autoradiographs. Approximately 50 nuclei per sample were photographed, and the negatives projected on white paper from a photographic enlarger for scoring. The number of grains per nucleus were counted in each cell and averaged for each experimental point.

Sister Chromatid Exchanges (SCEs)

The cells were subcultured into 4 P-100 dishes containing 10^{-5} M BrdUrd medium from 1 P-100 confluent culture, and grown for 2 rounds of cell replication. Colcemid was added to successive samples for 6-hr intervals before fixation as previously described (12). The cells were fixed by hypotonic method (13) and stained by the fluorescence-plus-Giemsa technique for the differential staining of sister chromatids (14). Sister chromatid exchanges were scored in the sample containing the peak number of second mitoses.

RESULTS

Figure 1A shows a comparison of the X-ray sensitivities of ataxia telangiectasia (AT), BCNS, and normal cell strains. The D_0 values for the AT, BCNS, and normal cell strains were approximately 50, 110, and 140 rads, respectively. In addition, the BCNS cell strains showed smaller shoulder regions to their survival curves than did the normal cells. Therefore, the D_{10} values were significantly different between AT (110 rads), BCNS (300), and normal (430 rads).

The D_0 values of normal and BCNS cells were similar after UV exposure (Fig. 1B) although the shoulder region appeared diminished in the BCNS cells as compared with normal strains. Similarly, the differences in D_0 doses for MMC (Fig. 1C) between normal (0.25 μg/ml) and BCNS (0.20 μg/ml) were also small as compared with Fanconi anemia (FA) (0.02 μg/ml). However, survival curves of BCNS cell strains again showed smaller shoulder regions than did normal cell strains.

There were no differences in X-ray-induced DNA single-strand breaks in normal and BCNS cells (Fig. 2). There was also no difference in MMC-induced cross-linking between normal and BCNS cells (Fig. 3A, 3B).

Although UDS measurements in 2 normal cell strains yielded linear response curves over a broad range, there was a marked difference in the relative amounts of UDS in each cell line as measured by the number of grains per cell (Fig. 4). Only very small amounts of UDS were observed in xeroderma pigmentosum cells from complementation groups A and C (XPA and XPC, respectively) following various doses of UV (up to 25 J/m^2) (Fig. 4).

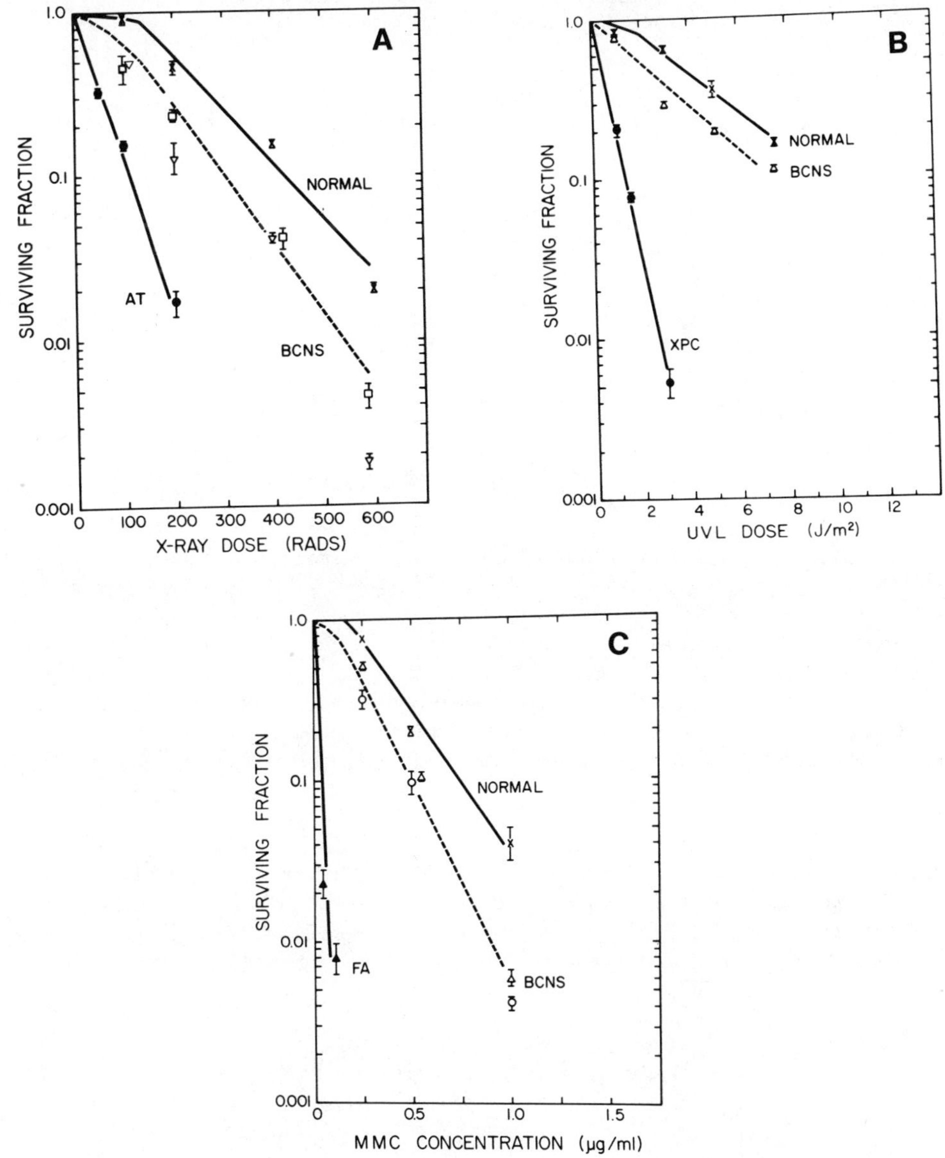

Fig. 1. A. The X-ray survival curves for normal, BCNS, and AT cells. B. The UV survival curves for normal, BCNS, and XPC cells. C. The MMC survival curves for normal, BCNS, and FA cells.

There was a distinct difference in the shape of the dose-response curve for the BCNS as compared with the normal cell strains (Fig. 4). The number of grains per cell in BCNS cultures increased linearly with doses up to 10 J/m^2, but they remained at the same level with increasing UV doses up to 30 J/m^2. The number of grains

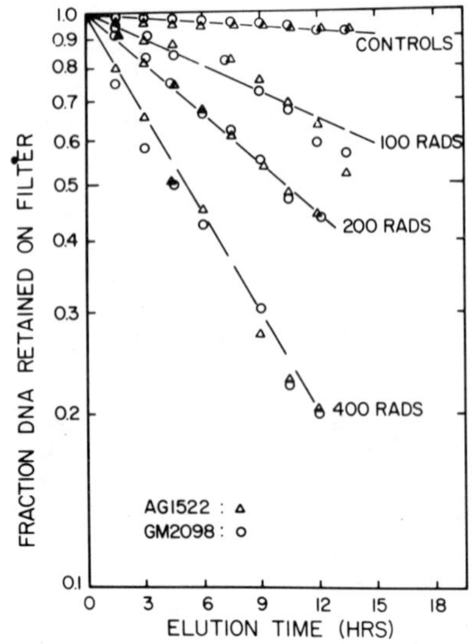

Fig. 2. X-ray-induced DNA single-strand breaks.

continued to rise linearly in normal cells exposed to similar doses. Furthermore, whereas a normal cell strain (AG1522) held in confluent growth for 24 hr after UV exposure showed little residual UDS, significantly higher UDS activity remained in a BCNS strain (GM 2098) treated and held under the same conditions (preliminary data).

There were no differences in the dose-response curves for UV-induced SCE between normal and BCNS cell strains (Fig. 5). When UV-irradiated normal cell cultures were held in confluent growth for 0-48 hr to allow for the repair of potentially lethal damage (PLD), the UV-induced SCE frequencies declined rapidly during the first 12 hr of PLD repair, reaching a plateau with longer repair periods

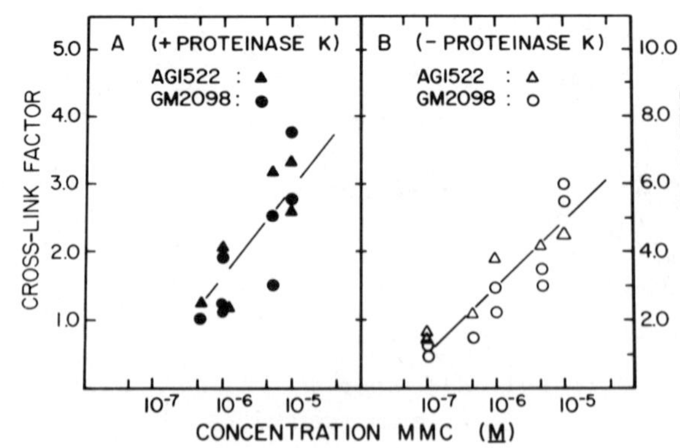

Fig. 3. MMC-induced cross-links. A. DNA-DNA cross-links. B. Total cross-links.

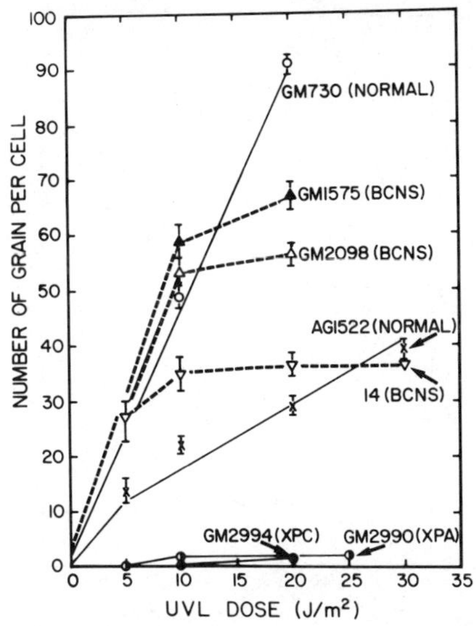

Fig. 4. UV-induced unscheduled DNA synthesis.

(Fig. 6). As can also be seen in Fig. 6, however, UV-induced SCE frequencies in BCNS cell strains declined very slowly during PLD repair periods of up to 48 hr.

DISCUSSION

Basal cell nevus syndrome fibroblasts showed no marked hypersensitivity to the cytotoxic effects of X-rays, UV, or MMC treatments, unlike AT, XP, and FA fibroblasts, respectively. However, the survival curves appear to indicate that BCNS cell strains are slightly hypersensitive to all 3 DNA-damaging agents (Fig. 1). Featherstone et al. (6) reported that D_0 values for BCNS and normal cell strains were within the same ranges (90-100 rads without shoulder regions) in either exponentially growing or confluent cultures. On the other hand, Chan and Little (7) reported BCNS cells were slightly hypersensitive to killing by X-rays. Interestingly, both groups obtained similar D_0 values for BCNS cell strains. These apparent differences might be related to the culture condition after X-irradiation, such as the presence (6) or absence (7) of feeder layers (15). Weichselbaum et al. (1980) reported that D_0 values for 6 normal cell strains were 140-152 rads, with appreciable shoulder regions, without feeder layers during the colony-forming assay after X-irradiation. Arlett and Harcourt (16) reported a wider range of D_0 values for normal cells (101-160 rads) without shoulders assayed in the presence of feeder layers.

Slightly higher frequencies of X-ray-induced CAs have been reported in G_0-irradiated lymphocytes from BCNS patients as compared

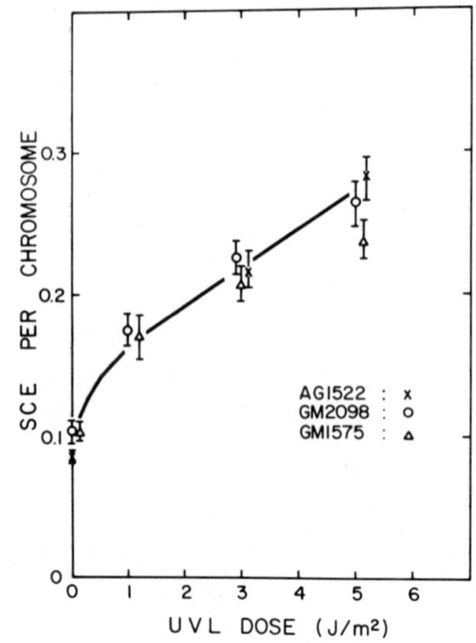

Fig. 5. The frequencies of UV-induced SCE between normal and BCNS cells.

with normal (6); however, there were no significant differences in BCNS fibroblasts (Nagasawa and Little, unpublished data).

In preliminary experiments, Yandell and Little (unpublished data) found that 30-50% of BCNS cells were irreversibly blocked in G_1 following irradiation with 400 rads in confluent cultures and subcultivation to low density. A similar but smaller X-ray-induced G_1 block has previously been observed in very slowly growing human diploid cell populations (17), in density-inhibited, confluent mouse 10T 1/2 cells (18) and normal human diploid fibroblasts (19). However, a much larger X-ray-induced G_1 block, similar to that in BCNS cells, was observed in X-ray-sensitive retinoblastoma (Rb) fibroblasts (19) and in AT-heterozygote cells (20). Rb and AT-heterozygote fibroblasts showed only normal levels of X-ray-induced CAs.

There were no differences in X-ray-induced DNA single-strand breaks and MMC-induced DNA-protein and DNA-DNA cross-links between normal and BCNS fibroblasts (Figs. 2, 3A, and 3B). However, UV-induced UDS showed 3 different types of dose-response relationships between normal, BCNS, and XP cell strains (Fig. 4). It is tempting to speculate that UV-induced DNA damage in BCNS cells is similar to that in normal cells, but that the excision repair process in BCNS cells might be limited in its capacity and thus take place more slowly when much damage is present. This would explain why significant amounts of UV-induced UDS still remained at 24 hr after irradiation in noncycling BCNS cells. Although the method of UDS determination differed, maximum DNA repair synthesis was reported to be about 25% lower in UV-exposed lymphocytes from BCNS patients than normal (21).

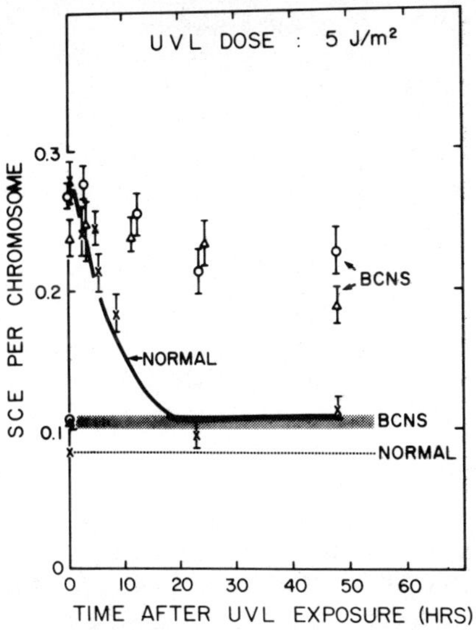

Fig. 6. UV-induced SCE frequencies during confluent holding periods.

A similar phenomenon was observed in measurements of the frequencies of UV-induced SCEs after varying periods of confluent holding. The dose-response relationships for immediately subcultured cells was the same as that for normal cell strains (Fig. 5), but UV-induced SCE frequencies declined very slowly as compared with normal cells during PLD repair (Fig. 6). Similar declining patterns of UV-induced SCE frequencies have previously been reported in mouse 10T½ cells and in normal human fibroblasts during PLD repair after UV exposure (22). Interestingly, the kinetics for the decline in SCEs with confluent holding were very similar to those for UV- and MMC-induced DNA protein cross-links studied under identical conditions (Ref. 23 and Targovnik et al., unpublished data).

These results suggest that BCNS cells may have a diminished capacity for the repair of some types of DNA damage. The specific DNA lesions and their molecular repair systems remain to be identified. Yet, our results suggest that this process may be common to the DNA damage induced by 3 different types of DNA-damaging agents: X-rays, UV light, and MMC. Such a process could be responsible for the degree of hypersensitivity to DNA-damaging agents reported in some other genetic disorders.

ACKNOWLEDGEMENTS

This research was supported by Training Grant CA-09078, Research Grant CA-33624, and Center Grant ES-00002 from the United States National Institutes of Health.

REFERENCES

1. Gorlin, R.J., and H.O. Sedano (1976) Cancer of the Skin, A. Andrade, S.L. Gumport, G.L. Popkin, and T.D. Rees, eds. W.B. Sanders, Philadelphia, pp. 883-898.
2. Strong, L.C. (1977) Genetic and environmental interactions. Cancer 40:1861-1866.
3. Strong, L.C. (1977) The Genetics of Human Cancer, J.J. Malvihill, R.W. Miller, and J.F. Fraumeni, Jr., eds. Raven Press, New York, pp. 401-414.
4. Hawkins, J.C., H.J. Hoffman, and L.E. Becker (1979) Multiple nevoid basal-cell carcinoma syndrome (Gorlin's syndrome): Possible confusion with metastatic medulloblastoma. J. Neurosurg. 50:100-102.
5. Meadows, A.T., S.J. D'Angio, V. Mike, A. Banfi, C. Harris, R.D.T. Jenkin, and A. Schwartz (1977) Patterns of second malignant neoplasms in children. Cancer 40:1903-1911.
6. Featherstone, T., A.M.R. Taylor, and D.G. Harnden (1983) Studies on the radiosensitivity of cells from patients with Basal Cell Naevus Syndrome. Am. J. Human Genet. 35:58-66.
7. Chan, G.L., and J.B. Little (1983) Cultured diploid fibroblasts from patients with the Nevoid Basal Cell Carcinoma Syndrome are hypersensitive to killing by ionizing radiation. Am. J. Pathol. 111:50-55.
8. Friedberg, E.C., U.K. Ehman, and J.I. Williams (1979) Advances in Radiation Biology, Vol. 8, J.T. Lett, H. Adler, and M. Zelle, eds. Academic Press, New York, pp. 85-174.
9. Chan, G.L., and J.B. Little (1976) Induction of oncogenic transformation in vitro by ultraviolet light. Nature (Lond.) 264:442-444.
10. Kohn, K.W., R.A.G. Ewig, L.C. Erickson, and L.A. Zwellig (1981) DNA Repair, Vol. 1, part B, E.G. Frieberg and P.C. Hanawalt, eds. Marcel Decker, New York and Basel, pp. 379-401.
11. Formace, A.J., Jr., and J.B. Little (1979) DNA-protein cross-linking by chemical carcinogens in mammalian cells. Cancer Res. 39:704-710.
12. Nagasawa, H., and J.B. Little (1979) Effect of tumor promoters, protease inhibitors, and repair processes on X-ray-induced sister chromatid exchanges in mouse cells. Proc. Natl. Acad. Sci., USA 76:1943-1979.
13. Hsu, T.C., and O. Klatt (1958) Mammalian chromosomes in vitro IX. On genetic polymorphism in cell population. J. Natl. Cancer Inst. 21:437-473.
14. Wolff, S., and P. Perry (1974) Differential Giemsa staining of sister chromatids and the study of sister chromatid exchanges without autoradiography. Chromosoma 48:341-353.
15. Weichselbaum, R.R., J. Nove, and J.B. Little (1980) X-ray sensitivity of fifty-three human diploid fibroblast cell stains from patients with characterized genetic disorders. Cancer Res. 40:920-925.
16. Arlett, C.F., and S.A. Harcourt (1980) Survey of radiosensitivity in a variety of human cell stains. Cancer Res. 40:926-932.

17. Little, J.B. (1969) Differential response of rapidly and slowly proliferating human cells to X-irradiation. Radiology 93:307-313.
18. Nagasawa, H., J.B. Robertson, C.S. Arundel, and J.B. Little (1984) Effect of X-irradiation on the progression of mouse 10T½ cells released from density inhibited cultures. Radiat. Res. 97:537-545.
19. Nagasawa, H., and J.B. Little (1983) Comparison of kinetics of x-ray-induced cell killing in normal ataxia telangiectasia and hereditary retinoblastoma fibroblasts. Mutat. Res. 109:297-08.
20. Nagasawa, H., S. Latt, and J.B. Little (1983) Effects of x-irradiation on cell cycle progression, induction of chromosomal aberrations and cell killing in homozygotic and heterozygotic ataxia-telagiectasia. Radiat. Res. 94:632-633.
21. Ringborg, N., B. Lambert, J. Landgren, and R. Lewensohn (1981) Decreased UV-induced DNA repair synthesis in peripheral leukocytes from patients with the Nevoid Basal Cell Carcinoma Syndrome. J. Invest. Dermatol. 76:268-270.
22. Nagasawa, H., A.J. Fornace, Jr., M.R. Ritter, and J.B. Little (1982) Relationship of enhanced survival during confluent holding recovery in ultraviolet-irradiated human and mouse cells to chromosome aberrations, sister chromatid exchanges, and DNA repair. Radiat. Res. 92:483-496.
23. Nagasawa, H., A.J. Fornace, Jr., J.B. Little, and J.R. Williams (1977) Potentially lethal damage repair:relationship to the induction of sister chromatid exchanges and DNA repair capacity. Radiat. Res. 70:706.

DNA INTERSTRAND CROSS-LINKING, REPAIR, AND SCE MECHANISM IN HUMAN CELLS IN SPECIAL REFERENCE TO FANCONI ANEMIA

Yoshisada Fujiwara, Yoshio Kano, and Yoko Yamamoto

Department of Radiation Biophysics
Kobe University School of Medicine
Kusunoki-cho 7-5-1, Chuo-ku, Kobe 650, Japan

ABSTRACT

The relation of DNA cross-linking and repair to sister chromatid exchange (SCE) formation was studied in normal human, Fanconi anemia (FA), and xeroderma pigmentosum (XP) cells. Despite a hypersensitive lethality response in FA cells, the SCE induction rates by mitomycin C (MMC), trimethylpsoralen (TMP)-light, cisplatin, and diepoxybutane were twice as high as in normal cells. For MMC, the induced SCE frequency in normal cells was reduced in a biphasic fashion with a repair incubation time (the first decline $t_{1/2}$ = 2 hr; the second $t_{1/2}$ = 14-18 hr) which corresponds exactly to the molecular kinetics of cross-link and monoadduct removal. However, FA cells lack the first half-excision process and exhibit a lack of the first rapid decline SCE component. The slow decline component is present, and a higher SCE frequency is observed 24 to 48 hr after treatment. By contrast, XP cells capable of the half-excision process reveal the first rapid decline component, followed by an extremely slow second-reduction component ($t_{1/2}$ = 48 hr) due to defective monoadduct repair. The endoreduplication-SCE method revealed that rates of both twin (first cycle) and single (second cycle) SCE formations by MMC and TMP-light were higher in FA cells than in normal cells. These results indicate that cross-links remaining unrepaired induce SCEs as do monoadducts. The probabilistic SCE induction occurs at a rate of 1 SCE per 35 MMC cross-links in FA cells. Further, a non-SCE-forming tolerance mechanism also operates in hypersensitive FA cells. These molecular and cytogenetic results allow us to construct a new probabilistic model for cross-link-induced SCE into which the replication-fork model, random cross-link transfer to both chromatids, and chromatid breakage-reunion are incorporated.

INTRODUCTION

Sister chromatid exchanges are a more sensitive indicator than chromosomal aberrations (CAs) of the S-phase-dependent clastogenic effects of mutagenic carcinogens (1-4). Sister chromatid exchanges that involve reciprocal double-strand recombination between homologous chromatid loci occur at replication forks blocked by a DNA lesion(s) during S phase (3-8). However, the precise molecular mechanism of SCE formation remains, as yet, uncertain. The nature of SCE-forming DNA lesions and their induction efficiency are still largely unknown because of the multiplicity of DNA lesions induced by even a single agent and the operation of different DNA repair mechanisms. Crosslinking agents are more effective in blocking DNA replication and inducing SCEs than are single-strand-damaging agents (1-4). Bifunctional agents produce both interstrand cross-link and monoadducts in different proportions in cellular DNA. The interpretation of whether cross-linking or monoaddition is responsible for SCE formation has not been possible based on the to-date SCE studies involving cross-linking agents (9-15). We have investigated the relation between cross-linking or monoaddition to SCE induction in normal human, FA, and XP complementation group A (XP-A) cells using MMC, trimethylpsoralen-plus-near-ultraviolet light (TMP-NUV), _cis_-diaminedichloroplatinum (_cis_-DDP), and deoxybutane (DEB). These agents are known to be supereffective in inducing CAs in FA (16,17). We have focused especially on MMC-induced SCEs, since FA and XP-A cells have clearly a specific defect in the half-excision of MMC cross-links and in the MMC-monoadduct repair, respectively (18-21). Based on these results, we have developed assumptive models for cross-linkinduced SCE and a non-SCE-forming tolerance.

SURVIVAL RESPONSES TO CROSS-LINKING AGENTS

Figure 1 shows chronogenic survival curves of NHSF6 (normal), FA17JTO, and XP6KO-A cells treated with the cross-linking agents. Analysis of mean lethal dose (Do) revealed 30- and 18-fold supersensitivities of FA17JTO cells to MMC and DEB killing, respectively, compared with NHSF6 cells (Figs. 1A and D) (12,18). Based on calculated Do values, TMP-NUV exerted only a twice greater lethal effect on FA17JTO and XP6KO-A cells than on NHSF6 cells (Fig. 1B) (21). This increase is consistent with the finding that FA and XP cells are deficient in the repair of psoralen cross-links (22). Similarly, _cis_-DDP exerted the same lethal effect on FA17JTO and XP6KO-A cells, which is twice greater than that on NHSF6 cells (Fig. 1C). The XP6KO cells were 6 times more sensitive to DEB than NHSF6 cells, but 3 times less sensitive than FA17JTO cells (Fig. 1D). Thus the killing efficiency, which reflects the FA and XP defects in cross-link and/or monoadduct repair differs greatly from agent to agent. Nevertheless, the lethal effects of these agents appear to be greater in FA cells and, in this regard, MMC is the agent selected to distinguish the defects in FA and XP-A cells.

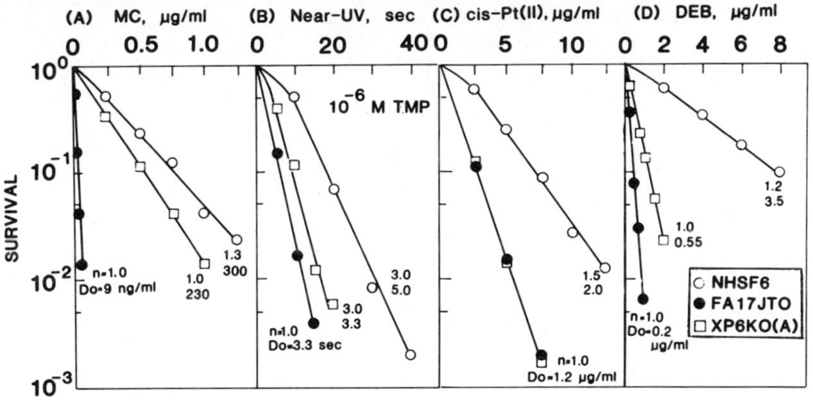

Fig. 1. Chronogenic survival curves. The appropriate numbers of plated cells were treated with MMC (A), 10^{-8} M TMP (B), cis-DDP (cis-Pt) (C), and DEB (D) for 1 hr in phosphate buffered saline at 37°C. In the case of TMP, cells were irradiated with NUV immediately after incubation. Following 2 washes, treated cells were incubated for 12-14 da in growth medium for clone survival. Extrapolation number (n) and mean lethal dose (Do) have been specified for each curve.

HALF-EXCISION REPAIR OF MITOMYCIN C CROSS-LINKS

Mammalian cross-link repair appears to consist of an excision step involving one-half of the cross-link followed by subsequent monoadduct repair (12,18,20). Thus, we studied the half-excision of MMC cross-links by determining the number of cross-links per dalton of purified DNA (19,20).

Figure 2 displays the typical half-excision kinetics. As described elsewhere (19-21), normal human cells manifest the first rapid half-excision characterized by the first-order kinetics of $t_{1/2}$ = 2-3 hr. However, several unrelated FA strains exhibit a unique defect in this process (Fig. 2) (18-21). A previous study showed the normal level of excision repair of monoadducts produced selectively by decarbamoyl mitomycin C (DMC) in FA cells, although the monoadduct removal was slow ($t_{1/2}$ = 14-18 hr) (12,18). By contrast, XP-A cells are normally proficient in the half-excision of MMC cross-links (Fig. 2) (18-21), but defective in the monoadduct repair (18). However, XP cells, like FA cells (22), have been shown to be defective in the removal of psoralen cross-links (22,23), a process which is much slower than that of MMC cross-links (see above). This observation suggests that there are differences in the repair kinetics and the unhooking specificity to types of cross-linking.

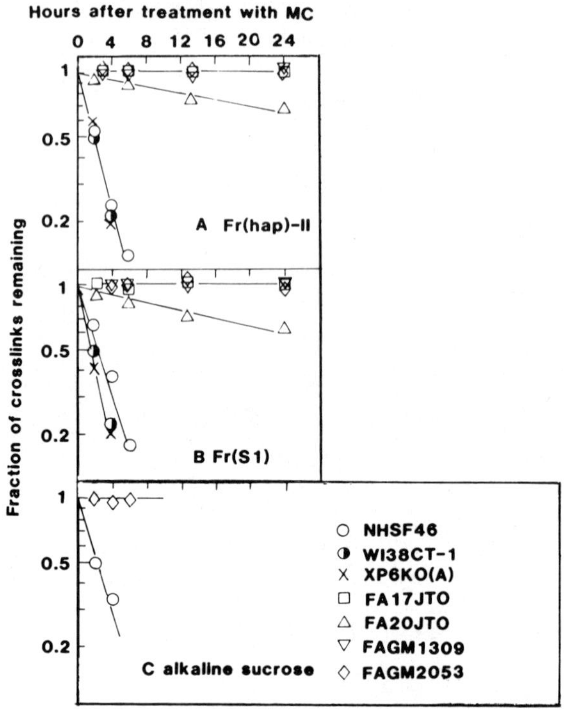

Fig. 2. Half-excision kinetics of MMC cross-links. Cells in log-growth phase were treated for 1 hr with 10 μg/ml of MMC (which produces 1.4 cross-links/10^8 daltons) (19,20), washed, and incubated in MMC-free medium for desired lengths of time as indicated. DNA was extracted and purified by use of proteinase K and RNases, followed by measurements of cross-links by hydroxyapatite chromatography (A) and S1-nuclease digestion (B) of denatured/renatured DNA, and alkaline sucrose sedimentation (C) as defined previously (19,20). Then, the number of cross-link units/dalton of purified DNA was computed (20). The fraction of cross-links remaining was plotted against incubation time. (Excerpted from Ref. 19 under permission by Elsevier Biomedical.)

Fornace et al. (24) failed to detect a defective cross-link repair in MMC-treated FA strain even at confluent state by the alkaline-elution method. On the contrary, our DNA study indicated the loss of cross-link repair activity in FA cells and in confluent, otherwise proficient, normal and XP cells (19,21). Using alkaline elution without digestion with proteinase K, Sognier and Hittelman (25) found no defect in the repair of MMC and nitrogen mustard cross-links in early-passage FA cells, but a deficiency only in late passages. Our results have so far shown no such culture-age dependent fluctuations on purified DNA (unpublished data). Further, a long treatment with alkali makes MMC cross-links, but not psoralen cross-links, labile (20). Therefore, the discrepancy is most likely to arise from long alkaline treatment (alkaline elution) and the proficient removal of DNA-protein cross-links.

ROLES OF INTERSTRAND CROSS-LINKING AND REPAIR IN SCE FORMATION

It has been suggested that cross-links may not form SCEs, but that monoadducts are exclusively responsible for their induction (9-11). Based on the molecular rationale of MMC-cross-link repair and the alternative repair defect of FA and XP cells (see above), we attempted to test the above notion and to elucidate the roles of MMC cross-linking and two-step repair in SCE formation.

Induction of SCEs by MMC

The induction rate of SCEs by MMC (treatment for 1 hr), as determined from the linear regression lines, was 1.7 times higher in FA17JTO cells than in NHSF46 (normal) cells (Tab. 1) (12,13). The same results were obtained in the other typical FA strains (12). The induction rate by monofunctional DMC was the same (0.5 SCEs/cell per 10 ng/ml) in both NHSF46 and FA17JTO cells (Tab. 1) (12,13). On the other hand, XP6KO-A cells possessing defective DMC-monoadduct repair (12) exhibited a 6-fold higher SCE rate for DMC, while only a 1.3-times higher rate for MMC than did NHSF46 cells (Tab. 1). Therefore, both cross-links and monoadducts by MMC are responsible for SCE formation.

Tab. 1. The rates of SCE formations in normal, FA, and XP cells induced by the cross-linking agents. Each cross-linking agent (1-hr treatment in phosphate buffered saline at 37°C) induced SCEs in any type of cells as a linear function of dose. SCE rate was determined from the regression line of each strain and expressed as SCEs/cell per unit dose. Unit dose: 10 ng/ml for MMC and DMC; 10 sec for NUV; 0.1 μg/ml for cis-DDP (cis-PT) and DEB.

	SCEs/cell per unit dose			Ratio	
Agent	NHSF46	FA17JTO	XP6KO	FA/NHSF46	XP/NHSF46
Bifunctional					
MC	2.6	4.3	3.3	1.7	1.3
TMP(10^{-8}M)-NUV	1.0	2.6	1.3	1.6	1.3
cis-Pt	2.6	7.5	4.5	2.9	1.7
DEB	8.4	15.1	9.3	1.8	1.1
Monofunctional					
DMC	0.5	0.5	3.0	1.0	6.0

SCE Inductions by TMP-NUV, cis-DDP, and DEB

More SCEs were induced by these agents in FA17JTO cells than in XP6KO-A and NHSF46 cells (Tab. 1). Near-UV irradiation at 10^{-8} M TMP, *cis*-DDP, and DEB were nearly twice as effective at inducing SCEs in FA17JTO cells than in XP6KO-A cells. This also suggests a possible retarded capacity of FA cells to half-excise cross-links by these agents. Further, DEB mimics MMC in the lethal (Fig. 1) and SCE (Tab. 1) effects, although DEB and MMC bind specifically to the N^7 and O^6 sites of guanine, respectively.

Differential Decline of MMC-Induced SCE Frequency with Repair Time

Change in MMC-induced SCE frequency as a function of repair time may clearly distinguish whether or not cross-links produce SCEs. Half-excision is rapid, but monoadduct excision is slow in proficient cells and FA and XP cells are alternatively defective in these 2 steps. The pertinent experiments and results have been described in detail previously (12,13). Briefly, NHSF46, FA17JTO, and XP6KO-A cells were treated with MMC or DMC for 1 hr and grown in normal medium between 0 and 48 hr before the 2-cycle incubation in BrdUrd-deoxycytosine (dCyt).

Figure 3 depicts time-dependent reductions in induced SCE frequencies. In NHSF46 cells (Fig. 3A), MMC-induced SCE frequency declines biphasically. The first SCE decline is rapid ($t_{1/2}$ = 2 hr), and the second component is much slower ($t_{1/2}$ = 14-18 hr). Such kinetics correspond exactly to the molecular 2 steps of cross-link repair. In NHSF46 cells, the second decline (Fig. 3A) parallels the decline curve of DMC-induced SCEs (Fig. 3B), reflecting the monoadduct repair. XP6KO-A cells capable of only half-excision do show the first rapid decline in MMC-induced SCE frequency (Fig. 3A), and a slower second decline ($t_{1/2}$ = 48 hr) (Fig. 3A) due to defective monoadduct repair. This latter curve runs parallel to that of DMC-induced SCEs (Fig. 3B). FA17JTO cells show the highest frequency of immediate SCEs by MMC and only a slow component of decline ($t_{1/2}$ = 14 hr) without the first component due to defective half-excision (Fig. 3A). Such a component perfectly fits the monotonic decline of DMC-induced SCEs in FA17JTO and NHSF46 cells (Fig. 3B). In the FA cells, further, the SCE frequency at later periods (24-48 hr) after treatment still remained high at a 50% level of the initial induced frequency (Fig. 3A). This SCE fraction arises from unrepaired cross-links, so that total cross-links and monoadducts in genomic DNA may have a 1:1 induction efficiency. Since the ratio of cross-linking to monoaddition by MMC is at least 1:10, cross-links are a more effective inducer of SCEs than are monoadducts. These clear-cut, differential results strongly indicate that cross-links induce SCEs.

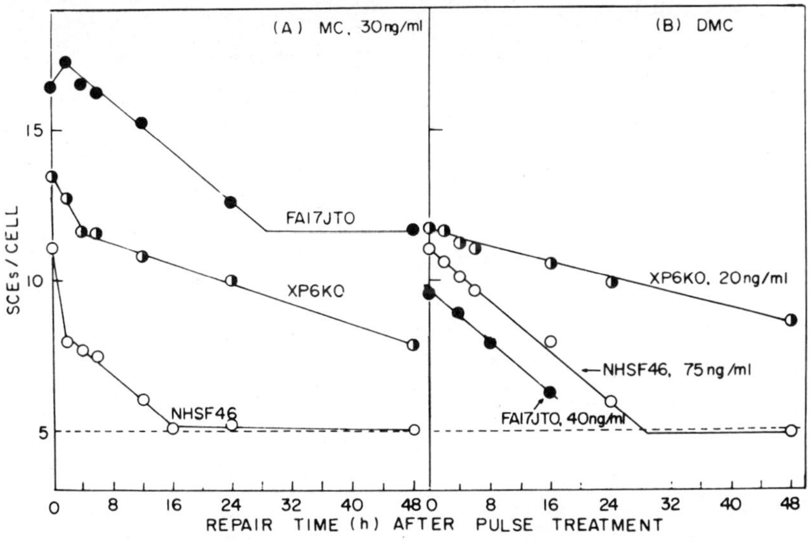

Fig. 3. The reduction kinetics of MMC- or DMC-induced SCEs as a function of post-treatment time. Exponentially growing NHSF46 (normal), FA17JTO, and XP6KO-A cells were treated with MMC at 30 ng/ml (A) or DMC at the indicated concentrations (B) for 1 hr at 37°C, washed, and incubated in the ordinary growth medium between 0 and 48 hr before the 2-cycle (72 hr in this case) incubation in BrdUrd-dCyt (10^{-5} M each) to detect SCEs. -----, base-line frequency. (Excerpted from Ref. 12 under permission by Elsevier Biomedical.)

MMC-Induced Single and Twin SCEs in Endoreduplicated Cells

Since endoreduplicated tetraploids contain pairs of daughter chromosomes, or diplochromosomes, the endoreduplication-SCE method, which we described earlier (14), allows the accurate identification of both twin and single SCEs.

Figure 4 illustrates our best protocol for endoreduplication of human fibroblasts and the scheme for twin and single SCEs. Exponentially growing cells (doubling time, 30-36 hr) were treated with MMC or TMP-NUV for 1 hr and incubated for 72 hr (2 cycles) in BrdUrd-dCyt (14). In the middle of the culture period (27 to 36 hr), the cells were treated with Colcemid (0.1 μg/ml) to induce endoreduplication and again for the final 5 hr to accumulate M2 metaphases (14). An exchange occurring only in the first S phase (S1) will result in a twin SCE at M2 (Fig. 4B), and an S2 exchange alone will produce a single SCE in one of diplochromosome (Fig. 4C).

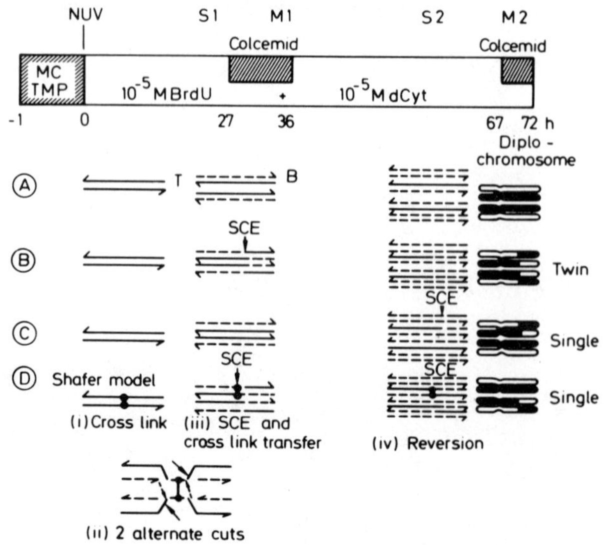

Fig. 4. The protocol for endoreduplication and the schemes for twin and single SCEs. The protocol for endoreduplication by Colcemid for NHSF46 and FA17JTO cells has been described in the text and Ref. 14. (A) Diplochromosome without SCE; (B) S1 exchange (twin SCE at M2); (C) S2 exchange (single SCE at M2); (D) cross-link-induced SCE predicted for diplochromosome from Shafer's replication bypass model (26): A potential SCE is formed at fork with replicative bypass of cross-link by alternative cuts (stage ii), and cross-link transfer to a chromatid (stage iii) in S1, and the inevitable reexchange by the cross-link persisting in one chromosome in S2 (stage iv) reverses the S1 exchange in that chromosome to result in only a single SCE at M2. (Excerpted from Ref. 14 under permission by Springer-Verlag.)

Figure 5 shows SCE induction in terms of SCEs/diploid cell in S1 and S2. The average frequencies of induced twin (S1) and single (S2) SCEs increased as a linear function of MMC concentration or NUV dose. The S1-exchange rates by MMC and TMP-NUV were 1.7 and 2.4 times higher, respectively, in FA17JTO cells than in NHSF46 cells. Since monoadduct repair is normal in the FA cells, the greater number of S1-exchanges arise mainly from defective half-excision. Similarly, the rates of single SCEs (S2 exchanges) in FA17JTO cells were much higher than those in NHSF46 cells. These SCEs must arise from persisting cross-links in S2. Therefore, such substantial evidence clearly indicates that unrepaired cross-links indeed create SCEs. In the FA cells, the S2-SCE frequency becomes less than that of S1-SCEs due to proficient monoadduct repair and cross-link dilution (see below).

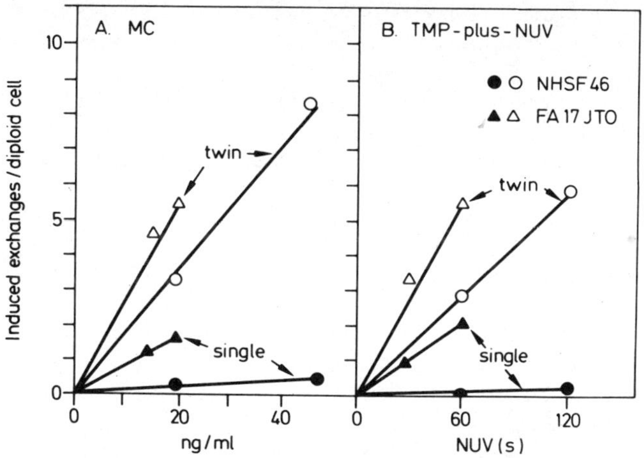

Fig. 5. Inductions of twin and single SCEs by MMC and TMP-NUV. Exponentially growing NHSF46 and FA17JTO cells were treated according to the protocol in Fig. 4. Induced twin (S1) and single (S2) SCE frequencies in terms of SCEs/diploid cell were plotted against MMC concentration or NUV dose in sec (dose rate = 8.79 J/m^2 per sec) at the fixed concentration of 10^{-8} M TMP. (Excerpted from Ref. 14 under permission by Springer-Verlag.)

THE NUMBER OF MMC CROSS-LINKS REQUIRED
FOR SCE FORMATION AND LETHAL HIT

In the typical FA strains, the SCE induction rate by MMC is about 4.3 SCEs/cell per 10 ng/ml (Tab. 1) (12-14), about a half of which (∿ 2 SCEs/cell) arises from unrepaired MMC cross-links and the remaining half from overwhelming monoadducts (see above). Our previous estimate for the MMC-cross-linking efficiency was 0.14×10^{-8} cross-link units/dalton per 1,000 ng/ml under the identical conditions (19,20). The approximate number of MMC cross-links per SCE in FA can be estimated as follows:

$0.14 \cdot 10^{-8}$ (efficiency/1,000 ng$\cdot$ml^{-1}) x $5 \cdot 10^{12}$ (genome size, dalton) x 10 ng/1,000 ng x 2^{-1} [SCE rate/10 ng)$^{-1}$] ≅ 35 cross-links/SCE.

Thus, induction by MMC cross-links occurs at a probabilistic rate of 1/35 and, therefore, MMC is a more potent SCE inducer than are single-strand-damaging agent, i.e., 1 SCE/10^{4-5} pyrimidine dimers (12,13). Unfortunately, the accurate estimation for normal cells is difficult due to the rapid first component of cross-link repair.

On the other hand, a Do = 9 ng/ml was obtained from MMC survival curves of FA17JTO (Fig. 1A) and other typical FA strains (12,20). If it is presumed that unrepaired cross-links and not repairable monoadducts have the superlethal effect on FA cells, then the number of MMC cross-links per lethal hit for an FA cell would be roughly 60 (= $0.14 \cdot 10^{-8} \times 5 \cdot 10^{-12} \times 9/1{,}000$). This number can be compared to a rough value of $\geq$ 3,000 cross-links/lethal hit for a normal cell. Therefore, supersensitive FA cells can still survive multiple cross-links by a tolerance mechanism independent of the SCE mechanism.

MODELING OF POSSIBLE MECHANISMS FOR CROSS-LINK-INDUCED SCE

Shafer's model of replication bypass (26) (see Fig. 4D legend) instructs the following: (a) each cross-link inevitably forms a potential S1-SCE by bypass through alternate parental-strand cuts and replication (Fig. 4D, stage ii); (b) the cross-link transferred to either sister chromatid (Fig. 4D, stage iii), unless repaired before S2, will cancel the twin SCE formation by its own reversing SCE in the cross-link-bearing chromosome in S2 (Fig. 4D, stage iv). Thus, only a single SCE will appear at M2 in response to a cross-link. In FA cells, therefore, the twin frequency is expected to be low due mainly to SCEs induced by repairable monoadducts (14,26,27). Such a phenomenon in FA cells has been observed when the continuous presence of cross-linking agents selects the minimally damaged cells (9,10). However, minimization of such a selection effect revealed a higher probabilistic induction of SCEs by cross-links in both S1 and S2 in FA than in normal cells (Fig. 5). Thus, this result does not support the Shafer model, as we (14) and Stekta (27) have discussed.

The replication-fork model (3-8) is currently acceptable for single-strand damage-induced SCE in that parental strands with ultimately 2 cuts at the replication fork are rejoined with newly formed daughter strands with polarity restraints. Double-strand cross-link may have some of the same characteristic features, as follows. Cross-links which block replication forks still cause a linear SCE induction as a function of dose (12-14) (Fig. 5), which allows us to utilize the replication-fork model. Unrepaired cross-links are randomly transferred, unless repaired, to both sister chromatids during S1 and induce S2-SCEs probabilistically as in FA cells (14) (Fig. 5). Cross-linking agents induce extensive CAs (16,17) and a high coincidence of chromatid breaks with SCEs (10,12), suggesting an involvement of chromatid breakage-reunion. Therefore, the model construction for SCEs by persisting cross-links should incorporate at least the replication-fork model, cross-link transfer to both chromatids and chromosome breakage-reunion. Figure 6 presents our assumptive models as described in detail previously (14).

S1-Exchange (Twin SCE)

In Fig. 6A, a persisting cross-link (stage i) blocks adjacent replication forks (stage ii). The 2 cuts in parental strands in one

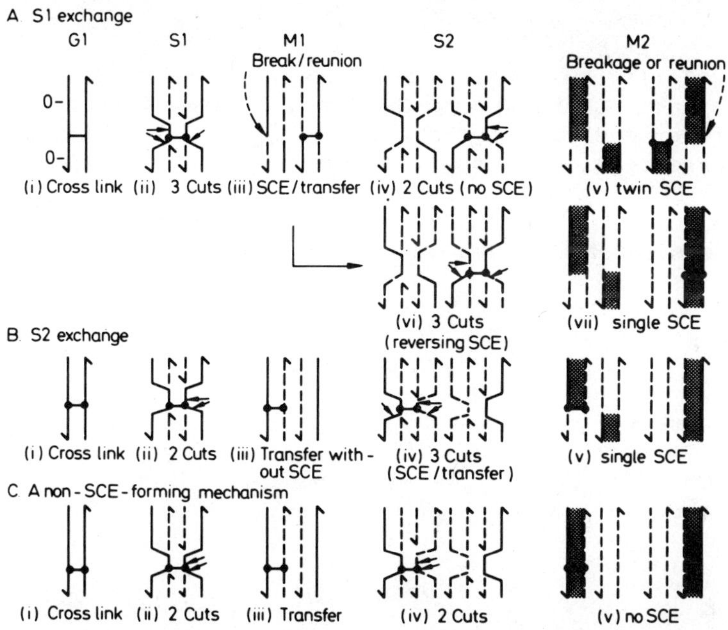

Fig. 6. Assumptive models for cross-induced twin and single SCE formations and for a non-SCE-forming tolerance mechanism. See the text and Ref. 14 for detailed explanation. (Excerpted from Ref. 14 under permission by Springer-Verlag.)

fork initiate SCE between parental and newly formed strands (replication model) (stage ii). One more break into the other side of (2 cuts across) the cross-link unit enhances unilateral crosslink transfer and exchange, leaving a chromatid break (stage ii-iii). Such a break will be reunited in survivors (stage iii). When a cross-link-bearing chromosome escapes from re-exchange in S2 by the 2-cut mechanism (stage iv), a twin SCE appears at M2 (stage v). Thus, the ordered sequence of the 3- and 2-cut mechanisms in S1 and S2 renders an S1-exchange twin in M2-diplochromosomes. If the 3-cut SCE/transfer mechanism works again in S2 (stage vi), the S1-SCE will be reversed to yield a single SCE at M2 (stage vii). However, the probability of this occurrence is suggested to be very low from our study (14).

S2-Exchange (Single SCE)

Figure 6B illustrates the sequence. If the 2-cut cross-link-transfer without an accompanying SCE in S1 (stage i-iii) and the 3-cut SCE/transfer mechanism in S2 (stage iv) occur, only a single SCE should appear at M2 (stage vii). We have observed the actual occurrence of single SCEs in MMC-treated FA cells as described above.

MODEL FOR A NON-SCE-FORMING TOLERANCE

The majority of MMC cross-links in an FA genome can be tolerated by a non-SCE-forming mechanism. Figure 6C suggests a possible cell tolerance, whereby the 2-cut cross-link-transfer mechanism with breakage-reunion is applied for 2 cycles of S1 and S2. As a result, no SCEs occur in spite of multiple cross-links, since a lethal hit even in FA cells requires about 60 cross-links (see above). Thus, such a mechanism is clearly important for cell tolerance.

ACKNOWLEDGEMENTS

This work was supported in part by Grants-in-Aid for Cancer and Scientific Research projects from the Ministry of Education, Science, and Culture, Japan. We also thank A. Katanoda for her skillful assistance.

REFERENCES

1. Perry, P., and H.J. Evans (1975) Cytological detection of mutagen-carcinogen exposure by sister chromatid exchange. Nature 250:121-125.
2. Wolff, S. (1977) Sister chromatid exchanges. Ann. Rev. Genet. 11:183-201.
3. Kato, H. (1977) Spontaneous and induced sister chromatid exchanges as revealed by the BUdR-labelling method. Int. Rev. Cytol. 49:55-97.
4. Latt, S.A. (1981) Sister chromatid exchange formation. Ann. Rev. Genet. 15:11-55.
5. Taylor, J.H., P.S. Woods, and W.L. Hughs (1957) The organization and duplication of chromosomes as revealed by autoradiographic studies using tritium-labelled thymidine. Proc. Natl. Acad. Sci., USA 43:122-128.
6. Ishii, T., and M.A. Bender (1980) Effects of inhibition of DNA synthesis on spontaneous and UV-induced sister chromatid exchanges in Chinese hamster cells. Mutat. Res. 79:19-32.
7. Kato, H. (1980) Evidence that the replication point is the site of sister chromatid exchange. Cancer Genet. Cytogenet. 2:69-77.
8. Painter, R.B. (1980) A replicative model for sister chromatid exchange. Mutat. Res. 70:337-341.
9. Carrano, A.V., L.H. Thompson, D.G. Stakta, J.L. Minkler, J.A. Mazrimas, and S. Fong (1979) DNA cross-linking, sister chromatid exchange and specific locus mutations. Mutat. Res. 63:175-183.
10. Latt, S.A., E. Steten, L.A. Juergens, A.R. Buchman, and P.S. Gerald (1975) Induction by alkylating agents of sister chromatid exchanges and chromatid breaks in Fanconi's anemia. Proc. Natl. Acad. Sci., USA 72:4066-4070.

11. Sahar, E., C. Kittrel, S. Feughum, M. Feld, and S.A. Latt (1981) Sister chromatid exchange induction in Chinese hamster ovary cells by 8-methoxypsoralen and brief pulses of laser light. Mutat. Res. 83:91-105.
12. Kano, Y., and Y. Fujiwara (1981) Roles of DNA interstrand crosslinking and its repair in the induction of sister chromatid exchange and a higher induction in Fanconi's anemia cells. Mutat. Res. 81:365-375.
13. Fujiwara, Y., Y. Kano, P. Paul, K. Goto, K. Yamamoto, and N. Miyazaki (1981) Excision and cross-link repair of DNA and sister chromatid exchange in human fibroblasts with different repair capacities. Gann Monogr. Cancer Res. 27:33-44.
14. Kano, Y., and Y. Fujiwara (1982) Higher inductions of twin and single sister chromatid exchanges by cross-linking agents in Fanconi's anemia cells. Human Genet. 60:233-238.
15. Bredberg, A., and B. Lambert (1983) Induction of SCE by DNA cross-links in human fibroblasts exposed to 8-MOP and UVA irradiation. Mutat. Res. 118:191-204.
16. Sasaki, M.S., and A. Tonomura (1973) A high susceptibility of Fanconi's anemia to chromosome breakage by DNA cross-linking agents. Cancer Res. 33:1829-1836.
17. Auerbach, A.D., and S.R. Wolman (1978) Carcinogen-induced chromosome breakage in Fanconi's anemia heterozygous cells. Nature 271:69-71.
18. Fujiwara, Y., M. Tatsumi, and M.S. Sasaki (1977) Crosslink repair in human cells and its possible defect in Fanconi's anemia cells. J. Mol. Biol. 113:634-649.
19. Fujiwara, Y. (1982) Defective repair of mitomycin C cross-links in Fanconi's anemia and loss in confluent normal human and xeroderma pigmentosum cells. Biochim. Biophys. Acta 699:219-225.
20. Fujiwara, Y. (1983) Measurement of interstrand cross-links produced by mitomycin C. In DNA Repair - A Laboratory Manual of Research Procedures, E.C. Friedberg and P.C. Hanawalt, eds. Marcel Dekker Inc., New York, Vol. 2, pp. 143-160.
21. Fujiwara, Y., and Y. Kano (1983) Loss of half-excision of mitomycin C cross-links in cycling Fanconi's anemia cells and in non-cycling normal and xeroderma cells. In Proc. 7th Int. Cong. Radiat. Res., J.J. Broerse, G.W. Barendesen, H.B. Kal, and A.J. Kogel, eds. Martinus Nijhoff Publ., Amsterdam, Vol. B., B3-9.
22. Greunert, D.C., and J.E. Cleaver (1983) Repair of photoactivated 8-methoxypsoralen damage in human cells. In Proc. 7th Int. Cong. Radiat. Res., J.J. Broerse, G.E. Barendesen, H.B. Kal, and A.J. Kogel, eds. Martinus Nijhoff Publ., Amsterdam, Vol. B., B2-14.
23. Kaye, J., C. Smith, and P.C. Hanawalt (1980) DNA repair in human cells containing photoadducts of 8-methoxypsoralen and angelicin. Cancer Res. 40:696-702.
24. Fornace, A.J., J. Little, and R.B. Weichselbaum (1979) DNA repair in Fanconi's anemia fibroblast cell strain. Biochim. Biophys. Acta 561:99-109.

25. Sognier, M.A., and W.N. Hittelman (1983) Loss of repairability of DNA interstrand cross-links in Fanconi's anemia cells with culture age. Mutat. Res. 108:383-393.
26. Shafer, D.A. (1977) Replicative bypass model of sister chromatid exchange and implications for Bloom's syndrome and Fanconi's anemia. Human Genet. 33:177-190.
27. Stekta, D.G. (1979) Further analysis of the replicative bypass model for sister chromatid exchange. Human Genet. 49:63-69.

HUMAN HEALTH SITUATION AND CHROMOSOME ALTERATIONS: SISTER CHROMATID EXCHANGE FREQUENCY IN LYMPHOCYTES FROM PASSIVE SMOKERS AND PATIENTS WITH HEREDITARY DISEASES

Kanehisa Morimoto, Kunihiko Miura, Tetsuya Kaneko, Kumiko Iijima, Mayumi Sato, and Akira Koizumi

Department of Public Health
Faculty of Medicine
University of Tokyo
Tokyo 113, Japan

INTRODUCTION

Many investigators have described strong correlations between the frequency of sister chromatid exchanges (SCEs) in blood lymphocytes and health situations of the blood donor. Recent studies have further shown that cigarette smoking as well as certain genetic factors, can potentiate the cellular response to chemical treatment. The chemical stress method might thus be a sensitive test for quantitating the cellular damage or defects that have been induced environmentally and genetically.

It is widely recognized that smoking habit strongly correlates with the incidence of lung cancer, and cigarette smoke condensate has already been shown to induce mutation or chromosomal aberrations (CAs) in vitro (1). These previous studies strongly suggest that cigarette smoke might cause some persistent damage(s) in peripheral lymphocytes (2-4), and several investigators are warning about the attribution of passive smoking to lung cancer in nonsmokers (5,6). So far, however, the influence of passive smoking has not yet been fully understood and no published data is available on the SCE frequencies in lymphocytes from passive smokers.

Fanconi amemia (FA) is an autosomal recessive disorder characterized by a high frequency of spontaneous CAs (7,8) and a hypersensitivity to bifunctional DNA-damaging agents such as mitomycin C (MMC) (9). Many studies have also been performed on the baseline and induced SCE frequencies in FA cells. However, the results of these studies are often inconsistent. For example, in FA cells exposed to MMC, Novotnå et al. (10) showed a slightly higher induction of SCEs in FA cells than in normal cells, whereas Latt et al. (11) and Cervenka et al. (12) reported only half the SCE induction in FA cells as in normal cells.

Alzheimer disease (Alz), a progeroid (premature aging) disease (13), has been suggested to be dominantly heritable (14). Lymphocyte chromosomes from patients with this syndrome have been known to show aneuploidy (15) which is thought to be one of the characteristics of aged cells (16). Aged cells are also suggested to show decreased abilities to repair DNA damage induced by chemicals (17), and baseline and chemically induced SCEs have been reported to decrease in aged cells which is thought to be the reflection of decreased DNA repair ability (18). However, there is no published data on the incidence of SCEs in lymphocytes from Alz patients when exposed to chemical mutagens.

Familial polyposis coli (FPC), among other adenomatosis coli, is another dominantly heritable disease (19) and cells from patients with this syndrome are known to be hyper-susceptible to viral or chemical transformation in vitro (20,21), cell killing (22), or CA induction (23). However, only few data are available on the SCE frequencies in FPC cells after exposure to DNA-damaging agents.

The differential staining of sister chromatids enabled us to investigate cell cycle kinetics as well as SCEs in human lymphocyte cultures (24,25). By these techniques, we show here that (i) lymphocytes from passive smokers show significantly higher SCE frequencies than cells from nonsmokers when treated with MMC; (ii) FA cells constantly show about 1.4 times higher frequencies of SCEs than normal cells in both MMC-treated and untreated cultures, while they show much longer cell proliferation delay and significantly higher frequencies of CAs than normal cells when exposed to MMC; (iii) in MMC-treated cultures, Alz cells show about 2 times higher SCE induction than normal cells, while they show several times smaller induction of SCEs compared to normal cells when treated with 4-nitroquinoline-N-oxide (4NQO): (iv) FPC and normal cells show about the same induction of SCEs when treated with MMC, 4NQO, or N-methyl-N'-nitro-N-nitrosoguanidine (MNNG).

MATERIALS AND METHODS

Blood Donor

Peripheral blood samples were drawn from 26 cigarette smokers, 16 passive smokers, and 9 nonsmokers. Samples were also drawn from patients with FA (sisters, aged 7 and 11 yr), their sister (9 yr), their father (34 yr), and their mother (36 yr), from patients with FPC [male, 18 yr; females, aged 24 yr (with Gardner syndrome) and

44 yr], and patients with Alz (males, 52 and 60 yr; females, 59 and 62 yr). Healthy, unrelated adults (males, 27, 35, 50, and 60 yr; females, 20, 51, and 58 yr) were used as controls.

Lymphocyte Culture

Whole blood from each donor (0.2 ml) was added to 5 ml RPMI 1640 medium (Gibco) containing 15% fetal bovine serum (Gibco) and 20 μM bromodeoxyuridine (BrdUrd) (Sigma). After stimulation by phytohemagglutinin (PHA-M) (Gibco) the cultures were incubated at 37°C for 72 hr (smokers, Alz, and FPC) or for 96 hr (FA). Colcemid (2×10^{-7} M; Calbiochem) was added to each culture before fixation for 3 to 6 hr. The cells were then collected by centrifugation, and air-dried chromosome preparations were made as previously reported (26). A modification of the fluorescence-plus-Giemsa method (27) was used to obtain differentially stained chromosomes. Cells dividing for the first (X1), second (X2), or third or more (X3+) time in culture can be determined in such preparations (25,28).

Chemical Treatment

For chemical treatments, MMC (Sigma) and MNNG (Sigma) were first dissolved in distilled water, 4NQO (Tokyo-Kasei) in dimethylsulfoxide, and after necessary dilutions in phosphate buffered saline (PBS), aliquots of freshly made solutions were added to each culture to give the appropriate final concentration for the last 48 hr of the 96 hr culture (FA), for 1 hr before PHA stimulation (FPC), or for 72 hr during incubation (smokers, Alz, and FPC).

Scoring

Two hundred metaphase cells were scored for CAs and cell cycle distribution, and 25 to 35 second-division metaphase cells were scored for SCEs per point per person.

RESULTS AND DISCUSSION

Baseline and MMC-Induced SCEs in Lymphocytes from Positive and Passive Smokers

SCE frequency in lymphocytes from cigarette smokers. Peripheral lymphocytes from cigarette smokers have already been reported to show higher SCE frequencies than those from nonsmokers in both chemically treated and nontreated cultures (2-4). In our own study the following relationship between the baseline SCE frequencies and the number of smoked cigarettes has been obtained:

$$y = 0.051x + 6.4$$

where y: SCE frequency per cell, and x: number of cigarettes smoked per da. In the MMC-treated cultures no significant differences in the mean SCE frequencies were observed between smokers and nonsmokers but the percentages of cells that show extremely higher SCE frequencies were increased in cells from smokers (data not shown, manuscript in preparation).

SCE frequency in lymphocytes from passive smokers. When lymphocytes from passive smokers were exposed to 3×10^{-8} M MMC, they showed significantly higher SCE induction than those from positive or nonsmokers (Tab. 1; $p < 0.05$, Student's t-test). The results suggest that SCE-leading DNA damage might be produced more in lymphocytes from passive smokers. The mean frequencies of baseline SCEs were within the same range among lymphocytes from positive smokers, passive smokers, and nonsmokers (Tab. 1). Mitomycin C-induced SCE frequencies in lymphocytes from both positive and passive smokers had significant correlations with hemoglobin concentration in blood. Mitomycin C has been reported to be activated through reduction (29) and the erythrocyte concentration in the whole blood cultures might have influence on this process. No significant correlations were observed among age/sex and SCE frequencies in both nontreated and MMC-treated cultures (data not shown).

Baseline and Chemical-Induced SCEs in Lymphocytes from Hereditary Diseases

SCEs and CAs in FA cells. Lymphocytes from FA patients showed consistently higher frequencies of SCEs than normal cells in MMC-treated and nontreated cultures (Fig. 1a). The ratio of the average SCE frequency in FA cells to that in normal cells was approximately 1.4 (Fig. 1b). Chromosomal aberrations were examined separately in X1, X2, and X3+ metaphase cells. In each division metaphase, FA cells showed several times higher frequencies of CAs than normal cells (Fig. 2a, 2b). Fanconi anemia cells, which are characterized

Tab. 1. Mean SCE frequencies on MMC-treated cells from positive smokers, passive smokers, or nonsmokers.

	Positive smokers	Passive smokers	Nonsmokers
Baseline	7.2 ± 0.42[a]	6.6 ± 0.23	6.6 ± 0.67
MMC-induced[b]	27.8 ± 0.44	27.6 ± 0.20	24.7 ± 0.98
Net induction	20.5 ± 1.23	20.3 ± 1.34	18.0 ± 1.39
Induct. ratio	4.0 ± 0.29	4.2 ± 0.26	4.0 ± 0.25

[a] Mean ± S.E.

[b] Cells were exposed to 3×10^{-8} M MMC for the entire culture period of 72 hr.

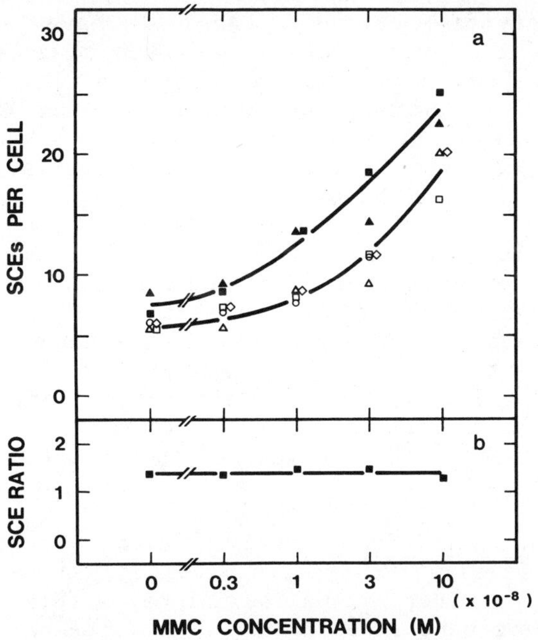

Fig. 1. a) Frequencies of SCEs in FA (solid symbols) and normal (open symbols) X2 cells exposed to various concentrations of MMC. b) Ratio of SCEs in FA cells to SCEs in normal cells.

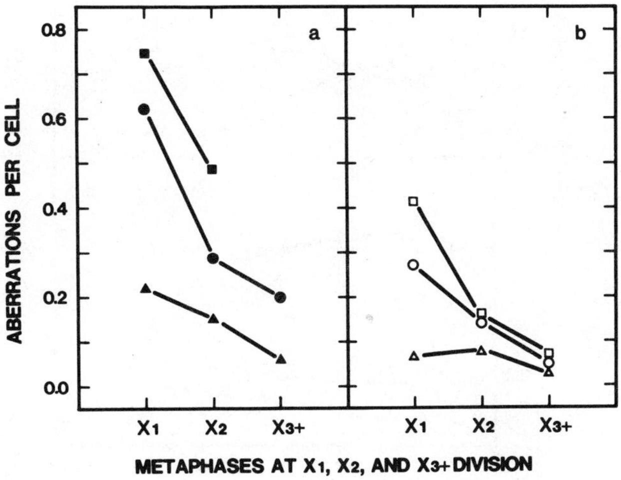

Fig. 2. a) Frequencies of CAs in (a) FA cells and (b) normal cells in X1, X2, and X3+ divisions in cultures exposed to various concentrations of MMC: ▲ Δ, 0; ● o, 3 x 10^{-9} M; ■ □, 3 x 10^{-8} M.

by a high frequency of spontaneous CAs (30), have been known to be hypersensitive to the production of CAs when treated with MMC (9) and these findings are consistent with the results in the present study. However, the results of previous studies on MMC-induced SCEs in FA cells have been inconsistent. About the same or a slightly higher induction of SCEs by MMC in FA lymphocytes or fibroblasts than in normal cells have been reported in several studies (31,32) and these findings are in agreement with our results over a wide range of MMC doses. However, Latt et al. (11) and Cervenka et al. (12) observed a reduced frequency of SCEs in MMC-treated FA cells, and these apparently contradictory results might be attributed to genetic heterogeneity in this syndrome. The exact mechanisms that bring about the interesting discrepancy between SCEs and CAs observed in MMC-treated FA cells are still to be clarified, but if FA cells had some repair deficiency, this discrepancy could, at least in part, be explained by the MMC-induced damage spectra that lead to SCEs or CAs.

Cell cycle kinetics in FA cells. When lymphocytes from untreated cultures fixed at 96 hr were examined, FA cells showed much slower proliferation kinetics than normal cells (Fig. 3a, 3b), manifested as higher frequencies of X1 and X2 metaphases: in cultures

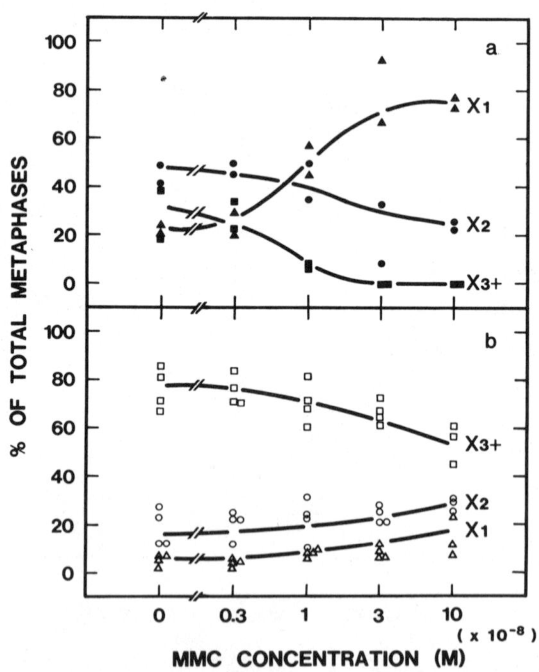

Fig. 3. Percentages of X1 (▲ △), X2 (● o), and X3+ (■ □) division metaphase cells from (a) FA patients and (b) control subjects in cultures exposed to various concentration of MMC.

fixed at 96 hr, FA cells showed relative frequencies of 20% X1, 35% X2, and 45% X3+ metaphases, whereas normal cells had a distribution of 5% X1, 15% X2, and 80% X3+ metaphases. In cultures exposed to increasing concentrations of MMC, both FA and normal cells showed a clearly dose-dependent delay in cell turnover times, with FA cells showing a much longer delay at all doses (Fig. 1). For instance, an exposure of FA cells to 3×10^{-8} M MMC gave an average distribution of 80% X1 and 20% X2 cells, whereas the identical treatment of normal cells showed a distribution of 10% X1, 20% X2, and 70% X3+ cells.

Our results are consistent with those of previous studies on cell cycle kinetics, in which FA lymphocytes were found to have longer cell cycle times than normal cells (10,33-35). The possible causes of longer cell cycle times in FA cells might be an increased intracellular thymidine pool (36-38), or defective or reduced ability to repair the oxygen-induced lesions during culture (39). Our present data, showing that MMC treatment induced much longer cell cycle delays in FA cells than in normal cells, indicate that FA cells are also hypersensitive in terms of cell cycle delays as well as CAs when exposed to MMC.

SCEs in Alz cells. When exposed to 20 ng 4NQO, Alz lymphocytes exhibited a several times smaller induction of SCEs than normal cells while they exhibited about 1.5 times higher induction of SCEs than normal cells in MMC-treated cultures (Tab. 2). In nontreated cultures, Alz and normal cells exhibited about the same baseline frequency of SCEs. These results strongly suggest some metabolic disorders and/or deficiencies in the repair ability to DNA damages in Alz cells. Recently, it has been shown that the DNA-binding form of 4NQO is 4-hydroxyaminoquinoline-N-oxide (4HAQO) which is formed

Tab. 2. SCE frequencies in Alz and normal cells in nontreated, MMC-, or 4NQO-treated cultures.

Subjects		Non-treated	MMC 3×10^{-8} M	4NQO 20 ng/ml
Alzheimer	1	11.8 ± 3.2[a]	41.8 ± 10.6	13.2 ± 3.6
	2	9.7 ± 2.3	-	12.2 ± 3.3
	3	9.4 ± 3.3	46.2 ± 10.5	11.6 ± 3.8
	4	9.5 ± 2.2	51.0 ± 11.0	15.1 ± 3.2
Mean ± S.E.[b]		10.1 ± 0.6	46.3 ± 2.7	13.0 ± 0.8
Control	1	10.2 ± 3.2	37.5 ± 4.7	34.7 ± 9.2
	2	9.1 ± 1.9	32.5 ± 4.5	32.8 ± 13.8
	3	9.6 ± 2.2	31.1 ± 6.6	36.8 ± 16.5
Mean ± S.E.		9.6 ± 0.3	33.7 ± 1.9	34.6 ± 1.2

[a] Mean ± S.E.
[b] Mean ± S.E. within the group.

through metabolic activation of 4NQO by reducing enzyme(s) (40,41). Our finding on the increased induction of SCEs by MMC treatment can be accounted for by some deficiencies of repair enzymes in Alz cells. The hypersensitivities to MMC might also be explained by hyper-membrane passage, increased metabolic activation of MMC, or anomalies in DNA/nuclei protein conformation that affects the accessibility of the active metabolite(s) to DNA.

SCEs in FPC cells. In any culture, SCE freqencies were about the same in FPC and normal cells, and no significant difference in the SCE induction was observed between FPC and normal cells (Tab. 3). Previous studies have reported FPC cells to be hypersensitive to MMC, MNNG, or 4NQO in colony formation, or in the induction of CAs. These contradictory results found in FPC cells might be attributed to the following three possibilities: i) heterogeneity in sensitivities in this syndrome, ii) differences in sensitivities in stages in the natural history of this disease, and/or iii) possible differences in the induction mechanisms between SCEs and CAs or cell death when treated with DNA-damaging chemicals. Taking into consideration the discrepancy between SCE induction and CA frequencies in FA data, the third possibility seems to be most plausible. The first possibility, however, cannot be excluded because cells from one patient exhibited about a 3 times higher SCE induction than normal cells. Although Little et al. (42) reported the hypersensitivity in cells from patients with Gardner syndrome, cells from the patient with Gardner syndrome did not show a significant difference in SCE induction from the other FPC cells in our study. The heterogeneity in sensitivities within this syndrome might be quite complicated.

SUMMARY

Lymphocytes from passive smokers, and patients with FA , Alz, or FPC were studied for SCEs in cultures treated with MMC, 4NQO, or MNNG. Fanconi anemia lymphocytes were also studied for cell cycle

Tab. 3. Mean SCE frequencies in FPC or normal cells.

Treatment			FPC	Control
MMC	3×10^{-6} M	1 h	43.4 ± 1.6[a]	40.2 ± 2.0
	3×10^{-8} M	72 h	32.3 ± 2.8	31.4 ± 0.6
MNNG	2 μg/ml	72 h	12.5 ± 0.4	10.8 ± 0.7
4NQO	6×10^{-7} M	1 h	11.0 ± 1.7	9.4 ± 0.9
	Non-treated		7.4 ± 0.4	7.8 ± 0.9

[a] Mean ± S.E.

kinetics, and CAs after completion of 1, 2, or 3 or more divisions in MMC-treated cultures. The results can be summarized as follows: (1) lymphocytes from passive smokers showed a slightly higher induction of SCEs than nonsmokers when exposed to MMC. (2) FA cells had about 1.4 times higher frequencies of SCEs than normal cells in both MMC-treated and untreated cultures while they showed several times higher frequencies of CAs in both cultures. Analyses of cell cycle kinetics by the sister chromatid differential staining method revealed that MMC treatments of FA and normal cells led to a clearly dose-related delay in cell turnover times, the duration of delay being much longer in FA than in normal cells. (3) Alz cells showed about 1.5 times higher induction of SCEs in MMC-treated cultures whereas they had only 10% as much SCEs as controls when exposed to 4NQO. Familial polyposis coli cells showed no significant difference in the induction of SCEs in untreated cultures and cultures treated with MMC, 4NQO, and MNNG.

REFERENCES

1. DeMarini, D.M. (1983) Genotoxicity of tobacco smoke and tobacco smoke condensate. Mutat. Res. 114:59-89.
2. Meiying, C., X. Jiujin, and Z. Xianting (1982) Comparative studies on spontaneous and mitomycin C-induced sister-chromatid exchanges in smokers and nonsmokers. Mutat. Res. 105:195-200.
3. Obe, G., H.-J. Vogt, S. Madle, A. Fahning, and W.D. Heller (1982) Double-blind study on the effect of cigarette smoking on the chromosomes of human peripheral blood lymphocytes in vivo. Mutat. Res. 92:309-319.
4. Livingston, G.K., and R.M. Fineman (1983) Correlation of human lymphocyte SCE frequency with smoking history. Mutat. Res. 119:59-64.
5. Hirayama, T. (1981) Non-smoking wives of heavy smokers have a higher risk of lung cancer: A study from Japan. Br. Med. J. 282:183-185.
6. Trichopoulos, D., A. Kalandidi, L. Sparros, and B. MacMahon (1981) Lung cancer and passive smoking. Int. J. Cancer 27:1-4.
7. Fanconi, G. (1927) Familiäre infantile perniziosaartige Anämie (perniziöses Blutbild und Konstitution). Jahrb. Kind. 117:257-281.
8. Schroeder, T.M. (1966) Cytogenetischer Befund und Ätiologie bei Fanconi-Anämie: Ein Fall von Fanconi-Anämie ohne Hexokinasedefekt. Humangenetik 3:76-81.
9. Sasaki, M.S., and A. Tonomura (1973) A high susceptibility of Fanconi's anemia to chromosome breakage by DNA cross-linking agents. Cancer Res. 33:1829-1836.
10. Novotnå, B., P. Goetz, and N.I. Surkova (1979) Effects of alkylating agents on lymphocytes from controls and from patients with Fanconi's anemia. Human Genet. 49:41-50.
11. Latt, S.A., G. Stetten, L.A. Juergens, G.R. Buchanan, and P.S.

Gerald (1975) Induction by alkylating agents of sister chromatid exchanges and chromatid breaks in Fanconi's anemia. Proc. Natl. Acad. Sci., USA 72:4066-4070.
12. Cervenka, J., D. Arthur, and C. Yasis (1981) Mitomycin C test for diagnostic differentiation of idiopathic aplastic anemia and Fanconi anemia. Pediatrics 67:119-127.
13. Bergener, M., and F.K. Jungklaas (1970) Genetische Befunde bei Morbus Alzheimer und seniler Demenz. Gerontol. Clin. 12:71-75.
14. Chase, G.A., M.F. Folstein, J.C.S. Breitner, T.H. Beaty, and S.G. Self (1983) The use of life tables and survival analysis in testing genetic hypotheses, with application to Alzheimer's disease. Am. J. Epidem. 117:590-597.
15. Jarvik, L.F. (1967) Senescence and chromosomal changes. Lancet 1:114.
16. Mattevi, M.S., and F.M. Salzano (1975) Senescence and human chromosome changes. Humangenetik 27:1-8.
17. Bowman, P.D., R.L. Meek, and C.W. Daniel (1976) Decreased unscheduled DNA synthesis in nondividing aged WI-48 cells. Mech. Aging Dev. 5:251.
18. Schneider, E.L., and R.E. Monticone (1978) Aging and sister chromatid exchange II. The effect of the in vitro passage level of human fetal lung fibroblasts on baseline and mutagen-induced sister chromatid exchange frequencies. Exp. Cell Res. 115:269-276.
19. Lipkin, M., S.J., Winawer, and P. Sherlock (1981) Early identification of individuals at increased risk for cancer of the large intestine Part I: Definition of high risk populations. Clin. Bul. 11:13-21.
20. Miyaki, M., N. Akamatsu, M. Rokutanda, T. Ono, H. Yoshikura, M.S. Sasaki, A. Tonomura, and J. Utsunomiya (1980) Increased sensitivity of skin fibroblasts from patients with adenomatosis coli and Peutz-Jegher's syndrome to transformation by murine sarcoma virus. Gann 71:797-803.
21. Miyaki, M., N. Akamatsu, T. Ono, A. Tonomura, and J. Utsunomiya (1982) Morphologic transformation and chromosomal changes induced by chemical carcinogens in skin fibroblasts from patients with familial adenomatosis coli. J. Natl. Cancer Inst. 68:563-571.
22. Barfknecht, T.R., and Little, J.B. (1982) Abnormal sensitivity of skin fibroblasts from Familial polyposis patients to DNA alkylating agents. Cancer Res. 2:1249-1254.
23. Hori, T., M. Murata, and J. Utsunomiya (1980) Chromosome aberrations induced by N-methyl-N'-nitro-N-nitrosoguanidine in cultured skin fibroblasts from patients with Adenomatosis coli. Gann 71:628-636.
24. Perry, P., H.J., Evans (1975) Cytological detection of mutagen-carcinogen exposure by sister chromatid exchange. Nature (Lond.) 258:121-125.
25. Tice, R., E.L. Schneider, and J.M. Rary (1976) The utilization of bromodeoxyuridine incorporation into DNA for the analysis of cellular kinetics. Exp. Cell Res. 102:232-236.

26. Morimoto, K., and S. Wolff (1980) Increase of sister chromatid exchanges and perturbations of cell division kinetics in human lymphocytes by benzene metabolites. Cancer Res. 40:1189-1193.
27. Goto, K., S. Maeda, Y. Kano, and T. Sugiyama (1978) Factors involved in differential Giemsa-staining of sister chromatids. Chromosoma 66:351-359.
28. Morimoto, K., and S. Wolff (1980) Cell cycle kinetics in human lymphocyte cultures. Nature (Lond.) 288:604-606.
29. Iter, V.N., W. Szybalski (1964) Mitomycins and Profiromycin: Chemical mechanism of activation and cross-linking of DNA. Science 145:55-58.
30. German, J., R. Archibald, and D. Bloom (1965) Chromosomal breakage in a rare and probably genetically determined syndrome of man. Science 148:506-507.
31. Sasaki, M.S. (1978) DNA Repair Mechanisms, P.C. Hanawalt, E.C. Friedberg, and C.F. Fox, eds. Academic Press, New York, pp. 285-313.
32. Kano, H., and Y. Fujiwara (1981) Roles of DNA interstrand crosslinking and its repair in the induction in Fanconi's anemia cells. Mutat. Res. 81:365-375.
33. Sasaki, M.S. (1975) Is Fanconi's anemia defective in a process essential to the repair of DNA cross links? Nature (Lond.) 257:501-503.
34. Dutrillaux, B., and A.M. Fosse (1976) Utilisation du BrdU dans l'étude du cellulaire de sujets normaux et anormaux. Ann. Genet. (Paris) 19:95-102.
35. Dutrillaux, B., A. Aurias, A.-M. Dutrillaux, D. Buriot, and M. Prieur (1982) The cell cycle of lymphocytes in Fanconi anemia. Human Genet. 62:327-332.
36. Galavazi, G., H. Schenk, D. Bootsma (1966) Synchronization of mammalian cells in vitro by inhibition of the DNA synthesis I. Optimal conditions. Exp. Cell Res. 41:428-437.
37. Elmore, E., and M. Swift (1975) Growth of cultured cells from patients with Fanconi anemia. J. Cell Physiol. 87:229-234.
38. Shoyab, M., M. Gunnell, and A. Lubiniecki (1981) Reduced uptake and incorporation of ^{3}H-thymidine in Fanconi anemia fibroblasts. Human Genet. 57:296-299.
39. Joenje, H., F. Arwert, A.W. Eriksson, H. Konnig, and A.B Oostra (1981) Oxygen-dependence of chromosomal aberrations in Fanconi's anemia. Nature (Lond.) 290:142-143.
40. Sugimura, T., K. Okabe, and M. Nagao (1966) The metabolism of 4-nitroquinoline-l-oxide, a carcinogen III. An enzyme catalyzing the conversion of 4-nitroquinoline-l-oxide to 4-hydroxyaminoquinoline-l-oxide in rat liver and hepatomas. Cancer Res. 26:1717-1721.
41. Tada, M., and M. Tada (1975) Seryl-tRNA synthetase and activation of the carcinogen 4-nitroquinoline-l-oxide. Nature (Lond.) 255:510-512.
42. Little, J.B., J. Nove, and R. Weichselbaum (1980) Abnormal sensitivity of diploid skin fibroblasts from a family with Gardner's syndrome to the lethal effects of X-irradiation, ultraviolet light and mitomycin-C. Mutat. Res. 70:241-250.

SENSITIVITIES OF PERIPHERAL LYMPHOCYTES FROM HEALTHY HUMANS AND CANCER PATIENTS TO INDUCTION OF SISTER CHROMATID EXCHANGES BY GENOTOXICANTS

Atsushi Oikawa, Setsu Sakai, Katsuhiko Horaguchi, Ryoko Sugawara, Kazunori Sato, Jun Tazawa, Hiroko Tohda, Masakazu Yokoyama, and Akira Wakui

Research Institute for Tuberculosis and Cancer
Tohoku University
Sendai 980, Japan

INTRODUCTION

The sister chromatid exchange (SCE) assay is a sensitive method for detection of genotoxic agents (1). On the other hand, the same method detects individuals whose cells are sensitive to any particular genotoxicant (2).

We surveyed spontaneous and chemically induced SCEs in cells from about 50 healthy persons ranging in age from 22 to 54 yr, living in and around Sendai City. Sendai City is located in the northern part of Honshu (Main Island) in Japan. Results have been reported briefly (2).

In this report we describe the above data in more detail and also include additional results from some cancer patients who were newly diagnosed and had not yet received any therapeutic treatments.

METHODS

Whole blood was diluted 16-fold with RPMI 1640 medium supplemented with 10% fetal bovine serum, and cultured in the presence of phytohemagglutinin and 5-bromodeoxyuridine (BrdUrd) for 4 da in the dark.

SCEs were induced by the "latter half treatment" with 50 μM methylmethanesulfonate (MMS) or 1 μM 4-nitroquinoline-1-oxide (4NQO), or by the "midterm 2-hour treatment" with 5 μM 3-amino-1-methyl-5H-pyrido[4,3-b]indole (Trp-P-2) (a pyrolytic product of tryptophan) in the presence of rat liver S-9 mix, as described previously (1). The concentration of chemicals were fixed as above, unless otherwise stated.

Microscopic slides for SCE scoring were prepared by a conventional method including the use of the fluorescence-plus-Giemsa technique (3).

Sister chromatid exchanges were scored in 30 metaphase cells in each preparation, and the mean and standard error were calculated as SCEs per cell for each individual. The induced SCE rate was defined as the mean number of SCEs per cell in the test culture minus that in the corresponding control with or without S-9 mix. The spontaneous SCE rate is the mean number of SCEs per cell in the culture without treatment. Cells treated with S-9 mix alone usually gave slightly more (but not more than one) SCEs per cell than did those without S-9 mix. Detailed procedures have been reported in previous publications (1,2).

RESULTS AND DISCUSSION

SCEs in Lymphocytes of Healthy Humans

Individual means of spontaneous and chemically induced SCEs in 30 metaphase lymphocytes for about 50 healthy donors are shown in Fig. 1. We have reported that age and sex differences were not significant (2), consistent with the findings of another study (4). It was noticed that a few donors had extremely high sensitivities to particular chemicals: M12 was sensitive to 4NQO; F3 to MMS; F18 to all 3 chemicals. These results were confirmed in repeat examinations carried out 3-12 mo after the first examinations (Fig. 2). The rare presence of persons whose lymphocytes are highly sensitive to particular chemicals as indicated by the increased SCE response could mean that cells in these individuals have a higher chance of suffering DNA damage from environmental genotoxicants. At present, our knowledge of SCEs does not allow us to predict any health consequences. A high spontaneous SCE of M5 (Fig. 1) will be discussed later.

No clear correlation between sensitivities in SCE induction by 2 of the 3 chemicals was found (Fig. 3), possibly because the 3 chemicals chosen in this study may not have the same metabolic properties, induce the same kind of DNA damage, or invoke the same DNA repair mechanisms, which result in SCE induction.

SCEs in Lymphocytes of Cancer Patients

Results of the studies are presented in Fig. 4. It is quite clear, although the number of cases is limited, that the frequency of patients (5/10) whose SCE sensitivity to Trp-P-2 exceeds 2 standard deviations above the control range of healthy humans is much higher than that of healthy persons (1/33). All 4 stomach cancer patients belong to this group with another patient bearing colon cancer. As shown in Tab. 1, MMS-sensitive persons seem also more frequent in cancer patients than those in the healthy population. Due to limited number, sensitivities in the spontaneous and 4NQO-induced SCEs could not be compared between 2 populations.

Although these facts are still to be confirmed with more patients, if the lymphyocytes from cancer patients, especially patients with stomach cancer, are really sensitive to SCE induction by Trp-P-2 as shown in Fig. 4, it is very intriguing to know whether the sensitivity is the cause or the consequence of carcinogenesis.

Induced SCEs in Individuals with High Spontaneous SCE Frequency

Patient P4 suffering from liver cancer showed an extraordinarily high spontaneous SCE. This was also true for a healthy human, M5. Both individuals were long-time smokers, but as can be seen in Tab. 2 and Figs. 1 and 4, and because P1, P3, P5, and P6 smoked more than P4, the very high frequency of spontaneous SCE in cells of M5 and P4 cannot be ascribed simply to their smoking habit. It is noted that their SCE frequencies induced by 3 chemicals were normal, in spite of their high spontaneous SCE frequency, suggesting that the induced SCE response is not related directly to the spontaneous level of SCEs.

SCE Sensitivity of Cell Lines Established from Sensitive Lymphocytes

Peripheral lymphocytes from healthy donors M12 and F18 were transformed with Epstein-Barr virus (EBV) by the method described previously (5). The established line from M12 was again sensitive to SCE induction by 4NQO, but that from F18 was not sensitive to any of 3 chemicals (Fig. 5). This suggests that lymphocytes are heterogeneous in their sensitivity to SCE induction. In this connection it should be noted that EBV transforms only B lymphocytes (5) and phytohemagglutinin activates T cells rather than B cells. Induction of SCE by 4NQO in the M12-derived line was dose-dependent and the sensitivity was between cell lines similarly derived from normal healthy donors and a xeroderma pigmentosum (XP) group A patient. The latter cell line needs only one-tenth the concentration of 4NQO to give a similar induced SCE level.

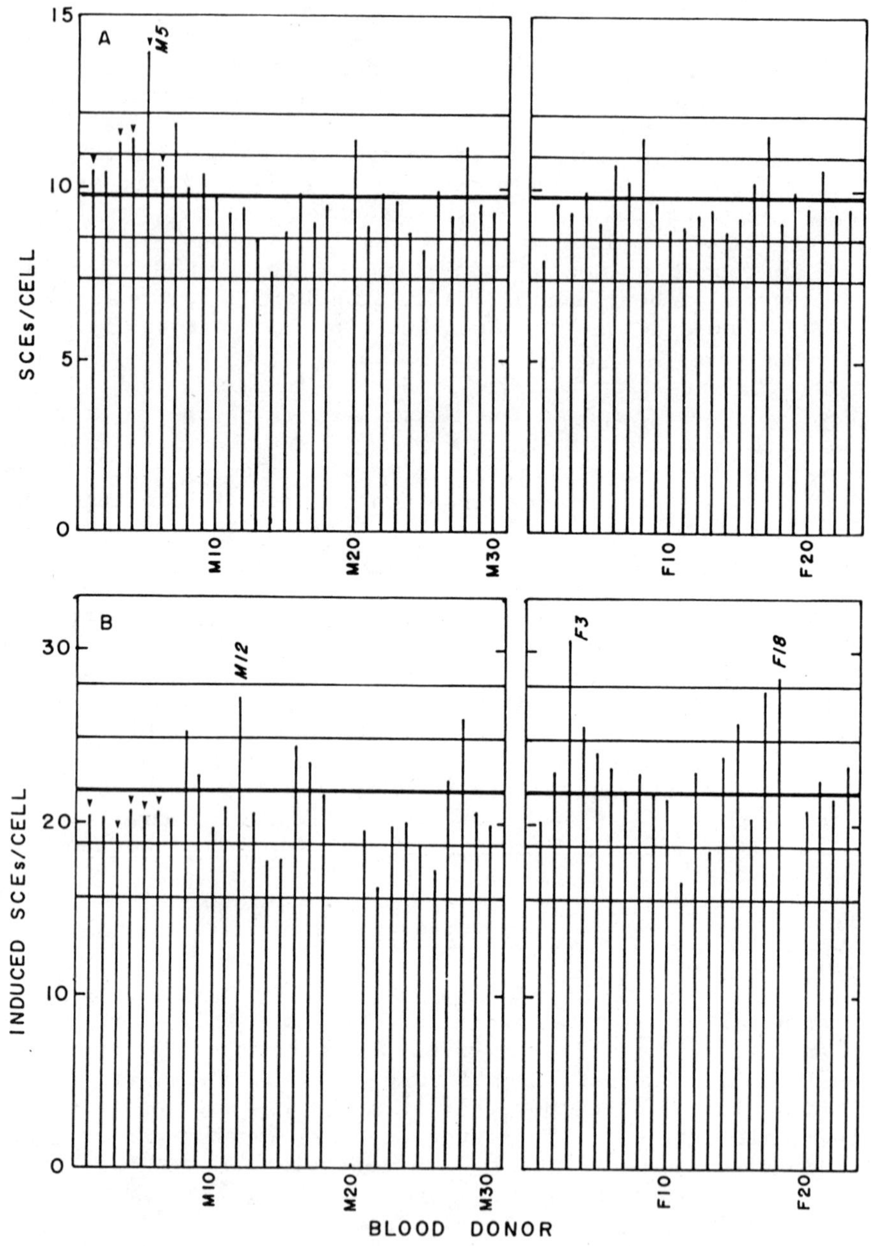

Fig. 1. Spontaneous and induced SCEs in peripheral lymphocytes from healthy donors. Sister chromatid exchanges were scored for 30 metaphase cells and expressed as a mean.

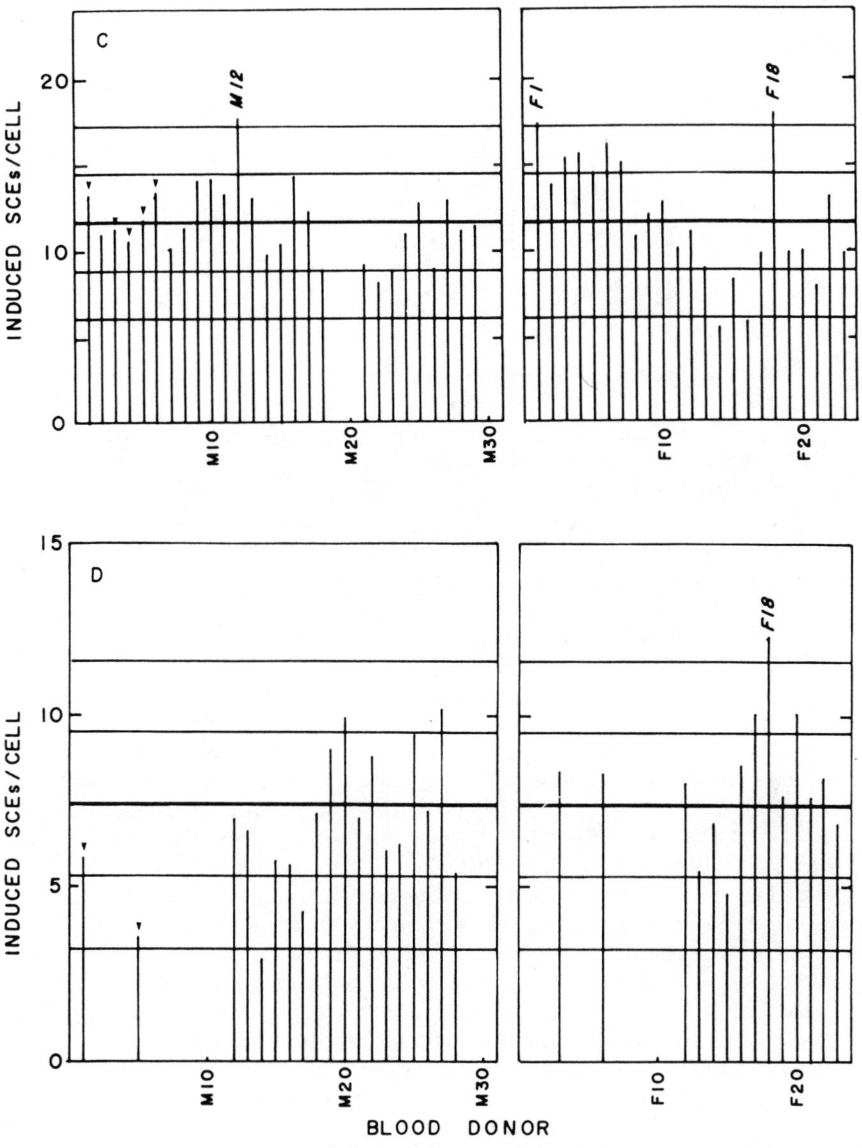

The mean induced SCE is defined in Methods. A, spontaneous SCE; B, C, and D, SCEs induced by 50 μM MMS, 1 μM 4NQO, and 50 μM Trp-P-2, respectively. The thick horizontal line is the mean of individual means, and thin lines are levels for ± 1σ and 2σ (S.D. unit). Arrowheads show smokers (see also Tab. 2).

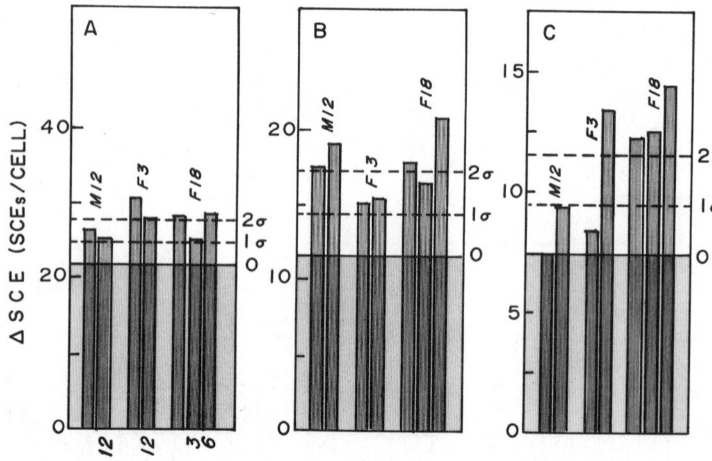

Fig. 2. Re-examination of SCEs induced by MMS (A), 4NQO (B), and Trp-P-2 (C). Horizontal lines 0 are the mean level for the total population in the first series of examinations. Figures at the bottom are number of months after the first examination.

General Discussion

A general correlation between the carcinogenic and genotoxic activities of chemicals has been recognized, although there are a few exceptions. This minor but practically important inconsistency seems partly due to species or individual differences in metabolic and repair activities of organisms. There are rare genetic disorders that are prone to cancer. Individuals with XP are specifically predisposed to skin cancer but probably only when exposed to ultraviolet light (UV). Cells of these patients are defective in the repair of UV-induced DNA damages (6). As shown in Tab. 3, lymphoblastoid cells from XP patients deficient in unscheduled DNA synthesis (UDS) are highly sensitive to the induction of SCEs by UV irradiation, irrespective of whether the individuals bear a tumor (7). This implies that a high sensitivity of a person to SCE induction by UV is related to tumor initiation in that person. However, cells from UDS-proficient XP patients are more sensitive than cells from normal individuals, although to a lesser degree than UDS-deficient cells, in SCE induction by UV but only when the donors are bearing a skin cancer (Tab. 3). In these instances, SCE sensitivity seems to be the result of the presence of a tumor. The facts that lymphocytes from breast cancer patients had a greater spontaneous SCE frequency than lymphocytes from normal individuals and that cells from women who had been successfully treated for breast cancer at least 5 yr previously had normal baseline SCE levels (4) seem to show a similar situation between spontaneous SCE and induced SCE.

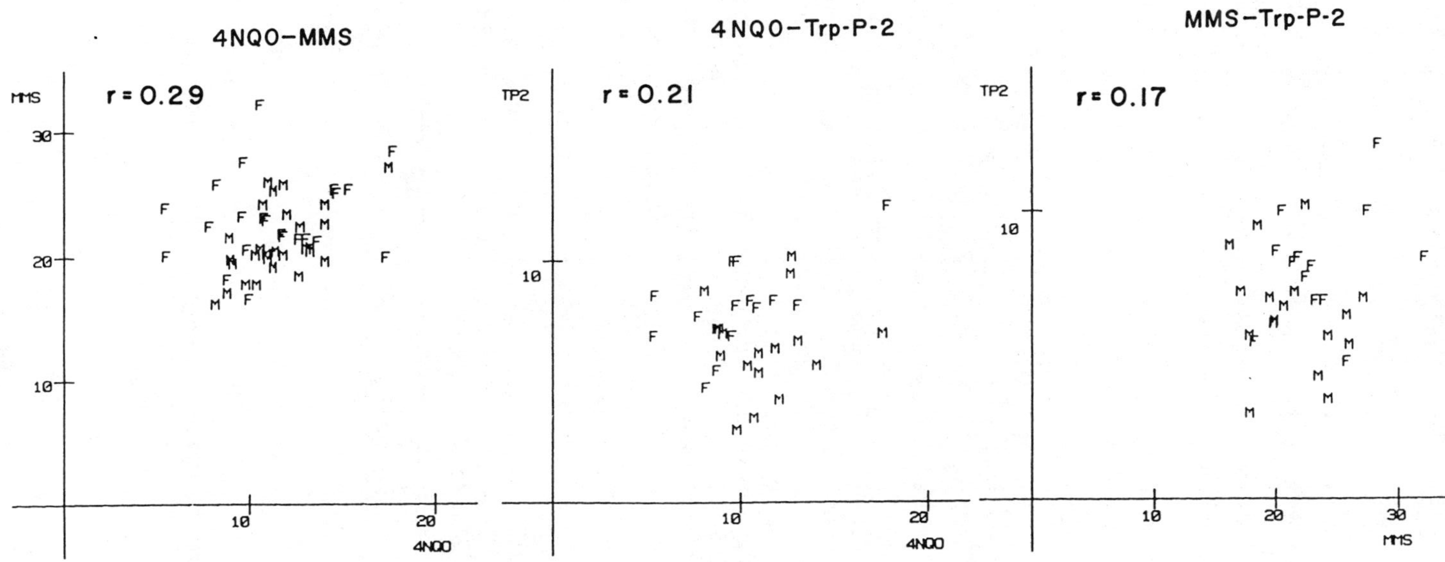

Fig. 3. Correlations between sensitivities to chemicals. A pair of SCE values induced by 2 chemicals were plotted as a point for each individual. Correlation coefficients (r) are given for respective pairs of chemicals.

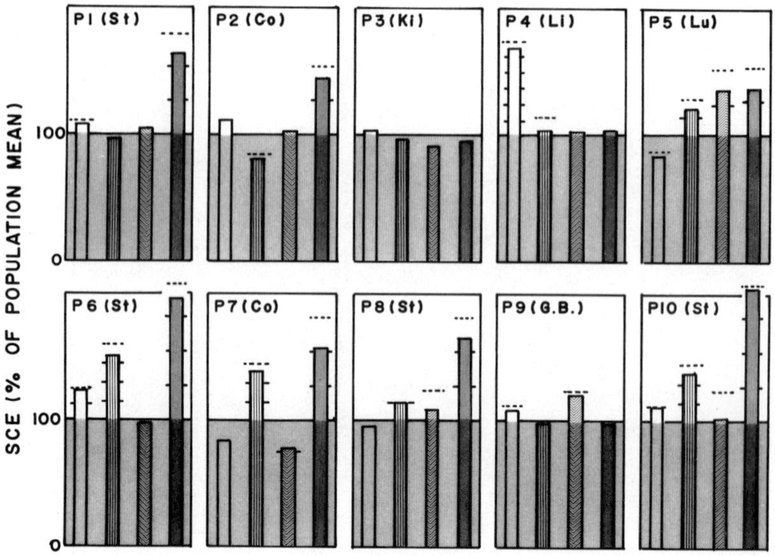

Fig. 4. Sister chromatid exchange sensitivities of lymphocytes from cancer patients. Ten patients (P1-P10) were examined for their SCE sensitivities with procedures identical to those for healthy population shown in Fig. 1. Sister chromatid exchange frequency is expressed as the percentage of the mean frequency of the healthy population. Four columns for each patient are, from left to right, spontaneous, MMS-, 4NQO-, and Trp-P-2-induced SCEs, and marked with the respective σ units. St, Co, Ki, Li, Lu, and G.B. are patients bearing stomach, colon, kidney, liver, lung, and gall bladder cancers, respectively.

Tab. 1. Frequencies of SCE-sensitive individuals in cancer patients and healthy humans.

	Cancer patients (A)	Healthy humans (B)	A/B	Difference between[a] A and B (P for A=B)
Spontaneous SCE	1 in 10	1 in 52	(5.2)	not signif. ($P>0.1$)
MMS-induced SCE	3 in 10	2 in 50	7.5	singif. ($P<0.01$)
4NQO-induced SCE	0 in 10	3 in 50	(0)	not singif. ($P>0.3$)
Trp-P-2-induced SCE	5 in 10	1 in 33	16.5	singif. ($P<0.001$)

Sensitive individual: Mean SCE frequency of peripheral lymphocytes exceeds more than 2σ above the mean of healthy humans (cf. Table 2).

a. χ^2-test.

Tab. 2. Induced SCEs in individuals with high spontaneous SCE levels.

Donor(sex, age)	Sister chromatid exchange (SCEs/cell)				Smoking habit[a]
	Spontaneous	MMS-induced	4NQO-induced	Trp-P-2-induced	
	(mean for 30 cells ± S.E.)				
M5(M, 50y)	14.0 ± 0.6	20.2 ± 1.4	11.9 ± 1.9	3.5	20 x 30
P4(F, 69y)	16.6 ± 1.1	22.9 ± 2.0	12.2 ± 1.9	7.8	13 x 25
Healthy humans	(mean of individual means ± S.D.)				
Total(M+F,55y>,>21y)	9.8 ± 1.2(52)[b]	21.8 ± 3.1(50)	11.7 ± 2.8(50)	7.4 ± 2.1(33)	
Smoker(M, >30y)[c]	11.3 ± 1.0(5)	20.6 ± 1.1(5)	12.0 ± 1.3(5)	6.0; 3.5	20 x (>10)
Non-smoker(M+F,>30y)	9.4 ± 0.6(5)	23.2 ± 5.2(5)	12.9 ± 2.3(5)	8.4	
Non-smoker(M+F,<30y)	9.5 ± 0.9(19)	22.7 ± 3.4(18)	11.4 ± 3.6(19)	7.4 ± 2.2(15)	

a. (Number of cigarettes/day) x (number of years).

b. Figures in parentheses are number of blood donors.

c. This includes M1, M3, M4, M5 and M6 in Fig. 1.

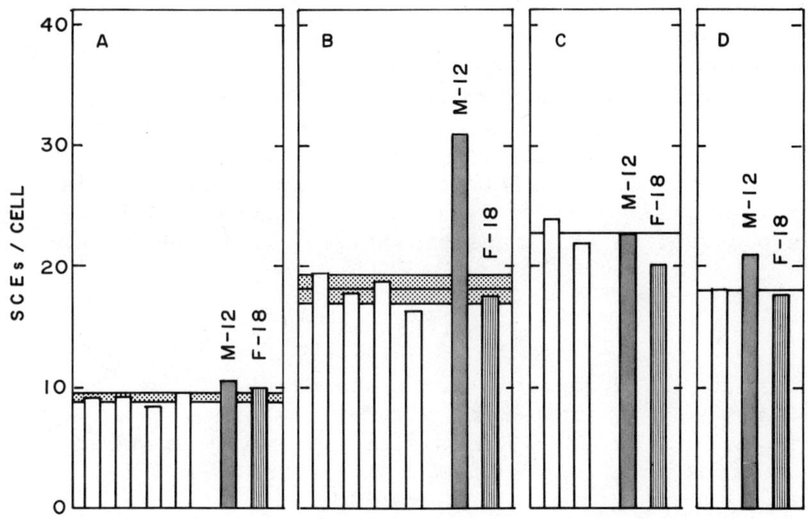

Fig. 5. Sensitivity to SCE induction in cell lines derived by EBV transformation from healthy donors M12 and F18. A, spontaneous SCE; B, C, and D, SCEs induced by 1 μM 4NQO, 50 μM MMS, and 10 μM Trp-P-2, respectively. Open columns are for cell lines from normal donors.

Tab. 3. Correlations among UDS, UV-induced SCE, and cancer contraction in XP patients.

Cell line	Sex	Age	UDS (% of normal)	UV-induced SCE[a] sensitivity	Skin cancer
UDS-deficient (< 70% of normal)					
XPL 12	M	1	7	++	-
XPL 15	M	7	5	+++	+
XPL 17	F	5	2	++	-
XPL 21	F	7	18	+++	-
XPL 23	F	2	6	++	-
XPL 24	F	21	63	+	+
XPL 26	M	15	31	++	-
UDS-proficient cells (≥ 80% of normal)					
XPL 5	M	36	80	+	+
XPL 18	M	18	83	normal	-
XPL 19	M	49	97	normal	-
XPL 20	M	20	89	+	+
XPL 22	M	32	85	+	+
XPL 25	M	38	83	normal	+

a. Degree of sensitivities higher than normal.

Whether the high sensitivity in SCE formation is related to the cause or the consequence of carcinogenesis should be solved by prospective observation of humans sensitive to SCE induction.

REFERENCES

1. Tohda, H., K. Horaguchi, K. Takahashi, A. Oikawa, and T. Matsushima (1980) Epstein-Barr virus-transformed human lymphoblastoid cells for study of sister chromatid exchange and their evaluation as a test system. Cancer Res. 40:4775-4780.
2. Oikawa, A., S. Sakai, K. Horaguchi, and H. Tohda (1983) Sensitivities of peripheral lymphocytes from healthy humans to induction of sister chromatid exchanges by chemicals. Cancer Res. 43:439-442.
3. Perry, P., and S. Wolff (1974) New Giemsa method for the differential staining of sister chromatids. Nature (Lond.) 251: 156-158.
4. Livingston, G.K., L.A. Cannon, D.T. Bishop, P. Johnson, and R.M. Fineman (1983) Sister chromatid exchange: Variation by age, sex, smoking, and breast cancer status. Cancer Genet. and Cytogenet. 9:289-299.

5. Tohda, H., A. Oikawa, T. Kudo, and T. Tachibana (1978) A greatly simplified method of establishing B-lymphyoblastoid cell lines. Cancer Res. 38:3560-3562.
6. Cleaver, J.E. (1968) Defective repair replication of DNA in xeroderma pigmentosum. Nature (Lond.) 218:652-656.
7. Tohda, H., and A. Oikawa (1983) Differential features of sister-chromatid exchange responses to ultraviolet radiation and caffeine in xeroderma pigmentosum lymphoblastoid cell lines. Mutat. Res. 107:387-396.

VALUE AND SIGNIFICANCE OF SCE IN HUMAN LEUKEMIA AND CANCER

Avery A. Sandberg, Reinhard Becher, and Zenon Gibas

Roswell Park Memorial Institute
Buffalo, New York 14263

INTRODUCTION

The introduction and refinement of sister chromatid exchange (SCE) techniques generated much enthusiasm, interest, and promise for their application in human neoplasia. In particular, the demonstration of high SCE levels in the cells of Bloom syndrome (BS) patients (1,2), a condition with a very high susceptibility to the development of cancer (primarily lymphoma or leukemia), and the application of SCE as a sensitive and specific index of carcinogenic and/or mutagenic activity (3,4), generated much excitement for the application of SCE studies in human neoplasia. It was hoped that unique changes, akin to those at the chromosome level, would characterize certain malignant conditions and states (5-7). Thus, the general concept was developed that individuals or families with high risk of cancer might show SCE changes characteristic of these groups and that cells involved in certain neoplastic conditions would display unique SCE patterns different from those of normal cells and allow for differential delineation of some neoplastic processes.

Whether the promise SCE studies held in unraveling specific alterations in human neoplasia has been fulfilled is a moot question. In this presentation we shall address the following areas: 1) SCE changes in cancer patients and their relatives, 2) SCE levels in affected cells of leukemia and cancer, both in vivo and in vitro, and 3) the value and significance of the SCE changes observed in human neoplasia to date.

Comments Regarding SCE Studies in Human Neoplasia

Before SCE studies in human neoplasia are discussed, it is important that some of the bothersome and limited aspects of these studies be discussed.

Studies based on PHA-stimulated lymphocytes. Generally, studies on SCE levels in cancer patients and their relatives have been performed on peripheral blood lymphocytes stimulated by phytohemagglutinin (PHA). Thus, the bulk of the SCE data is based on observations on such cells which, for all practical purposes, represent populations of T cells. Though examination of these cells may be appropriate for certain conditions (e.g., T-cell lymphoma), data are egregiously lacking on SCE levels on other cells, particularly those in the tissues or organs affected by the neoplasia under study. For example, though it may be assumed that SCE levels in PHA-stimulated T lymphocytes of lung cancer patients reflect a general condition in the body, that may not be necessarily true. Another example in that regard is the study of SCE levels of lymphocytes in smokers when, in fact, in order to obtain more meaningful data, the SCE levels in lung cells would represent a much more useful and appropriate parameter to examine. Even in lymphomas, which are preponderantly of B cell origin, the SCE data published to date are almost exclusively based on PHA-stimulated cells (T cells). To a large extent, the same applies to studies in leukemia.

When studying SCE levels in lymphocytes of patients with treated leukemia or cancer, it is essential that a sufficient period of time be allowed to elapse to make sure that the effects of the therapy no longer exist. Practically and hypothetically, this may never be the case, since of the stimulated lymphocytes many still carry within them the long-term effects of such therapy. Thus, it is best to study lymphocytes of untreated subjects.

The systemic effects of tumor growth and the tumor per se cannot be ignored, since these may manifest themselves in parameters affecting SCE levels (changes in immune area, inappropriate or ectopic production of substances by the tumor, metabolic derangements).

Studies based on neoplastic cells. The shortcomings alluded to above can be further extended to the situation in leukemia where the blood lymphocytes may not represent the cell type involved by the leukemic process in the bone marrow. Also, when examining SCEs in bone marrow cells, the metaphases usually observed consist of dividing erythroblasts or granulocytic cells at certain stages of maturation and seldom reflect the more immature types of cells, such as myeloblasts, which are commonly involved in leukemia. In addition, comparing SCE levels in the cells of normal marrow with those in leukemic marrows presents a further complication in that, in certain acute leukemias, the preponderant number of the cells in the marrow is of a particular variety (lymphoblasts or myeloblasts), whose normal SCE levels cannot be established easily due to the fact that they constitute a small number of the metaphase in normal marrow and, furthermore, they cannot be identified as such. Thus, caution must be used in comparing and interpreting data obtained on circulating T cells, normal fibroblasts, or bone marrow cells with those obtained in conditions such as leukemia and various cancers. In the case of bone marrow cells, it is the cellular heterogeneity in the normal condition, and in the case of cancers, the inability to examine the cells of the affected tissue which constitute important obstacles and shortcomings in the interpretation of SCE results. Optimally, control SCE values for comparison with those obtained on affected cells should be established on the cells of the normal tissues from which the cancers originate (e.g., in the case of malignant melanoma, normal melanocytes should serve as control cells and in the case of lung cancer, either epithelial or fibroblastic cells of the lung).

In vitro and life style effects. In interpreting the results to be presented, it should be also kept in mind that in vitro conditions [e.g., concentration of bromodeoxyuridine (BrdUrd)], type of medium, type of and sensitivity to mitogenic agents) may in themselves affect the levels of SCEs and not necessarily reflect those which would be observed under in vivo conditions. Furthermore, certain aspects of an individual's life style may be reflected in the SCE levels, such as the well-established effects of smoking and those of certain drugs (e.g., estrogenic compounds) (7), parameters which often are not taken into consideration when establishing SCE data on so-called control populations. There is little doubt in our minds that exposure to substances in the environment, diet, and working conditions may affect the SCE levels; often these considerations are not given due to attention when establishing control populations for comparison with studies of SCE levels in patients with leukemia and cancer. The effects of various forms of chemotherapy on SCEs have been well documented and need not be discussed here, although, again, these could play an important role in affecting some of the results published to date.

SCE Changes in PHA-stimulated Lymphocytes of Cancer Patients

A number of studies have now demonstrated the ability of carcinogenic agents to induce high SCE levels in vitro (Fig. 1-3) and in vivo in a variety of cell systems (4). It was these observations that led to an examination of SCE levels in cancer patients and their families as a means of establishing criteria for

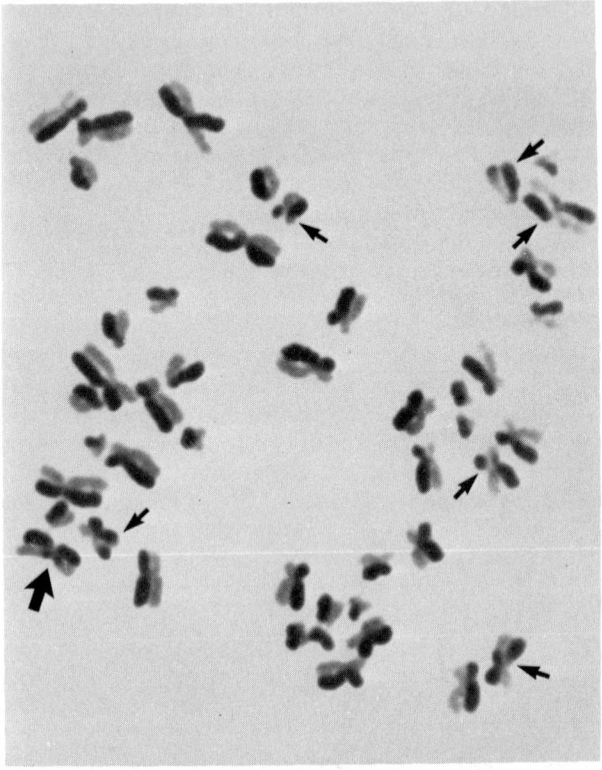

Fig. 1. SCE in a normal lymphocyte. Small arrows point to 6 SCEs and the thick arrow to a pericentric exchange.

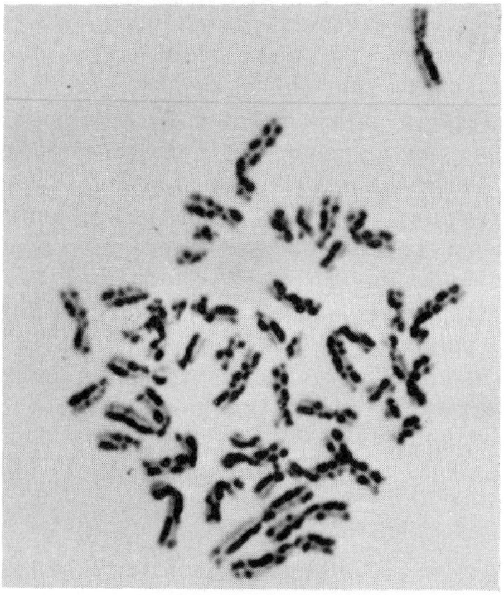

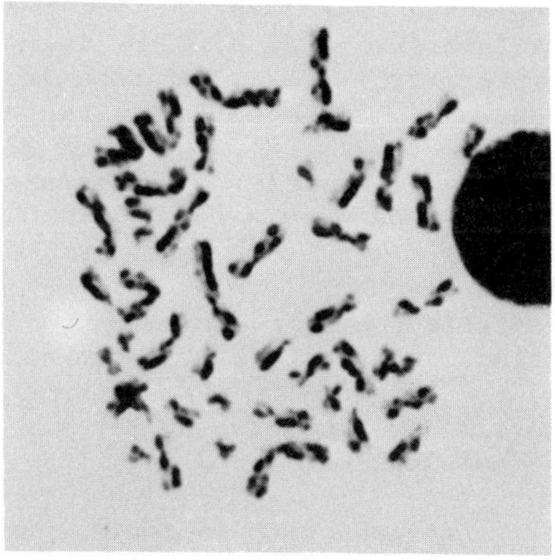

Figs. 2, 3. Very high levels of SCEs observed in lymphocytes exposed to mitomycin C. Such changes can be observed in the lymphocytes of patients receiving certain forms of chemotherapy or in subjects exposed to carcinogenic agents.

predicting possible susceptibility to neoplasia. In addition to baseline SCE levels, the in vitro effects of substances capable of raising the SCE levels were evaluated in order to ascertain whether a higher sensitivity existed in cancer-prone populations. Here, again, the problem of control populations and possible in vitro effects may play an important part and make comparison of data rather difficult.

Patients with cancers. The results to date regarding the SCE levels in cancer patients and their relatives are conflicting and have led to inconsistent conclusions. Some workers have found increased baseline SCE levels in some patients (8-15), whereas others have not (16-25). Some have found an increased sensitivity to SCE-inducing agents (26) and others have not (27,28). Thus, the picture is confusing and uncertain, at least as far as the SCE levels in cancer patients or in cancer-prone individuals and families are concerned. As indicated above, examination of lymphocytes and fibroblasts may be far removed from the area of the primary tumors, cells of which would constitute more appropriate material for examination. Unfortunately, such examinations cannot be undertaken due to a number of practical reasons.

It may be worthwhile to discuss a few specific conditions. In retinoblastoma, examination of blood lymphocytes or fibroblast cultures has shown normal baseline SCE levels in affected individuals (22,26), with one group of authors stating that retinoblastoma is not associated with chromosome instability (22). In another study, though, a higher sensitivity of the cells to mitomycin C (MMC) was demonstrated (26), the authors concluding that DNA repair efficiency may be decreased in the lymphocytes of retinoblastoma patients and that the changes may be due to the retinoblastoma gene or to another factor acting on the degree of expressivity of the disease in gene carriers. In patients with lung cancer, the SCE levels have been shown to be normal in 2 studies (14,23,27), including the response of the cells to benzo-(a)pyrene (BP) (27). The situation in breast cancer and cancer of the uterine cervix remains confusing, since in each condition the results of studies are conflicting, with some studies finding increased SCE levels (10,12) and others, normal ones (19,24). Variable results have been obtained in a study of malignant melanoma patients (11), most of them of the familial variety, with a few of the family members showing definitely increased SCE levels (as high as 28 per cell) and others showing essentially normal levels.

Patients with leukemia. Studies of SCE levels in circulating lymphocytes, predominantly of the T cell variety, in various leukemias have shown normal levels in subjects with chronic myelocytic leukemia (15), chronic lymphocytic leukemia (16), and lymphoma (18). Several studies, however, have shown significantly increased lymphocyte SCE levels in patients with acute lymphoblastic leukemia (9) or lymphoma (8). An exact interpretation of the conflicting results in these cases is, at present, difficult, particularly in acute leukemia or lymphoma patients in whom a variety of systemic effects and changes could readily account for or contribute to the increased SCE levels. Such systemic effects could be due to infections with bacteria or viruses, from metabolites or substances originating from the tumor or leukemic cells per se, or a changed response of the lymphocytes in vitro resulting from immunologic derangements. For example, stimulated human phagocytes have the ability to elevate the SCE levels in cultured cells exposed to various concentrations of these stimulated leukocytes (29,30). The authors have postulated that the generation by these cells of oxygen radicals could account for the increased levels of SCEs (30). It is possible to visualize a situation of that nature occurring in patients with leukemia and lymphoma, wherein, under in vivo conditions, such radicals could readily affect the SCE levels observed in the stimulated cells of these patients. Complicating the interpretation further is a recent publication in which lower SCE values were found in the stimulated lymphocytes of patients with both acute and chronic leukemia, with the authors stating that the lower SCE levels may be related to the leukemic process (25).

Evaluation of SCE studies on lymphocytes and fibroblasts in cancer patients or cancer-prone subjects. In viewing the results obtained to date on lymphocytes and fibroblasts of patients with various forms of neoplasia, it appears to us that the last word has not been said in this area, although it is unlikely that future results will unravel facets to this approach which will prove to be a useful or sensitive test for leukemia or cancer risk. The reasons for this pessimistic view have been given above; unless major breakthroughs in methodology and specificity in this area occur, the results to date indicate a confusing picture with little application to the clinical situation.

In fact, in one study with lymphocytes, the authors could not find increased baseline levels of SCE in the cells of patients with lung cancer or in their response to BP (27); but when fibroblasts were examined, a slight increase in SCE was observed in the cultured cells, particularly in those with the familial variety of lung cancer (28). The response of fibroblasts to BP, though, was less than that observed in those of the control group. Thus, the results, at best, are borderline and indicate that SCEs will probably not be a useful approach in establishing susceptibility to lung cancer or, for that matter, to other cancers.

In a number of the studies it has been demonstrated that chemotherapy of one type or another was capable of increasing the SCE levels and, occasionally, to extremely high levels (7). In one study of an acute lymphocytic leukemia (ALL) patient with Down syndrome, the trisomic cells showed definitely increased levels of SCE both prior to and following chemotherapy, findings which differed from those observed in the diploid cells of the same patient (31). This recurring theme of heteroploid cells being more sensitive to SCE induction will become more apparent as we discuss SCE levels in established cell lines of lymphomas and cancers.

Patients with X-linked proliferative disease had normal SCE levels in their lymphocytes (20) and, furthermore, lymphoblastoid lines established from the lymphocytes of these subjects had significantly lower SCE levels than those observed in PHA- or pokeweed mitogen-stimulated lymphocytes (20).

SCE Changes in Affected Cells of Leukemia, Lymphoma, and Cancer

Leukemia. The results on SCE levels in affected cells of leukemia and cancer, established either in vivo or in vitro, have been of some interest but not consistent. For example, results in various leukemias have generally indicated SCE levels which are somewhat lower than those observed in normal bone marrow cells (5-7), although, as discussed above, a comparison of SCE levels of myeloblasts or lymphoblasts, which are the predominant cells in acute leukemia, with SCE findings in normal bone marrow cells consisting primarily of dividing erythroblasts and granulocytes, has shortcomings and limitations. Nevertheless, it appears that generally the leukemic marrow cells either in chronic myelocytic leukemia (32-35) or acute nonlymphocytic leukemia (36-38) are characterized by SCE levels which tend to be lower than those seen in normal bone marrow cells and certainly much lower than those in normal lymphocytes. (The SCE levels in normal bone marrow cells have been shown consistently to be lower than those of PHA-stimulated lymphocytes.) In contrast, one group has reported definitely elevated SCE levels in the cells of childhood leukemia whether the bone marrow was examined directly or following stimulation with PHA (39). Elevated SCE levels have also been found in the bone marrow cells of patients with multiple myeloma (13), but it would be difficult to be certain that the myeloma cells were, in fact, the ones examined, since the authors did not indicate whether the metaphases were aneuploid or not.

The results obtained on bone marrow cells reflecting in vivo conditions in leukemia, also point to the possibility that transformed cells are not necessarily accompanied by high SCE levels, at least at a stage when the neoplasia is well established. In fact, in one study it was shown that progression of the cells in chronic myelocytic leukemia from the chronic stage to the acute (blastic) phase was accompanied by an actual decrease in the levels of SCEs (35), indicative that progression in malignancy is not necessarily accompanied by elevated SCE levels.

The interpretation of the decreased SCE levels in leukemic cells observed by most workers raises the question as to the possible relationship between SCEs, cell cycle length, and the efficiency of DNA repair in such cells. These aspects have been discussed in another chapter in this volume and thus will not be taken up at this point.

Lymphoma or cancer cell lines. Studies on established cell lines from lymphomas (40-42) or cancers (43-45) have generally revealed higher SCE levels than those observed in diploid cell lines or in the lymphocytes of the affected patients. Such cell lines have been established from cancers of the breast, colon, malignant melanoma, and lymphomas. The general opinion has been that the elevated SCE rates in these cell lines seem to correlate with the number of chromosome markers and the degree of karyotypic instability of each line. Furthermore, the SCE levels appear to be proportional to the number of chromosomes and, hence, a more proper way of expressing results in these lines would be to show them as SCE per chromosome rather than per cell. In fact, because of the wide variation in chromosome sizes observed in cells with high chromosome number and/or morphologic changes, a more accurate way of expressing SCEs in such cells (as well as in normal cells) would be to do so on the basis of chromosomal unit length or amount of DNA. On the other hand, other authors (44) have stated that the karyotypic instability in tumor cells is not necessarily associated with increased baseline SCE levels or induced SCEs per chromosome, indicative that not all cell lines have increased SCEs.

A related observation is that seen in one of our patients with erythroleukemia who had at least 2 populations of cells, one in the near-diploid and another in the near-tetraploid range (46). The SCE levels in near-diploid marrow cells were comparable to those seen in normal marrow, whereas they were definitely elevated in the tetraploid cells. Allusion has been made above to increased SCE levels following therapy in the trisomic cells of a patient with Down syndrome and acute leukemia, as compared to the normal SCE levels in the diploid cells (31). Thus, one of the definite conclusions that can be reached from studies of in vitro material of human cancer or lymphoma origin is that cells which are aneuploid, particularly those with high chromosome counts and many morphologic changes, are more apt to have high SCE levels, either as a baseline or following exposure to various substances, than are diploid cells. This obviously indicates a genomic instability, as well as a deranged DNA repair mechanism, in aneuploid cells as compared to those of normal cells. It is possible that similar mechanisms are responsible for the increased SCE rates observed in the bone marrow cells of patients with refractory anemia, a condition associated with disturbed replication (47).

Sister chromatid exchange studies on noncultured cancer cells are essentially nonexistent. In a recently completed study, we investigated the SCE levels in 4 malignant melanomas directly. Chen and Li et al. (44,45) had reported a slightly increased SCE frequency in 6 permanent malignant melanoma cell lines. This, however, was not the case in our study. Inasmuch as the number of chromosomes is subject to considerable variability in tumor cells (e.g., in the 4 melanomas studied by us the modal chromosomal number ranged from pseudodiploid to more than triploid, with a wide range of chromosome numbers within each melanoma), comparison with SCE rates per cell in normal tissues is to be avoided. Thus, we calculated the SCE frequency per chromosome which ranged in the 4 melanomas from 0.111 to 0.168 SCEs per chromosome. This is compatible with an SCE frequency of 5.1 to 7.7 per diploid metaphase (Tab. 1, Fig. 4) of normal lymphocytes at the same BrdUrd concentrations.

Evaluation of SCE Changes in Leukemic and Cancer Cells

Even though a statistically more frequent involvement by SCE of some chromosomes has been published in a few conditions (10,12) to date, chromosomal or intrachromosomal specificity of SCE changes has not been observed in any human neoplastic entity. Thus, the use of SCE patterns in uniquely delineating some cancers and leukemias has fallen short of expectations, with the bulk of

Tab. 1. SCE frequency in malignant melanoma metaphases.

Case #	Culture time (h)	Total number of metaphases	Total number of chromosomes	Total number of SCE	Mean SCE/metaphase	Mean SCE/chromosome
1	72	23	1414	160	6.96	0.113
2	48	58	4221	603	10.4	0.143
3	72	20	1401	236	11.8	0.168
4	72	17	1133	126	7.4	0.111

the results to date being reflected in statistical analyses. The latter leave much to be desired in applicability. Also, as long as the exact mechanisms for generating SCEs continue to be unknown, the changes observed defy interpretation and lack specificity.

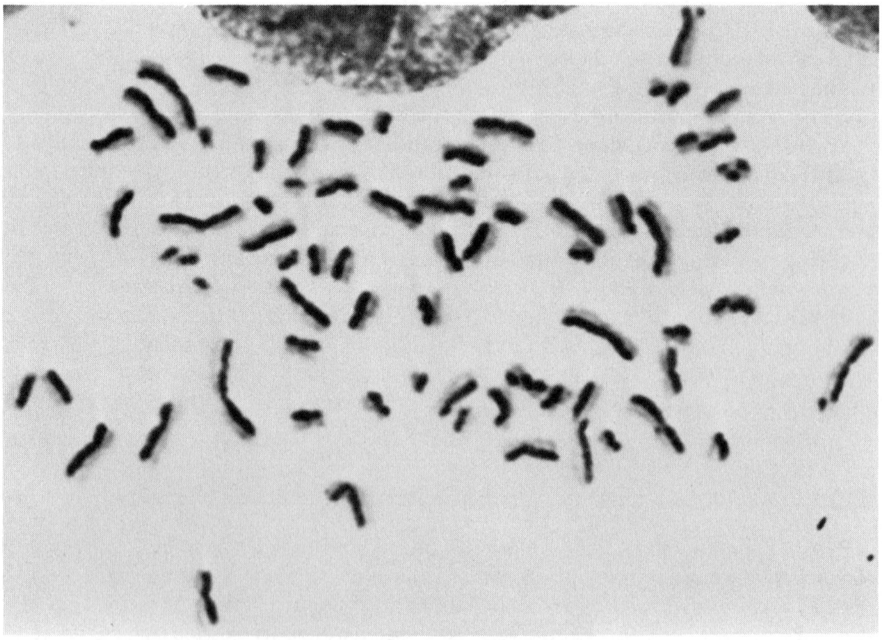

Fig. 4. An aneuploid malignant melanoma cell with 67 chromosomes, 10 SCEs, and 0.15 SCEs/chromosome. The latter is within the normal range.

Even though it has been established that various chemotherapeutic agents used in the therapy of cancer and leukemia induce high levels of SCEs (7), specificity regarding each agent has not been demonstrated. The elevated SCE levels observed during and following chemotherapy are not permanent, with several groups reporting normal levels of SCE after such therapy has been discontinued, including long-term survivors with acute leukemia (7,12, 48,49).

The higher SCE values, expressed per metaphase or chromosome, observed in aneuploid cells, particularly those in established cultures, are of interest. It is possible that these changes are related to a general instability of the genome. In contrast, normal to low values have been observed in aneuploid cells of chronic myelocytic leukemia and acute leukemia. These findings indicate that the in vitro environment of established cell lines may predispose aneuploid cells to generally higher SCE levels than in diploid cells. Obviously, we must learn much more about this facet of SCE formation, particularly in relation to aneuploidy.

The determination of SCE levels in lymphocytes of patients with cancer or leukemia has not proven to date to be an approach providing us with a deep insight into a better understanding of the genesis and origin of neoplasia or, for that matter, a useful means of monitoring predisposing backgrounds to neoplasia, either genetic or ecologic (50).

REFERENCES

1. Chaganti, R.S.K., S. Schonberg, and J. German (1974) A manifold increase in sister chromatid exchanges in Bloom's syndrome lymphocytes. Proc. Natl. Acad. Sci., USA 71:4508-4512.
2. Shiraishi, Y., A.I. Freeman, and A.A. Sandberg (1976) Increased sister chromatid exchange in bone marrow and blood from Bloom's syndrome. Cytogent. Cell Genet. 17:163-173.
3. Perry, P., and H.J. Evans (1975) Cytological detection of mutagen-carcinogen exposure to sister chromatid exchange. Nature 258:121-125.
4. Abe, S., and M. Sasaki (1982) SCE as an index of mutagenesis and/or carcinogenesis. In Sister Chromatid Exchange, A.A. Sandberg, ed. Alan R. Liss, Inc., New York, pp. 461-514.
5. Sandberg, A.A. (1980) Some comments on sister chromatid exchange (SCE) in human neoplasia. Cancer Genet. Cytogenet. 1:197-206.
6. Shiraishi, Y., and A.A. Sandberg (1980) Sister chromatid exchange in human chromosomes, including observations in neoplasia. Cancer Genet. Cytogenet. 1:363-380.
7. Sandberg, A.A. (1982) Sister chromatid exchange in human states. In Sister Chromatid Exchange, A.A. Sandberg, ed. Alan R. Liss, Inc., New York, pp. 619-651.

8. Kurvink, K., C.D. Bloomfield, K.M. Keenan, S. Levitt, and J. Cervenka (1978). Sister chromatid exchange in lymphocytes from patients with malignant lymphoma. Human Genet. 44:137-144.
9. Otter, M., C.G. Palmer, and R.L. Baehner (1979). Sister chromatid exchange in lymphocytes from patients with acute lymphoblastic leukemia. Human Genet. 52:185-192.
10. Mitra, A.B., V.V.V.S. Murty, and U.K. Luthra (1982) Sister chromatid exchanges in leukocytes of patients with cancer of cervix uteri. Human Genet. 60:214-215.
11. Ghidoni, A., E. Privitera, E. Raimondi, D. Rovini, M.T. Illeni, and N. Cascinelli (1983) Malignant melanoma: Sister chromatid exchange analysis in three families. Cancer Genet. Cytogenet. 9:347-354.
12. Livingston, G.K., L.A. Cannon, D.T. Bishop, P. Johnson, and R.M. Fineman (1983) Sister chromatid exchange: Variation by age, sex, smoking, and breast cancer status. Cancer Genet. Cytogenet. 9:289-299.
13. Singh, J., G.R. Malik, A.K. Malhotra, and I.C. Verma (1984) Increased sister chromatid exchanges in the bone marrow in multiple myeloma. Int. J. Cancer (in press).
14. Hollander, D.H., M.S. Tockman, Y.W. Liang, D.S. Borgaonkar, and J.K. Frost (1978) Sister chromatid exchanges in the peripheral blood of cigarette smokers and in lung cancer patients; and the effect of chemotherapy. Human Genet. 44: 165-171.
15. Hopkin, J.M., and H.J. Evans (1980) Cigarette smoke-induced DNA damage and lung cancer risks. Nature (Lond.) 283:388-390.
16. Cheng, W.-S., J.J. Mulvihill, M.H. Greene, L.W. Pickle, S. Tsai, and J. Whang-Peng (1979) Sister chromatid exchanges and chromosomes in chronic myelogenous leukemia and cancer families. Int. J. Cancer 23:8-13.
17. McDonald, M.A., and P.H. Fitzgerald (1979) Sister chromatid exchange and cell cycle progression in cultured lymphocytes from patients with chronic lymphatic leukemia. J. Natl. Cancer Inst. 62:1169-1171.
18. Husum, B., H.C. Wulf, and E. Niebuhr (1981) Sister-chromatid exchanges in lymphocytes in women with cancer of the breast. Mutat. Res. 85:357-362.
19. Crossen, P.E., P.H. Fitzgerald, and B.M. Colls (1981) Normal sister chromatid exchange and cell cycle progression in cultured lymphocytes from patients with malignant lymphoma. J. Natl. Cancer Inst. 67:821-824.
20. Barnabei, V.M., K. Sakamoto, J.K. Seeley, and D.T. Purtilo (1982) Chromosomal breakage and sister chromatid exchange in peripheral blood lymphocytes and lymphoblastoid cell lines in the X-linked lymphoproliferative syndrome. Cancer Genet. Cytogenet. 6:313-321.
21. Bazopoulou-Kyrkanidou, E., A.P. Angelopoulos, J. Garas, and F. Zervou-Valvi (1983) Lymphocyte reactivity in patients with cancer of the oral cavity measured by sister chromatid

differential staining. Cancer Genet. Cytogenet. 10:43-49.

22. Takabayashi, T., M.S. Lin, and M.G. Wilson (1983) Sister chromatid exchanges and chromosome aberrations in fibroblasts from patients with retinoblastoma. Human Genet. 63:317-319.
23. Knight, L.A., H.A. Gardner, and B.L. Gallie (1983) Normal rates of sister chromatid exchange in lymphocytes from retinoblastoma patients. J. Clin. Hematol. Oncol. 13:73-80.
24. Adhvaryu, S.G., R.C. Vyas, B.J. Dave, A.H. Trivedi, and B.N. Parikh (1984) Spontaneous and induced sister chromatid exchange frequencies and cell cycle progression in lymphocytes of patients with carcinoma of uterine cervix. Cancer Genet. Cytogenet. (in press).
25. Larripa, I., M.L. de Vinuesa, I. Slavutsky, and S.B. de Salum (1984) Sister chromatid exchanges in leukemic patients. Cancer Genet. Cytogenet. (in press).
26. Der-Sarkissian, H., C. Bonaiti-Pellie, M.-L. Briard-Guillemot, and J.-M. Zucker (1982) Sister chromatid exchanges in patients with retinoblastoma. Cancer Genet. Cytogenet. 7: 73-77.
27. Schonwald, A.D., C.R. Bartram, and H.W. Rudiger (1977) Benzpyrene-induced sister chromatid exchanges in lymphocytes of patients with lung cancer. Human Genet. 36:261-264.
28. Rudiger, H.W., W. Harder, P. Maack, F.-V. Kohl, and U. Schmidt-Preuss (1981) Decreased rate of benzo(a)pyrene-induced sister chromatid exchange in fibroblast cultures from patients with lung cancer. J. Cancer Res. Clin. Oncol. 102: 169-175.
29. Weiss, S.J., M.B. Lampert, and S.T. Test (1983) Long-lived oxidants generated by human neutrophils: Characterization and bioactivity. Science 222:625-628.
30. Weitberg, A.B., S.A. Weitzman, M. Destrempes, S.A. Latt, and T.P. Stossel (1983) Stimulated human phagocytes produce cytogenetic changes in cultured mammalian cells. New Eng. J. Med. 308:26-30.
31. Heidemann, A., B. Schmalenberger, and H. Zankl (1984) SCE-frequency and lymphocyte proliferation in a Down-syndrome-mosaic developing an acute lymphoblastic leukemia. Cancer Genet. Cytogenet. (in press).
32. Kakati, S., S. Abe, and A.A. Sandberg (1978) Sister chromatid exchange in Philadelphia chromosome (Ph^1)-positive leukemia. Cancer Res. 38:2918-2921.
33. Becher, R., C.G. Schmidt, G. Theis, and D.K. Hossfeld (1979) Sister chromatid exchange in Ph^1-positive chronic myelocytic leukemia. Int. J. Cancer 24:713-716.
34. Abe, S., S. Kakati, and A.A. Sandberg (1980) Growth pattern and SCE incidence in Ph^1-positive cells of CML. Cytobios 28:137-149.
35. Stoll, C., F. Oberling, and M.-P. Roth (1982) Sister chromatid exchange and growth kinetics in chronic myeloid leukemia. Cancer Res. 42:3240-3243.
36. Abe, S., S. Kakati, and A.A. Sandberg (1979) Growth rate and sister chromatid exchange (SCE) incidence of bone marrow

cells in acute myeloblastic leukemia (AML). Cancer Genet. Cytogenet. 1:115-130.

37. Abe, S., and A.A. Sandberg (1980) Sister chromatid exchange and growth kinetics of marrow cells in aneuploid acute non-lymphocytic leukemias. Cancer Res. 40:1292-1299.
38. Becher, R., G. Zimmer, and C.G. Schmidt (1981) Sister chromatid exchange and growth kinetics in untreated acute leukemia. Int. J. Cancer 27:199-204.
39. Heerema, N.A., C.G. Palmer, and R.L. Baehner (1982) Elevated sister chromatid exchange and cell cycle analysis in bone marrow in childhood ALL. Cancer Genet. Cytogenet. 6:323-330.
40. Fonatsch, C., M. Schaadt, and V. Diehl (1979) Sister chromatid exchange in cell lines from malignant lymphomas (lymphoma lines). Human Genet. 52:107-118.
41. Fonatsch, C., M. Schaadt, H. Kirchner, and V. Diehl (1980) A possible correlation between the degree of karyotype aberrations and the rate of sister chromatid exchanges in lymphoma lines. Int. J. Cancer 26:749-756.
42. Fonatsch, C., H.H. Kirchner, M. Schaadt, M. Gunzel, and V. Diehl (1981) Sister chromatid exchange in lymphoma lines and lymphoblastoid cell lines before and after heterotransplantation into nude mice. Int. J. Cancer 28:441-445.
43. Reisbach, G., E. Gebhart, and R. Cailleau (1982) Sister chromatid exchanges and proliferation kinetics of human metastatic breast tumor cell lines. Anticancer Res. 2:257-260.
44. Li, S., W.W. Au, R.L. Schmoyer, Jr., and T.C. Hsu (1982) Baseline and mitomycin-C induced sister chromatid exchanges in a melanoma and a colon tumor cell line. Cancer Genet. Cytogenet. 6:243-248.
45. Chen, T.R. (1981) Frequencies of sister chromatid exchanges in heteroploid cell lines of human melanoma origin. J. Natl. Cancer Inst. 66:273-277.
46. Gibas, Z., R. Becher, and A.A. Sandberg (1983) Chromosomes and causation of human cancer and leukemia. LIII. Comprehensive cytogenetic analysis of an erythroleukemia. Cancer Genet. Cytogenet. 10:241-253.
47. Carbone, P., G. Barbata, G. Granata, G. Margiotta, and F. Caronia (1981) Sister chromatid exchange distribution in bone marrow cell chromosomes from patients with refractory anemia. Acta Haematol. 65:177-182.
48. Abe, S., and A.A. Sandberg (1980) Growth pattern and sister chromatid exchanges of bone marrow cells in acute lymphoblastic leukemia. Cytobios 29:165-173.
49. Inoue, S., L. Brown, Y. Ravindranath, and M.J. Ottenbreit (1982) Normal sister chromatid exchange frequency in long-term survivors with acute leukemia. Cancer Res. 42:2906-2908.
50. Sandberg, A.A. (1983) Chromosomes and carcinogenesis: Public health application. In Biology of Cancer (2), 13th International Cancer Congress, Part C, Alan R. Liss, Inc., New York, pp. 295-309.

SCE AND CELL CYCLE STUDIES IN LEUKEMIA

R. Becher,[1,2] C. G. Schmidt,[1] and A. A. Sandberg[2]

[1]Innere Universitätsklinik (Tumorforschung)
Westdeutsches Tumorzentrum
Hufelandstr. 55
D-4300 Essen 1, Federal Republic of Germany

INTRODUCTION

During recent years an increasing number of papers has been published on sister chromatid exchanges (SCEs) in neoplasia. These papers can be divided roughly into 2 major groups, i.e., SCEs in the circulating lymphocytes of cancer patients or their relatives and SCEs in affected cells. A considerable number of authors have investigated SCEs in phytohemagglutinin (PHA)-stimulated lymphocytes of cancer patients, which is comprehensively discussed by Sandberg et al. (1). Here, we want to report on SCE frequencies and cell cycle-specific patterns of the leukemic cells themselves. As SCE induction is a sensitive parameter of DNA damage, the available data have to be carefully interpreted regarding the effects of previous cytostatic treatment and the length of the therapy-free interval before samples are collected.

After the detection of high levels of SCEs in subjects with Bloom syndrome, interpreted as a parameter of genetic instability, the high incidence of malignant disease in these patients led to the presumption that genetic instability might play a key role in cancer etiology (2). The high rate of chromosomal abnormalities, generally also found in malignant cells, was frequently associated with genetic instability. Consequently, many authors suggested a close relationship, not only between high SCE levels and the susceptibility to cancer, but also between the origin of chromosomal abnormalities and the induction of SCE frequency (3).

---

[2]Roswell Park Memorial Institute, 666 Elm Street, Buffalo, New York.

The finding of low SCE frequency in leukemic cells indicated that reality was different in regard to the relationship between SCEs, genetic stability, and chromosomal aberrations, and it clarified some of the misinterpretations of the past.

As metaphases show a typical sister chromatid differentiation (SCD) pattern with respect to the number of S phases during which bromodeoxyuridine (BrdUrd) is incorporated, a cell cycle-specific pattern can be established which allows an easy estimation of the rate of cell proliferation. It has been suggested that the length of the cell cycle may influence final SCE rates. Therefore, we have also included in this presentation cell cycle studies, particularly in leukemic cells.

EVALUATION CRITERIA

It is well established that chromosomes of leukemic cells frequently display a higher degree of contraction than those of normal lymphocytes after identical culture and preparation procedures. We have recently shown (4) that evaluation procedures avoiding SCEs in the centromere might bias results. Centromeric and pericentromeric SCE events, which can be easily distinguished in long chromosomes, may appear as centromeric ones only in chromosomes of a more contracted morphology. In our study we differentiated between centromeric and noncentromeric SCEs and related our findings to the extent of chromosomal contraction. Surprisingly, the percentage of centromeric exchanges appeared to increase with the contraction state of the chromosomes. Obviously, as noted before, exchanges close to the centromere cannot be differentiated from centromeric ones in shorter chromosomes, resulting in a spuriously higher percentage of exchanges appearing to occur at the centromere.

Because of this finding, we propose, in agreement with Tice et al. (5) but for different reasons, that centromeric exchanges should be counted if chromosome material of different nature and contraction grade is studied for SCE frequency.

The nomenclature utilized for the evaluation of cell cycle-specific patterns (CCSP) after SCD, according to the number of cell cycles, was adopted from a previous publication (6).

MO/1 Metaphases

This group of cells is able to incorporate BrdUrd during a single S phase. Metaphases without any incorporation of BrdUrd (MO metaphases) cannot be clearly differentiated from M1 metaphases and consequently also appear in this category, which can therefore be named MO/1.

M2 Metaphases

This group contains cells which incorporated BrdUrd during 2 S phases and display a typical differential staining of sister chromatids. M2 metaphases are used for the evaluation of SCE frequency.

M3 Metaphases

These metaphases incorporated BrdUrd during 3 S phases and contain chromosomes with double-stranded, BrdUrd-substituted DNA in both sister chromatids. Cells incorporating BrdUrd for more than 3 cell cycles were not observed in cultures exposed to BrdUrd for 48 to 50 hr, the normal culture duration used for leukemic cells. (See Fig. 1 for M1 and M2 metaphases.)

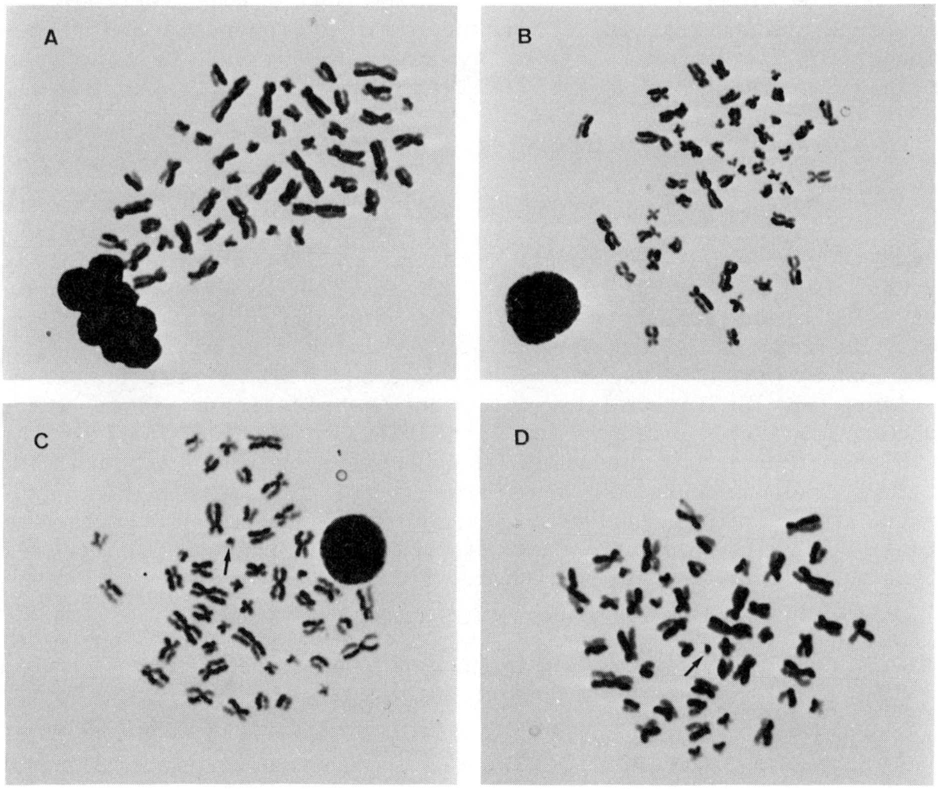

Fig. 1. M2 (A) and M3 (B) metaphases of normal bone marrow. C and D display Ph^1-positive M2 metaphases.

SCEs IN NORMAL BONE MARROW CELLS

Knowledge of SCE frequencies in normal bone marrow cells is a prerequisite for the interpretation of studies in leukemia. Data obtained on mitogen-stimulated lymphocytes are not suitable for comparison with those of leukemic cells, as they are differentiated cells which divide only after stimulation by a mitogen. These cells could be considered only as control populations in lymphocytic leukemias.

Normal bone marrow cells in our study were obtained from donors for bone marrow transplantation. In comparing SCE frequencies of PHA-stimulated lymphocytes with those of normal bone marrow cells, we found a significant decrease in the SCE frequency in unstimulated marrow cells, as shown in Tab. 1 (7). Our first impression (8) has now been confirmed by a larger number of analyses. Apart from a lower SCE rate, bone marrow metaphases are also characterized by a more contracted chromosomal morphology (9,10).

Another interesting feature is that while the distribution of metaphases with identical SCE numbers is heterogeneous and probably bimodal in lymphocytes, it is typically monomodal in bone marrow cells (7). This finding is likely to be related to the heterogeneity of lymphocytic subpopulations with regard to T and B lymphocytes and their subsets in the circulation.

The data reported on SCE frequency in normal bone marrow are shown in Tab. 2. Differences in SCE frequency might be partly related to various BrdUrd concentrations and laboratory-specific features. Compared to average values reported in the literature on PHA-stimulated lymphocytes (16), the level of SCEs in bone marrow cells is remarkably low. Most of the cases studied by Abe et al. (12) and by Honeycombe et al. (13) were not healthy persons but were patients who suffered from different neoplastic diseases without bone marrow involvement. These conditions might explain the slightly higher rates of spontaneous SCEs observed by these authors. Most of the cases reported by Knuutila et al. (11) were patients with mental retardation, the majority of which had been treated with drugs. Only the control cases reported by Stoll et al. (14) and those studied by us were healthy persons.

Tab. 1. SCE frequencies in lymphocytes and bone marrow of 5 healthy donors grown concomitantly.

SCE $\bar{X} \pm SD$	1	2	3	4	5
bone marrow	2.66 ± 1.65	3.62 ± 1.84	4.18 ± 2.21	4.22 ± 1.94	4.27 ± 1.93
lymphocytes	3.86 ± 2.o7	6.68 ± 3.o8	6.52 ± 2.59	4.4o ± 1.98	6.32 ± 3.22

Tab. 2. SCE frequencies in normal bone marrow reported in the literature.

Author/Year	No. of cases	SCE $\bar{x}$	BrdU concentration (μg/ml)
Knuutila et al., 1978 (11)	8	4.69	8.o
Abe et al., 1979 (12)	8	4.65	1.o
Honeycombe, 1981 (13)	17	5.68	3.o7
Stoll et al., 1982 (14)	1o	4.118	3.o7
Becher et al., 1979 (9)	9	4.16	3.o7
Becher, 1983 (15)	29	3.87	3.o7

SCEs IN PH^1-POSITIVE CHRONIC MYELOCYTIC LEUKEMIA (CML)

For the study of SCEs in neoplasia, it is preferable to study those diseases in which the leukemic or cancerous metaphases can be identified by means of their distinct marker chromosomes. This is the case in Ph^1-positive CML. The Ph^1-chromosome can be found in more than 90% of CML cases (17). Generally, the Ph^1 chromosome originates from a translocation of part of the long arm of chromosome 22 (break point q11) to the terminal long arm of chromosome 9, forming the so-called standard translocation t(9;22). The Ph^1 chromosome can be identified easily without chromosome banding in SCD-stained metaphases. The first investigations on SCEs in Ph^1-positive cells were published by Kakati et al. (19) and by Knuutila et al. (11) in 1978. These studies were based on material from patients treated with different cytostatic drugs. The SCE frequency was rather high in some of the cases, reflecting the interaction of cytostatics with the DNA of leukemic cells. Kakati et al. (19) concluded that the Ph^1-chromosome did not seem to influence SCE rates either in the chronic phase or during the blast crisis of CML. In our own data on untreated patients with Ph^1-positive CML in the chronic phase (9), we found significantly lower SCE rates as compared to those of normal bone marrow cells. Figure 1 shows Ph^1-positive M2 metaphases of untreated patients with CML.

As the mean age of donors of normal bone marrow was lower than that of CML patients, we performed a regression analysis of our results in order to reveal a possible relationship between SCE frequency and age. Such a relationship was not found either in normal bone marrow or in Ph^1-positive cells (9).

Another question was whether the decrease of SCE in leukemic cells was related to specific groups of chromosomes. Therefore, our evaluation included an analysis of the chromosome groups involved in SCE events. The final analysis showed that all chromosome groups were equally involved (9). Shortly after our publication of the results in untreated cases, Abe et al. (19) reported on SCE frequencies in 17 CML patients in the chronic as well as in the active phase of the disease and in 3 cases in blast crisis. As all patients were pretreated with different chemotherapy regimens, conclusions regarding spontaneous SCE rates in CML were not possible. In 1981, Honeycombe (20) reported on the effects of busulfan treatment on SCEs in Ph^1-positive cells and added spontaneous SCE rates of 11 untreated patients to this study. Spontaneous SCE rates in Ph^1-positive cells were significantly lower than in PHA-stimulated lymphocytes (20). Stoll et al. (14) recently published data on SCEs in untreated Ph^1-positive CML and compared them with those of normal bone marrow. These authors confirmed our findings of low SCE rates in Ph^1-positive cells (9). Stoll et al. (14) investigated 16 cases and found mean SCE values in the range of 2.16 to 4.04 per metaphase. The mean value in the whole group was 2.97 SCE per metaphase, as compared to 4.12 SCE per metaphase in normal bone marrow, a frequency based on 10 cases. SCE rates in Ph^1-positive CML published up to now are summarized in Tab. 3.

Stoll et al. (14) also reported on SCEs in untreated blast crisis of CML. It was surprising that SCE rates were even more decreased. Data on untreated blast crises of CML are difficult to obtain, as most patients receive cytostatic treatment during the accelerated phase. In the 16 cases investigated, extremely low SCE levels, ranging from 1.16 to 2.32 per metaphase, were found. This finding is of interest as additional abnormalities appear in more than 60% of the CML cases at the onset of blast crisis (22). Obviously, these karyotypic changes do not indicate instability of the genome, at least not a form of instability which can be monitored by SCEs. The gain of numerical and/or structural chromosomal aberrations may, thus, be more a reflection of an active ability of neoplastic cells to increase the amount of specific genetic material capable of providing growth advantages to the cells (23).

Tab. 3. SCE frequency in Ph^1-positive CML during the chronic phase of the disease.

Author	No. of cases	Status	SCE range	SCE ($\bar{x}$)
Kakati et al. (1978)	8	treated	3.4-6.4	4.4o
Knuutila et al. (1978)	3	treated	3.6-4.o	3.89
Becher et al. (1979)	12	untreated	2.32-3.43	3.o3
Abe et al. (198o)	17	untreated	2.38-1o.62	5.11
Honeycombe (1981)	11 3	untreated treated	4.17-5.73 3.63-11.o3	- -
Stoll et al. (1982)	16	untreated	2.16-4.o4	2.97
Becher (21)	37	untreated	2.32-5.72	3.37

ACUTE LEUKEMIAS

In contrast to CML, acute leukemias represent a rather heterogeneous group of diseases. Roughly, myelocytic and lymphocytic leukemias and erythroleukemias can be differentiated. A further subclassification is possible, using cytogenetic findings; it is based on the presence of specific chromosomal abnormalities. One group of leukemias without chromosomal abnormalities (NN) can be differentiated from another one characterized by normal and abnormal karyotypes (AN) and from a third one in which only abnormal cells (AA) are present. A more detailed classification utilizes specific chromosomal aberrations (24).

Overall, "the" acute leukemia does not exist and, thus, it can be presumed that the SCE data would also be less homogeneous than they have been shown to be in the Ph^1-positive CML.

Acute Myelocytic Leukemia

The first SCE analysis in acute myelocytic leukemia (AML) was published by Knuutila et al. (11) in 1978. In 2 cases, material was used from relapsing leukemias and 2 others were investigated during remission. A karyotypic description of the leukemic cells was not included. The SCE frequency was in the range of 4.4 to 6.8 SCE per metaphase. Three of their cases were intensively pretreated and no information was given on the fourth one.

In 1979, Abe et al. (12) published a comprehensive SCE analysis on 11 cases of AML which were repeatedly studied as many as 10 times during overt leukemia. Six other patients were already in remission when the SCE analyses were performed, excluding the presence of leukemic cells at the time of analysis. Six of the 11 leukemic patients were studied before cytostatic therapy was started. It was noted by these authors that the SCE frequency before chemotherapy was signficantly lower than that observed after chemotherapy. In the following year, the same group (25) published a second paper on SCEs in nonlymphocytic leukemia, citing additional cases. In this new study they investigated 10 leukemias classified as AN cases, in which a leukemic aneuploid popluation of cells coexists with those remaining from normal myelopoiesis. With one exception, all patients were pretreated. The authors found that the SCE frequency in aneuploid leukemic cells was lower than that in the normal marrow cells.

In 1981 we published our findings on 7 untreated myeloid leukemias and 1 undifferentiated leukemia (10). Our study was based on euploid as well as aneuploid leukemias. As compared to normal bone marrow cells, SCE rates in acute leukemic cells were significantly lower, ranging from 2.11 to 2.97 SCEs per metaphase. Therefore, the low spontaneous SCE rates already described in Ph^1-positive CML were also shown to be a characteristic feature of AML.

Acute Lymphocytic Leukemia

To date, 2 studies have been reported on acute lymphocytic leukemia (ALL). Heerema et al. (26) reported on SCEs in the bone marrow cells of 8 cases of childhood ALL which were studied before treatment was started. These authors found notably higher SCE frequencies in the leukemic cells as compared to control values. These findings are contrary to those described by Abe and Sandberg in 1980 (27). As a matter of fact, in spite of previous cytostatic treatment, the latter authors found lower SCE rates than those reported by Heerema et al. (26). Further studies are necessary to clarify the reason for these conflicting data.

GENERAL PROBLEMS ENCOUNTERED WITH SCE STUDIES IN LEUKEMIA

SCE induction has been shown to be a very sensitive parameter which can be influenced by a variety of factors. The degree of chromosomal spiralization and its impact on the evaluation procedure with regard to centromeric exchanges has already been mentioned (4). In addition, the age of the patients has to be considered. Therefore, a regression analysis should be made in all evaluations in order to exclude bias due to age (8).

Another important point is that laboratory-specific features and individual handling of material are able to influence SCE frequencies. Therefore, each laboratory should use its own control values in order to prevent misinterpretations.

COMPARISON OF SCE FREQUENCIES IN NORMAL AND LEUKEMIC BONE MARROW

It has been shown that SCE frequencies are significantly lower in normal bone marrow cells as compared to those of normal lymphocytes (7,13). All studies so far reported on untreated cases of Ph^1-positive CML have shown that there is a further decrease in SCE frequency from normal bone marrow to Ph^1-positive cells (9,14). The clonal evolution of Ph^1-positive cells at the onset of the blast crisis seems to induce a further drop in SCE frequency (14), very similar to reports on SCEs in acute leukemia. The only exception to this relationship is in ALL, for which the data are still conflicting. A graphic illustration of SCE rates reported by us shows a progressive decrease in SCE frequency from lymphocytes to normal bone marrow cells to CML and, finally, to the acute leukemias (Fig. 2). This observation raises the suspicion that spontaneous SCEs could be related to cellular differentiation stages, i.e., the SCE frequency seems to drop as cells are less differentiated. Leukemic blasts, which are most closely related to bone marrow stem cells, reveal the lowest SCE frequency ever observed in human cells. Cairns (28) suggested that the SCE frequency in stem cells might be extremely low, as each DNA strand exchange involved in SCE formation

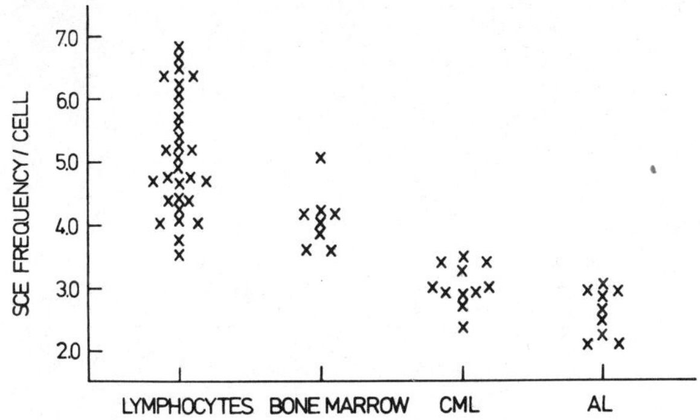

Fig. 2. SCE frequency in lymphocytes, normal bone marrow cells, Ph^1-positive CML, and AML (6,8,9,56). A significant decrease of SCE frequency could be found as cells become less differentiated.

could easily lead to misrepair and, consequently, to point mutations, thus challenging genetic stability. Therefore, low SCE frequencies in acute leukemia might be simply an expression of cellular immaturity due to the fact that these cells are more closely related to stem cells as compared to normally differentiated bone marrow cells. Chromosomal aberrations appearing during clonal evolution of Ph^1-positive CML, therefore, do not seem to indicate genetic instability and must, more probably, be interpreted as an active process of malignant cells. Similarly, leukemic cells express cell surface markers, which are normal in an early ontogenetic phase and are not primarily related to their leukemic character. Therefore, low SCE frequencies in leukemic cells might be a feature generally associated with cells which are closely related to bone marrow stem cells.

CELL CYCLE STUDIES BY MEANS OF SCD

As metaphases display a typical pattern with regard to the number of S phases during which BrdUrd was incorporated into the DNA, metaphases can be discriminated as to the number of cell cycles completed (M1, M2, and M3). This technique allows a unicellular measurement of cell cycle division. An important advantage of this method, as compared to autoradiography, is the fact that neoplastic metaphases can be rigorously identified by their chromosomal aberrations, which allow an analysis of heteroclonal cell populations. Such a differentiation is easily possible in Ph^1-positive CML with residual normal myelopoiesis in the bone marrow. Until recently, autoradiography was the only method allowing cell cycle studies by means of the incorporation of ^3H-thymidine into the DNA during the S phase. The incorporation of ^3H-thymidine, however, is liable to interfere with cell division (30-34).

Lymphocytes

Concomitant cultures of PHA-stimulated lymphocytes and normal bone marrow cells of donors regularly revealed a higher percentage of M2 and M3 metaphases in normal bone marrow cells (7). At first glance, therefore, it could be concluded that lymphocytes proliferate at a significantly slower rate than marrow cells. Yet, according to investigations of Bender and Prescott (38), Michalowski (39), and Cooper et al. (40), lymphocytes need up to 24 hr after PHA stimulation before they start measurable DNA synthesis. Crossen and Morgan (41) calculated a mean cell cycle time of 12 hr for PHA-stimulated lymphocytes, which is in agreement with our observations (7). The individual variability in the pattern of M1, M2, and M3 metaphases is rather high. The heterogeneity of PHA-stimulated lymphocytes might be responsible for this finding, as PHA, in addition to stimulating T lymphocytes, may stimulate a small percentage of B lymphocytes. Santesson et al. (42) found different SCE frequencies in T and B lymphocytes. The more slowly proliferating T lymphocytes seemed to express higher SCE rates.

Normal Bone Marrow Cells

Normal bone marrow cells grow in vitro without stimulation by mitogens. Due to this fact, the cell cycle time can be calculated over the whole culture period. The majority of metaphases is able to divide twice during a culture time of 50 hr (9,14). Up to 20% of cells pass through 3 cell cycles. A maximum cell cycle time of 16 hr can be calculated for M3 metaphases. This fast rate of cell division has so far been reported only for myeloblasts (43,44) and for proerythroblasts using autoradiographic methods (45). Unfortunately, no discrimination can be made between erythroid and myeloid bone marrow elements. Thus, we are not able to establish whether more erythroid or myeloid precursors are found among the fast or slowly proliferating cells, respectively. The first cell cycle data using SCD published by us in 1981 (10) were recently confirmed in an investigation of Stoll et al. (14).

Cell Cycle Studies in Leukemia

An important problem in cell cycle studies using autoradiographic methods is that many leukemias present a heteroclonal system, with normal cells frequently coexisting with the leukemic ones. Cytogenetic analysis allows a discrimination of both populations by chromosomal markers. Our cell cycle studies were obtained from untreated Ph^1-positive CML (6) and acute leukemia (10) cases using normal bone marrow cells as controls. The data discussed here are based on cell cycle studies in untreated leukemias, as cell cycle division is severely disturbed by chemotherapy depending on the interval from its application, as clearly shown by Abe et al. (1979). In our study of untreated cases (6,10), cell cycle-specific patterns in CML and acute leukemia were very similar, though a high individual variability could be found in each of the groups. CML and acute leukemias were characterized by a relative increase in the frequency of slowly growing cells (M1 metaphases). In CML, this finding corresponds quantitatively to the findings of Killman et al. (46), Ogawa (47), and Gavosto (48), utilizing autoradiographic methods. As red cell precursors (45) have a rather short generation time and erythropoiesis is reduced in CML, the finding of long cell cycle times in CML could partly be related to the deficiency of red cell precursors--a fact which could also apply to acute leukemia (49).

Figure 3 shows a summary of our cell cycle studies based on 20 normal bone marrows, 21 cases of Ph^1-positive CML and 15 cases of acute leukemia (50). Stoll et al. (14) also published cell cycle studies in untreated CML and confirmed our findings. In this study, in addition to CML, the blast crisis was also studied, revealing a further prolongation of cell cycle duration.

SCE AND CELL PROLIFERATION

Even though the biological meaning of SCEs is still unknown, a close relationship is suggested between increased levels of SCE and effects of DNA-damaging agents (51,52).

It was suggested that the period of time during which cells are able to repair DNA damage might be responsible for, or at least have an important bearing on, the final SCE frequency. In fact, Gibas and Limon (53) found different SCE frequencies in lymphocytes treated with mutagens depending on whether they passed 2 or 3 cell cycles during a given culture time. The spontaneous SCE levels

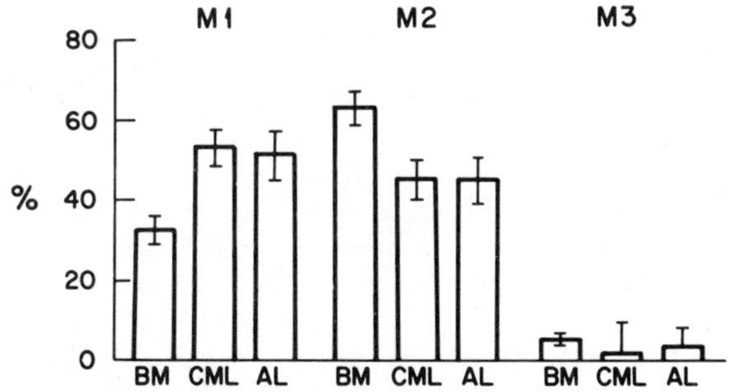

Fig. 3. Percentage of M1, M2, and M3 metaphases in normal bone marrow, Ph^1-positive CML and acute leukemia. The number of M1 metaphases was significantly higher in leukemias. The percentage of M2 metaphases, however, decreased in leukemias, indicating a longer cell cycle for the neoplastic cells.

(54), however, were shown to be identical in M2 and M3 metaphases, suggesting that the cell cycle time for the measurement of spontaneous SCE might be of little significance. Thus, longer cell cycle alone does not seem to be the only cause of low SCE frequencies in leukemic cells.

CONCLUSIONS

The finding of significantly lower SCE rates and slower cell proliferation in leukemic cells, as compared to those of non-leukemic cells, might be used as an indicator of the neoplastic nature of bone marrow cells. In acute leukemia, so far, only 60% of cases have shown clonal abnormalities. Even though Yunis et al. (55) challenged this dogma using high-resolution banding, their proposal so far has to be confirmed by others and in a larger series of cases. Thus, in cases without chromosomal abnormalities, low SCE rates and long cell cycle times might be regarded as indicators of the leukemic nature of the cells. The majority of leukemic cells, even after 48 hr, are not able to complete 2 cell cycles.

REFERENCES

1. Sandberg, A.A., R. Becher, and Z. Gibas (1984) Value and significance of SCE in human leukemia and cancer. In Sister Chromatid Exchange: 25 Years of Experimental Research, R.R. Tice and A. Hollaender, eds. Plenum Press, New York.
2. Ray, J.H., and J. German (1981) The chromosome changes in Bloom's syndrome, ataxia-telangiectasia and Fanconi's anemia. In Genes, Chromosomes and Neoplasia, F.E. Arrighi, P.N. Rao, and E. Stubblefield, eds. Raven Press, New York.
3. Preisler, H.D., and A. Raza (1982) Chronic myelocytic leukemia: Comments on new approaches to therapy. Cancer Treat. Rep. 66:1073-1076.
4. Becher, R., and A.A. Sandberg (1983) Sister chromatid exchange in the centromere and centromeric area. Human Genet. 63:358-361.
5. Tice, R., J. Chaillet, and E.L. Schneider (1975) Evidence derived from sister chromatid exchanges of restricted rejoining of chromatid subunits. Nature (Lond.) 256:642-644.
6. Becher, R., and C.G. Schmidt (1982) Sister chromatid differentiation and cell-cycle-specific patterns in chronic myelocytic leukemia and normal bone marrow. Int. J. Cancer 29:617-620.
7. Becher, R., and A.A. Sandberg (1984) Sister chromatid exchange levels and cell cycle time in human bone marrow cells and lymphocytes. Cancer Genet. Cytogenet. 11:19-23.
8. Becher, R., C.G. Schmidt, G. Theis, and D.K. Hossfeld (1979) The rate of sister chromatid exchanges in normal human bone marrow cells. Human Genet. 50:213-216.

9. Becher, R., C.G. Schmidt, G. Theis, and D.K. Hossfeld (1979) Sister chromatid exchange in Ph[1]-positive chronic myelocytic leukemia. Int. J. Cancer 24:713-716.
10. Becher, R., G. Zimmer, and C.G. Schmidt (1981) Sister chromatid exchange and growth kinetics in untreated acute leukemia. Int. J. Cancer 27:199-204.
11. Knuutila, S., E. Helminen, P. Vuopio, and A. de la Chapelle (1978) Sister chromatid exchanges in human bone marrow cells. Hereditas 88:189-196.
12. Abe, S., S. Kakati, and A.A. Sandberg (1979) Growth rate and sister chromatid exchange (SCE) incidence of bone marrow cells in acute myeloblastic leukemia (AML). Cancer Genet. Cytogenet. 1:115-130.
13. Honeycombe, J.R. (1981) Spontaneous and busulphan induced sister chromatid exchange (SCE) frequencies in haematologically normal human bone marrow and lymphocytes. Mutat. Res. 84:399-407.
14. Stoll, C., F. Oberling, and M.-P. Roth (1982) Sister chromatid exchange and growth kinetics in chronic myeloid leukemia. Cancer Res. 42:3240-3243.
15. Becher, R. (unpublished data).
16. Crossen, P.E. (1982) SCE in lymphocytes. In Sister Chromatid Exchange, A.A. Sandberg, ed. Alan R. Liss, Inc., New York, pp. 175-193.
17. Sandberg, A.A. (1980) The cytogenetics of chronic myelocytic leukemia (CML): Chronic phase and blastic crisis. Cancer Genet. Cytogenet. 1:217-228.
18. Kakati, S., S. Abe, and A.A. Sandberg (1978) Sister chromatid exchange in Philadelphia chromosome (Ph[1])-positive leukemia. Cancer Res. 38:2918-2921.
19. Abe, S., S. Kakati, and A.A. Sandberg (1980) Growth pattern and SCE incidence in Ph[1]-positive cells of CML. Cytobios 28:137-149.
20. Honeycombe, J.R. (1981) The cytogenetic effects of busulphan therapy on the Ph[1]-positive cells and lymphocytes from patients with chronic myeloid leukemia. Mutat. Res. 81:81-102.
21. Becher, R. (unpublished data).
22. Sandberg, A.A. (1980) The Chromosomes in Human Cancer and Leukemia. Elsevier North-Holland, Inc., New York.
23. Becher, R. (1984) Chromosomale Befunde und Chemotherapieresistenz. Regensburger Symposium, 9-11, June, 1983, Bd. 18. Contributions to Oncology, Vol. 18. Kasper, Basel, München, Paris, London, New York, Tokyo, Sydney (in press).
24. Sandberg, A.A. (1983) Chromosomes in human neoplasia. In Current Problems in Cancer, R.C. Hickey, ed. Vol. 8, No. 2, pp. 1-52. Year Book Medical Publishers, Inc., Chicago-London.
25. Abe, S., and A.A. Sandberg (1980) Sister chromatid exchange and growth kinetics of marrow cells in aneuploid acute nonlymphocytic leukemias. Cancer Res. 40:1292-1299.
26. Heerema, N.A., C.G. Palmer, and R.L. Baehner (1982) Elevated

sister chromatid exchange and cell cycle analysis in bone marrow in childhood ALL. Cancer Genet. Cytogenet. 6:323-330.
27. Abe, S., and A.A. Sandberg (1980) Growth pattern and sister chromatid exchanges of bone marrow cells in acute lymphoblastic leukemia. Cytobios 29:165-173.
28. Cairns, J. (1975) Mutation selection and the natural history of cancer. Nature (Lond.) 255:197-200.
29. Becher, R. (unpublished data).
30. Drew, R.M., and R.B. Painter (1959) Action of tritiated thymidine on the clonal growth of mammalian cells. Radiat. Res. 11:535-544.
31. Cleaver, J.E. (1967) Thymidine Metabolism and Cell Kinetics. Frontiers of Biology, Vol. 6, A. Neuberger and E.L. Tatum, eds. North-Holland, Amsterdam. Wiley Interscience Division, New York.
32. Ehmann, U.K., J.R. Williams, W.A. Nagle, J.A. Brown, J.A. Belli, and J.T. Lett (1975) Perturbations in cell cycle progression from radioactive DNA precursors. Nature (Lond.) 258: 633-636.
33. Marz, R. J.M. Zylka, P.G.W. Plagemann, J. Erbe, R. Howard, and J.R. Sheppard (1976) G_2 + M arrest of cultured mammalian cells after incorporation of tritium labeled nucleosides. J. Cell. Physiol. 90:1-8.
34. Pollack, A., C.B. Bagwell, and G.L. Irvin (1979) Radiation from tritiated thymidine perturbs the cell cycle progression of stimulated lymphocytes. Science 203:1025-1027.
35. Morimoto, K., and S. Wolff (1980) Cell cycle kinetics in human lymphocyte cultures. Nature (Lond.) 288:604-606.
36. Tice, R., E.L. Schneider, and J.M. Rary (1976) The utilization of bromodeoxyuridine incorporation into DNA for the analysis of cellular kinetics. Exp. Cell Res. 102:232-236.
37. Schneider, E.L., H. Sternberg, and R.R. Tice (1977) In vivo analysis of cellular replication. Proc. Natl. Acad. Sci., USA 74:2041-2044.
38. Bender, M.A., and D.M. Prescott (1962) DNA synthesis and mitosis in cultures of human peripheral leukocytes. Exp. Cell Res. 27:221-229.
39. Michalowsky, A. (1963) Time course of DNA synthesis in human leukocyte cultures. Exp. Cell Res. 32:609-612.
40. Cooper, E.H., and P. Barkhan (1963) Observations on the proliferation of human leukocytes cultured with phytohaemagglutinin. Brit. J. Haemat. 9:101-111.
41. Crossen, P.E., and W.F. Morgan (1977) Analysis of human lymphocyte cell cycle time in culture measured by sister chromatid differential staining. Exp. Cell Res. 104:453-457.
42. Santesson, B., K. Lindahl-Kiessling, and A. Mattsson (1979) SCE in B and T lymphocytes. Possible implications for Bloom's syndrome. Clin. Genet. 16:133-135.
43. Cronkite, E.P., and P.C. Vincent (1970) Granulocytopoiesis. In Symposium on Hemopoietic Cellular Proliferation, Boston, Nov. 5-6, 1969, F. Stohlman, Jr., ed. Grune & Stratton, New York, pp. 211-228.

44. Boll, I.T., and G. Fuchs (1970) A kinetic model of granulocytopoiesis. Exp. Cell Res. 61:147-152.
45. Dörmer, P. (1973) Kinetics of erythropoietic cell proliferation in normal and anemic man. A new approach using quantitative ^{14}C-autoradiography. In Progress in Histochemistry and Cytochemistry, W. Graumann, Z. Lojda, A.G.E. Pearse, and T.H. Schiebler, eds. Fischer, Stuttgart, Vol. 6, No. 1, pp. 1-83.
46. Killmann, S.A., E.P. Cronkite, J.S. Robertson, T.M. Fliedner, and V.P. Bond (1963) Estimation of phases of the life cycle of leukemic cells from labeling human beings in vivo with tritiated thymidine. Lab Invest. 12:671-684.
47. Ogawa, M., J. Fried, Y. Sakai, A. Strife, and B.D. Clarkson (1970) Studies of cellular proliferation in human leukemia. VI. The proliferative activity, generation time, and emergence time of neutrophilic granulocytes in chronic granulocytic leukemia. Cancer 25:1031-1049.
48. Gavosto, F. (1974) Granulopoiesis and cell kinetics in chronic myeloid leukemia. Cell Tissue Kinet. 7:151-163.
49. Vincent, P.C. (1974) Cell kinetics of the leukemias. In Leukemia, F. Gunz and A.G. Baikie, eds. Grune & Stratton, New York, 3rd ed., pp. 189-221.
50. Becher, R. (unpublished data).
51. Kato, H. (1977) Mechanisms for sister chromatid exchanges and their relations to the production of chromosomal aberrations. Chromosoma 59:179-191.
52. Wolff, S. (1977) Sister chromatid exchange. Ann. Rev. Genet. 11:183-201.
53. Gibas, Z., and J. Limon (1979) The induction of sister-chromatid exchanges by 9-aminoacridine derivatives. I. The relation between the yield of SCE induction and cell kinetics in cultured human lymphocytes. Mutat. Res. 67:93-96.
54. Giulotto, E., A. Mottura, R. Giogi, L. De Carli, and F. Nuzzo (1980) Frequencies of sister chromatid exchanges in relation to cell kinetics in lymphocyte cultures. Mutat. Res. 70:343-350.
55. Yunis, J.J., C.D. Bloomfield, and K. Ensrud (1981) All patients with acute nonlymphocytic leukemia may have a chromosomal defect. New Engl. J. Med. 305:135-139.
56. Becher, R., G. Zimmer, C.G. Schmidt, and A.A. Sandberg (1983) Sister chromatid exchange and proliferation pattern after ultrasound exposure in vivo. Am. J. Human Genet. 35:932-937.

SISTER CHROMATID EXCHANGE IN PHYTOHEMAGGLUTININ-STIMULATED LYMPHOCYTES OF NONFAMILIAL CUTANEOUS MALIGNANT MELANOMA PATIENTS

A. Ghidoni,[1] E. Privitera,[1] E. Raimondi,[1]
D. Rovini,[2] M.T. Illeni,[2] and N. Cascinelli[2]

[1]Università di Milano
Sezione di Genetica e Microbiologia
Dipartimento di Biologia
Milan, Italy

The relationship between sister chromatid exchange (SCE) and DNA damage or carcinogenesis has been the subject of numerous investigations (1-6). Altered frequencies of SCEs have been reported in cancer patients (7-13), but their meaning with regard to cancer development has not yet been elucidated.

SCEs in phytohemagglutinin (PHA)-stimulated peripheral blood lymphocytes (normal cells) were examined in 36 nonfamilial cutaneous malignant melanoma (CMM) patients. In 27 cases, a close relative (1st or 2nd degree) was also examined. All of the patients had undergone surgical treatment 2 mo to 1 yr before blood sampling, but had undergone no chemotherapy nor radiotherapy. Cultures were set up with RPMI medium supplemented with 15% calf serum (Labtek), 0.2 mg/ml PHA (M Form, Difco), and 10 µg/ml bromodeoxyuridine (BrdUrd), and kept in the dark at 37°C for 72 hr.

The mean SCE rates were found to be significantly higher among the CMM patients as compared with the group of close relatives or with the group of unrelated subjects. Higher SCE rates were apparently a characteristic of only about one-third of the CMM patients and of a few of their close relatives as well (Tabs. 1 and 2). No effect of the clinical progression of the neoplasm on SCEs was observed. Most patients with relatively higher SCE rates (9 or more SCE/cell) also had an altered (diminished) proliferation pattern of PHA-stimulated lymphocytes (Tab. 3 and Fig. 1).

---

[2]Istituto Nazionale per lo Studio e la Cura dei Tumori, Milan, Italy.

Tab. 1. Mean SCE rates (30 cells per subject) in lymphocytes.

	A) CMM PATIENTS		B) CLOSE RELATIVES OF CMM PATIENTS		C) UNRELATED SUBJECTS	
	SCE/cell (and S.E.)	No. of subjects	SCE/cell (and S.E.)	No. of subjects	SCE/cell (and S.E.)	No. of subjects
D) SMOKERS	9.0 (.2)	14	8.1 (.6)	8	7.0 (.2)	11
E) NON-SMOKERS	7.7 (.2)	22	7.2 (.3)	19	6.8 (.2)	9

Analysis of variance:

A vs. (B+C)	P : <.01
B vs. C	P : <.05
D vs. E	P : <.01
INTERACTION	P : n.s.

Tab. 2. Distribution of subjects in relation to SCE rates.

SCE/cell	A (CMM Patients)	B (close relatives)	C (unrelated subjects)
up to 6	1	3	2
6.1- 7	3	7	7
7.1- 8	12	9	10
8.1- 9	12	4	1
9.1-10	6	3	0
over 10	2	1	0
	36	27	20

χ^2 VALUES

A vs. (B+C)	14.4 (d.f.=5)	P = 0.02 ~ 0.01
B vs. C	5.1 (d.f.=5)	P = n.s.

Tab. 3. Frequency of subjects with 20%, or over, M1 cells.

15/36	CMM PATIENTS
8/10	CMM PATIENTS WITH 9 or more SCE/cell
6/27	CLOSE RELATIVES OF CMM PATIENTS
4/20	UNRELATED SUBJECTS

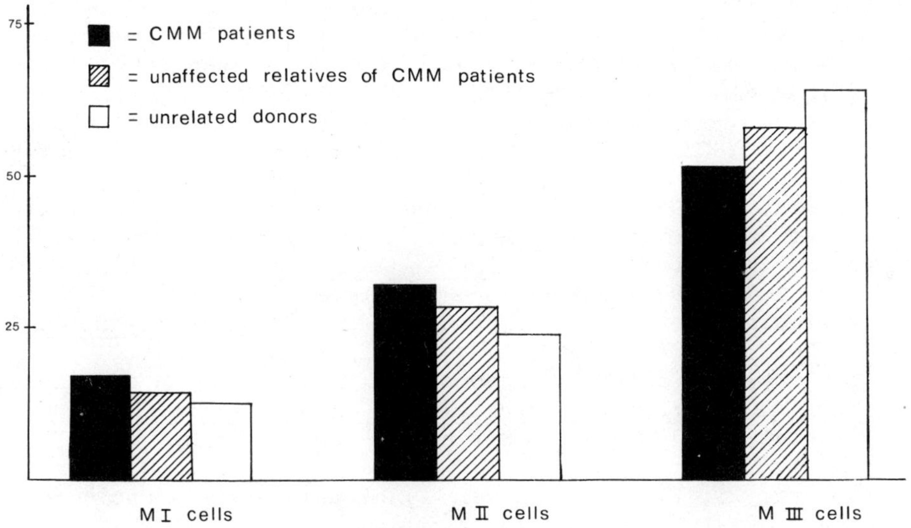

Fig. 1. Relative frequencies of cells in metaphase of 1st, 2nd, and 3rd division, [M1 (MI), M2 (MII), M3 (MIII)], respectively. Lymphocytes were stimulated with PHA and cultures were arrested at 72 hr.

These observations indicate, similar to previous ones on familial malignant melanoma (FMM) patients (13), that SCEs may have an unknown relationship, in some CMM patients, with the mechanism of onset of this neoplasy.

REFERENCES

1. Latt, S.A. (1974) Sister chromatid exchanges, indices of human chromosome damage and repair: Detection of fluorescence and induction by mitomycin C. Proc. Natl. Acad. Sci., USA 71:3162-3166.
2. Perry, P., and H.J. Evans (1975) Cytological detection of mutagen-carcinogen exposure by sister chromatid exchange. Nature (Lond.) 258:121-124.
3. Stetka, D.G., and S. Wolff (1976) Sister chromatid exchange as an assay for genetic damage induced by mutagen-carcinogens. Mutat. Res. 41:333-342.
4. Carrano, A.V., L.H. Thompson, P.A. Lindl, and J.L. Minkler (1978) Sister chromatid exchange as an indicator of mutagenesis. Nature (Lond.) 271:551-553.
5. Kinsella, A.R., and M. Radman (1978) Tumor promoter induces sister chromatid exchanges: Relevance to mechanisms of carcinogens. Proc. Natl. Acad. Sci., USA 75:6149-6153.
6. Bradley, M.O., I.C. Hsu, and C.C. Harris (1979) Relationship between sister chromatid exchange and mutagenicity, toxicity and DNA damage. Nature (Lond.) 282:318-320.
7. Kakati, S., S. Abe, and A.A. Sandberg (1978) Sister chromatid exchange (SCE) in Philadelphia-positive leukemia. Cancer Res. 38:2918-2921.
8. Heerema, N.A., C.P. Palmer, and R.L. Baehner (1982) Elevated sister chromatid exchange and cell cycle analysis in bone marrow in childhood ALL. Cancer Genet. Cytogenet. 6:323-330.
9. Chen, T.R. (1981) Frequencies of sister chromatid exchanges in heteroploid cell lines of human melanoma origin. J. Natl. Cancer Inst. 66:273-277.
10. Cervenka, J. (1981) Chromosomal changes associated with neoplasia. In Results and Problems in Cell Differentiation, R.G. McKinnell et al., eds. Springer Verlag, Berlin, Vol. 11, pp. 93-101.
11. Lambert, B., U. Ringborg, and A. Lindblad (1979) Prolonged increase of sister chromatid exchanges in lymphocytes of melanoma patients after CCNU treatment. Mutat. Res. 59:295-300.
12. Turleau, C., M.O. Cabanis, and J. de Grouchy (1980) Augmentation des échanges de chromatides dans les fibroblastes d'un enfant atteint de del(13)-rétinoblastome. Ann. Génét. 23:169-170.
13. Ghidoni, A., E. Privitera, E. Raimondi, D. Rovini, M.T. Illeni, and N. Cascinelli (1983) Malignant melanoma: Sister chromatid exchange analysis in three families. Cancer Genet. Cytogenet. 9:347-354.

SCE INDUCTION BY CYTOSTATICS AND ITS RELATION TO IATROGENIC LEUKEMOGENESIS

Tibor Raposa

Department of Medicine
Semmelweis University Medical School
1121 Budapest, Eötvös u. 12, Hungary

INTRODUCTION

There is considerable information currently available about the genotoxic effect, as measured by an alteration in the frequency of sister chromatid exchanges (SCEs), of various environmental chemical compounds (1,13). Evidence is also accumulating on the role of these compounds in carcinogenesis (10) and their ability to induce cytogenetic changes [i.e., chromosomal aberrations (CAs) and SCEs (1,22)]. Considerable information has also been acquired in the past 10 years about CAs in malignancies, especially hemoblastoses, and there seems to be a close relationship between the location of the break points in specific chromosomal rearrangements and cellular oncogen localization (20). This relationship seems to provide further support to the importance of CA studies in detecting carcinogens, although the role of CAs in tumors and the relationship between the nature of CAs induced by carcinogens and the ones observed in cancer cells remains to be explored. The same lack of understanding, as far as the process of carcinogenesis is concerned, holds true for the relevance of SCEs. Malignant process per se seems not to have an influence on the baseline frequency of SCEs (18), although it is well known that many environmental carcinogens induce SCEs (1) at much lower concentrations of the compounds than is needed for the induction of CAs (16). These marked differences between SCEs and CAs (i.e., their occurrence in malignancies and circumstances of inducibility) underline the fact that SCEs and CAs are different cellular phenomena as far as carcinogenesis and the detection of the genotoxic effects of various compounds are concerned. However, before making any decision as to the relative importance of these two cytogenetical endpoints for mutagenesis and carcinogenesis, one has to realize that most of the studies on CAs have been performed on the tumor tissue, while the information on SCEs in the target tumor tissue is, as yet, very limited (7,17,18).

What then is the meaning of SCEs in regard to the process of carcinogenesis? What do SCEs mean in regard to the genotoxicity versus the carcinogenicity of a compound? Until we know the exact mechanism(s) involved in the formation of SCEs, the nature of the cellular lesions involved in the induction of SCEs and of those leading to malignancies as well as their relationship to each other, these questions remain to be answered.

Comparative genotoxicity, mutagenicity, and carcinogenicity studies can be made, however. By comparing data from the epidemiology of malignancies in which the role of environmental chemical compounds is suspected with that obtained from SCE studies involving the same compounds, a step towards understanding more about the possible role of SCE induction in carcinogenesis may be possible. To perform such comparative mutagenesis/carcinogenesis/genotoxicity studies is all the more important since the relationship between the different cytogenetic endpoints such as CAs and SCEs, as well as the role of various cytogenetic changes in malignancies and their usefulness for detecting mutagenic and carcinogenic compounds is a matter of much debate and awaits further elucidation. An appropriate comparative study will be faced immediately with at least two important obstacles:

i) the validity of the comparison of the results of a simple test such as SCE analysis in comparison with the complexity of the biological event involved in a process like carcinogenesis, and

ii) the exposure of individuals to a mixture of suspected mutagens and carcinogens, most of which are unknown.

If we want to make experiments of this kind relevant to man, the following requirements have to be met:

i) it should be a human model;

ii) it should involve known mutagenic-carcinogenic compounds;

iii) it should provide extensive data on the SCE-inducing characteristics of the suspected mutagenic-carcinogenic compounds in man.

Based on these prerequisites, a selected but important range of malignancies associated with exposure to chemical carcinogens (i.e., the malignancies following medical treatment with cytostatics and irradiation) seems to be a very reasonable model for this purpose. From a practical point of view it would be advantageous to take into consideration only acute leukemias (AL) following the treatment of solid tumors and lymphomas, as well as immunopathological disorders with cytotoxic agents (i.e., the so-called therapy-related AL). The peculiar advantage of this system is that the cytological subtype classification of AL is reliably settled (2). Furthermore, it provides an opportunity for studying the cytological subtypes of AL in relationship to the type of inducing carcinogens. The following reasons also prompted us to use the therapy-related leukemias as a model system for attempting to assess the relationship between SCEs and carcinogenesis:

i) the accumulation of therapy-related AL during the past 10 years (Tab. 1).

ii) the extensive studies and knowledge on the SCE-inducing characteristics of different cytostatic agents used for the treatment of our patients and implicated in the therapy of the primary diseases later followed by an AL.

The purpose of this paper is to determine further the predictive value of the SCE test for the carcinogenic or noncarcinogenic activity of cytostatics. The following questions were asked:

(a) To what extent is the inducibility of SCEs in man by various cytostatics related to their leukemogenic activity?

(b) Which are the most hazardous cytostatics from the point of view of iatrogenic leukemogenesis?

(c) How can the results of SCE studies on cytostatics be utilized in planning new therapeutical protocols and for testing new cytostatics with a lower carcinogenic activity?

To attempt to provide an answer to these questions, we present our data on:

(1) SCE frequencies in the phytohemagglutinin (PHA)-stimulated lymphocytes of patients receiving cytostatic therapy.

Tab. 1. Rate of increase of therapy-related acute leukemias during the period 1930-1980.

	HD	NHL	MM	ST	IP	
1930-70	41	4	1	28	2	total AL cases
1970-80	194	80	123	181	103	
1930-70	1,0	0,09	0,024	0,68	0,048	AL cases/year
1970-80	17,63	7,27	11,18	13,9	9,39	
rate of increase	17,63	80,77	465	20,44	195	

(2) Details of the leukemogenic activity of different cytostatics based on an extensive literature survey of induced AL during the period 1930-1980.

(3) A comparison of the genotoxicity of the cytostatics as measured by the analysis of SCEs (point 1 plus literature data) and their leukemogenic potential in man (point 2).

MATERIALS AND METHODS

SCE Analysis of Patients Under Cytostatic Therapy

The frequency of SCEs was determined in PHA-stimulated peripheral blood lymphocytes in patients undergoing cytostatic therapy (59 patients) at different intervals in relation to the administration of the cytostatic therapy. An SCE analysis was conducted before the therapy started in 14 of the 59 patients, and was regularly performed during the drug administration at 1-7 da intervals and/or after the cessation of cytostatic therapy at different intervals. A total of 254 SCE analyses were performed in PHA-stimulated lymphocyte cultures, 75 of which were performed on normal cells exposed to the plasma of the patients (i.e., to reveal the ability of the plasma of the patients treated with cytostatics to induce SCEs in normal cells). In these cases, the plasma taken from the patients under cytostatic therapy was cultured with the lymphocytes of healthy donors. These experiments are designated as SCE_n, whereas those with the patients' own plasma and lymphocytes are designated as SCE_p. A total of 140 SCE analyses were made after monochemotherapy and 114 following polychemotherapy. The monotherapy consisted of cyclophosphamide (CP), n = 30; melphalan (MP), n = 7; busulfan (BUS), n = 14; chlorambucil (CB), n = 4; lycurim, n = 1; 5-fluorouracil (5FU), n = 2; vincristine (VCR), n = 9; hydroxyurea (HU), n = 5; cytosine arabinoside (AraC), n = 62; Acyclovir, n = 2; Tamoxifen, n = 4; azathioprine (Aza), n = 3 SCE analyses. The combination chemotherapy involved CP + VCR + procarbazide + prednisolon (COPP), n = 21; CP + VCR + AraC + prednisolon (COAP), n = 13; CP + VCR, n = 18; CP + adriamycin (ADM) + VCR + prednisolon + bleomycin (CHOP-Bleo), n = 8; CP + ADM, n = 11; AraC + 6-thioguanine (6TG), n = 21; CP + 5FU + Methotrexate (MTX), n = 6 SCE analyses. The schedules of the administration of the cytostatic agents in the combination protocols are given in Tab. 2.

In each SCE analysis, 20 mitoses were analyzed for SCE frequency. In follow-up studies in the course of treatment at least 4 SCE analyses were made from each patient; the number of mitoses analyzed in these cases amounted to approximately 100/case.

The following questions were asked:

Tab. 2. Treatment schedule for combination chemotherapy in cancer patients. n = number of SCE analyses. Arrows indicate the administration of the cytostatics in the course of the combination protocols.

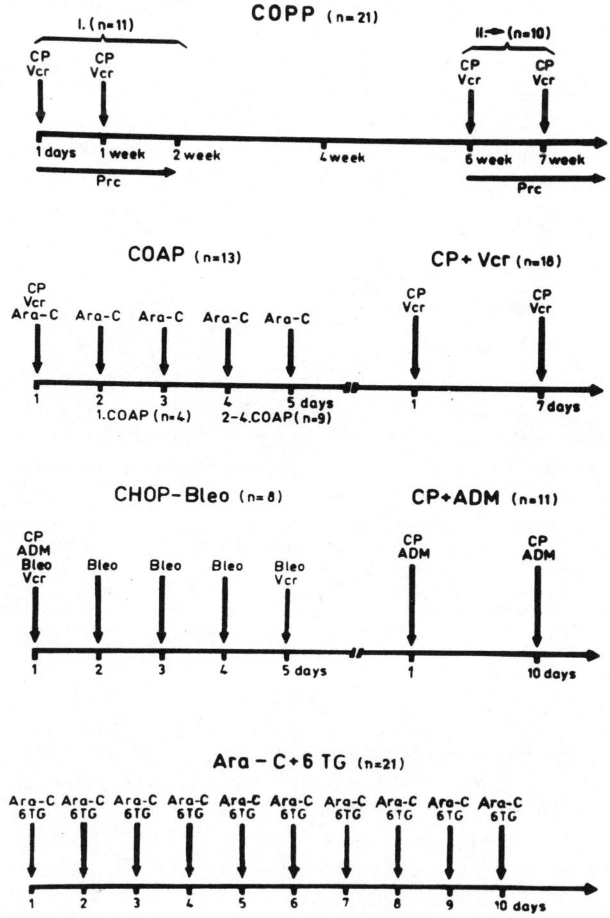

(1) Is there any alteration in the frequency of SCEs after the administration of the drug(s) to the patients as measured in peripheral blood lymphocytes?

(2) What is the tendency of the elevation in SCE frequency in PHA-stimulated blood lymphocytes to return to a normal level after the therapy had been stopped?

(3) Is there any change in SCEs in the peripheral blood lymphocytes of cancer patients before a therapy started?

The changes in SCE frequencies following the administration of the various cytostatics are plotted against time and the tendency of the SCE values to normalize to control frequencies is demonstrated relative to the baseline values of SCEs as estimated on the basis of SCE frequencies determined in 14 cancer patients before therapy (SCE mean/cell ± S.D. in these patients was: 6.05 ± 1.85). All figures contain pooled data from the SCE analyses on different patients but having the same therapeutical protocols (although with various drug doses), carried out at the same intervals after the start or cessation of the therapy. The least square test was used in some cases to calculate a best fitting curve to demonstrate the tendency of the higher frequencies of SCEs to normalize to control values.

Assessment of the Leukemogenic Potential of Different Cytostatic Protocols in Man

An extensive literature survey was made for therapy-related AL from the period of 1930 until 1980. Acute leukemias following chemotherapy and/or irradiation therapy for primary diseases such as Hodgkin's disease (HD), non-Hodgkin lymphoma (NHL), multiple myeloma (MM), solid tumors (ST), and immunopathological disorders (IP) were considered if the AL developing in these conditions followed at least 6 mo later the administration of the cytotoxic protocols. As many as 746 cases of AL were found based on the reports of 114 authors (26-140). All published cases were reviewed individually and analyzed for several parameters: a) the type of the therapy - i) monochemotherapy, ii) combined chemotherapy, iii) irradiation, and iv) irradiation and chemotherapy; b) type of the AL induced in relationship to the therapy administered; c) the cytogenetic changes observed; d) the relative participation of different therapeutic modalities in inducing secondary AL; e) the mean latency period to AL from the start of the cytotoxic therapy for the primary disease. Out of the total of 746 AL cases, 616 cases contained sufficient information regarding the type of the therapy. Monochemotherapy was given for the primary disease in 164 cases, irradiation in 77, and combined chemotherapy (CHT) and irradiation in the remaining 375 cases. The relative participation of each cytostatic agent in the treatment history of an AL was calculated for the different groups of cytotoxic drugs. The groups of cytostatics were constructed on the basis of the SCE-inducing ability of various cytotoxic compounds as known in human cell systems, especially after their administration for cancer patients (Refs. 12,17,21, and this chapter). Three different groups of cytotoxic agents were constructed: cytostatics with a) strong, b) weak, and c) negative SCE-inducing capability. Then, the different combinations of the cytotoxic therapies in the combination CHT group and the monochemotherapy group were placed in one of the above categories.

RESULTS

SCE Studies in Peripheral Blood Lymphocytes of Humans Exposed to Cytostatic Treatment

The SCE studies have been performed in 2 different systems:

a) The lymphocytes of the patients receiving cytotoxic therapy were cultured in the presence of autologous plasma from the patients (179 SCE analyses).

b) In as many as 75 cases of SCE analyses, normal lymphocytes were cultured in the presence of the treated plasma of the patients. In these experiments, the presence of the Y chromosome in the mitoses scored for SCEs was a discriminating marker to disclose contamination with the patients' cells. White blood cells from a healthy female were cultured in the presence of the plasma from a male patient, and vice versa.

Figures 1-5 show that the malignant process itself does not influence the frequency of SCEs in the peripheral blood lymphocytes of 14 patients. The cytostatic therapy, however, induced an increase in the frequencies of SCEs which was influenced by: a) the

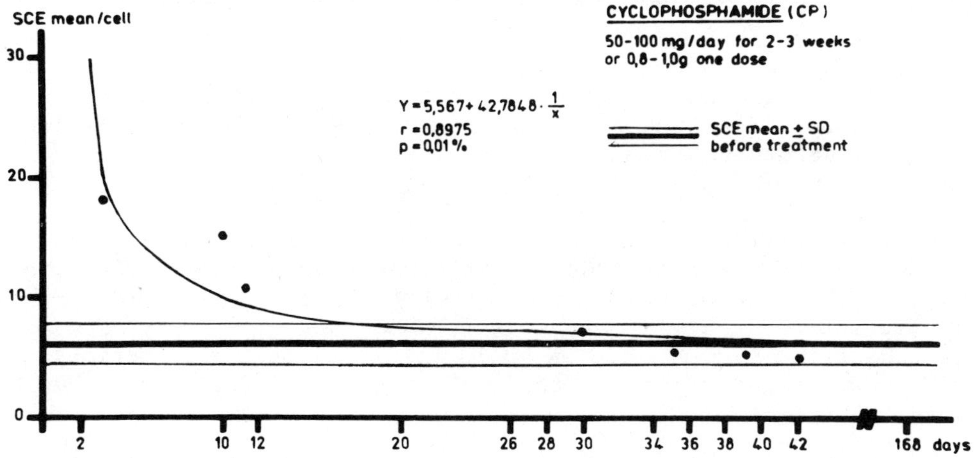

Fig. 1. The alteration of the SCE frequency in PHA-stimulated peripheral blood lymphocytes of patients treated with CP alone. SCE is plotted against time (da) after the cessation of the therapy. Dose ranges of the cytostatic(s) are given. Horizontal lines indicate SCE mean ± S.D. in 14 cancer patients before treatment.

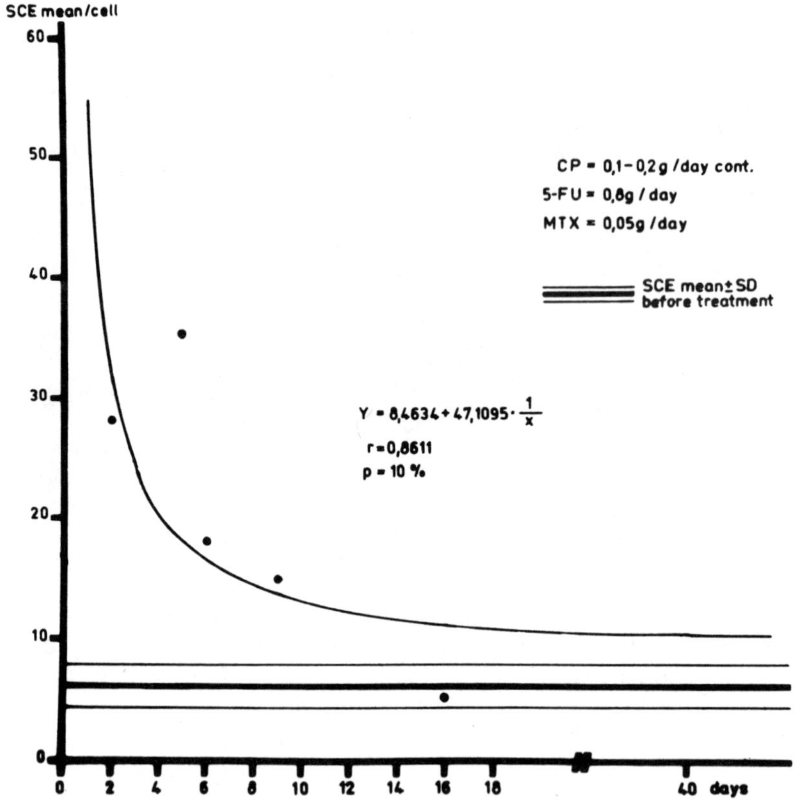

Fig. 2. Sister chromatid exchange changes in peripheral blood lymphocytes in patients treated with melphalan. Horizontal lines indicate SCE mean ± S.D. in 14 cancer patients before treatment.

type of cytostatic, b) the time elapsing between the administration of the drug and when the SCE analysis was conducted and, most probably, c) the dose of the cytostatic(s). The study of various doses of cytostatics on the frequencies of SCEs was not our primary aim. Even the clinical type studies would not allow such distinctions, so only dose ranges in which the pooled data are collected can be given.

There are cytostatics which do not increase the frequencies of SCEs even under continuous administration (17). These are: AraC, VCR, 5FU, HU, Tamoxifen, Acyclovir, and Immuran(azathioprine). On the contrary, CP and MP induced a long-lasting elevation of SCEs which tended to return to the baseline level at about 3-4 wk after the cessation of the therapy (Figs. 1 and 2). The combination cytostatic schedules containing CP and one or more SCE noninducing compounds (1,17) show a similar tendency toward normalization of the

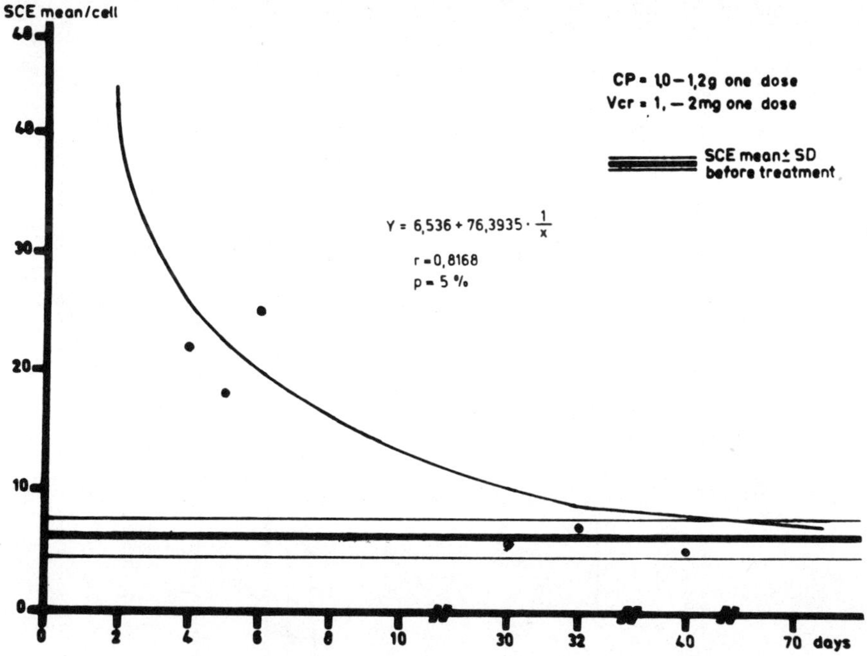

Fig. 3. Sister chromatid exchange changes in PHA-stimulated lymphocytes of patients treated with a combination of CP + 5FU + MTX (see also Tab. 2). Horizontal lines indicate SCE mean ± S.D. in 14 cancer patients before treatment.

elevation in SCE frequency to control values as does CP alone (Figs. 3 and 4). However, COPP protocols induce an elevation of SCEs in the peripheral blood lymphocytes of the patients (Fig. 5), which tends to show a somewhat more retarded rate of normalization than CP alone or in other combination [i.e., CP + VCR or CP + 5FU + MTX (Fig. 6)]. These results indicate that it is either the addition of procarbazide (PCR) to the combination protocol or the repetition of the COPP cycles which may lie behind this phenomenon. The role of PCR to induce SCEs could not be excluded. On the contrary, it was evident that repeated COPP cycles induced a significantly higher level of SCEs 1-7 da after the termination of the therapy (Fig. 7). This indicates an additive effect of the repeated doses of the cytostatics as far as the normalization of the elevated level of SCEs to control levels after the cessation of the therapy. The combination of AraC + 6TG was never found to induce an increase in the frequencies of SCEs if blood sampling was made 2-4 hr after the cessation of the therapy. This combination did, however, if blood sampling was made soon (10-30 min) after the administration of a bolus iv (17).

An increase in SCE frequency was also observed if blood samples were taken from patients with no history of cytotoxic treatment and

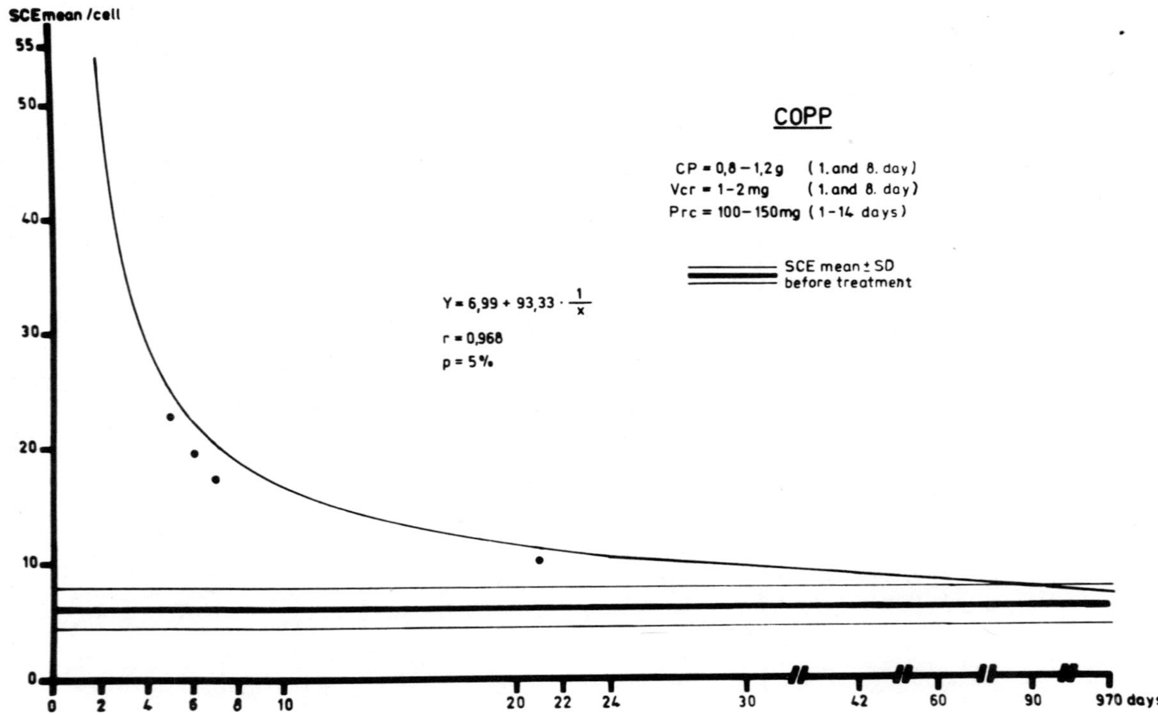

Fig. 4. Sister chromatid exchange changes in PHA-stimulated lymphocytes of patients treated with a combination of CP and VCR (see also Tab. 2). Horizontal lines indicate SCE mean ± S.D. in 14 cancer patients before treatment.

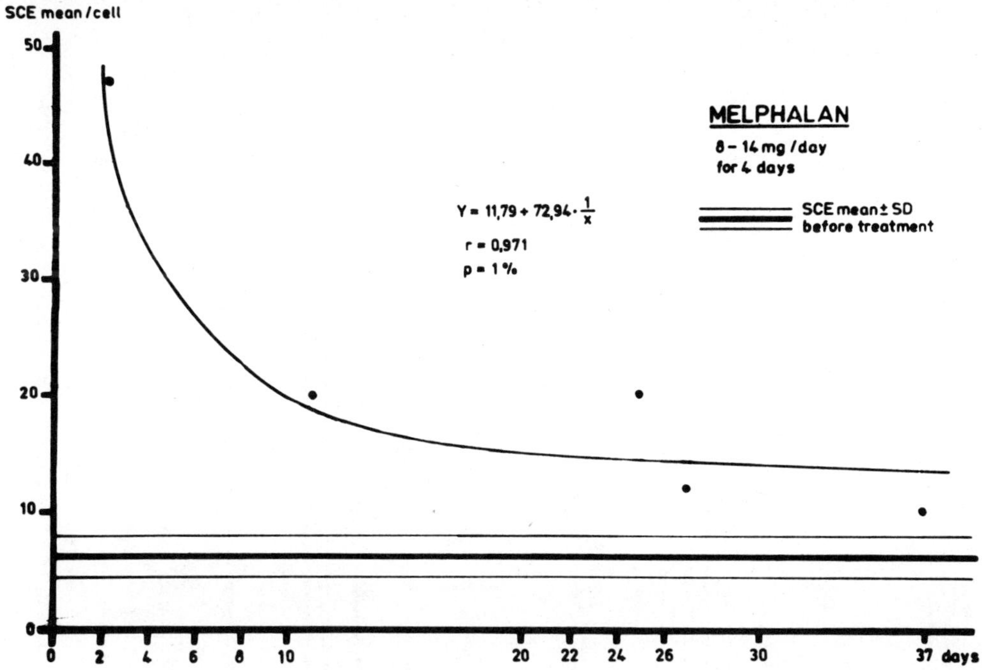

Fig. 5. Sister chromatid exchange changes in PHA-stimulated lymphocytes in patients treated with COPP protocols (see also Tab. 2). Horizontal lines indicate SCE mean ± S.D. in 14 cancer patients before treatment.

cultured in the presence of the sera taken from patients treated with CP, MP, COPP, CHOP-Bleo, and COAP. This effect suggests that treated plasma from patients under CP + ADM + BLM therapy contains active metabolites which induce SCEs in normal PHA-stimulated lymphocytes (SCE_n) (Fig. 8). The rate of normalization to control values of SCEs induced under these in vitro conditions is quite similar to that found in the SCE_p experiments. This implies that in patients treated with cytostatics, the SCE elevation in the peripheral blood lymphocytes is not only influenced by damage to the DNA at G_0, but also to the presence of metabolites in the plasma which are active in the induction of SCEs at the time of the genotoxicity study. The genotoxicity of the treated plasma taken from the patients under cytostatic therapy may open a new approach to the assessment of the effective concentration of a cytostatic available for the biotransformation of the cells in a given patient.

Leukemia-Inducing Ability of Different Cytostatics

From a total of 616 secondary AL following the treatment of lymphomas (HD, NHL, MM), solid tumors (lung, ovarian, gastrointestinal, etc.), and immunopathological disorders, 164 AL were found

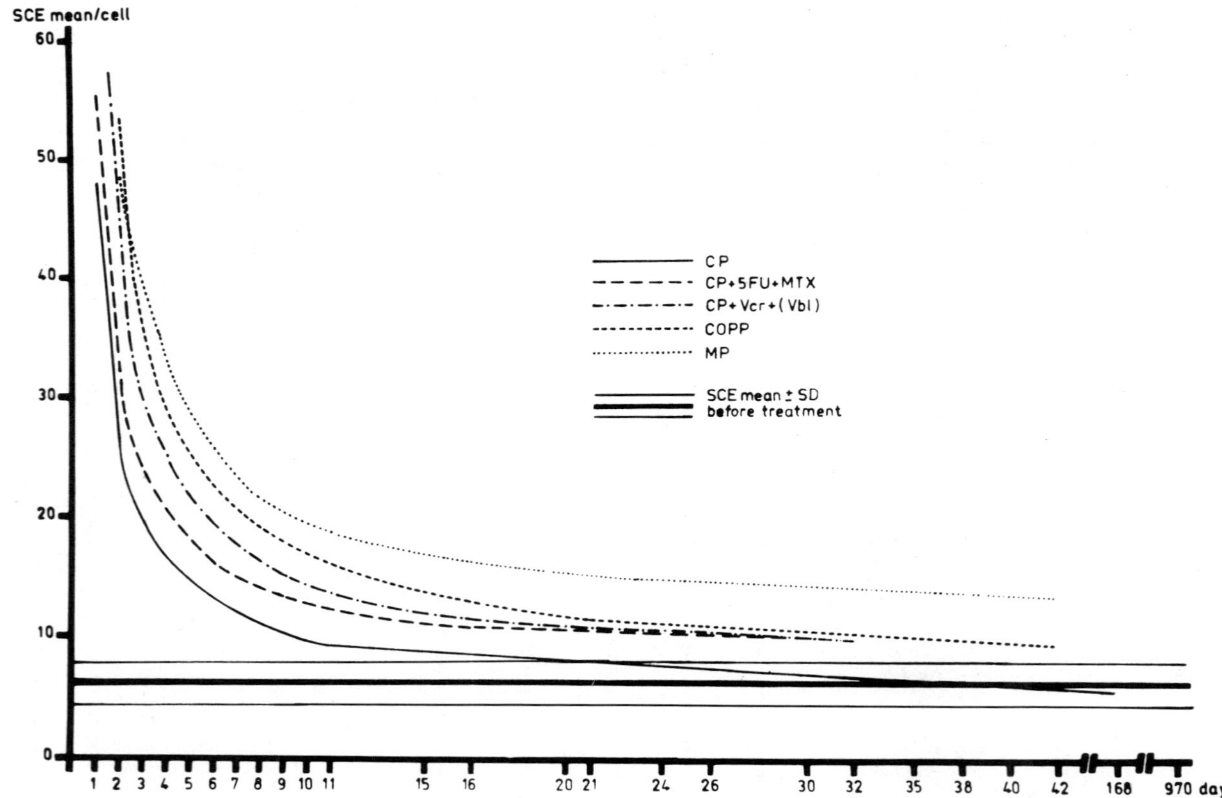

Fig. 6. Comparison of the rate of normalization of the SCE frequencies in PHA-stimulated lymphocytes of cancer patients treated with 1) CP; 2) CP + 5FU + MTX; 3) CP + VCR; 4) COPP; 5) MP (see also Figs. 1-5). Note that there is a similar rate of normalization of SCE levels after CP and CP + VCR + MTX and CP + VCR. After the administration of COPP and MP there is a somewhat retarded rate of normalization of the higher levels of SCE (about 7-8 wk).

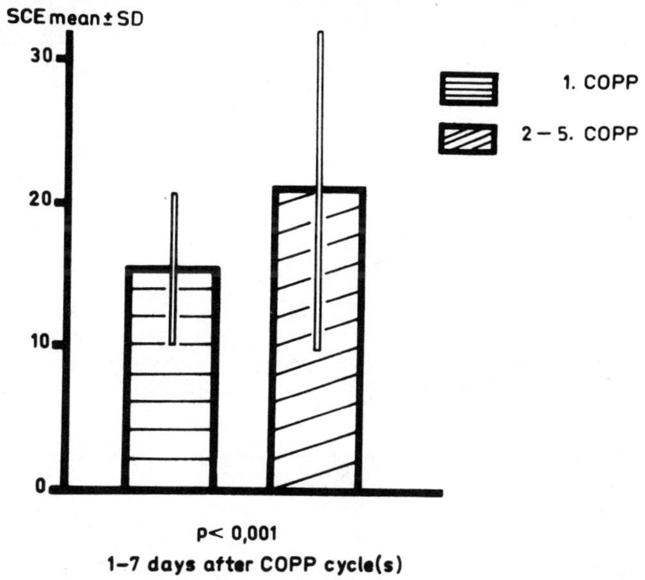

Fig. 7. Sister chromatid exchange frequency in PHA-stimulated lymphocytes of patients following the first COPP cycle compared to repeated (2-5) COPP cycles. Sampling for SCE analysis was made 1-7 da after the cessation of the appropriate schedule (pooled data).

where the chemotherapy for the primary disease consisted only of monochemotherapy. Seventy-seven AL cases were preceded only by irradiation, and the remaining 375 AL by the administration of combination CHT and/or plus irradiation. The relative participation of leukemia induction by the different drugs in the monochemotherapy group is of special interest.

Of 164 AL induced in this group, 151 followed the administration of MP (64 AL), CP (18 AL); HN_2 (1 AL), BUS (5 AL), treosulfan (6 AL), N,N',N"-triethylenethiophosphoramide (ThioTEPA) (8 AL), 1-(2-chloroethyl)-3-(4-methylcyclohexyl)-1-nitrosourea (mCCNU) (1 AL). These chemotherapy agents are all potent inducers of SCEs (1,12,17). Streptozotocin, aminopterin, MTX, 5-(3,3-dimethyl-1-triazeno)imidazole-4-carboxamide (DTIC), and Aza induced a total of 13 AL and are known to be negative or weak inducers of SCE (1,12, 17). As far as the combination chemotherapy group is concerned, it is especially evident that only those cytostatics which are positive in the SCE test induce the overwhelming majority of secondary AL. Drugs which are negative or uncertain SCE inducers caused only one AL after their administration (Fig. 9). Thus, of the 616 AL following the treatment of lymphomas, solid tumors, and IP disorders, 523 (84.9%) were induced by cytostatic agents which are positive in SCE tests and only 8 (1.29%) by those agents not able or uncertain inducers of SCEs.

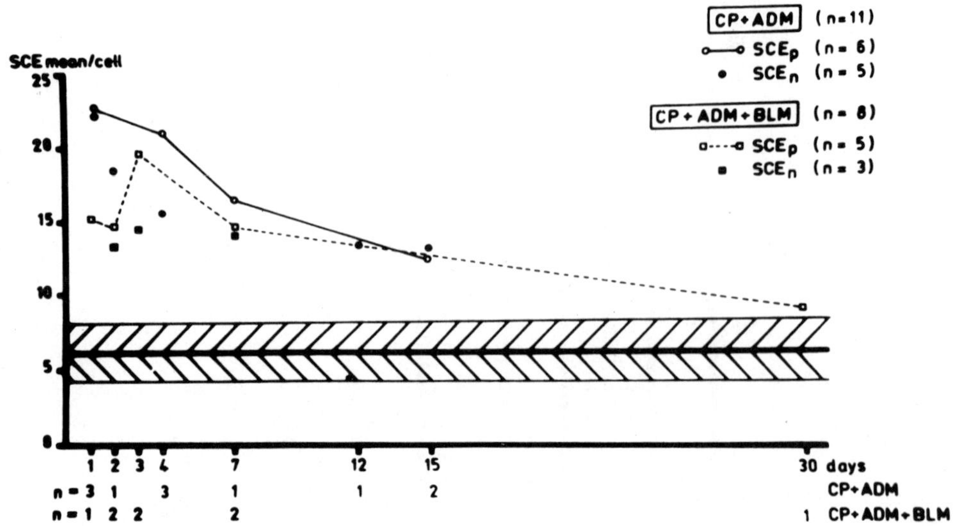

Fig. 8. Sister chromatid exchange in PHA-stimulated human lymphocytes after the exposure to a combination of cytostatics (CP + ADM; CP + ADM + BLM; see also Tab. 2). SCE_p designates the experiments in which the lymphocytes of the treated patients are cultured in the presence of autologous plasma. SCE_n designates experiments in which the plasma taken from the patients at various intervals after the cessation of the cytostatic therapy is cultured with lymphocytes derived from healthy donors. Note that treated plasma induces a long-lasting elevation in SCEs in PHA-stimulated lymphocytes of healthy donors.

DISCUSSION

The principal aim of this study is to consider the probable relationship between SCEs and carcinogenesis. Such a correlation, if it exists, assumes to find a relationship between lesions in the DNA responsible for cancer initiation and those capable of inducing SCEs. Furthermore, this study hopes to better understand the relationship between various cytogenetic endpoints (SCEs and CAs) and their relation to the complex process of carcinogenesis. Although much information has been acquired about the characteristics of SCE induction by various compounds, the data can be extrapolated to man only with great caution. Due to rapidly increasing research involving human systems on the SCE-inducing characteristics of different compounds, there is now a body of information which is worth consideration from the point of view of the complex process of carcinogenesis.

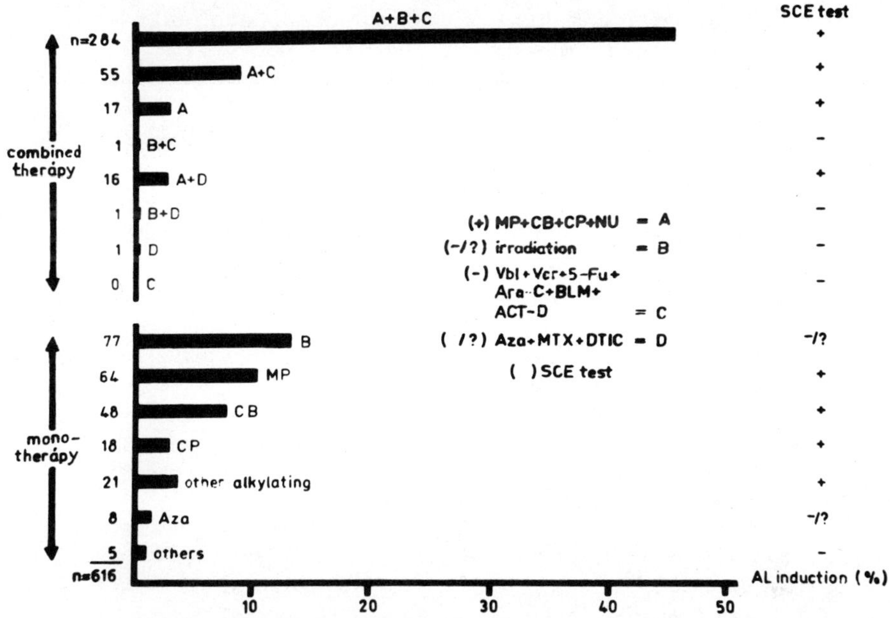

Fig. 9. Comparison of the SCE-inducing ability of different cytostatics and their relative participation in the treatment of the primary diseases (lymphomas and solid tumors and immunopathological disorders) which were later followed by an acute leukemia (a total of 616 acute leukemia from the literature during the period 1930-1980). A = group of drugs which are unequivocally positive in the SCE test (see Results and Refs. 1,12,17). NU = nitrosoureas; B = irradiation; C = group of drugs which are negative in the SCE test (see Results and Refs. 1,12,17). VBL = vinblastine; BLM = bleomycin; D = group of drugs with weak SCE-inducing ability (Refs. 1,12,17).

Why Secondary Acute Leukemias as a Model for Carcinogenesis?

Generally, it is well recognized that various environmental chemical compounds contribute to the emergence of malignancies (Tab. 3). From the middle of the last decade it has also been increasingly realized that cytotoxic agents (i.e., irradiation and chemotherapy) are implicated in the etiology of malignancies following primary tumors after chronic exposure to these agents (15,19). One of the most often observed malignancies after the administration of cytotoxic therapy is the appearance of acute leukemias, mainly a variant of the myeloid ones (4). These so-called therapy-related, or therapy-induced, or secondary AL are now considered as a specific

Tab. 3. List of environmental chemical compounds suspected to be carcinogenic in man (modified from Ref. 10).

CHEMICAL	TISSUE SPECIFICITY
Occupational exposure	
aromatic amines	bladder
arsenic	skin, bronchus
asbestos	bronchus, pleura peritoneum
bis/chloromethlyl/ether	
bis/chloroethyl/sulfide	respiratory tract
chrome ores , coke ovens,	
nickel ores, soots, tars, wood dust	
vinyl chloride	liver
Medical drugs	
alkylating agents	bladder, hemopoietic
anabolic steroids	liver
chlornaphasine	bladder
diethylstilbestrol	vagina
immunsuppressive drugs	lymphoid tissue
phenacetin	renal pelvis
oral contraceptives	liver
Miscellaneous	
aflatoxins	liver
tobacco smoke	respiratory, gastro-intestinal

/ Modified from Farber, 1981 /

clinicopathological entity among ALs (11) which are different from the so-called de novo AL in many respects [i.e., cytochemical (9) and cytogenetic (23) as well as for a poor response to therapy (11)].

Although the exact mechanism of the therapy-related leukemia remains to be identified, there are several indications which suggest that it is the cytotoxic therapy which is of importance for the emergence of these malignancies:

(i) The mean interval of emergence of secondary AL is about 5.52 yr.

(ii) The accumulation of secondary AL started 4-5 yr after the large-scale introduction of cytostatic therapy for malignant disorders and immunopathological dyscrasias.

(iii) There are markedly different cytogenetic alterations in therapy-related acute nonlymphocytic leukemias from that of the de novo ones (Fig. 10).

Most of the cancer epidemiology and environmental mutagenesis studies are hampered by the complexity of carcinogens to which the affected individuals are suspected to have been exposed (25). This prompted us to restrict ourselves to iatrogenic leukemogenesis, especially acute leukemias following the treatment of lymphomas, solid tumors, and immunopathological disorders. This system is especially useful in two respects: a) it takes into account malignancies with well-defined and comparable histology, and b) involves well-known mutagens and carcinogens (i.e., therapeutically administered cytostatics).

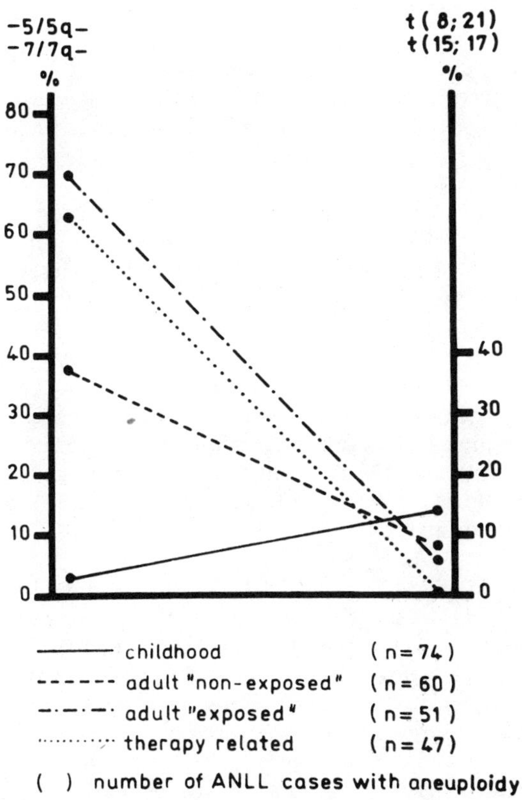

Fig. 10. The relative occurrence of chromosomal abnormalities (such as -5/5q-, -7/7q-, and t(8;21) and t(15;17) in therapy-related acute nonlymphocytic leukemia (ANLL) versus childhood ANLL and de novo ANLL in adult with and without occupational exposure [pooled data from Golomb et al. (1982) Blood 60:404-412; Kaneko et al. (1982) Blood 60:389-399; and present review (see References)].

The use of this system has already led to the recognition of some peculiar cytogenetic changes (i.e., the remarkable differences in the nature of CAs between therapy-related and de novo leukemias (20,23).

Thus, the iatrogenetic leukemias induced by cytostatics in man seem to be a relevant model of carcinogenesis in which the genotoxicity of different cytotoxic agents and their carcinogenic potentials can be tested.

Relationship Between the Carcinogenicity-Mutagenicity of Various Drugs and Different Cytogenetic Parameters

This relationship is not clear and it is due to the fact that different systems in vitro and in vivo with different cell types, only a few of which involve human cells, are available for genotoxicity as well as carcinogenicity and mutagenicity studies.

If we examine the 379 compounds tested for their ability to induce SCEs, there are 219 compounds which have also been tested for CA. Among these agents there is a 77% correlation between the SCE and CA results (1).

If only medical drugs and related compounds are considered (a total of 164 drugs tested so far), there is a higher agreement (86.9% correlation) between SCEs and CAs, which is even higher (97.05%) if only data on human cells are taken into consideration (Fig. 11).

If the same compounds are evaluated for their carcinogenic (C) and mutagenic (M) ability and cellular transformation (T) in relationship to SCE and CA induction, then a positive SCE response remarkably correlates with M, C, and T. However, if a compound is negative in the SCE test, the available data suggest that the compound may still be carcinogenic.

The same relationship is valid for CA and M/C/T (1). This correlation clearly indicates that a positive SCE response in the group of chemical compounds with medical relevance tested so far reflects carcinogenic potency. However, the data also leave no doubt that a negative result in the SCE test does not necessarily mean that a compound is not carcinogenic. The data presented here shows circumstantial evidence about the meaning of SCE data in relationship to malignancies in man. Most of the data cited above originate from different cell types for cytogenetic studies, for bacterial mutagenesis systems as far as mutagenesis, and only to a limited extent from human cells as far as transformation assays are concerned (for review see Ref. 1). If we want to understand the meaning of these tests in man and their relationship to human carcinogenesis, real human systems should be used. The system of known mutagen/carcinogen-induced ALs and their relation to SCEs can be a step in this direction.

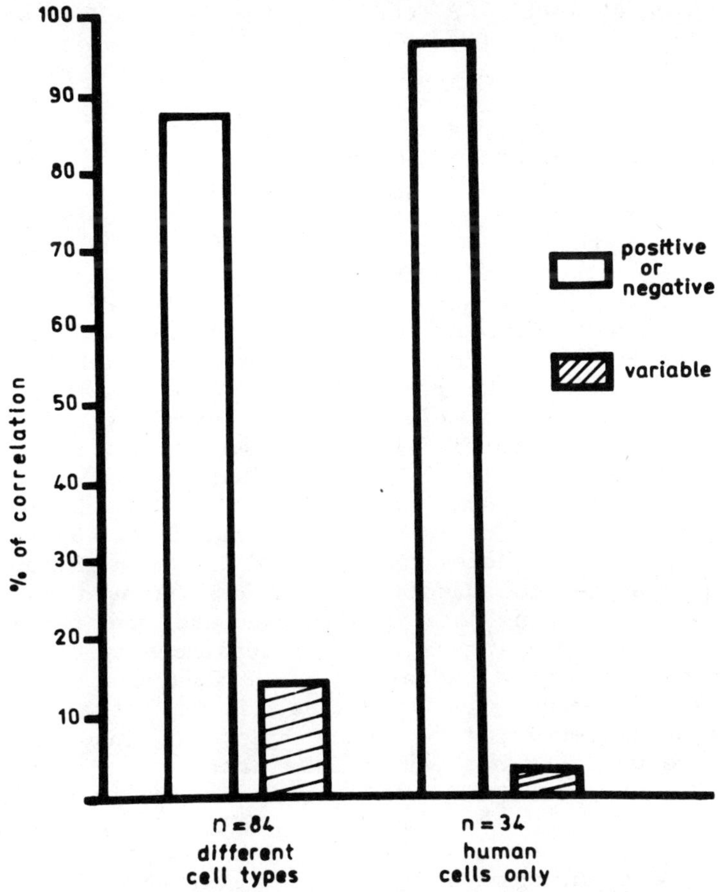

Fig. 11. Correlation between SCEs and CAs.

SCE in Patients with Malignant Disorders Treated with Cytostatics

Most of the SCE studies on cytostatics in man have been made in PHA-stimulated peripheral blood lymphocytes of patients receiving cytostatic therapy. The cytostatics can very clearly be grouped into 3 different categories: i) strong inducers of SCEs, ii) negative for SCE induction, and iii) weak or uncertain in SCE induction (12,17,21). The data presented here are in agreement with previous reports demonstrating that cytostatics which directly act on DNA, such as alkylating and intercalating agents, induce an elevation in SCEs, whereas other agents such as spindle poisons [VCR, vinblastine (VBL)], compounds inhibiting purine and pyrimidine synthesis (5FU, HU, AraC), or like Acyclovir or Tamoxifen do not induce SCE. The dynamics of the elimination of the higher frequencies of SCEs after the therapeutic administration of alkylating cytostatics, such as CP and MP or BUS, and combination chemotherapy protocols containing alkylating agents, mostly CP [COPP or CP + VBL + PCR + prednisolon (CVPP); COAP; CP + VCR; CP + 5FU + MTX; CHOP-Bleo] is of special interest (12,17).

After the administration of combinations such as CP + VCR or CP + 5FU + MTX (i.e., containing one or more cytostatics except for CP which are negative inducers of SCEs in human peripheral blood lymphocytes), the rate of normalization of SCEs to control values is very similar to that of CP alone. The high frequency of SCEs returns to the control SCE level about 5-6 wk after the administration of CP, CP + VCR, or CP + 5FU + MTX. After MP, however, the curve shows a more elongated tendency to approach the baseline range (Fig. 6). The SCE data exhibited considerable variability at different intervals after the termination of the therapy. This may i) indicate interindividual differences in lymphocyte SCE response to the cytostatics due to differences in metabolism, or to subpopulations of lymphocytes, or ii) be attributable to the different doses of the cytostatics administered to the patients. For CP this dose range is 0.8-1.2 gm/da in different protocols. By knowing the dose-range dependent SCE elevations caused by known mutagens-carcinogens, it is suggested that SCE levels can be used as tentative models to estimate "equal dose mutagen exposure" when an individual is exposed to unknown mutagens (i.e., occupationally or accidentally). The cytostatic agents studied for SCE induction in vivo in the peripheral blood lymphocytes of humans can be categorized into 3 different classes according to their ability to induce SCEs. For example, drugs which do not induce SCEs are: AraC, actinomycin D, BLM, HU, 5FU, MTX, 6MP, VCR, VBL, Acyclovir, and Tamoxifen. The strong inducers of SCEs are: ADM, CP, CB, CCNU, MP alone and in combination with each other or with drugs with otherwise no SCE-inducing ability. Weak inducers of SCEs are: irradiation, Aza, MTX, and DTIC.

This subclassification of cytostatics indicates that it is the direct action of the cytostatics on the DNA which is important for SCE induction (see below). It will not provide any information concerning any particular type of lesions introduced in the DNA.

Treatment for the Primary Disease Preceding Secondary AL and SCE Induction by Cytostatics

A very striking correlation between the SCE-inducing ability of cytostatics and the induction of iatrogenic leukemias was found. Of the 616 AL in the literature between 1930-1980, 164 were preceded by the administration of monochemotherapy, 77 by irradiation alone, and 375 by combined CHT and/or irradiation. After subclassifying these induced AL according to the capability of the therapeutic protocol to induce SCEs, a strong correlation was noted between the SCE as well as the secondary leukemia-inducing ability of the same cytotoxic agent(s). For the monochemotherapy group, 92.07% (i.e., 151 of the secondary leukemias induced) were preceded by the administration of a drug which is positive in the SCE test and only 13 (about 8%) by the administration of those drugs which are not able to induce SCEs in vivo in the human lymphocyte system (Fig. 9). For the total number of secondary AL, 523 of the 616 (84.9%) were induced by therapy involving drugs capable of inducing SCEs and only 8 (1.29%) by drugs which do not appear to induce SCEs. The remaining 85 AL (13.79%) were induced by those cytotoxic agent(s) for which an SCE response is uncertain. In this group, 77 (12.5%) AL

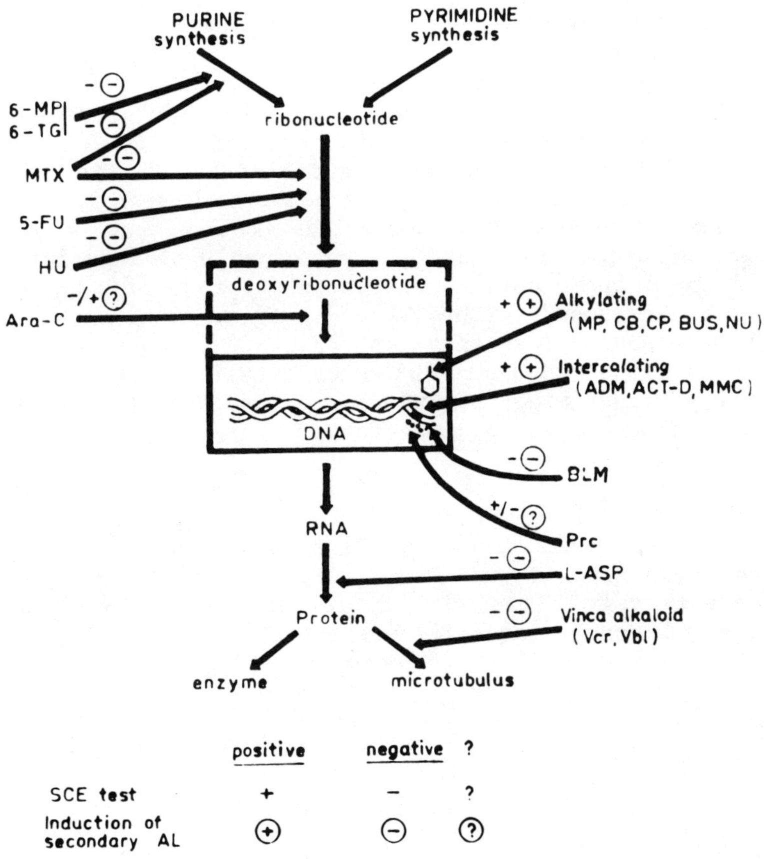

Fig. 12. Scheme of DNA synthesis, the interaction of cytostatics with the DNA, and their relation to SCE induction and iatrogenic leukemogenesis. (6MP = 6-mercaptopurine; MMC = mitomycin C; L-asp = L-asparaginase). "+" means a compound induces SCE; "-" means a compound does not induce SCE. "⊕" means a compound is involved in iatrogenic leukemogenesis; "⊖" means a compound is not involved in iatrogenic leukemogenesis.

were preceded by the administration of irradiation alone, the consideration of which does not strictly belong to the subject of this chapter. These data show a very striking correlation between an SCE positive response and carcinogenesis in man.

The Interference of the Cytostatics with DNA Synthesis and its Relation to SCE Induction and Secondary Leukemias

In terms of the action of various cytostatics on DNA synthesis, the data on SCEs in the PHA-stimulated lymphocyte system in patients treated with cytostatics clearly shows almost exclusively that the cytostatics capable of directly introducing some kind of lesion(s) into the DNA will be able to induce SCEs (i.e., those agents which alkylate and intercalate into the DNA but not those which block purine, pyrimidine, or tubulin synthesis). One has to keep in mind, however, that the number and types of lesions in the DNA introduced by various agents, including cytostatics, are quite large. A visible SCE induction might not only be influenced by the type (5,14) but also by the location (8) and the number of these lesions (6) in the DNA. As a tentative hypothesis one may say that only the type of lesions introduced into the DNA and capable of inducing SCEs are relevant to iatrogenic carcinogenesis, while those lesions which are not directly introduced to the DNA are not important either for SCEs or for the emergence of secondary AL (Fig. 12). Thus, one may argue that those lesions induced by the cytostatic compounds, whatever different kinds they might be, which are important for the induction of SCEs, are also of crucial importance for secondary AL.

The recognition of this relationship may have implications in clinical practice. It may help in the planning of new therapeutical trials involving cytostatics with less carcinogenic potency (3,24) and in the selection of new drugs with low leukemogenic potential for the treatment of lymphomas and other diseases with high cure rates and a subsequent long life expectancy.

APPENDIX: THERAPY-INDUCED AL - LITERATURE SURVEY

in Hodgkin 's disease: 26./ Mitrou,P.S.,et al.:Dtsch.Med.Wsch., 99,1596,1974; 27./ Toland,D.M.,et al.:Blood,46,1013,1975; 28./ Rosenberg,S.A.,et al.: Cancer,35,55,1975; 29./ Rosner,F., et al.: Am.J.Med.,58,339,1975; 3o./ Raich,P.C.,et al.: Am.J.Med. Sci.,269,237,1975; 31./ Lundh,B.,et al.: Scand.J.Haemat.,14, 3o3,1975; 32./ Canellos,G.P.,et al.: Lancet,1,947,1975; 33./ Jacquillat,C.I.,et al.: Proc.Am.Assoc.Cancer Res.,17,247,1976; 34./ Parmley,R.T.,et al.: Cancer,38,1188,1976; 35./ Rowley,J.D., et al.: Blood,5o,759,1977; 36./ Larsen,J.,et al.:Scand.J.Haemat., 18,197,1977; 37./ Cavalli,F.,et al.: Dtsch.Med.Wsch.,1o2,1o19, 1977; 38./ Cadman,E.C.,et al.:Cancer 4o,128o,1977; 39./ Coleman,C.N.,et al.:N.Engl.J.Med.,297,1249,1977; 4o./ Kuse,R.,et al.: Dtsch.Med.Wsch.,49,1824,1977; 41./ Brody,R.S.,et al.: Cancer, 4o,1917,1977; 42./ Cavallin-Stähl,E.,et al.: Scand.J.Haemat., 19,273,1977; 43./ Preisler,H.D.,et al.: Am.J.Hemat.,3,2o9,1977; 44./Li,P.F.: Cancer,4o,1899,1977; 45./Vardiman,J.W.,et al.: Cancer, 42,229,1978; 46./Neufeld,H.,et al.: JAMA,239,247o,1978; 47./ Kong-O O Goh,et al.; Am.J.Med.Sci.,276,189,1978; 48./ Powers,J.S.,et al.: J.Pediatr.,93,323,1978; 49./ Jouet,J.P.,et al.:Nouv.Press.Med.,8,613,1979; 5o./ Auclerc,G.,et al.:Cancer, 44,2o17,1979; 51./Foucar,K.,et al.: Cancer,43,1285,1979; 52./ Papa,G.,et al.: Scan.J.Haemat.,23,339,1979; 53./Kithara,M.,et al.: Ann.Intern.Med.,92,625,198o; 54./ Rosenstranch,K.J.C.J., et al.: S.Afr.Med.J.,58,454,198o; 55./ Valagussa,P.,et al.: Br.Med.J.,28o,216,198o; 56./ Kapadia,S.B.,et al.: Cancer,45, 1315,198o; 57./ Baccarini,M.,et al.: Cancer,46,1735,198o; 58./ Gulati,S.C.,et al.: Cancer,46,725,198o; 59./ Lederlin,P.,et al.: Ann.Med.Nancy Est.,119,59,198o; 6o./Rowley,J.D.,et al.: Blood, 58,759,1981;

in Non-Hodgkin's Lymphoma : 61./ Anderson,T.,et al.:Cancer.Treat. Rep.,61,1o57,1977; 62./ Cavalli,F.,et al.: Acta Haemat.,6o,25o, 1978; 63./ Vardiman,J.W.,et al.:Cancer,42,229,1978; 64./ Beltran, G.,et al.: Blood,52/suppl/ 329,1978; 65./ Strauss,D.J.,et al: Blood/suppl/, 52,277,1978; 66./Zarrabi,M.H.,et al.: Cancer,44, 1o7o,1979; 67./ O'Donell,J.P.,et al.: Cancer,44,193o,1979; 68./Lönqvist,B.,et al.:N.Engl.J.Med.,3oo,367,1979; 69./ Foucar,K.,et al.: Cancer,43,1285,1979; 7o./ Casciato,D.A., et al.: Medicine/Baltimore/,58,32,1979; 71./ Carabell,S.C., et al.:Cancer,43,994,1979; 72./ Risdall,R.,et al.: Cancer, 44,529,1979; 73./ Harris,R.E.,et al.: Med.Pediatr.Oncol., 7,3o3,1979; 74./ Zarrabi,M.H.,et al.: Semin.Oncol.,7,34o, 198o; 75./ Pucket,J.B.,et al.: Ann.Intern.Med.,93,51o,198o; 76./Kapadia,S.B.,et al.: Cancer,45,1315,198o; 77./ Lederlin,P. et al.: Ann.Med.Nancy Est,119,59,198o; 78./ Harousseau,J.L., et al.: La Nouv.Press.Med.,9,3513,198o;

(Continued)

APPENDIX (Continued)

in solid tumors: 79./ Reimer,R.R.,et al.: N.Eng.J.Med., 297,177,1977; 8o./ Youness,E.,et al.: Cancer,Treat.Rep.,62, 1513,1978; 81./ Einhorn,N.: Cancer,41,444,1978; 82./ Vogl,S.E.: Cancer,41,333,1978; 83./ Kapadia,S.B.,et al.: Cancer,41,1676, 1978; 84./ Smithon,W.A.,et al.: Mayo Clin.Proc.,53,757,1978; 85./ Wiggans,R.G.,et al.: Blood,52,659,1978; 86./ Rosner,F., et al.: Am.J.Hemat.,4,151,1978; 87./ Zarrabi,M.H.,et al.: Am.J.Hemat.,7,357,1979; 88./ Casciato,D.A.,et al.: Medicine /Baltimore/,58,32,1979; 89./ Auclerc,G.,et al.: Cancer,44, 2o17,1979; 9o./ Ortonne,J.P.,et al.: Ann.Dermatol.Venereol., 1o6,251,1979; 91./ Reimer,R.R.,et al.: Ann.Intern.Med.,9o, 989,1989; 92./Kapadia,S.B.,et al.: Cancer,45,1315,198o; 93./ Pedersen-Bjergaard,J.,et al.:Cancer,45,19,198o; 94./ Stewart, A.L.,et al.: J.Cancer Res.Clin.Oncol.,1oo,1o9,1981; 95./ Shimp,W.,et al.: N.Eng.J.Med.,3o4,845,1981; 96./ May,J.T.,et al.: Oncology,38,134,1981; 97./ Schmitt-Gräff,A.,et al.: J. Cancer Res.Clin.Oncol,1o2,93,1981; 98./ Green,M.R.,et al.: Cancer,47,1963,1981; 99./ Rowley,J.D.,et al.: Blood,58,759, 1981; 1oo./ Pizzolo,G.,et al.: Acta Haemat.,65,128,1981; 1o1./ Nowell,P.,et al.: Cancer,48,667,1981;

in multiple myeloma: 1o2./ Drivsholm,Aa.,et al.: Reported at Hematologisk Vårmøte in Oslo,June,1971; 1o3./Rosner,F.,et al.: Am.J.Med.,57,927,1974; 1o4./ Auclerc,G.,et al.:Cancer,44,2o17, 1979; 1o5./Rosner,F.,et al.: New York State Journal of Medicine,8o,558,198o; 1o6./ Kapadia,S.B.,et al.: Cancer,45, 1315,198o; 1o7./ Swolin,B.,et al.: Blood,58,986,1981; 1o8./ Rowley,J.D.,et al.: Blood,58,759,1981;

in immunpathological diseases:1o9./ Witz,F.,et al.: Nouv.Press. Méd.,7,2392,1978; 11o./ Fiere,D.Et al.: Nouv.Press.Méd.,7,756, 1978; 111./ Carcassone,Y.,et al.: Cancer Treat.Rep.,62,111o, 1978; 112./ Grünwald,H.W.,et al.: Arch.Intern.Med.,139,461,1979; 113./ Khan,M.F.,et al.: Nouv.Press.Méd.,8,1393,1979; 114./ Auclerc,G.,et al.: Cancer,44,2o17,1979; 115./ Lebranchu,Y.,et al.: Lancet,I.,649,198o; 116./ Aymard,J.P.,et al.: Acta Haemat., 63,283,198o; 117./ Sheibani,K.,et al.: Hum.Pathol.,11,175,198o; 118./ Kapadia,S.B.,et al.: Cancer,45,1315,198o; 119./ Wheeler, G.E.: Ann.Intern.Med., 94,361,1981;

Cytogenetical investigations in secondary AL :

12o./ Durant,J.R.,et al.: Am.J.Med.Sci.,254,824,1967; 121./ Novell,P.G.: in Proc.of the International Conference on Leukemia-Lymphoma, Ed.C.J.D. Zarafonetis. Philadelphia,Lea and Feiberg,1968,pp.47; 122 ./ Ezdinli,E.Z.,et al.: Ann.Int.Med., 71,1o97,1969; 123./ Steinberg,M.H.,et al.: Arch.Intern.Med., 125,496,197o; 124./ Bennett,J.M.; Proc.Am.Assoc.Cancer Res., 12,16,1971; 125./ Khaleeli,M.,et al.:Blood,41,17,1973; 126./ Weiden,P.L.,et al.:Blood,42,571,1973; 127./ Rodrigez,V.,et al: Arch.Int.Med.,132,874,1973; 128./Rutten,F.J.,et al.: Br.J.

Haemat.,26,391,1974; 129./ Dahlke,M.B.,et al.: Br.J.Haematol, 31,111,1975; 13o./ Hossfeld,D.K.,et al.: Cancer.Res. 35,28o8, 1975; 131./ Oshimura,M.,et al.: Cancer,38,748,1976; 132./ Meytes,D.,et al.: Acta Haematol.,55.,358,1976; 133./ Gonzales,F et al.: Ann.Intern.Med.,86,44o,1977; 134./ Swolin,B.,et al.: Helsinki Chromosome Conference,1977, Abstract p.214; 135./ Whang-Peng,J.,et al.: Leukemia Res.,1,19,1977; 136./ Rowley, J.D.; Symposium on Chromosomes et Hémopathies Malignes,Poitiers,1978; 137./ Boetius et al.: Br.J.Haemat. 37,1o1,1977; 138./ Whang-Peng,J.,et al.: Cancer,44,1592,1979; 139./Berger, R.,et al.: Blood Cells,7,293,1981; 14o./ Berger,R.,et al.: Nouv Rev.Fr.Hematol.,23,275,1981; ... and also under the following references : 3o,31,32,35,4o,42,43,47,51,52,56,57,58,6o,66, 73,35,87,93,96,1o1,1o7,117.

REFERENCES

1. Abe, S., and M. Sasaki (1982) SCE as an index of mutagenesis and/or carcinogenesis. In *Progress and Topics in Cytogenetics Vol. 2. Sister Chromatid Exchange*, A.A. Sandberg, ed. Alan R. Liss, Inc., New York, pp. 461-514.
2. Bennett, J.M., D. Catovsky, M.T. Daniel, C. Flandrin, D.A.G. Galton, H.R. Gralnick, and C. Sultan (1976) Proposals for the classification of acute leukemias. *Br. J. Haemat.* 33:755-760.
3. Berger, M., M. Habs, and D. Schmähl (1983) Noncarcinogenic chemotherapy with a combination of vincristine, methotrexate and 5-fluorouracil (VMF) in rats. *Int. J. Cancer* 32:131-137.
4. Borum, K. (1980) Increasing frequency of acute myeloid leukemia complicating Hodgkin's disease. A review. *Cancer* 46:1247-1252.
5. Bredberg, A., and B. Lambert (1983) Induction of SCE by DNA cross-links in human fibroblasts exposed to 8-MOP and UVA irradiation. *Mutat. Res.* 118:191-204.
6. Cassel, D.M., and S.A. Latt (1980) Relationship between DNA adduct formation and sister chromatid exchange induction by (^{3}H)-8-methoxypsoralen in Chinese hamster ovary cells. *Exp. Cell Res.* 128:15-22.
7. Der-Sarkissian, H., C. Bonaiti-Pellié, M.L. Briard-Guillemot, and J.M. Zacher (1982) Sister chromatid exchanges in patients with retinoblastoma. *Cancer Genet. Cytogenet.* 7:73-77.
8. Duncan, A.M.V., and H.J. Evans (1982) Molecular lesions involved in the induction of sister chromatid exchange. *Mutat. Res.* 105:423-427.
9. Etiemble, J., C. Picat, and P. Boivin (1982) Enzymopathies induced by drugs and acute leukemia. *Br. J. Haemat.* 52:165-166.
10. Farber, E. (1981) Chemical carcinogenesis. *New Eng. J. Med.* 305:1379-1389.

11. Foucar, K., R.W. McKenna, C.D. Bloomfield, T.K. Bowers, and R.D. Brunning (1979) Therapy-related leukemia. A panmyelosis. Cancer 43:1285-1296.
12. Lambert, B., A. Lindblad, K. Holmberg, and D. Francesconi (1981) The use of sister chromatid exchange to monitor human populations for exposure to toxicologically harmful agents. In Sister Chromatid Exchange, S. Wolff, ed. John Wiley and Sons, New York, pp. 149-182.
13. Latt, S.A., J. Allen, S.E. Bloom, A. Carrano, E. Falke, D. Kram, E. Schneider, R. Schreck, R. Tice, B. Whitfield, and S. Wolff (1981) Sister-chromatid exchanges. A report of the gene-tox program. Mutat. Res. 87:17-62.
14. Natarajan, A.T., A.A. van Zeeland, and F. Degratti (1982) UV-induced SCE and effects of photoreactivation. In Progress and Topics in Cytogenetics Vol. 2. Sister Chromatid Exchange, A.A. Sandberg, ed. Alan R. Liss, Inc., New York, pp. 315-325.
15. Penn, I. (1978) Malignancies associated with immunosuppression or cytotoxic therapy. Surgery 83:492-502.
16. Perry, P., and H.J. Evans (1975) Cytological detection of mutagen-carcinogen exposure by sister chromatid exchange. Nature (Lond.) 258:121-125.
17. Raposa, T. (1982) SCE and chemotherapy of noncancerous and cancerous conditions. In Progress and Topics in Cytogenetics Vol. 2. Sister Chromatid Exchange, A.A. Sandberg, ed. Alan R. Liss, Inc., New York, pp. 579-617.
18. Reischbach, G., E. Gebhart, and R. Cailkan (1982) Sister chromatid exchanges and proliferation kinetics of human metastatic breast tumor cell lines. Anticancer Res. 2:257-260.
19. Rosner, F., and H. Grünwald (1975) Hodgkin's disease and acute leukemia. Report of eight cases and review of the literature. Am. J. Med. 58:339-353.
20. Rowley, J.D. (1983) Human oncogene locations and chromosome abberations. Nature (Lond.) 301:290-291.
21. Sandberg, A.A. (1982) Sister chromatid exchange in human states. In Progress and Topics in Cytogenetics Vol. 2. Sister Chromatid Exchange, A.A. Sandberg, ed. Alan R. Liss, Inc., New York, pp. 619-651.
22. Sieber, S.M., and R.H. Adamson (1975) Toxicity of antineoplastic agents in man: Chromosomal aberrations, antifertility effects, congential malformation, and carcinogenic potential. In Advances in Cancer Research, Vol. 22. Academic Press, New York, pp. 57-155.
23. Smadja, N., M. Krulik, and J. Debray (1982) Etude cytogenétique des ét t -réleucémiques et des leucémiés aiqués secondariés a un traitment par chimotherapié et/ou radiotherapié. Revue de la litérature de 102 observations. Path. Biol. 30:775-783.
24. Valgussa, P., A. Santoro, R. Kenda, F.F. Bellani, and A. Banfi (1980) Second malignancies in Hodgkin's disease: A complication of certain forms of treatment. Br. Med. J. 280:216-220.
25. Wolff, S. (1982) Difficulties in assessing the human health effects of mutagenic carcinogens by cytogenetic analyses. Cytogenet. Cell Genet. 33:7-13.

SISTER CHROMATID EXCHANGE FREQUENCY AND CELL CYCLE KINETICS IN CANCER PATIENTS TREATED WITH CYTOSTATIC DRUGS

Narendra P. Singh and Steven M. D'Ambrosio

Ohio State University
Department of Radiology
College of Medicine
Columbus, Ohio 43220

ABSTRACT

The ability of various cytostatic drugs to induce sister chromatid exchanges (SCE) and to alter the progression of cells through mitosis was analyzed in lymphocytes cultured from cancer patients following various in vivo chemotherapy treatments. Control individuals exhibited 4.87 ± 0.08 SCEs per metaphase. Patients being treated with cyclophosphamide (CP), mitomycin C (MMC), and/or cis-platinum (CPT) in combination with other drugs exhibited 3- to 5-fold greater levels of SCEs. Other cytostatic drugs: the plant alkaloids; vincristine (VCR); and homoharringtonine (HHT); the antibiotic, adriamycin (ADM); the folic acid antagonist, methotrexate (MTX); and the nucleotide analogue, dihydroazacytidine (HAC) did not appear to induce SCE levels significantly above controls. Most of the cytostatic drugs used in cancer patients appeared to delay the progression of cells through mitosis. All the drug protocols which included a known DNA-damaging agent induced SCE. There was no relationship between SCE and cell cycle kinetics. Thus, SCE appears to be a sensitive assay for monitoring the in vivo exposure of individuals to genotoxic agents.

INTRODUCTION

Many chemotherapeutic drugs have been shown to elicit long-term genotoxic effects, especially those classes that have the potential to interact with the DNA. These drugs have been shown by various in vitro and in vivo systems to be mutagenic, carcinogenic, and/or teratogenic. It is thought that individuals being given these types of drugs are at risk for developing neoplasms many years following treatment. In the few studies conducted to assess this possibility, secondary cancers, unrelated to the treated primary cancer, appeared an average of 6.5 yr after termination of the initial treatment (1). Thus, it is of interest to evaluate the in vivo genotoxic effects of current chemotherapeutic protocols, utilizing assays which can be used to monitor human exposure to genotoxic agents.

Sister chromatid exchanges have been used as a sensitive assay to detect the genotoxic effects of various mutagenic and carcinogenic agents in vitro and in vivo (2-12). This assay system has also been used to evaluate the genotoxicity of various types of cytostatic drugs, particularly those used in the treatment of cancer (13-19). These studies, using lymphocytes treated in culture or lymphocytes obtained from cancer patients on chemotherapy, indicate that cytostatic drugs damage the DNA and induce SCEs.

Another parameter which also has been studied extensively in relation to SCE and one which may be of importance in evaluating the toxicity of cytostatic agents in cancer patients is how these drugs affect cellular proliferation. Studies (20,21) have shown that chemotherapeutic agents markedly inhibit the progression of cells through mitosis. As such, the determination of proliferation rates in lymphocyte cultures should be a useful and sensitive indicator of the cellular toxicity of chemotherapeutic agents.

In this paper we report on the effect of various clinical cytostatic drug protocols on the induction of SCE and the perturbation of the cell cycle in lymphocytes isolated after in vivo administration of chemotherapy. The data were compared to lymphocytes isolated from healthy individuals not exposed to any known genotoxic agent, and to an individual being treated with cytostatic drugs for idiopathic thrombocytopenic purpura (ITCP). The chemotherapy-treated cancer patients were given single and combination treatment protocols of various classes of cytotoxic drugs, including an investigational phase I drug, Homoharringtonine.

MATERIAL AND METHODS

Lymphocyte Cultures and Chromosome Preparations

Venous blood was collected from healthy individuals, cancer patients, and an individual having ITCP. The cancer patient and ITCP patient were under prescribed chemotherapy drug protocols. These protocols included the following chemotherapeutic agents: ADM, bleomycin (BLM), cytarabine (CA), CPT, CP, HAC, HHT, fluorouracil (FU), MMC, prednisone (PRD), and vincristine (VCR). Lymphocytes were isolated using standard Ficoll Hypaque techniques. Approximately 0.5×10^6 lymphocytes were cultured in T75 (Corning) tissue culture flasks in 10 ml RPMI 1640 medium buffered with 15 mM HEPES (pH 7.2) containing 20% fetal bovine serum and 2 mM L-glutamine (GIBCO), 10 μg/ml 5'-bromo-2'-deoxyuridine (BrdUrd) and 10 mg/ml phytohemagglutinin (Sigma). Cultures were incubated at 37°C in a shaking water bath for 76 hr. This incubation time allowed us to have more second-division metaphases, as lymphocytes from cancer patients treated with cytostatic drugs showed marked retardation of cell cycle. Four hr before harvesting, 0.1 μg/ml (final concentration) of colchicine (Sigma) was added to the cultures. Cultures were collected in 15 ml polyproylene screw cap tubes (Corning) and centrifuged at 750 rpm for 15 min in a TJ-6 centrifuge (Beckman). Cells were put into a hypotonic solution of 0.075 M KCl for 10 min and centrifuged at 750 rpm for 10 min. The supernatant was aspirated, leaving 0.5 ml. Cells were suspended gently in this volume and fixed by adding dropwise 2 ml of 3:1 fixative (3 methanol:1 glacial acetic acid). After 30 min, cells were pelleted and the supernatant was aspirated, leaving 0.5 ml. Cells were resuspended in this volume and again fixed by adding 2 ml fresh fixative. One or 2 more changes of such fixative were given. Finally, cells were suspended in a volume of fixative to give the appropriate cell density. Air-dried chromosomal preparations were made on precleaned glass slides.

Sister Chromatid Differential Staining

Sister chromatid differential staining was achieved by the method of Goto et al. (22). Briefly, slides were stained in Hoechst 33258 (10 mg/ml in distilled water) for 10 min. Slides were mounted in McIlvaine buffer, pH 8.00 (0.2 M Na phosphate, 0.1 M Na citrate) and placed on a slide warmer at 40°C. They were irradiated with short-wave UV for 35 min at 18 $J/m^2/sec$ using 2 germicidal lamps (General Electric Company) mounted parallel, and 4 cm apart. Slides were placed in 2 SSC (0.3 M NaCl 0.03 M Na citrate) at 58-60°C for 70 min. Staining was accomplished by using 1% stock Giemsa solution in 0.006 M phosphate buffer, pH 6.8, for 5 min with continuous stirring. Permanent chromosomal preparations were made by mounting slides in Permount (Fisher), and SCEs were scored under oil immersion (1,000 X). Data are presented as mean and standard deviation from at least 25 metaphases per lymphocyte culture for SCE. At least 100 metaphases were analyzed for cell cycle analysis.

RESULTS AND DISCUSSION

The effects of various cytostatic drugs, administered to cancer patients, on the induction of SCE in lymphocytes and the in vitro proliferative rate of cells was determined. Table 1 describes the clinical diagnosis, and Tab. 2 compares the drug treatment protocols to the number of SCEs induced and the number of metaphase cells in M1, M2, and M3. As shown in Tab. 2, the frequency of SCE in lymphocytes obtained from 6 healthy donors, ranging in age between 22 and 82 yr, were fairly consistent, i.e., 4.4 to 5.6 SCEs per cell.

Tab. 1. Clinical characteristics of individuals evaluated.

Subject	Age (yr)	Sex	Clinical diagnosis[a]
NF	27	M	Control
KK	34	M	Control
MF	22	F	Control
MS	82	M	Control
SD	33	M	Control
HE	75	M	Control
KL	18	F	adenocarcinoma of the breast
WW	68	M	Non-Hodgkins lymphoma, abdomen
CW	61	M	Large cell carcinoma of the lung
CW	61	M	Large cell lung carcinoma
DH	43	M	Adenocarcinoma of the colon
FS	73	M	Adenocarcinoma of the prostrate
PO	57	M	Adenocarcinoma of the prostrate
RM	51	M	Adenocarcinoma of the maxillary sinus
KE	19	M	Acute lymphatic leukemia
RD	82	F	Acute myeloid leukemia
SG	72	M	Idiopathic thrombocytopenic purura
RS	45	F	Adenocarcinoma of the colon

[a]Based upon histo-pathological diagnosis.

There was a larger variation in the percentage of cells in mitosis among these controls. The percentage of cells in M1 varied from 9.3 in the 33-yr old male, SD, to 41.5 in the 82-yr old male, MS. Patient HE, 75 yr of age, also exhibited a high percentage of cells (30.8) in M1. The number of cells in M2 was more consistent among the individuals treated. The lymphocytes from 3 of the 6 controls indicated that approximately 75% of the cells were in M2, while the other 3 controls showed approximately 55%. These data on cell cycle kinetics agree closely with those reported by Crossen (23) and Abdel-Fadil et al. (20).

Three of the 10 cancer patients were being treated with 150 to 800 mg CP in combination with other drugs (MTX, FU, CA, ADM, and/or CPT). Patients KL, WW, and CW exhibited the highest levels of SCE, 15-23 SCEs per cell. Patient CW, given the highest single dose of CP and well as ADM and CPT, exhibited the highest number of SCEs. Also, CW had been given a combination of FU, VCR, and MMC 22 da

Tab. 2. Induction of SCE and effect on cell cycle kinetics following chemotherapy. (ADR = ADM, CTX = CP in the text.)

Subject	Treatment[a] drug	dose	days	Assay[b] time post treatment	SCE per Cell mean ± S.D.	Percentage of cells in[3] M_1	M_2	M_3
NF	none				4.40±1.32	18.2	75.7	6.1
KK	none				5.55±1.25	21.8	70.4	7.8
MF	none				5.31±1.75	19.5	72.9	7.6
MS	none				3.44±1.39	41.5	54.7	3.8
SD	none				5.56±2.54	9.4	58.0	32.6
HE	none				4.96±1.75	30.8	52.6	16.6
KL[d]	CTX	150 mg	14	12 hr	20.00±8.25	52.2	43.8	3.9
	MXT	100 mg	2					
	FU	10 mg	2					
WW	CTX	690 mg	1	4 da	15.28±4.50	78.7	21.3	0
	CA	80 mg	1	18 hr				
CW[e]	CTX	800 mg	1	24 hr	23.30±4.42	80.1	19.9	0
	ADR	70 mg	1					
	CPT	100 mg	1					
CW	FU	500 mg	4	22 da	12.68±3.79	69.8	29.8	0.40
	VCR	2 mg	4	22 da				
	MMC	17.5 mg	4	22 da				
DH	HHT	8.0 mg	5	2 hr	5.46±1.20	75.2	24.8	0
FS	HHT	5.7 mg	2	2 hr	4.64±1.17	37.1	56.2	6.7
PO	HHT	3.8 mg	2	2 hr	4.92±1.52	32.3	64.9	2.8
RM	CPT	160 mg	2	24 hr	16.79±2.65	84.2	15.8	0
	MXT	80 mg	2					
	BLM	30 mg	2					
KE	VCR	3.8 mg	6	2 hr	6.12±2.83	13.5	32.0	54.5
	MXT	12 mg	6	2 hr				
	PRD	40 mg	6	2 hr				
RD	ADR	60 mg	1	24 hr	5.02±1.64	26.3	67.1	6.6
SG	VCR	7.7 mg	1	24 hr	5.87±1.59	72.4	27.6	0
RS	HAC	4 g	3	24 hr	6.89±2.54	76.1	23.9	0

[a]The drugs were given to the patients for the number of days indicated.

[b]Blood samples for assay were drawn at the time indicated following the last treatment.

[c]M_1, M_2, and M_3 are the metaphases in first, second, and third mitosis.

[d]Patient KL was given MXT and FU on day 6 and on day 14 of therapy while CTX was given every day for 14 days.

[e]Patient CW was treated with FU, VCR and MMC 22 days prior to the CTX, ADR and CPT therapy.

prior to the CP, ADM, and CPT treatment. Patient CW exhibited a high SCE level, 16 SCEs per cell, when assayed 2 hr prior to CW's second treatment protocol. This is similar to other studies (4) showing that SCE levels persist for long periods following MMC treatment. Receiving the lowest single dose of CP but for the longest duration, KL also exhibited high levels of SCE. The difference in SCE levels between WW and CW, who received similar doses of CP, is probably related to CW being given ADM and CPT, and/or that WW was assayed 4 da post CP treatment. All 3 patients showed a marked increase in the number of cells in M1 and a decrease in the number of cells in M2. No cells were observed in M3, indicating that this treatment protocol greatly inhibited cell progression through mitosis. Earlier work using FU, MTX, and CA (24,25) indicated that these drugs do not induce SCEs. Thus, the SCEs observed in the present study in patients KL and WW are probably due to CP and not due to FU, MTX, and CA. A known mutagen (26), a suspected carcinogen (27), and a cross-linking agent (28,29), CPT has been shown to be a potent inducer of SCE in cell culture systems (27). Our data (patients RM and CW) are consistent with Wiencke et al. (30), who showed that this compound was a very potent inducer of SCEs in vivo in humans.

Three patients were being treated only with a Phase I investigational drug, HHT. This drug, a plant alkaloid, is one of a group of cephalotoxine esters and has been shown to exhibit good antileukemia activity (31). The level of SCE in the lymphocytes isolated from these patients was similar in number to the control group, indicating that HHT does not induce SCE in vivo. Homoharringtonine in one of the patients, DH, did alter the progression of cells into M2. Like that observed in the patients described above for the CP combination protocol, a marked decrease in the number of cells in M2 and M3 were noted. No abnormal perturbations of the cell cycle were observed in the other 2 patients, FS and PO, who were given HHT. The reason for this difference may be related to dose and duration of treatment, since DH was given 8.0 mg for 5 da, while FS and PO were given 5.7 and 3.8 mg, respectively, for 2 da. This is the first report on the effect of HHT on SCE and cell cycle kinetics.

Patient RM, being given a combination of CPT, MTX, and BLM, showed elevated levels of SCE, while patient KE, on VCR, MTX, and PRD, did not. These data would suggest that MTX, VCR, and PRD by themselves do not induce SCE. Other studies (5,16,32) have indicated that these drugs do not change SCE levels significantly. The SCE observed in RM is probably due to the CPT. Bleomycin does not induce SCE in vitro (5), while CPT does in vitro and in vivo (12). Patient RM also showed a marked delay in the progression of cells through mitosis. Patient KE, on the other hand, appeared to have a greater number of cells progressing to M3. This could be due to the acute lymphocytic leukemia or to the PRD.

The other cancer patients were on single drug treatments. Treated with ADM, RD did not show an elevated level of SCE nor cell cycle kinetics different from controls. This is different from previous observations (24) showing that ADM increases SCEs over control levels. This difference may be due to the higher doses of ADM (800 mg) used in their patients. Patient RS, treated with HAC, also did not show elevated levels of SCEs. However, the progression of cells into M2 and M3 was markedly delayed. We were able to obtain lymphocytes from a noncancer patient (SG) being treated with VCR for ITCP. The drug VCR did not induce SCEs above controls but it did markedly retard cell cycle progression.

It appears that those drug protocols which include an agent that can interact strongly with DNA, i.e., CP (15), CPT (29), and MMC (32), are potent inducers of SCEs. Other drugs which do not interact with the DNA (i.e., FU, VCR, CA, MXT, and HAC) did not induce SCE above the controls. Most of the drug protocols, regardless of whether the drug could interact with DNA, delayed the cell cycle. Also, there did not appear to be any relationship between the inhibition of the cell cycle and induction of SCEs. These data presented here should provide new and additional information on the genotoxicity of in vivo cytostatic drug treatments, and the usefulness of SCEs in monitoring humans exposed to genotoxic agents.

ACKNOWLEDGEMENTS

We wish to thank Dr. M. Grever and Ms. D. Derocher for helping to obtain the cancer patients' blood samples. This work was supported by U.S. E.P.A. cooperative agreement CR-807268.

REFERENCES

1. Rosner, F. (1976) Acute leukemia as a delayed consequence of cancer chemotherapy. Cancer 37:1033-1036.
2. Perry, P.E., and H.J. Evans (1975) Cytological detection of mutagen-carcinogen exposure by sister chromatid exchange. Nature 258:121-125.
3. Perry, P.E. (1980) Chemical mutagens and sister chromatid exchange. In Chemical Mutagens, Vol. 6, F.J. de Serres and A. Hollaender, eds. Plenum Press, New York, pp. 1-40.
4. Stetka, D.G., J. Minkler, and A.V. Carrano (1978) Induction of long-lived chromosome damage, as manifested by sister chromatid exchange, in lymphocytes of animals exposed to mitomycin C. Mutat. Res. 51:383-396.
5. Littlefield, L.G., S.P. Colyer, and R.J. DuFrain (1980) Comparison of sister chromatid exchanges in human lymphocytes after Go exposure to mitomycin C in vivo vs. in vitro. Mutat. Res. 69:191-197.

6. Brown, R.L., and P.E. Crossen (1976) Increased incidence of sister chromatid exchanges in Rauscher leukemia virus infected mouse fibroblast. Exp. Cell Res. 103:418-420.
7. Nichols, W.W., C.I. Bradl, L.H. Toji, M. Godley, and M. Segawa (1978) Induction of sister chromatid exchanges by transformation with SV40. Cancer Res. 38:960-964.
8. Brown, E.H., and C. Basilico (1982) Induction of sister chromatid exchange by polyoma large viral tumor antigen in transformed rat fibroblasts. Cancer Res. 42:1909-1912.
9. Abe, S., and M. Sasaki (1977) Chromosome aberrations and sister chromatid exchanges in chinese hamster cells exposed to various chemicals. J. Natl. Cancer Inst. 58:1635-1641.
10. Carrano, A.V., L.H. Thompson, P.A. Linde, and J.L. Minkle (1978) Sister chromatid exchange as an indicator of mutagens. Nature 271:551-553.
11. Kato, H., and H. Shimada (1975) Sister chromatid exchanges induced by mitomycin C, a new method of detecting DNA damge at chromosomal level. Mutat. Res. 23:459-464.
12. Popescu, N.C., D. Turnbull, and J.A. DiPaolo (1977) Sister chromatid exchange and chromosome aberration analysis with the use of several carcinogens and non-carcinogens brief communication. J. Natl. Cancer Inst. 59:289-293.
13. Ohtsuru, M., Y. Ishii, S. Takai, H. Higashi, and G. Kosaki (1980) Sister chromatid exchanges in lymphocytes of cancer patients receiving mitomycin treatment. Cancer Res. 40:477-480.
14. Banerjee, A., and W.F. Benedict (1979) Production of sister chromatid exchanges by various chemotherapeutic agents. Cancer Res. 39:797-799.
15. Benedict, W.F., A. Banerjee, and N. Venkatesan (1978) Cyclophosphamide induced oncogenic transformation, chromosomal breakage and sister chromatid exchange following microsomal activation. Cancer Res. 38:2922-2924.
16. Raposa, T. (1978) Sister chromatid exchange studies for monitoring DNA damage and repair capacity after cytostatics in vitro and in lymphocytes of leukaemic patients under cytostatic therapy. Mutat. Res. 57:241-251.
17. Parkes, J.J.G., and S. Scott (1982) A quantitative comparison of cytogenetic effects of anti-tumor agents. Cytogenet. Cell Genet. 33:27-34.
18. Nevstad, N.P. (1977) Sister chromatid exchange and chromosomal aberrations induced in human lymphocytes by the cytostatic drug adriamycin in vivo and in vitro. Mutat. Res. 57:253-258.
19. Ohisuru, M., Y. Ishii, S. Takai, and G. Kosake (1982) Sister chromatid exchanges in lymphocytes of a patient treated with cyclophosphamide and vincristine for non-Hodgkins lymphoma. Gann 73(3):433-438.
20. Abdel-Fadil, M.R., C.G. Palmer, and H. Heerema (1982) Effect of temperature variation on sister chromatid exchange and cell cycle duration in cultured human lymphocytes. Mutat. Res. 104:267-273.

21. Snope, A.J., and J.M. Rary (1979) Cell cycle duration and sister chromatid exchange frequency in cultured human lymphocytes. Mutat. Res. 63:345-349.
22. Goto, K., S. Maeda, Y. Kano, and T. Sugiyoma (1978) Factors involved in differential Giemsa staining of sister chromatids. Chromosoma 66:351-359.
23. Crossen, P.E. (1982) Variation in sensitivity of human lymphocytes to DNA-damaging agents measured by sister chromatid exchange frequency. Human Genet. 60:19-23.
24. Musilova, J., K. Michalova, and J. Urban (1979) Sister chromatid exchanges and chromosomal breakage in patients treated with cytostatics. Mutat. Res. 67:289-294.
25. Littlefield, L.G., S.P. Colyer, A.M. Sayer, and R.J. DuFrain (1979) Sister chromatid exchanges in human lymphocytes exposed during Go to four classes of DNA damaging chemicals. Mutat. Res. 67:259-269.
26. Andersen, K.S. (1979) Platinum (II) complexes generate frameshift mutations in test strains of Salmonella typhimurium. Mutat. Res. 67:209-214.
27. Turnbull, D., N.C. Popescu, J.A. DiPaolo, and B.C. Myhr (1979) Cis-platinum (II) diamine dichloride causes mutation, transformation and sister chromatid exchanges in cultured mammalian cells. Mutat. Res. 66:267-275.
28. Zwelling, L.A., M.O. Bradley, N.A. Sharkey, T. Anderson, and K.W. Kohn (1979) Mutagenicity, cytotoxicity and DNA crosslinking in V-79 Chinese hamster cells treated with cis and trans-Pt(II) diamminedichloride. Mutat. Res. 67:271-280.
29. Oredsson, S.M., D.F. Dean, and L.J. Marton (1982) Decreased cytotoxicity of cis-diamminedichlor-platinum (II) by α-difluoromethylornithine depletion of polyamines in 9L rat brain tumor cells in vitro. Cancer Res. 42:1296-1299.
30. Wiencke, J.K., J. Cervenka, B.J. Kennedy, J. Prlina, and R. Gorlin (1982) Sister-chromatid exchange induction by cis-platinum/adriamycin cancer chemotherapy. Mutat. Res. 104:131-136.
31. Cephalotaxus Research Coordinating Group (1976) Cephalotaxine esters in the treatment of acute leukemia: A preliminary clinical assessment. Chinese Med. J. 2(4):263-272.
32. Ishii, Y. (1981) Nature of the mitomycin C induced lesion causing sister chromatid exchange. Mutat. Res. 91:51-55.

MONITORING PATIENTS ON LONG-TERM DRUG THERAPY FOR GENOTOXIC EFFECTS

I. A. Erskine, J. M. Mackay, and D. P. Fox

Department of Genetics University of Aberdeen
2 Tillydrone Avenue
Aberdeen, AB2 2TN, Scotland

ABSTRACT

The problems associated with the design and conduct of experiments involving surveys of human population sister chromatid exchange (SCE) frequencies are discussed. It is suggested that the problems of variation between culture occasions may be overcome by the rigid control of experimental conditions and the inclusion of the same negative controls (herein called "base controls") on all culture occasions. In addition, all experimental subjects are cultured, as far as possible, at the same time as controls drawn from the same population and matched for age (± 5 yr) and sex. Many factors in such surveys remain uncontrolled but the collection of data on potential environmental mutagen exposure in all subjects is suggested as a mechanism to measure and evaluate such factors.

Data are reported on SCE frequencies in lymphocytes from patients receiving 4 separate drugs for chronic conditions. Using a square root transformation of SCE frequencies and the analysis of variance, there is clear evidence for a rise in SCE frequency in patients receiving sulphasalazine (SASP) and azathioprine (Aza). On the other hand, patients receiving atenolol or chlorpropamide show no evidence of a rise in SCE frequency.

SCE AND POPULATION MONITORING

The justification in using SCE analysis in population monitoring rests upon there being a relationship between SCE and gene mutation on the one hand, and SCE and carcinogenesis on the other. Several experimental studies do show such relationships (1-3) but the ratio of one effect to another varies widely between chemical compounds and various physical agents. As a result, raised SCE levels in populations, while probably indicating DNA damage and a qualitatively increased risk of mutagenesis and carcinogenesis, cannot generally be equated with a quantitative estimate of risk. However, population estimates of SCEs do have the advantage of being a single endpoint which may be produced by many different chemical and physical agents, or a complex mixture of these agents at realistic dose levels and subject to the full panoply of human metabolic modifications.

TECHNICAL PROBLEMS

The final level of SCEs observed in an experiment is the product of many different interacting factors, some of which are technical. Bromodeoxyuridine (BrdUrd) concentration in the culture medium (4), light exposure of the BrdUrd-substituted DNA (5), and culture media used (6) may all have such effects. We have tried to overcome these problems by keeping rigidly to the same experimental protocol on all culture occasions. In addition to using the same concentration of BrdUrd (30.4 μM) and minimizing light exposure by wrapping all cultures in black plastic bags, we also keep to the same batches of chemicals, particularly tissue culture components, for the same series of experiments. However, tissue culture media generally have a storage life of only ∿3 mo, thus, some changes of media batches are usually required. In addition, to limit the possibility of repair of lesions in G_0 lymphocytes (7,8), we transport and store blood samples at 4-8°C as far as possible, and always initiate culture within 24 hr of blood collection.

CULTURE CONDITIONS AND SLIDE STAINING

Venuous blood samples (10 ml) were collected in lithium heparin. Each culture contained the following: nutrient mixture F-10 Ham (1X) with L-glutamine (Gibco) 4.4 ml, RPMI 1640 (1X) with L-glutamine (Gibco) 4.4 ml, newborn calf serum (Tissue Culture Services) 2.2 ml, ∿150 μg/ml phytohemagglutinin (PHA) (Wellcome), 128 μg/ml streptomycin sulphate (Glaxo), 80 units/ml Crystapen (Glaxo), and 9.32 μg/ml BrdUrd. To this mixture 0.8 ml whole blood was added, giving a final volume of 11.8 ml.

Cultures were incubated at 37°C for 72 hr with gentle mixing every 24 hr. Colcemid solution (0.2 ml), giving a final concentration of 0.133 μg/ml) was added 90 min prior to harvesting.

Slides were stained using a modification of the technique of Weber and Hoegerman (9). Slides were first stained in Hoechst 33258 (Calbiochem) 20 μg/ml at 20°C for 60 min and then exposed in a shallow covering of 2 x SSC, face up, to a UV germicidal lamp (Philips

TUV15W) at a distance of 45 cm for 60 min. The slides were subsequently incubated in 2 x SSC at 75°C for 15 min, washed in hot tap water for 5 min, and stained in 3% Giemsa-in-phosphate buffer (pH 6.8) for 15 min.

EXPERIMENTAL DESIGN

Availability of Subjects

In any large human population study, all subjects may not be available at the same time or, if available, it may be impossible for logistical reasons to deal with them concurrently. Hence, it is often necessary to culture individuals drawn from the same population on several occasions with the possibility that this will introduce a significant between-occasions variance to the results. We have attempted to overcome this complication in our experimental design by several measures, including: 1) Culture conditions are kept rigidly constant. 2) On every culture occasion, blood from the same donor (or every member of a group of donors) is cultured as a negative control. We refer to these cultures as base controls. This allows some check on the constancy of culture conditions between occasions (see below). 3) For every experimental subject, a matched control is cultivated, as far as possible, on the same culture occasion.

Matched Controls

We have used age (± 5 yr) and sex-matching of controls, although there is little evidence that either factor (10,11) has much influence upon baseline SCE frequency. However, this procedure may help to equalize other (unknown) factors which do influence SCE frequency. The one factor which we have consistently found to be important in influencing SCE and chromosomal aberration (CA) frequency in our control subjects is smoking (12). Perhaps in the future it would be better to match as far as possible for this factor. Further problems arise in choosing a source for the matched controls. In the case of matched controls for hospital patients, it is probably better to find these from within the hospital population rather than from among colleagues, i.e., from patients with the same condition as the experimental subjects but prior to treatment, or patients in the hospital for minor surgery.

Other Potential Mutagen Exposures

Population surveys seldom compare well with controlled laboratory-based experiments because so many potentially important variables are involved which cannot be controlled. By imposing further conditions upon the choice of matched controls, one would find it increasingly difficult to obtain those controls, especially if they were to be cultured on the same occasion as the patient samples with which they are matched. At the start of this project we made an arbitrary decision to match age and sex and, in addition, to record potential mutagen exposure from the following sources: alcohol consumption, tobacco consumption, drugs, radiations, industrial chemical exposure, and recent viral infections. These data are collected using a standardized questionnaire and, at the same time, the aims of the project may be explained to the participant and informed consent may be obtained. Analysis of the data is usually confined to comparing groups of patients with their matched controls, for the frequency and extent of consumption or exposure in relation to each parameter being recorded, and analysis is important in excluding these agents as the cause of any observed difference.

Positive Controls

We also include positive controls in our experiments to show that the detection system, including the slide reader, is operative. Initially, this involved a 10^{-7} M mitomycin C (MMC) initial addition to blood cultures from a base control donor. This gives a mean SCE frequency of ∿90/cell. More recently, we have reduced the MMC concentration to give an SCE frequency of ∿30/cell, a level which is less easily recognized as coming from a positive control during blind scoring of randomized slides.

Slide Reading

Ideally, all slides from one experiment should be coded and randomized together before scoring. In this way, variations in scoring techniques will be randomized and not confounded with real changes in SCE frequency. This ideal is not easily achieved when samples are collected over several months or years. The normal practice we adopt is to keep the same scorer(s) throughout one survey and to randomize groups of slides from blood samples collected sequentially. We do have some evidence for variation between scorers in recorded SCE frequencies, probably relating to telomeric SCEs involving small pieces of chromosome (see Tabs. 1 and 3). This may be an important factor if scorers are not randomized within an experiment, for example, due to changes in personnel. We score 20 cells from each culture, 4 from each of 5 slides. Only cells with 46 chromosomes are included.

CHOICE OF DRUGS

In this study we deliberately chose drugs which are taken in relatively high doses and often indefinitely, since they are generally used for treating chronic conditions. However, we deliberately avoided chemotherapeutic agents (13) which were likely to be highly mutagenic. The drugs were as follows: 1) Chlorpropamide (94-20-2), an oral hypoglycemic agent used to treat maturity-onset diabetes. There is clear evidence in the literature for an in vitro drug-induced rise in SCE frequency (14) and CAs have been reported in the lymphocytes of patients taking the drug (15). 2) Sulphasalazine (599-79-1), an anti-inflammatory agent which probably acts locally in the bowel to suppress inflammation (16). It is the preferred treatment to prevent relapses in patients suffering from ulcerative colitis (17). There appear to be no reports in the literature indicating that SASP is directly mutagenic, although it is known to induce reversible male sterility in man (18). 3) Atenolol (29122-68-7) is a β-blocker used in the treatment of hypertension. There is no evidence that this drug is mutagenic in vitro or in vivo (19). (4) Azathioprine (446-86-6) is an immunosuppressive drug used to treat a variety of inflammatory conditions as well as to effect the suppression of the immune response following kidney transplantation. There is abundant evidence that azathioprine is mutagenic, both in vivo and in vitro, causing an increase in CAs or micronuclei (20-23). There is some evidence for an effect on SCEs (20) although some workers have obtained negative results (24,25).

RESULTS AND DISCUSSION

Base Controls

Three base controls were used in the reported experiments: BC1 and BC2 for Aza, atenolol, and chlorpropamide, and BC3 for SASP. A summary of the results for the 12 culture occasions involving BC1 and BC2 and the 6 culture occasions involving BC3 is contained in Tab. 1. Sister chromatid exchange frequency distributions (not shown) are typically non-normal with a longer tail on the upper side caused by outlier cells with a high SCE level. By using a square root transformation (26,27) of the data, the distributions become more normal and the variances more equal. This allows the use of second-degree statistical analyses such as the analysis-of-variance. (AOV). The AOV tables for the base controls are shown in Tab. 2. In all 3 cases the between-occasions variance is not significantly greater than the between-cells variance within occasions. The between-occasions variance contains 2 components; technical variation due to differing conditions and biological variation due to a changed response of the same source of cells with time. Clearly, in these 3 subjects, under our rigidly controlled culture conditions, both between-occasions variance components must be minimal compared to the between-cell variance, within occasions. Thus we may compare subjects cultured on different occasions with confidence. This is an important point since in most surveys samples are collected and cultured on a number of occasions, usually over several months.

Patients Taking Drugs

Group mean SCE values (untransformed data) for patients receiving each of the 4 drugs and their matched controls are contained in Tab. 3. These data were subsequently transformed to square roots

Tab. 1. Base control SCE frequencies for each culture occasion.

OCCASION	MEAN SCE/CELL ± S.D. BC 1	BC 2	BC 3
1	9.05 ± 3.69	9.15 ± 3.31	12.35 ± 4.30
2	8.95 ± 4.03	9.20 ± 2.31	12.55 ± 3.19
3	8.90 ± 4.70	9.85 ± 2.37	11.45 ± 3.79
4	8.50 ± 4.21	7.25 ± 3.37	12.50 ± 3.52
5	9.45 ± 2.72	9.45 ± 2.95	10.10 ± 4.22
6	8.20 ± 2.82	7.75 ± 3.18	11.00 ± 4.38
7	9.20 ± 3.32	8.35 ± 2.98	
8	8.40 ± 2.84	9.40 ± 3.59	
9	8.80 ± 3.04	8.55 ± 3.41	
10	10.00 ± 3.42	8.90 ± 3.13	
11	9.70 ± 3.18	8.65 ± 3.51	
12	8.10 ± 2.57	8.65 ± 2.52	
Scorer	1	1	2

Tab. 2. AOV tables for the 3 base controls.

ITEM	DF	MS	VR	P
AZATHIOPRINE, ATENOLOL AND CHLORPROPAMIDE				
BC 1				
Between occasions	11	0.2112	<1	>0.2
Within occasions	228	0.3288		
Total	239			
BC 2				
Between occasions	11	0.4018	1.3608	0.2-0.1
Within occasions	228	0.2953		
Total	239			
SULPHASALAZINE				
BC 3				
Between occasions	5	0.5314	1.4194	>0.2
Within occasions	114	0.3744		
Total	119			

Tab. 3. Group mean SCE frequencies for patients receiving drugs and their matched controls.

Group		Group Mean SCE/Cell ± S.D.			
		Sulphasalazine	Azathioprine	Atenolol	Chloropropamide
Patients		15.87 ± 2.47	12.46 ± 2.00	8.82 ± 1.90	8.91 ± 1.21
Matched Controls		11.59 ± 1.26	9.90 ± 1.74	9.10 ± 1.62	9.71 ± 0.94
Scorer		2	1	1	1
Base Control	1	-		8.94 ± 0.59	
	2	-		8.76 ± 0.74	
	3	11.66 ± 0.99		-	

and the analysis of variance performed (Tab. 4). A number of generalizations may be made concerning these results: 1) The between-cells variance within individuals is very similar to that observed in the base controls (Tab. 2). 2) there is always a significantly greater variance between individuals within groups, than between cells within individuals, and this variance is also greater than the between-occasions variance noted in the base controls. It indicates that there are significant biological differences between individuals within groups. This difference remains when the analysis is confined to the matched controls and is not solely the product of the drug treatment and/or the disease state. 3) Using the between-individuals within groups variance as the error variance, no significant difference is found between patients taking chlorpropamide and their matched controls or between patients taking atenolol and their matched controls. On the other hand, patients taking Aza have a moderately significant increase in SCE frequency over their matched controls, and patients taking SASP have a highly significant rise in SCE frequency over their matched controls.

Retrospective Studies

Retrospective studies of the type reported here can clearly indicate that patients taking a drug for some particular condition have a raised SCE frequency. However, a major drawback of such a design is that 2 factors, the drug treatment and the disease state, will usually be confounded. For example, all the patients taking SASP also had ulcerative colitis. In the case of the patients taking chlorpropamide, the matched controls were maturity-onset diabetics being treated by diet alone. Had there been an SCE rise in this

Tab. 4. AOV tables for the 4 groups of patients receiving drug therapy and their matched controls.

Item	DF	MS	VR	P
SULPHASALAZINE				
Between groups	1	49.2131	37.5960	<0.001
Between individuals within groups	28	1.3090	3.4420	<0.001
Between cells within individuals	570	0.3803		
Total	599			
AZATHIOPRINE				
Between groups	1	12.5649	7.5086	0.05-0.01
Between inds. within groups	14	1.6734	3.4290	<0.001
Between cells within individuals	303	0.4880		
Total	318			
ATENOLOL				
Between groups	1	0.2750	<1	>0.2
Between inds. within groups	20	1.8815	4.8567	<0.001
Between cells within individuals	418	0.3874		
Total	439			
CHLOROPROPAMIDE				
Between groups	1	1.5082	2.3336	0.2-0.1
Between inds. within groups	16	0.6463	2.6952	0.01-0.001
Between cells within individuals	342	0.2398		
Total	359			

case, it could have been attributed to the drug and not the disease state, but the experimental design still could not eliminate the disease state itself as a cause of SCEs.

Prospective Studies

The obvious alternative to retrospective studies is to conduct prospective studies, investigating the same patients after diagnosis but before treatment, and again at intervals after treatment has been implemented. However, this strategy also has problems associated with it. Inevitably, collection of first samples, i.e., after diagnosis but before treatment is sporadic with many culture occasions containing usually only one such blood sample. Second and subsequent samples may be grouped (e.g., due to several patients attending on the same clinic date). Although we did not detect a significant between-occasions variance in our base controls, the SCE technique has the potential to generate such an effect, and we must be careful to guard against confounding variation between culture occasions with variation between first and second (or subsequent) samples. A situation in which such confounding might most easily arise is the clinical trial where all first (or second, or subsequent) samples are likely to be cultured simultaneously. The best way to avoid these difficulties is by culturing matched controls or base controls on all occasions that the experimental samples are cultured. In this way, the disease state and the drug may have their effects measured separately.

REFERENCES

1. Carrano, A.V., L.H. Thompson, P.A. Lindl, and J.L. Minkler (1978) Sister chromatid exchanges as an indicator of mutagenesis. Nature (Lond.) 271:551-553.
2. Nishi, Y. (1984) Interrelationships of SCE, mutation at the HGPRT locus and toxicity in Chinese Hamster V79 cells. In Sister Chromatid Exchanges: 25 years of Experimental Research, R.R. Tice and A. Hollaender, eds. Plenum Press, New York.
3. Parodi, S. (1984) Quantitative predictivity of carcinogenesis for SCE in vivo. In Sister Chromatid Exchanges: 25 years of Experimental Research, R.R. Tice and A. Hollaender, eds. Plenum Press, New York.
4. Schvartzman, J.B., and R.R. Tice (1982) 5-Bromodeoxyuridine and its role in the production of sister chromatid exchange. In Progress and Topics in Cytogenetics, Vol. 2., Sister chromatid Exchange, A.A. Sandberg, ed. Alan R. Liss, Inc., New York, pp. 123-134.
5. Ikushima, T., and S. Wolff (1974) Sister chromatid exchanges induced by light flashes to 5-bromodeoxyuridine and 5-iododeoxyuridine substituted Chinese Hamster chromosomes. Exp. Cell Res. 87:15-19.
6. Morgan, W.F., and P.E. Crossen (1981) Factors influencing sister chromatid exchange rate in cultured human lymphocytes. Mutat. Res. 81:395-402.
7. Lambert, B., M. Sten, and D. Hellgren (1984) Removal and persistence of SCE-inducing damage in human lymphocytes in vitro. In Sister Chromatid Exchanges: 25 years of Experimental Research, R.R. Tice and A. Hollaender, eds. Plenum Press, New York.
8. Littlefield, L.G., S.P. Colyer, A.M. Sayer, and R.J. DuFrain (1984) Persistence of SCE-inducing lesions after G_0 exposure of human lymphocytes to differing classes of DNA-damaging chemicals. In Sister Chromatid Exchanges: 25 years of Experimental Research, R.R. Tice and A. Hollaender, eds. Plenum Press, New York.

9. Weber, K.E., and S.F. Hoegerman (1980) Timing of endoreduplication induced by colcemid or radiation in BudR-labelled human lymphocytes. Exp. Cell Res. 128:31-39.
10. Morgan, W.F., and P.E. Crossen (1977) The incidence of sister chromatid exchanges in cultured human lymphocytes. Mutat. Res. 42:305-312.
11. Crossen, P.E., M.E. Drets, F.E. Arrighi, and D.A. Johnston (1977) Analysis of the frequency and distribution of Sister Chromatid Exchange in cultured human lymphocytes. Human Genet. 35:345-352.
12. Fox, D.P., F.W. Robertson, T. Brown, and J.D.M. Douglas (1984) Chromosome aberrations in divers. Undersea Biomedical Research (in press).
13. Littlefield, L.G., S.P. Colyer, A.M. Sayer, and R.J. DuFrain (1979) Sister chromatid exchanges in human lymphocytes exposed during G_0 to four classes of DNA-damaging chemicals. Mutat. Res. 67:259-269.
14. Brown, R.F., and Y. Wu (1977) Induction of sister chromatid exchanges in Chinese Hamster cells by chlorpropamide. Mutat. Res. 56:215-217.
15. Watson, W.A.F., J.C. Petrie, D.B. Galloway, I. Bullock, and J.C. Gilbert (1976) In vivo cytogenetic activity of sulfonylurea drugs in man. Mutat. Res. 38:71-80.
16. Goldman, P. and M.A. Peppercorn (1975) Sulfasalazine. New Engl. J. Med. 293:20-23.
17. Dissanayake, A.S., and S.C. Truelove (1973) A controlled therapeutic trial of long-term maintenance treatment of ulcerative colitis with sulphasalazine (Salazopyrin). Gut 14:923-926.
18. Birnie, G.G., T.I.F. McLeod, and G. Watkinson (1981) Incidence of sulphasalazine-induced male infertility. Gut 22:452-455.
19. Presta, M., C. Mazzocchi, S. Ziliani, and G. Ragnotti (1983) In vitro and in vivo DNA damage of male and female rat liver nuclei by oncogenic and non-oncogenic beta blockers. J. Natl. Cancer Inst. 70:747-752.
20. Kučerová, M., V. Kočandrle, V. Matoušek, Z. Polivková, and I. Reneltova (1982) Chromosomal aberrations and sister chromatid exchanges (SCEs) in patients undergoing long-term Imuran therapy. Mutat. Res. 94:501-509.
21. Van Went, G.F. (1978) Mutagenicity testing of 3 hallucinogens: LSD, psilocybin and Δ^9-THC, using the micronucleus test. Experientia 34:324-325.
22. Matter, B.E., P. Donatsch, R.R. Racine, B. Schmid, and S. Suter (1982) Genotoxic evaluation of cyclosporin A, a new immunosuppressive agent. Mutat. Res. 105:257-264.
23. Van Went, G.F. (1979) Investigation into the mutagenic activity of azathioprine (Imuran[R]) in different test systems. Mutat. Res. 68:153-162.
24. Apelt, F., J. Kolin-Gerresheim, and M. Bauchinger (1981) Azathioprine, a clastogen in human somatic cells? Analysis of chromosome damage and SCE in lymphocytes after exposure in vivo and in vitro. Mutat. Res. 88:61-72.

25. Schinzel, A., and W. Schmid (1976) Lymphocyte chromosome studies in humans exposed to chemical mutagens. The validity of the method in 67 patients under cytostatic therapy. Mutat. Res. 40:139-166.
26. Whorton, E.B., Jr. (1984) Statistical design, analysis and inference issues in studies using SCE. In Sister Chromatid Exchanges: 25 years of Experimental Research, R.R. Tice and A. Hollaender, eds. Plenum Press, New York.
27. Wulf, H.C. (1984) Guidelines for the statistical evaluation of SCE. In Sister Chromatid Exchanges: 25 years of Experimental Research, R.R. Tice and A. Hollaender, eds. Plenum Press, New York.

PRACTICAL APPLICATIONS OF THE SCE STUDIES FOR GUIDING AND IMPROVING CHEMOTHERAPY

A. Kourakis, J. Dozi-Vassiliades, P. Hatzitheodoridou, J. Tsiouris,[1] and D. Mourelatos[2]

Department of General Biology
Faculty of Medicine
Aristotelian University
Thessaloniki, Greece

ABSTRACT

The effect of diphylline (DP) or (1,2-dihydroxy-3-propyl)-theophylline and theobromine (TB) on sister chromatid exchange (SCE) rates induced in vitro by cytosine arabinoside (AraC) was studied in normal human lymphocytes. The combined treatments with AraC plus DP or TB showed the potentiating ability of the latter drugs.

In a combined in vivo and in vitro study, lymphocytes taken from 14 patients suffering from various types of cancer who had been given Cytoxan (5 patients) or AraC (9 patients) by injection 3 hr before and then treated with DP or TB in vitro were found to have synergistically increased exchange rates. This has implications for interpreting the repair processes involved and for monitoring drug combinations that synergistically damage DNA in vivo and in vitro.

INTRODUCTION

Tumors that grow most rapidly are often the most sensitive to chemotherapy. This might be obvious for chemotherapeutic regimens utilizing antimetabolites that effect DNA synthesis, but it is not

---

[1]and Pediatric Clinic, Saint Sophia University Hospital, Thessaloniki, Greece.

[2]Presenter at the International Sister Chromatid Exchange Symposium.

apparent why other regimens are also more effective against the rapidly proliferating cancers. Whatever the source of this differential activity, it probably involves the induction of DNA damage in cells during S phase. Therefore, agents that enhance the S phase-dependent damage of existing chemotherapeutics may be of interest. The desirable increase in the cells' sensitivity to S phase damage might be obtained by increasing the toxicity of sublethal damage. This increased toxicity might be achieved by increasing or prolonging the intracellular level of active drug or by inhibiting the capacity of the cell to repair the drug's damage. Evidence is presented in this paper to indicate that methylated xanthines (MXs) may be useful in increasing the toxic response by inhibiting repair. Many antitumor drugs exert their cytotoxic effect principally as a result of the chemical alteration of DNA of precisely the kind which can be repaired by the DNA repair mechanisms.

We used MX with AraC in vitro, as well as MX with either AraC or Cytoxan in a combined in vivo and in vitro study, to determine whether SCE rates are an extremely sensitive indicator of mutagen exposure (1). Antitumor treatments provide models of controlled mutagen human exposure which facilitate the comparison of the sensitivity and specificity of various mutagen-assay systems (1,2). Sister chromatid exchanges in vivo have been proposed as a method of evaluating chemotherapy (3,4). Methylated xanthine has been reported to enhance the effect of several SCE-inducing agents, including antitumor drugs, in cultured normal human lymphocytes (5-8).

MATERIALS AND METHODS

Lymphocyte cultures were prepared by adding 4 drops of whole blood from normal subjects to 4 ml of chromosome medium 1A (Gibco). The cultures were incubated for 72 hr at 37°C, and metaphases were collected during the last 2 hr with colchicine at 0.5 μg/ml. For SCE demonstration, 5-bromodeoxyuridine (BrdUrd) was present at 4 μg/ml for 2 rounds of replication. To obtain a sufficient number of mitoses, treatment with AraC was given 48 hr after initiation of culture. When required, MX was present immediately after the introduction of AraC into the culture. Throughout, all cultures were maintained in the dark to minimize photolysis of BrdUrd. For studying the effect of cytostatics on the frequency of SCEs in vivo, phytohemagglutinin (PHA)-stimulated peripheral blood cultures were set up from 14 patients suffering from cancer [8 from acute lymphocytic leukemia (ALL), 1 from acute myelomonoblastic leukemia (AML), 2 from neuroblastoma (NBL), 1 from nephroblastoma (NPBL), 1 from non-Hodgkin lymphoma (NHL), and 1 from ovary's teratoma (OTER)]. The ages of the patients varied from 3 to 17 yr. The effect of DP or TB in vitro on SCEs induced by AraC (Upjohn) or Cytoxan (Asta Werke) in vivo was also studied. The culture conditions were the same as above, except that no cytostatics were added to the cultures. Air-dried preparations were stained by the fluorescence-plus-Giemsa (FPG) procedure (9) and scored for SCEs. The slides were coded and a minimum of 20 well-spread metaphases with clear differentiation between chromatids were scored from each culture.

RESULTS

The results presented in Tab. 1 indicate that, when DP or TB was added to the cultures exposed to AraC the effect of either DP or TB was synergistic. The SCE rate was consistently much higher than that expected by the simple addition of the effects of AraC and MX. Methylated xanthine alone, at the concentration tested throughout these experiments (120 μg/ml), induced only a small increase in SCEs which varied from donor to donor (Tabs. 1-3). Tables 2 and 3 present the results obtained after the combined study in vivo and in vitro. The blood samples were taken from the patients before the initiation of the therapy and 3 hr after administration of a single dose of AraC (doses varied from patient to patient within the range of 28 to 480 mg) or Cytoxan (dose variation: 130 to 1,080 mg). The 2 drugs acted synergistically with MX on the induction of SCEs in patients' lymphocytes. The frequency of SCEs in the patients' own lymphocyes with and without exposure to MX was determined before the cytostatic therapy, and was used as a control for later comparison in each individual case. In all patients the number of SCEs per cell fluctuated largely between mitoses in the same sample. This phenomenon was also observed by Raposa (10), and may be attributed to the effect of the cytostatics and/or by the cell cycle characteristics of circulating lymphocytes (11). The relative ability of the MXs at the concentration of 120 μg/ml to potentiate the production of SCEs induced in vivo by AraC appears to be DP > TB (Tab. 2). The synergism between Cytoxan taken in vivo and either DP or TB given subsequently in vitro appears to be equally strong (Tab. 3). The synergism between Cytoxan and DP or TB is greater than the one between AraC and DP or TB (Tabs. 2 and 3).

Tab. 1. Comparison of the effect of AraC on SCE rates in the presence and absence of a nontoxic concentration of a methylxanthine.

Ara-C (ng/ml)	Without Methylxanthine Mean SCE/cell ± S.E.M.	Methylxanthine Mean SCE/cell ± S.E.M.
		DP, I20 μg/ml
200	I0.8 ± I.04	I4.6 ± 0.72
Control	9.3 ± 0.48	9.8 ± 0.76
		TB, I20 μg/ml
200	I0.8 ± 0.64	I5.2 ± 0.76
Control	9.7 ± 0.45	II.7 ± I.02

Tab. 2. SCE induction by AraC taken in vivo by leukemic patients and the potentiating effect by MX given subsequently in vitro.

Patient (type of leukemia)	Without Methylxanthine SCE/cell ± S.E. (range)	DP, I20 μg/ml SCE/cell ± S.E. (range)	TB, I20 μg/ml SCE/cell ± S.E. (range)
I (ALL)	Before treatment		
	Control: I0.6 ± 0.8	II.7 ± 0.7	I2.I ± I.I
	(5–I4)	(8–20)	(7–I9)
	After in vivo		
	treatment: I3.4 ± 0.9	I9.3 ± I.6	I8.2 ± I.3
	(8–23)	(7–49)	(6–26)
2 (ALL)	Before treatment		
	Control: 9.9 ± I.2	I0.2 ± I.2	II.4 ± 2.I
	(2–I7)	(4–30)	(7–I9)
	After in Vivo		
	treatment: I4.3 ± 0.9	I8.9 ± I.5	I8.5 ± I.2
	(8–23)	(6–35)	(I3–29)
3 (ALL)	Before treatment		
	Control: I0.I ± 0.6	I2.2 ± I.I	I2.8 ± 0.9
	(6–I5)	(8–I8)	(9–I6)
	After in vivo		
	treatment: I0.7 ± I.I	I5.I ± I.I	I5.0 ± I.I
	(7–24)	(8–20) R	(9–I8)
4 (ALL)	Before treatment		
	Control: I2.I ± I.I	I3.2 ± I.I	I3.7 ± I.2
	(6–22)	(6–22)	(8–26)
	After in vivo		
	treatment: I5.3 ± I.6	24.7 ± 2.6	I9.I ± I.3
	(8–30)	(I0–57)	(8–33)
5 (ALL)	Before treatment		
	Control: I0 ± 0.9	II.4 ± 0.7	II.2 ± 0.7
	(5–I6)	(6–I5)	(4–I4)
	After in vivo		
	treatment: I2.4 ± I.I	I6.7 ± I.0	I7.9 ± 0.6
	(8–26)	(I2–29)	(6–28)
6 (ALL)	Before treatment		
	Control: 8.4 ± 0.7	8.7 ± 0.8	
	(5–I7)	(2–20)	
	After in vivo		
	treatment: 8.9 ± 0.7	I2.8 ± 0.9	
	(2–I4)	(2–33)	
7 (AML)	Before treatment		
	Control: 8.3 ± 0.9	I0.6 ± I.4	
	(2–I9)	(3–22)	
	After in vivo		
	treatment: I0.3 ± I.3	I6.9 ± I.9	
	(5–2I)	(3–32)	
8 (ALL)	Before treatment		
	Control: I2.2 ± 0.8	I2.6 ± I.2	
	(7–2I)	(4–20)	
	After in vivo		
	treatment: I3.5 ± 0.8	20.2 ± 2.I	
	(7–I8)	(I8–29)	
9 (ALL)	Before treatment		
	Control: I0.9 ± I.I	II.6 ± I.8	
	(5–I7)	(7–20)	
	After in vivo		
	treatment: I0.8 ± 0.5	I5.9 ± I.0	
	(6–I4)	(6–72)	

Tab. 3. SCE induction by Cytoxan taken in vivo by patients and the potentiating effect by methylxanthine given subsequently in vitro.

Patient (type of cancer)	Without Methylxanthine SCE/cell ± S.E. (range)	DP, I20 µg/ml SCE/cell ± S.E. (range)	TB, I20 µg/ml SCE/cell ± S.E. (range)
I (NHL)	Before treatment		
	Control: I0.5 ± 0.7 (7–I9)	II.9 ± 0.8 (9–23)	I2.7 ± 0.9 (5–I8)
	After in vivo treatment: I4.3 ± 0.7 (I0–23)	20.8 ± I.3 I0–29)	22.0 ± I.4 (6–28)
2 (OTER)	Before treatment		
	Control: I3.2 ± I.0 (I0–27)	I6.3 ± I.3 (7–30)	I7.2 ± I.3 (5–I8)
	After in vivo treatment: 25.4 ± I.I (6–35)	39.4 ± 2.3 (30–50)	36.7 ± 2.0 (I4–48)
3 (NBL)	Before treatment		
	Control: 9.I ± 0.9 (2–20)	9.I ± 0.6 (4–I6)	
	After in vivo treatment: I2.3 ± 0.6 (5–I7)	I7.8 ± I.3 (I0–29)	
4 (NBL)	Before treatment		
	Control: 8.8 ± 0.9 (2–I5)	I0.9 ± 0.8 (3–26)	
	After in vivo treatment: I4.4 ± 0.8 (8–2I)	20.7 ± 0.9 (II–34)	
5 (NPBL)	Before treatment		
	Control: 9.6 ± 0.7 (6–I4)	I3.8 ± I.4 (6–I7)	
	After in vivo treatment: 22.5 ± 2.3 (5–32)	32.4 ± 2.4 (I0–40)	

DISCUSSION

Successful excision repair, prior to S phase, removes damage which might otherwise give rise to SCEs. Even a small effect in the efficiency of this repair system may be sufficient to bring about an increase in the number of incompletely repaired lesions at the time the cells reach S phase. Subsequently, they may give rise to SCEs (5,6). In human lymphocytes, interference by MXs with the excision repair of cytostatic-induced DNA damage would lead to an increase in the number of incompletely repaired lesions at the time the cells reach S phase in vitro, lesions which may subsequently give rise to SCEs (Tabs. 1-3). The synergistic increase in SCE frequency induced by MXs may be similar to the effect of 3-aminobenzamide (3AMB) which acts synergistically with methylmethanesulfonate in the induction of SCEs (12). Like 3AMB, MX is an inhibitor of poly-(ADP- ribose)-polymerase, albeit a weak one (13). Poly-(ADP-ribose)-polymerase participates in DNA excision repair (14).

The synergistic interaction between alkylating agents and poly-(ADP-ribose)-polymerase inhibitors may be of use in the treatment of human cancer (7,8,15). The main action of caffeine in its enhancement of the antitumor effects of X-rays and alkylating agents in the hamster (16) and in the diminution of 4-nitroquinoline-1-oxide-induced lung tumor production in the mouse is inhibition of DNA repair (17). In combination with vitamin A, caffeine significantly enhanced the antitumor effect of 1,3-bis(2-chloroethyl)-1-nitrosourea against murine leukemia and this antitumor action was also attributed to the ability of caffeine to inhibit DNA repair (18).

Cytosine arabinoside, which acts not by base alkylation but by inhibition of DNA replication due to its effect on DNA and RNA polymerases, induces SCEs (10). AraC and TB or DP, present during the second cell cycle, synergistically increased the frequency of SCE in vitro (Tab. 1). We have already reported that caffeine synergistically increased the frequency of SCEs induced by AraC during the second cell cycle (19). The results from the in vivo and in vitro study indicate (Tabs. 2 and 3) that cytostatics induce lesions in the DNA of G_0 lymhpocytes in vivo that can subsequently be expressed as SCEs after culture in vitro. The data from the 14 patients show that damage induced in vivo by Cytoxan or AraC can be potentiated subsequently by MX in vitro (Tabs. 2 and 3). These results may be used for further consideration in trying to improve and guide the cytostatic therapy. This proposition is supported by Raposa's findings which indicate that the duration of the higher frequencies of SCEs in lymphocytes of leukemic patients treated with AraC was longer when the therapeutic response to the drug was clinically better (4). In view of the suggestion of an increased DNA excision repair as the pathogenesis of chronic lymphocytic leukemia (20), clinical studies with cytostatics and MXs may be expected. Sister chromatid exchange analysis may be useful in identifying individual differences in the capacity for repair of chemically induced DNA damage (10). Improvement of cancer chemotherapy could result from better combinations of existing agents, or from a more effective use of already established single agents.

In the future it may be envisaged that the ability of MXs (or other drugs that interfere with DNA repair mechanisms in man) to amplify the in vitro SCE response of human lymphocytes to cytostatics may be of value as a test system for enhancing the therapeutic ratio of combinations of drugs. By selecting cytostatics dependent on active cellular transport for cell penetration and using appropriate MX scheduling, an increase in the clinical effectiveness of these cytostatics might be achieved (7,15). The partial substitution of antitumor therapeutic regimens by reduced therapeutic regimens combined with a nontoxic concentration of an MX may be preferable in view of the side effects which may arise after antineoplastic therapy (8,19). MXs at the concentrations tested in these experiments (7,8,15,19) may be found to be especially suitable for initial trials in the future because of their low toxicity (16).

REFERENCES

1. Perry, P., and H.J. Evans (1975) Cytological detection of mutagen carcinogen exposure by SCE. Nature (Lond.) 258:121-125.
2. Nevstad, N.P. (1978) SCEs and chromosome aberrations induced in human lymphocytes by the cytostatic drug adriamycin in vivo and in vitro. Mutat. Res. 57:253-258.
3. Ohtsuru, M., Y. Ishii, S. Takai, S. Higashi, and G. Kosaki (1980) SCEs in lymphocytes of cancer patients receiving MMC treatment. Cancer Res. 40:477-480.
4. Raposa, T. (1981) Some practical applications of the SCE studies in human leukemic patients. Mutat. Res. 85:239-240.
5. Faed, M.J.W., and D. Mourelatos (1978) Enhancement by caffeine of SCE frequency in lymphocytes from normal subjects after treatment by mutagens. Mutat. Res. 49:437-440.
6. Ishii, Y., and M. Bender (1978) Caffeine inhibition of prereplication repair of MMC-induced DNA damage in human lymphocytes. Mutat. Res. 51:419-425.
7. Mourelatos, D. (1979) Enhancement by caffeine of SCE frequency induced by antineoplastic agents in human lymphocytes. Experientia 35:822-824.
8. Mourelatos, D., J. Dozi-Vassiliades, and A. Granitsas (1982) Antitumor alkylating agents act synergistically with methlyxanthines on induction of SCEs in human lymphocytes. Mutat. Res. 104:243-247.
9. Perry, P., and S. Wolf (1974) New Giemsa method for the differential staining of sister chromatids. Nature (Lond.) 251:156-158.
10. Raposa, T. (1978) SCE-studies for monitoring DNA damage and repair capacity after cytostatics in vitro and in lymphocytes of leukemic patients under cytostatic therapy. Mutat. Res. 57:241-251.
11. Beek, B., and G. Obe (1974) The human lymphocyte test system. II. Different sensitivities of subpopulations to a chemical mutagen. Mutat. Res. 24:395-398.
12. Morgan, W.F., and J.E. Cleaver (1982) 3-Aminobenzamide synergistically increases SCEs in cells exposed to methylmethanesulfonate but not to ultraviolet light. Mutat. Res. 104:361-366.
13. Purnell, M., and W. Whish (1980) Novel inhibitors of poly(ADP-ribose)synthetase. Biochem. J. 185:775-777.
14. Durkacz, B., O. Omidiji, D. Gray, and S. Shall (1980) (ADP-ribose)$_n$ participates in DNA excision repair. Nature (Lond.) 283:593-596.
15. Byfield, J.E., J. Murnane, J.E. Ward, P. Calabro-Jones, M. Lynch, and F. Kulhanian (1981) Mice, man, mustards and methylated xanthines: The potential role of caffeine and related drugs in the sensitization of human tumours to alkylating agents. Br. J. Cancer 43:669-683.
16. Gaudin, D., K.L. Yieldin, A. Stabler, and G. Brown (1971) The effect of DNA repair inhibitors on the response of tumours treated with X-ray and alkylating agents. Proc. Soc. Exp. Biol. Med. 137:202-206.

17. Nomura, T. (1976) Diminution of tumorigenesis initiated by 4-NQO by post-treatment with caffeine. Nature (Lond.) 260:547-549.
18. Cohen, M.H. (1972) Enhancement of antitumour effect of 1,3-bis-(2-chloroethyl)-1-nitrosourea by vitamin A and caffeine. J. Natl. Cancer Inst. 48:927-932.
19. Mourelatos, D., J. Dozi-Vassiliades, V. Tsigalidou-Balla, and A. Granitsas (1983) Enhancement by methylxanthines of SCE frequency induced by cytostatics in normal and leukemic human lymphocytes. Mutat. Res. 121:147-152.
20. Huang, A.T. (1975) Increased excision DNA repair as pathogenesis of a Human Leukemia. In Molecular Mechanisms for Repair of DNA, P. Hanawalt and R.B. Setlow, eds. Plenum Publishing Corporation, New York, pp. 805-806.

GENOTOXICITY OF ANTIAMEBIC, ANTHELMINTIC, AND ANTIMYCOTIC DRUGS IN HUMAN LYMPHOCYTES

P. Ostrosky-Wegman, G. García, L. Arellano, J. J. Espinosa, R. Montero, and C. Cortinas de Nava

Instituto de Investigaciones Biomédicas
Universidad Nacional Autónoma de México
Apartado Postal 70228
04510 México, D.F., México

INTRODUCTION

The use of amebicidal, anthelmintic, and antimycotic drugs in Mexico is widespread. Few studies have evaluated the mutagenic activity of the most commonly used antiparasitic drugs. Of these, the majority of the mutagenic evaluations have been conducted in microbial test systems and have yielded inconsistent results (1-5). Experiments in our laboratory using the Salmonella typhimurium microsomal test system have shown that niclosamide (anthelmintic) and dehydroemetine (amebicide) induce frameshift mutations following activation by microsomal enzymes (6). While neither chloroquine diphosphate (amebicide) nor ketoconazole (antimycotic) display mutagenic activity on the standard Ames test (unpublished observations for ketoconazole), the former induces frameshift mutations in the fluctuation test. Mutagenic activity has also been detected in the urine of mice treated with niclosamide (6). The capacity of chloroquine diphosphate to bind DNA is well documented (7), and it has also been reported to induce clumping in nuclei, as well as to produce chromosomal abnormalities in a bone marrow cell line (8). In view of these data we decided to evaluate the genotoxic effects produced in vitro in human lymphocytes, cultured with or without the addition of an auxiliary metabolic test system, by 4 antiparasitic drugs: chloroquine diphosphate, deyhdroemetine, niclosamide, and ketoconazole. Furthermore, we monitored chromosomal changes in lymphocytes of 2 patients treated chronically with ketoconazole.

MATERIALS AND METHODS

Drugs and Chemicals

The following drugs were a generous gift of pharmaceutical companies located in Mexico: chloroquine diphosphate (Winthrop), dehydroemetine (Roche), and niclosamide (Bayer). Ketoconazole was used as the pharmaceutical preparation (Nizoral, Janssen Pharmaceutica). All solutions were freshly prepared before each experiment. Chloroquine diphosphate and dehydroemetine were dissolved in distilled water, and the two other drugs in dimethylsulfoxide (DMSO). Glucose 6-phosphate and NADP (cofactors) were obtained from Sigma; DMSO from Merck; cyclophosphamide from Schering Mexicana; and Aroclor 1254 from Analabs, Inc.

Cell Culture

Aliquots of whole heparinized blood (0.4 ml) were cultured following standard procedures (9) at 37°C in 5 ml McCoy's 5a modified medium (Microlab) or RPMI 1640 medium (Gibco) to which had been added 50 I.U. penicillin-50 μg streptomycin (Gibco), 0.2 ml phytohemagglutinin (Gibco or Microlab), and 30 μg 5-bromodeoxyuridine (Sigma). No serum was added to the culture medium. Cell cultures were harvested after 48 or 72 hr. Colcemid (2 μg; Gibco) was added to the cultures 2 hr before harvesting.

The percentage of cells in the first, second, and third cycle of division was determined by analyzing 100 metaphases per sample. The frequency of chromosomal aberrations (CAs) (including gaps and pulverizations, each scored as one anomaly) was analyzed using 100 first-division cells, while the number of sister chromatid exchanges (SCEs) was determined in 25 second-division cells per sample.

In Vitro Treatment

Without auxiliary metabolic activation system. Blood from 2 healthy volunteers was used in these experiments. Various concentrations of dehydroemetine (6-48 ng/ml) and chloroquine diphosphate (0.2-20 μg/ml) were added at the initiation of the cultures, and were incubated with the lymphocytes for 48 or 72 hr. Niclosamide (2-8 μg/ml) and ketoconazole (3-50 μg/ml) were added to the cultures after 24 hr of incubation. The samples containing niclosamide were centrifuged after 2 hr, the supernatant was replaced with fresh growth medium, and the cells were incubated until the end of a 72-hr growth period. Cultures with ketoconazole were incubated in the presence of this compound until the end of a 48-hr growth period. Controls were carried out using the vehicle (distilled water or DMSO) alone.

With the exception of chloroquine diphosphate, higher doses of the drugs could not be tested because of their inhibitory effects on cell growth.

With auxiliary metabolic activation system. Blood obtained from 4 healthy volunteers was used in these experiments. Twenty-four hr after culture initiation, cofactors and liver homogenate containing microsomal enzymes from male rats stimulated with Aroclor 1254 (S-9 mix) (10) were added to the lymphocytes together with various concentrations of niclosamide (2-8 µg/ml) dissolved in DMSO. Incubation was continued for 2 hr, the samples were centrifuged, the supernatant was replaced with fresh growth medium, and the cells were cultured for a total of 72 hr. Cyclophosphamide (4 µg/ml) was employed as a positive control for this system since it displays cytogenetic activity in the presence of S-9 mix. The conditions of incubation were the same as for niclosamide. Controls were also carried out with DMSO alone.

In Vivo Treatment

Ketoconazole. Two patients were followed in this study, a mother (age 44) and son (age 10). Ketoconazole (200 and 100 mg, respectively) was administered daily for 509 and 211 da, respectively, to treat a mycosis of the nails provoked by *Trichophyton rubrum*. Blood lymphocytes from both patients were studied before (control), during, and after therapy.

RESULTS

Dehydroemetine was able to induce structural CAs as well as a marked inhibition in the mitotic cycle, reflected by an increase of cells in the first-generation and a reduction of second-generation cells (Fig. 1). Due to this toxic effect, we were not able to determine its capacity to induce SCEs at higher doses. The other amebicide, chloroquine diphosphate, also increased the frequency of CAs, but even at the highest dose (20 µg/ml) did not induce SCEs or alter the cell cycle (Fig. 2).

In a nonactivated system, in vitro administration of niclosamide induced a small clastogenic effect in 1 of the blood sample donors (2) at the highest dose which still permitted cell division (4 µg/ml), while in 2 other blood samples (3 and 4) it inhibited cell division and had a cytotoxic effect. In contrast, in the presence of the S-9 metabolic activation system, the anthelmintic exhibited a marked clastogenic effect and weak SCE activity in lymphocytes from donor 1 (Fig. 3). The response of lymphocytes from donors 2, 3, and 4 to niclosamide, tested in 2 experiments using this system, varied; the only increase in clastogenicity was found in blood from donor 2 (Fig. 3 and Tab. 1). We also observed a great deal of variability in the growth of the lymphocytes from different donors (Fig. 4). The results using cyclophosphamide demonstrated that the test system was functioning adequately (Tab. 1).

In lymphocytes exposed to ketoconazole in vitro, an increase in CAs (mainly gaps and chromatid breaks) was observed as the dose was

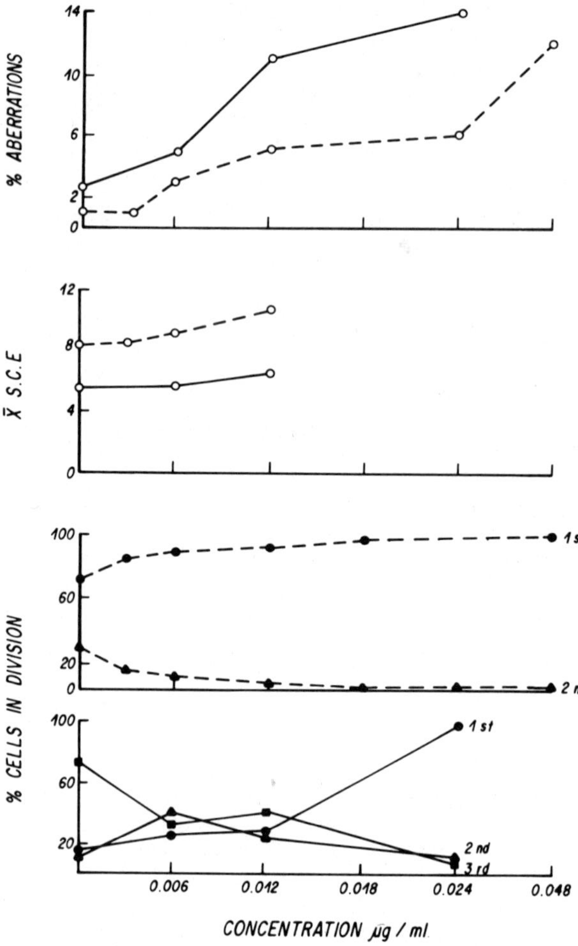

Fig. 1. Effects of various concentrations of dehydroemetine on human lymphocytes incubated in vitro with the amebicide for 48 hr (dashed lines) and 72 hr (solid lines).

increased to 6 µg/ml, and then remained constant at higher concentrations (Fig. 5). An increase in first-division cells, as well as a reduction in second-generation cells, was also detected. The frequency of SCEs did not differ from the control values.

In a study of 2 patients under prolonged treatment with ketoconazole, CAs were observed to increase after 140 da of treatment. The frequency peaked 274 da after treatment was initiated in patient 1, and 84 da after suspension of treatment in patient 2 (Fig. 6). We suspect that this effect could be associated with an undiagnosed viral or bacterial infection. The increase in CAs was maintained in both patients even after treatment was terminated. The average number of SCEs showed little variation throughout the study.

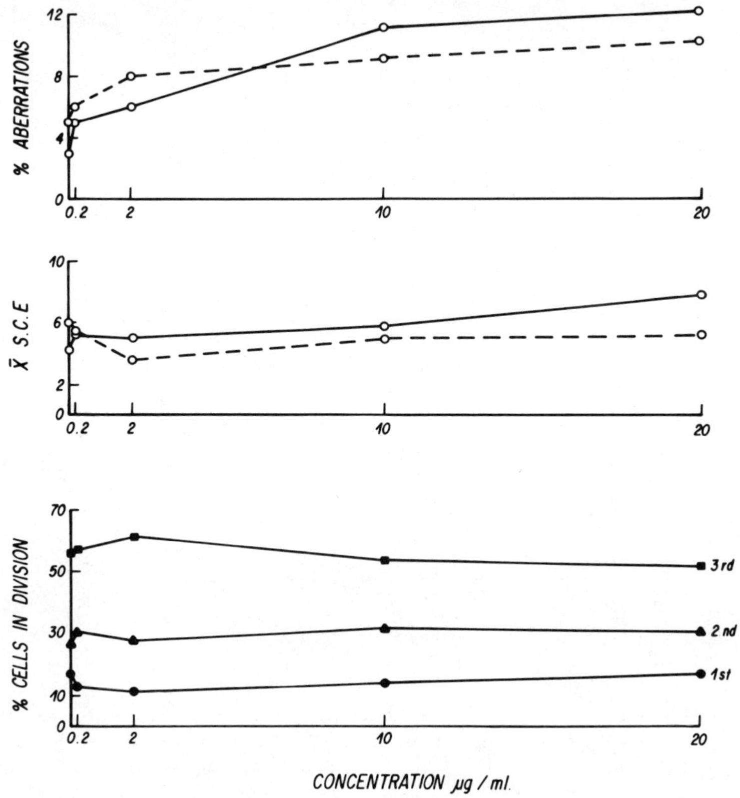

Fig. 2. Effects of various concentrations of chloroquine diphosphate on human lymphocytes incubated in vitro with the amebicide for 48 hr (dashed lines) and 72 hr (solid lines).

DISCUSSION

Two amebicides (dehydroemetine and chloroquine diphosphate), an anthelmintic (niclosamide), and an antimycotic drug (ketoconazole) were found to be very poor inducers of SCEs in human lymphocytes, or to have no effect. Nevertheless, both of the amebicides induced chromatid gaps and breaks in cultured lymphocytes. Previously, we have shown that dehydroemetine and chloroquine diphosphate behaved as intercalant mutagens in the Salmonella typhimurium test system, but the former was active only in the presence of rat liver microsomal enzymes (S-9 mix) (6). As dehydroemetine inhibited cell proliferation in lymphocytes in the absence of S-9 mix, we plan to study its effects on SCE in the presence of this metabolic activation system. Higher doses of chloroquine diphosphate should also be tested to firmly establish that it does not induce SCEs.

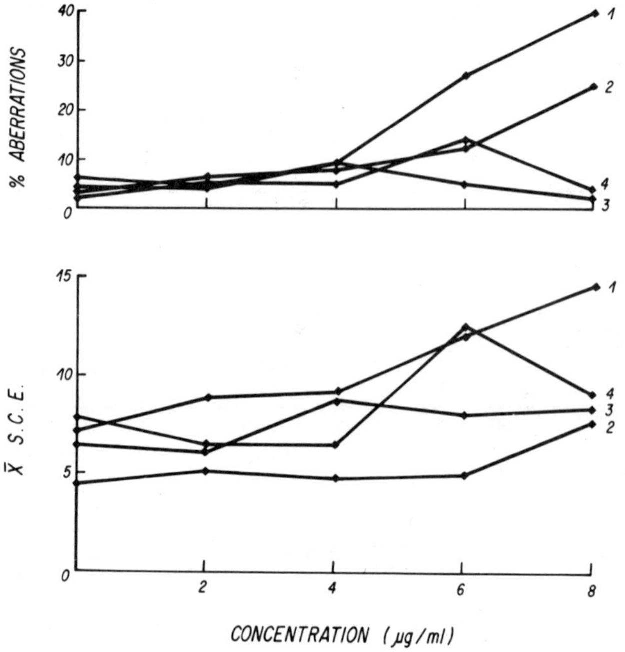

Fig. 3. Effects of various concentrations of niclosamide on the frequency of CAs (including pulverizations) and SCEs in human lymphocytes from 4 healthy donors (traces 1-4). The cells were incubated in vitro in the presence of an S-9 metabolic activation system. Following 24 hr of culture, the anthelmintic was added for 2 hr, the cells were centrifuged, and fresh medium was added. The lymphocytes were cultured for a total of 72 hr. The data represented here are those obtained for donors 2, 3, and 4 in experiment 2, shown in Tab. 1.

In agreement with recent studies showing that lymphocytes from different subjects vary in their response to chemical mutagens (Ref. 11; see Oikawa et al. in this volume), niclosamide was found to induce dissimilar effects on lymphocytes from 4 donors incubated in vitro with the antiparasitic drug in the presence of S-9 mix. Only 2 of the blood sample donors (1 and 2) showed an increase in CAs, and a small increase in SCEs was found in donor 1. Even though small variations were found in the response of 3 blood sample donors (2, 3, and 4) to cyclophosphamide, there was a definite increase in both the frequency of CAs and SCEs in all samples. These results suggest differences in lymphocyte susceptibility to the genotoxic effects of therapeutic drugs, and also indicate the need to study samples from at least 3 donors in the evaluation of weak mutagens.

Tab. 1. Cytogenetic effects of niclosamide and cyclophosphamide on human lymphocytes cultured in vitro.

Blood Sample	Treatment + S-9*	No. Aberrations/100 Cells			No. Pulverizations/100 Cells			SCEs**		
		Experiment 1	2	Average	Experiment 1	2	Average	Experiment 1	2	Average
2	Control	4	3	3.5	0	0	0	6.9	4.4	5.6
	Niclosamide									
	2 μg/ml	24	6	15.0	16	0	8.0	11.0	5.1	8.0
	4	8	8	8.0	0	0	0	6.8	4.8	5.8
	6	8	13	10.5	7	0	3.5	9.6	5.0	7.3
	8		10			15			7.8	
	Cyclophosphamide									
	4 μg/ml	24	12	18.0	11	4	7.5	30.6	26.5	28.5
3	Control	5	4	4.5	0	0	0	7.6	6.4	7.0
	Niclosamide									
	2 μg/ml	2	4	3.0	0	0	0	7.0	6.1	6.5
	4	5	9	7.0	0	0	0	8.0	8.9	8.4
	6	3	5	4.0	0	0	0	5.0	8.2	6.6
	8		2			0			8.4	
	Cyclophosphamide									
	4 μg/ml	17	25	21	0	0	0	37.2	29.8	33.5
4	Control	10	6	8.0	0	0	0	7.0	7.8	7.4
	Niclosamide									
	2 μg/ml	9	5	7.0	0	0	0	7.7	6.5	7.1
	4	9	6	7.5	1	0	0.5	8.2	6.5	7.3
	6	10	9	9.5	1	4	2.5	7.0	12.6	9.8
	8		4			0			9.3	
	Cyclophosphamide									
	4 μg/ml	22	14	18.0	1	0	0.5	32.1	27.8	29.9

* Metabolic activation system mediated by male rat liver homogenate plus added cofactors.
** Mean of sister chromatid exchanges (SCEs) for 25 second division cells.

After more than 140 da of treatment with ketoconazole, lymphocytes from 2 patients showed only a small clastogenic effect. In contrast, this antimycotic drug had a marked clastogenic effect on lymphocytes exposed in vitro. In the Salmonella typhimurium assay, it did not induce mutations with or without the addition of S-9 mix (unpublished results).

The greater degree of damage observed when lymphocytes are exposed to mutagenic agents in vitro as compared to in vivo has previously been reported in various studies (e.g., see Refs. 11,12). The low values obtained in in vivo lymphocyte studies should nevertheless be considered as an indication of potential risk, in particular when genotoxic effects are clearly evident in vitro. More comparative studies are needed to establish the relationship between the genotoxic effects induced in human lymphocytes in vitro and those induced in lymphocytes of subjects exposed medically, occupationally, or environmentally to chemical agents.

In agreement with Gebhart (14), our data also suggest that SCEs and structural CAs are different cytogenetic endpoints, and that both need to be evaluated in any study of genotoxic substances.

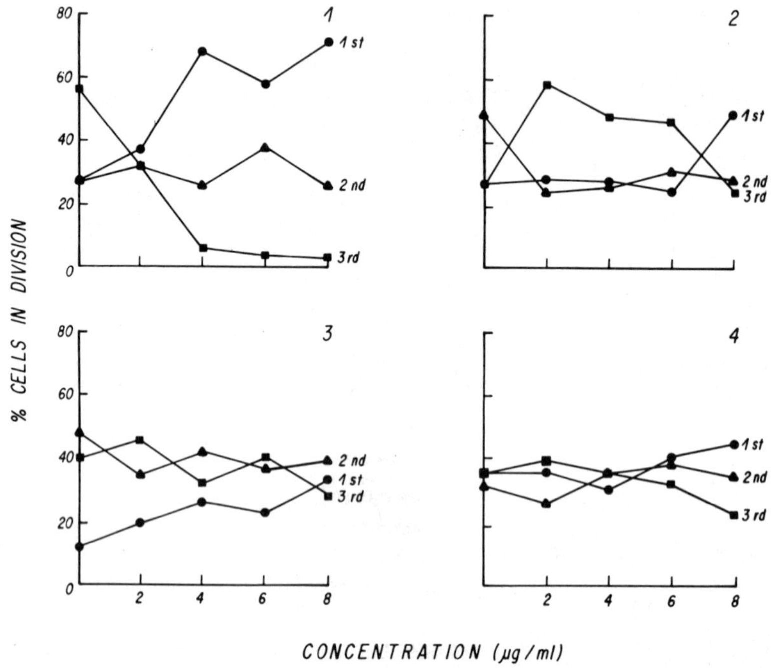

Fig. 4. Effect of various concentrations of niclosamide on cell division in human lymphocytes from 4 healthy donors (panels 1-4). The cells were incubated in vitro in the presence of an S-9 metabolic activation system. The conditions were the same as those described in Fig. 3.

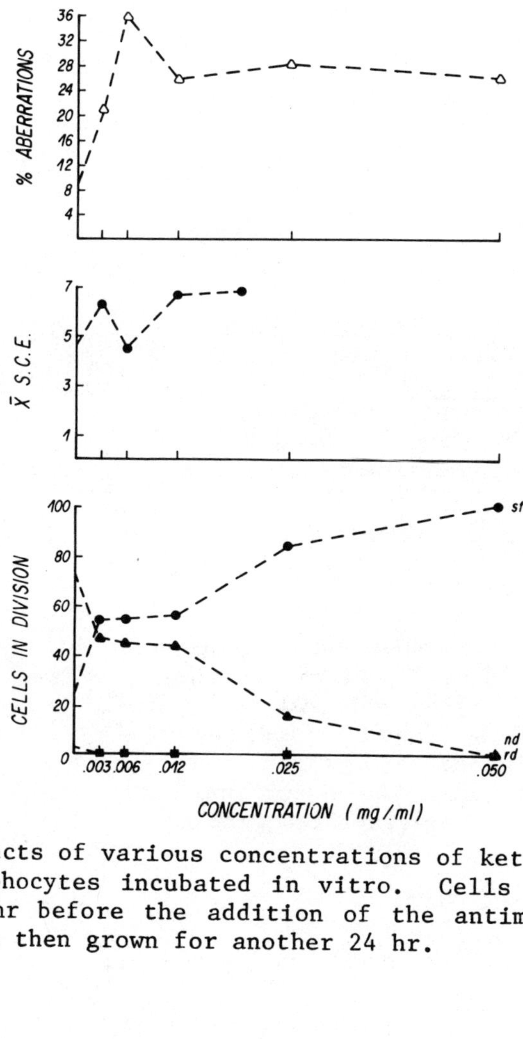

Fig. 5. Effects of various concentrations of ketoconazole on human lymphocytes incubated in vitro. Cells were cultured for 24 hr before the addition of the antimycotic agent, and were then grown for another 24 hr.

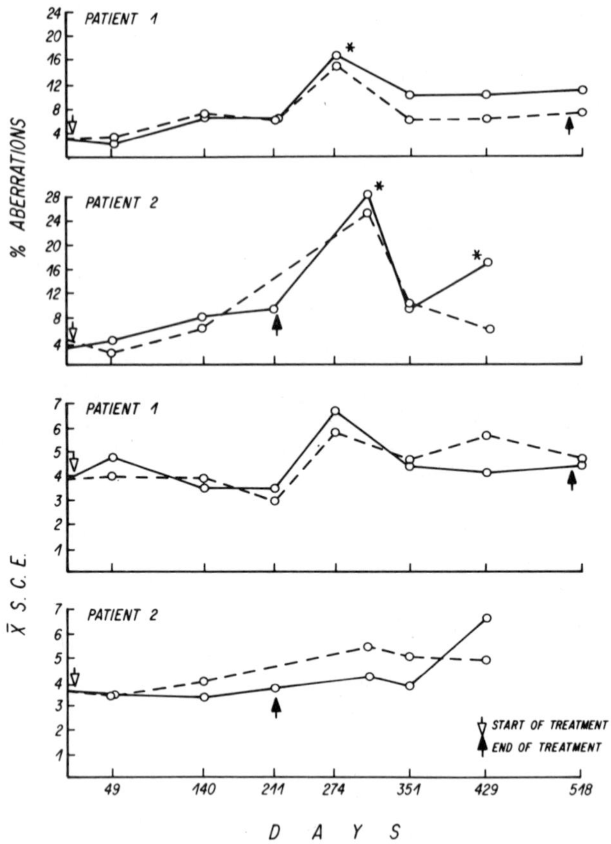

Fig. 6. Effects of chronic administration of ketoconazole on the lymphocytes of 2 patients (patient 1: female, age 44; patient 2: male, age 10). The cells were cultured for either 48 hr (dashed line) or 72 hr (solid line). The initiation and termination of treatment are indicated for each patient. The asterisks indicate that the marked peaks could be associated with microbial contamination.

ACKNOWLEDGEMENTS

This work was partially supported by the "Programa Universitario de Investigación Clínica, U.N.A.M." We wish to thank Mrs. Yolanda Sánchez, Mr. Enrique Vázquez, and Mr. José Avilés for their assistance, and Ms. Marcella Vogt for editorial advice. We also want to express our gratitude to Dr. Raymond Tice for several valuable discussions.

REFERENCES

1. Tutikawa, K., N. Shimoi, and Y. Yagi (1978) Mutagenicity of the products generated by a reaction between chloroquine and nitrite. Mutat. Res. 54:230.
2. MacPhee, D.G., and D.M. Podger (1977) Mutagenicity tests on anthelmintics: Microsomal activation of viprynium embonate to a mutagen. Mutat. Res. 48:307-312.
3. Schüpbach, M.E. (1979) Mutagenicity evaluation of the two antimalarial agents chloroquine and mefloquine, using a bacterial fluctuation test. Mutat. Res. 68:41-49.
4. Rosenkranz, H.S., B. Gutter, and W.T. Speck (1976) Mutagenicity and DNA-modifying activity: A comparison of two microbial assays. Mutat. Res. 41:61-70.
5. Epstein, S.S., J. Andrea, W. Bass, and I. Bishop (1972) Detection of chemical mutagens by dominant lethal assay in mouse. Toxicol. Appl. Pharmacol. 23:288-325.
6. Cortinas de Nava, C., J. Espinosa, L. García, A.M. Zapata, and E. Martínez (1983) Mutagenicity of antiamebic and anthelmintic drugs in the Salmonella typhimurium microsomal test system. Mutat. Res. 117:79-91.
7. Hahn, F.E. (1968) Interaction of antimalarials with DNA. Hoppe Seylers Z. Physiol. Chem. 349:955-956.
8. Siddique, K.M., and J. Biesele (1974) Chromosomal and mitotic damages by chloroquine diphosphate, an antimalarial and antirheumatic drug, on cultured human cells. Tex. J. Sci. 21:125-126.
9. Moorhead, P.S., P.C. Nowell, W.J. Mellman, D.M. Battips, and D.A. Hungerford (1960) Chromosome preparations of leukocytes cultured from human peripheral blood. Exp. Cell Res. 20:613-616.
10. Ames, B.N., J. McCann, and E. Yamasaki (1975) Methods for detecting carcinogens and mutagens with the Salmonella mammalian-microsome mutagenicity test. Mutat. Res. 31:347-364.
11. Crossen, P.E. (1982) Variation in the sensitivity of human lymphocytes to DNA-damaging agents measured by sister chromatid exchange frequency. Human Genet. 60:19-23.
12. Stevenson, A.C., and C. Patel (1973) Effects of chlorambucil on human chromosomes. Mutat. Res. 18:333-351.
13. Weinstein, D., I. Mauer, and H.M. Solomon (1972) The effect of caffeine on chromosomes of human lymphocytes. In vivo and in vitro studies. Mutat. Res. 16:391-399.
14. Gebhart, E. (1981) Sister chromatid exchange (SCE) and structural chromosome aberration in mutagenicity testing. Human Genet. 58:235-254.

SISTER CHROMATID EXCHANGE INDUCTION BY CIGARETTE SMOKE

Julian M. Hopkin

Department of Medicine
University of Birmingham
Edgbaston, Birmingham, England

SUMMARY

This paper presents evidence that cigarette smoke condensate (CSC) is a potent inducer of sister chromatid exchanges (SCEs) in cultured human lymphocytes; that benzo(a)pyrene (BP) contributes very little to this activity; that smokers have higher SCE rates than nonsmokers; that smokers with untreated lung cancer have consistently higher basal and CSC-induced SCE rates than their matched heavy smoking controls; and that, on a weight-for-weight basis, CSCs from different tar categories of cigarettes induce similar numbers of SCEs.

These results are in keeping with the evidence that many smokers die of lung cancer, and that the basis of malignant transformation may be an alteration in cellular DNA. The results raise questions about possible innate differences in individuals' responses to cigarette smoke, and about the "safeness" of lower tar cigarettes.

INTRODUCTION

The collection of our data stemmed from the question, given that we accept the formidable epidemiological evidence that cigarette smoking causes lung cancer (1), what factors allow some genuinely heavy smokers to escape this fate? Environmental and occupational factors are known to influence risk but do not account for the outcome in the majority of smokers. Community-based studies on the families of individuals with lung cancer and smoking matched controls show that there is a very significant familial influence on the risk of developing lung cancer which compounds the risk associated with cigarette smoking (2). This relationship raises the possibility that genetic factors may be important in determining the carcinogenic response to cigarette smoke. Attempts to account for individual variation in response in terms of differences in BP metabolism have not produced consistent results (3). Based on the steadily accruing evidence that malignant transformation involves a change in cellular DNA (4) and that induction of SCEs offers a method of detecting mutagen exposure in human cells (5,6), we elected to test the SCE-inducing properties of cigarette smoke. Our study set out to ask the questions: was cigarette smoke a potent inducer of SCEs in human lymphocytes? Did lymphocytes from different individuals, when exposed to similar amounts of cigarette smoke, produce different SCE rates? Were any such differences related to the risk of developing lung cancer?

MATERIALS AND METHODS

Cigarette smoke condensates were prepared on an automatic smoking machine that simulated human smoking in terms of puff volume and frequency (7). Condensates were dissolved in dimethylsulfoxide (DMSO). Cultures were set up with lymphocytes from heparinized venous blood using a whole-blood microculture technique with RPMI 1640 medium with fetal calf serum (16%), glutamine (0.8%), phytohemagglutinin (1%), penicillin (100 U), streptomycin (100 μg) and 5-bromodeoxyuridine (BrdUrd) (25 μM). To a series of 10 ml cultures, 20 μl of DMSO alone and DMSO containing different dosages of smoke condensate were added at the start of incubation. All cultures were grown for 72 hr in the dark (each receiving Colcemid solution for the final 3 hr), collected, fixed, and air-dried preparations made. SCE staining was carried out according to the BrdUrd-Hoechst 33258-Giemsa technique (8) except that the preparations were exposed to 20 w black light tubes at 5 cms for 30 min and incubated in 2 x SSC at 60°C for 20 min before staining in 2% Giemsa at pH 6.8. The number of SCEs were recorded in 20 second-generation metaphase cells from each culture.

RESULTS AND DISCUSSION

Our results are shown in the accompanying figures. We found that CSC, produced by a cigarette smoking machine, increased SCE frequency and that this increase seemed to occur at surprisingly low dosages. The SCE frequency was detectably increased by 0.1 mg of condensate and nearly doubled by 0.5 mg, these dosages representing 1/400 and 1/80, respectively, of a single high tar cigarette (Fig. 1). In terms of equivalence of effect, the yield of SCEs following exposure to 0.5 mg of condensate was equal to that obtained in our exposure of human lymphocytes to the powerful mutagens

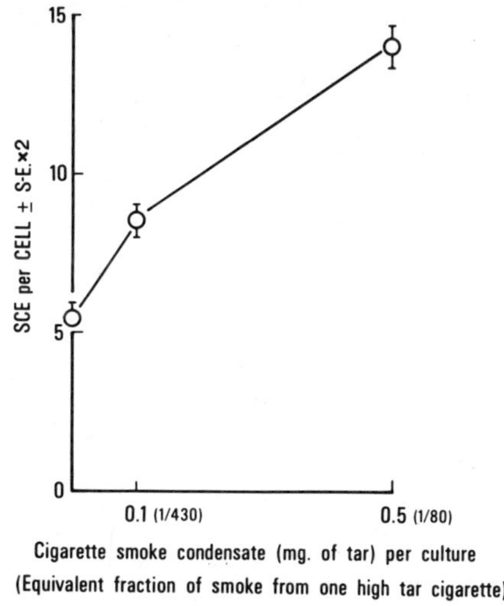

Fig. 1. Pooled data on SCE frequencies (± S.E. x 2) from 2 donors exposed to smoke condensates from each of 3 brands of cigarettes.

aflatoxin B_1 and mitomycin C at concentrations of 10^{-4} M and 5 x 10^{-8} M, respectively. Although BP has been regarded as the most potent carcinogen in cigarette smoke, we found that 2.5 µg of BP, an amount equivalent to that in the smoke of 50-250 cigarettes, was required to produce a near-doubling rate in SCEs in vitro. These results certainly suggested that BP contributed very little to the SCE-inducing properties of cigarette smoke, and were confirmed by further experiments performed on Chinese hamster ovary (CHO) cells and human cells (see Figs. 2, 3, and 4) (9). In CHO cells, BP required the presence of microsomal enzymes in the form of S-9 mix in order to produce SCEs, whereas cigarette smoke did not. In human cells, we found that an inhibitor of the microsomal metabolism of BP, α-naphthoflavone (ANF), inhibited the SCE-inducing properties of BP but not cigarette smoke. The possibility that BP was the important constituent of smoke but that its effect only became obvious when potentiated by the other constituents of cigarette smoke was refuted by our results in Fig. 4, where we found no such effect. These results with BP may to some extent explain the difficulties that have arisen in trying to relate differences in BP metabolism in humans to risk of developing lung cancer. While our results with CSC and lymphocytes in vitro cannot predict the response of bronchial epithelium in vivo, they do show that the levels of agents that can react with cellular DNA must be particularly high in the respiratory airways of inhaling cigarette smokers. This in vitro lymphocyte system did offer an opportunity to examine variation in individuals' responses to smoke, and to attempt to relate this variation to risk of developing lung cancer.

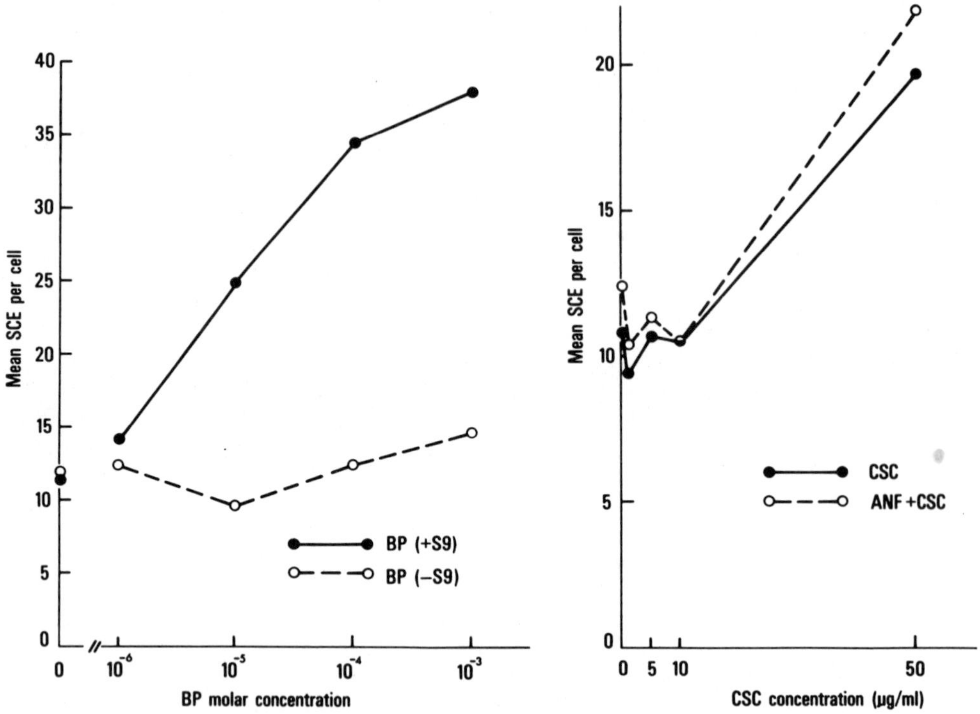

Fig. 2. SCE induction in CHO cells by (a) BP in the presence and absence of S-9 mix, (b) CSC in the presence and absence of 5 μg/ml ANF.

Our next study (10), therefore, used the same methods for examining 4 groups of individuals: A) 10 healthy smokers, none of whom to our knowledge had been unduly exposed to any mutagen other than cigarette smoke; B) 10 healthy nonsmokers who were examined in pairs, matched for age and sex with members of group A; C) 12 patients with histologically proven but untreated, nondisseminated anaplastic or squamous cell bronchial carcinomas, in whom diagnostic biopsies had been obtained by fiberoptic bronchoscopy under local anaesthetic. Ten of these patients were continuing to smoke at the time of testing (C_1), but 2 had stopped smoking 2 and 5 yr previously (C_2). D) 10 control patients with disorders, but none with any form of cancer. These individuals were on average 10 yr older and had smoked similar numbers of cigarettes per day, but for 10 yr longer than our lung cancer patients. All individuals in groups C and D were inhaling smokers and smoked their cigarettes to comparable butt lengths. Only 1 individual had been exposed to a recognized occupational risk factor (asbestos). All were hospital inpatients, none of whom was acutely ill, and individuals from both groups had received similar exposure to X-rays and drugs. Our results (shown in Tabs. 1 and 2 and summarized in Fig. 5), show that

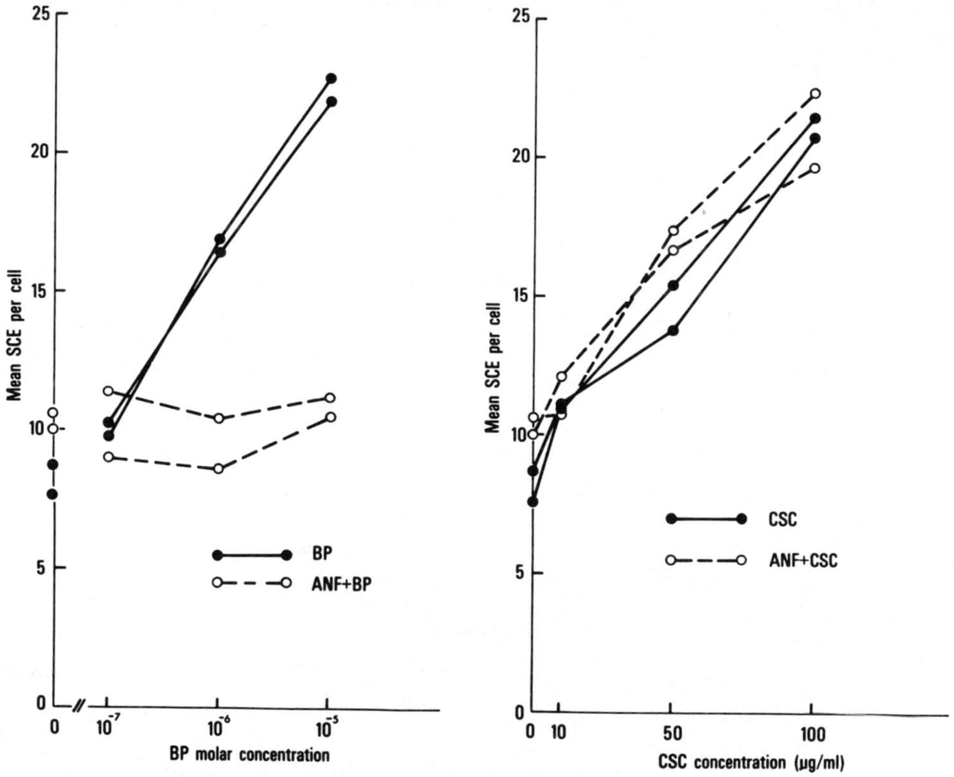

Fig. 3. SCE induction in human lymphocytes by (a) BP, (b) CSC in presence and absence of ANF.

CSC produced dose-related increases in SCEs in lymphocytes from all individuals. Healthy smokers (A) had, relative to healthy nonsmokers (B), higher mean basal and also higher induced SCE rates and the heaviest smokers had the highest SCE rates. A simple study of 2 smokers and 2 nonsmokers showed that these differences were not dependent on the presence of autologous plasma in the whole blood. This finding suggests that the greater number of SCEs produced by condensate in smokers' cells in comparison with nonsmokers must reflect some cellular alteration following exposure to smoke in vivo. For patients with lung cancer who were continuing to smoke (C_1), SCE rates were significantly higher at all levels of testing than for the smoking matched cancer controls (D), and this result was unaffected when allowances were made for the differing number of plain and filter cigarette smokers in the 2 groups. The 2 ex-smokers with lung cancer (C_2) had responses similar to those of healthy nonsmokers, suggesting that the high SCE rates present in our lung cancer group were not simply secondary to development of that disease.

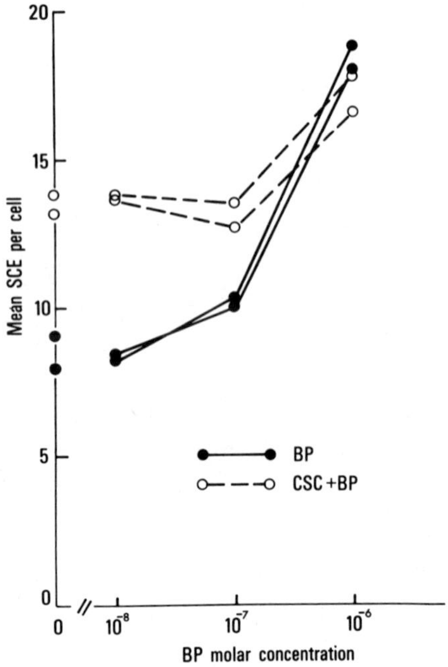

Fig. 4. SCE induction in human lymphocytes by BP in the presence and absence of 50 μg/ml of CSC.

Although we have shown that cigarette smoking and the number and type of cigarettes smoked significantly influenced SCE rates, it is also clear that smokers seemingly exposed to similar amounts of cigarette smoke in vivo can have significantly different SCE rates (groups C_1 and D). The differences raise the possibility that genetic factors may have influenced the extent of DNA damage suffered. This would not be an unlikely result in a species such as man where rates of DNA and protein polymorphism are high and would also be in keeping with the findings that there is a significant familial risk of lung cancer which compounds that due to cigarette smoking alone (2). These conclusions are, however, somewhat uncertain since the

Tab. 1. Frequency of SCEs in cultured lymphocytes from healthy smokers and nonsmokers exposed to CSC.

		CONDENSATE DOSE mg	
CIGARETTE TYPE	CONTROL	0.1	0.5
BRITISH HIGH TAR	7.2	8.8	13.6
BRITISH MIDDLE TAR		8.4	14.1
BRITISH LOW TAR		8.6	14.2
AMERICAN MIDDLE TAR		9.8	13.0

Tab. 2. Frequency of SCEs in cultured lymphocytes from patients with lung cancer and matched smoking controls exposed to CSC.

	Age	Sex	Cigarettes/ Day	Dose of CSC**				
				0	0.1	0.5	1.0	
B Non-smokers	47	M		7.5	9.4	11.6	13.6	
	45	F		7.2	8.4	11.1	16.8	
	29	F		7.1	8.0	12.7	15.8	
	24	M		7.8	9.5	11.6	16.0	
	21	M		7.5	7.6	12.7	15.8	
	75	M	(ex-smoker of 17 years)	7.0	9.9	12.2	18.3	
	73	F		7.0	9.1	12.7	19.4	
	59	M		7.2	8.2	15.1	18.0	
	48	M		8.5	9.5	12.8	15.2	
	54	F		7.5	9.1	14.0	17.9	
	47.6			7.43(.14)	8.87(.24)	12.65(.38)	16.68(.55)	Mean (SE) of individual means
A Smokers	47	M	15	6.6	9.7	13.4	15.4	
	54	F	20	7.3	11.7	14.1	17.7	
	32	F	20	8.5	11.9	15.3	19.0	
	24	M	15	8.2	10.5	14.0	16.5	
	20	M	15	6.8	10.0	12.8	15.8	
	70	M	10	8.0	13.8	17.0	22.3	
	75	F	12	10.5	11.9	16.2	-	
	64	M	10	6.5	7.5	11.1	16.9	
	53	M	35	9.4	12.0	15.1	17.5	
	54	F	30	12.0	14.6	17.5	20.1	
	49.3		18.3	8.38(.57)	11.36(.65)	14.62(.64)	17.91(.74)	Mean (SE) - overall
				9.30(.89)	12.55(.65)	15.5 (.72)	18.53(.60)	Mean (SE) ≥ 20 cigarettes/day
				7.76(.62)	10.57(.79)	14.08(.89)	17.38(1.26)	Mean (SE) < 20 cigarettes/day

* Mean of 20 cells per culture.

** CSC mg per 10 ml culture containing 10^6 lymphocytes (diluent, dimethylsulphoxide 20 µl)

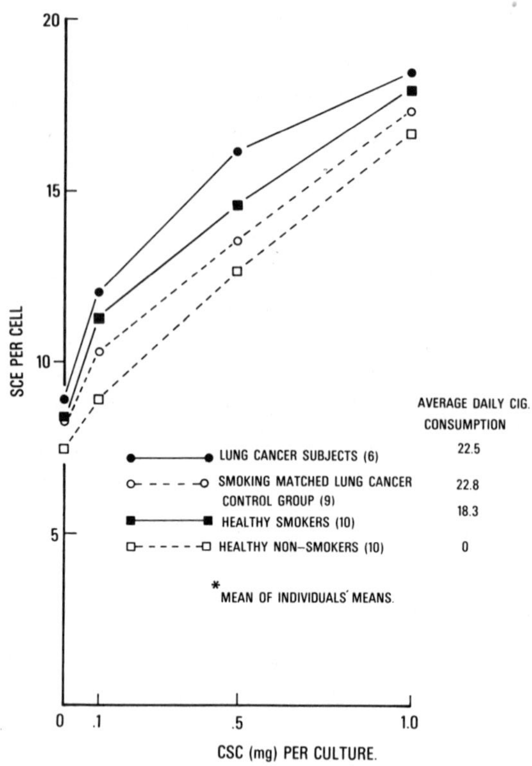

Fig. 5. Summary of basal and CSC-induced SCE rates from the 4 groups of individuals shown in Tabs. 1 and 2. Only filter cigarette smokers are included.

differences in the SCE rates may simply have resulted from discrepancies in actual amounts of tobacco smoked by these smokers. There is no doubt that all smoking histories describe only alleged amounts and that varying patterns of inhalation will also influence the amount of carcinogen delivered to the bronchial mucosa and peripheral lymphocytes. In either case, our observations may provide some rational experimental basis for the epidemiological findings that many, but not all, chronic cigarette smokers ultimately die of lung cancer.

Finally, we have been able to use our system of SCE induction by CSC in vitro to compare the mutagenicity of condensates from different categories of tar cigarettes. The production of the "safer" lower tar cigarette has been one of the main strategies for diminishing the risk of disease associated with cigarette smoking, but recent studies showing that smokers tend to compensate or extract more smoke from lower tar cigarettes have raised cautions about these products (11). One important issue is whether the quality of smoke from these products is different since there are grounds to

Tab. 3. Mean SCE frequencies from lymphocytes of healthy individuals exposed to different CSC.

	Age	Sex	Cigarettes/ Day	Cigarette Type & Smoking Duration		Dose of CSC				
				Plain	Filter	0	0.1	0.5	1.0	
C_1 Lung cancer subjects	60	M	40	47	-	12.4	14.1	16.5	21.6	
	51	M	25	36	-	9.6	12.2	16.8	20.4	
	66	M	20	58	-	12.4	15.8	21.0	24.6	
	51	M	25	36	-	12.3	14.0	20.3	20.1	
	63	M	30	36	20	8.5	12.4	15.5	19.8	
	55	F	20	23	15	7.3	13.2	14.3	16.3	
	63	F	20	25	15	9.9	10.8	15.9	20.4	
	65	M	25	39	12	8.3	13.8	19.0	20.3	
	53	F	20	18	20	9.8	11.4	16.5	17.8	
	45	M	20	-	25	9.8	11.3	15.6	16.5	
	56.9		24,5	42.3 yrs.		10.03(.57)	12.9(.49)	17.14(.70)	19.73(.78)	Mean (SE) overall
						11.7 (.69)	14.03(.74)	18.65(1.17)	21.67(1.02)	Plain cigarette smokers at testing
						8.93(.44)	12.10(.40)	16.13(.64)	18.43(.75)	Filter cigarette smokers at testing
C_2 Lung cancer subjects (ex-smokers of 2 and 5 yrs)	57	M	30	20	20	7.3	9.5	10.4	12.7	
	56	M	20	20	11	7.5	9.1	12.0	14.4	
D Smoking matched cancer control group	56	M	30	42	-	10.6	13.9	18.0	19.2	
	65	M	20	40	12	8.7	10.8	13.8	-	
	85	M	20	55	20	8.6	10.1	12.9	-	
	73	M	30	43	20	7.2	10.8	13.9	17.0	
	61	F	20	28	15	9.4	11.2	13.2	16.7	
	64	F	20	34	15	8.6	10.6	15.9	21.9	
	73	M	25	44	15	7.7	8.3	13.1	18.4	
	58	F	20	33	10	8.6	11.4	13.1	15.0	
	71	M	30	56	2	6.9	8.6	12.2	15.7	
	53	M	20	17	20	8.8	11.1	13.8	16.8	
	65.9		23.5	52.1 yrs.		8.51(.33)	10.68(.49)	13.99(.54)	17.59(.77)	Mean (SE) overall
						8.28(.27)	10.30(.38)	13.54(.34)	17.35(.85)	Filter cigarette smokers

consider that the altered tobacco packing, the differences in wrapping paper, and the presence of a filter (all part of the production of the lower tar cigarette) may influence the quality of the smoke, perhaps by changing combustion temperature. Table 3 shows our results and demonstrates that, on a weight-for-weight basis, the condensates from the low-, middle-, and high-tar cigarettes produce similar mutagenic effects detected by induced SCEs. We found an identical result with an assay for cytotoxicity--a process which is likely to form the cellular basis for the development of severe emphysema amongst smokers (12). These findings, taken with the evidence that smokers extract more smoke from lower tar cigarettes to compensate for low nicotine yields, suggest that the health dangers associated with smoking these "safer" products are seriously underestimated. A critical appraisal of mortality statistics in Britain (13) supports this conclusion.

ACKNOWLEDGEMENTS

This work was done in collaboration with John Evans and Paul Perry at the Medical Research Council Clinical and Population Cytogenetics Unit in Edinburgh, Scotland.

REFERENCES

1. Doll, R., and R. Peto (1976) Mortality in relation to smoking. Brit. Med. J. 2:1525-1536.
2. ToKuhata, G.K. (1964) Familial factors in lung cancer and smoking. Amer. J. Public Health 54:24-32.
3. Gelboin, H.V. (1977) Cancer susceptibility and carcinogen metabolism. N. Eng. J. Med. 297:384-386.
4. Rigby, P.W.J. (1982) The oncogenic circle closes. Nature 297: 451-453.
5. Perry, P., and H.J. Evans (1975) Cytological detection of mutagencarcinogen exposure by sister chromatid exchanges. Nature 258:121-125.
6. Carrano, A.V., C.H. Thompson, P.A. Lindl, and J.L. Minkler (1978) Sister chromatid exchange as an indicator of mutagenesis. Nature 271:551-553.
7. Hopkin, J.M., and H.J. Evans (1979) Cigarette smoke condensates damage DNA in human lymphocytes. Nature 279:241-242.
8. Perry, P., and S. Wolff (1974) A new method for the differential staining of sister chromatid exchanges. Nature 251:156-158.
9. Hopkin, J.M., and P. Perry (1980) Benzpyrene does not contribute to the SCEs induced by cigarette smoke. Mutat. Res. 77: 377-381.
10. Hopkin, J.M., and H.J. Evans (1980) Cigarette smoke induced DNA damage and lung cancer risks. Nature 283:388-390.

11. Lenfant, C. (1983) Are low yield cigarettes really safer. N. Eng. J. Med. 309:181-182.
12. Hopkin, J.M., and D. Gorecka (1983) Tough cells and old age. Lancet 2:1170-1173.
13. Office of Population Censuses and Surveys (1981) Cancer statistics, Her Majesty's Stationery Office, London.

THE SCE TEST AS A TOOL FOR CYTOGENETIC MONITORING OF HUMAN EXPOSURE TO OCCUPATIONAL AND ENVIRONMENTAL MUTAGENS

Toshiaki Watanabe and Akira Endo

Department of Hygiene and Preventive Medicine
Yamagata University School of Medicine
Zao-Iida, Yamagata 990-23, Japan

SUMMARY

The sister chromatid exchange (SCE) test system using human peripheral lymphocytes is proposed as a valuable tool for the cytogenetic monitoring of exposure to potential chemical mutagens in the occupational environment. We examined the SCE frequencies in organic solvent-exposed workers and reviewed the effects of occupational and environmental chemicals on SCE frequencies. The results obtained in these exposed populations are rather contradictory, which may be related to confounding factors, such as personal life style (smoking, drinking, and drugs) of the examinees, tissue culture conditions and the number of subjects examined. We discussed some practical problems for exposure estimation and sample size determination and are led to the following conclusions. 1) The possible combined effects of potential mutagens and cigarette smoking should be taken into consideration when using the SCE test system. 2) Cell cycle kinetic analysis by differential chromatid staining would provide valuable information as a biological indicator for the monitoring of the workers exposed to xenobiotics. 3) By appropriately setting the number of examinees, monitoring would become more efficient in detecting an increased SCE frequency in the exposed populations. Further studies are also required to enhance the sensitivity of the SCE test system for monitoring purposes.

INTRODUCTION

Sister chromatid exchanges are considered to be a very sensitive marker for the measurement of DNA damage and repair (1). Recently, the SCE test system has been extensively utilized in investigations of the mutagenic and carcinogenic properties of chemicals. The SCE test system using human peripheral lymphocytes is proposed as an especially valuable tool for the cytogenetic monitoring of exposure to potential chemical mutagens in the occupational environment. To date, many studies on the effects of occupational chemicals on SCE frequencies have been conducted [reviewed by Brøgger (2)]. However, as shown in Tab. 1 the results obtained in these exposed populations are often contradictory. The contradictions may be related to confounding factors such as personal life style (e.g., cigarette smoking, alcohol drinking, and drugs), cell culture conditions and the number of subjects examined. These factors, which may be responsible for variation in the population baseline SEC frequency, should be clearly delineated.

We have been examining the SCE frequency and the cell cycle kinetics of lymphocytes cultured from workers occupationally exposed to various organic solvents, such as benzene, tetrachloroethylene, and styrene (3-6). Our observations suggest that styrene induced a more severe effect in cigarette smokers than in nonsmokers and, also, that the simultaneous analysis of both SCE frequency and cell cycle kinetics is useful for the monitoring of workers exposed to xenobiotics. The purpose of this contribution is to discuss some practical problems for exposure estimation and sample size determination.

SCE STUDY IN BENZENE-EXPOSED WORKERS

We examined the frequencies of SCEs in cultured peripheral lymphocytes from benzene-exposed female workers (varnishers and line drawers) (3). None of the individuals had a history of cigarette smoking. The SCE frequencies in the benzene-exposed workers were 9.6 for workshop 1 and 10.5 for workshop 2, as compared with 11.4 for the control group (Tab. 2). The decrease in SCE frequencies appeared to depend on the intensity of the past exposure to benzene as measured in terms of concentration. The benzene concentrations in workroom air were in the range of 3-50 ppm for workshop 1 and less than 9 ppm for workshop 2. However, no difference in cell proliferation index or mitotic index was found among groups, suggesting that cell viability was not affected.

Funes-Cravioto et al. (7) have shown that organic solvent-exposed technicians in chemical laboratories and the children of these technicians who had worked during pregnancy had significantly increased frequencies of SCEs in cultured lymphocytes, whereas no increase in SCE frequency was detected in technicians who had past exposure to benzene and were currently exposed to toluene only.

These findings may imply that, at relatively low levels, certain chemicals such as benzene or benzene epoxide mainly affect DNA repair at the replication point, inhibiting the formation of SCEs. At higher concentrations, the chemicals induce DNA damage, such as

Tab. 1. (Part 1) Sister chromatid exchange frequency in subjects occupationally or environmentally exposed to potential mutagens.

Chemicals	Exposed groups	SCE frequency[a]	References
Styrene	reinforced plastic industry	-	[11]
	reinforced plastic industry	-	[12]
		-(children)	
	plastic boat factory	-(nonsmorkers)	[13]
		-(smokers)	
	reinforced plastic industry	-(nonsmokers)	[14]
		-(smokers)	
	plastic boat factory	+	[15]
	FRP-boat factory	-	[5]
	polyester-resin board factory	-	[5]
	FRP-boat factory	-(nonsmokers)	[6]
		+(smokers)	
	reinfoced plastics industry	+	[16]
Toluene	paint industry	-	[54]
	rotogravure plants	+(nonsmokers)	[14]
		-(smokers)	
	rotogravure plants	+(nonsmokers)	[55]
		+(smokers)	
	rotogravure printing plants	-(nonsmokers)	[56]
		-(smokers)	
	glue production	-(nonsmokers)	[57]
		-(smokers)	
Toluene/Xylene	paint industry	-	[54]
Tilene/Benzene	chemical laboratory	-*	[7]
Benzene	pottery workshop	-*	[3]
		-*	
Tetrachloro-ethylene	degreasing workshop	-	[4]
	(support department)	-	
Vinyl chloride	PVC factory	+	[58]
	PVC factory	-	[59]
Organic solvents	chemical laboratory	+	[7]
	chemical laboratory	+(nonsmokers)	[60]
		-(smokers)	
	chemical laboratory	+(nonsmokers)	[27]
		+(smokers)	
Pentachlorophenol	PCP producing factory	+(smokers)	[61]
Petrol products	bus drivers	-(nonsmokers)	[57]
		-(smokers)	

Tab. 1. (Part 2) Sister chromatid exchange frequency in subjects occupationally or environmentally exposed to potential mutagens.

Chemicals	Exposed groups	SCE frequencies	References
Anaesthetics	operating room	-*	[62]
	hospital	-	[29]
Halothane	hospital	-	[63]
Enflurane	hospital	-	[63]
Ethylene oxide	sterilization plants	-(nonsmokers)	[27]
		+(smokers)	
	hospital laboratory	+(nonsmokers)	[14]
		-(smokers)	
	hospital	-(nonsmokers)	[57]
		-(smokers)	
Epoxy resins	gluing metal and rubber	-	[64]
	gluing metal and rubber	+(nonsmokers)	[57]
		-(smokers)	
SO_2	alminium industry	-	[65]
Lead	lead smeltery	-(nonsmokers)	[56]
		+(smokers)	
	contaminated area	-	[66]
Nickel	nickle refinery factory	-	[29,67]
2,4-D/MCPA	forestry	-	[68]
(U.V)	hospital	-	[69]
(X-ray)	hospital	-	[29]
	hospital(patients)	-	[70]

[a]+: increased frequency of SCE. -: no change.

-*: decreased frequency of SCE.

cross-linking or base-chemical adducts, increasing the formation of SCEs. The above-mentioned observation suggests that the apparently reversed finding of a decreased SCE frequency in benzene-exposed individuals may be a manifestation of the subtler effect of chemicals on DNA replication mechanisms. Alternatively, the benzene-induced decrease in SCE frequency may be explained in another way. Moszczynski (8) demonstrated that T lymphocyte counts were diminished in workers with histories of a long exposure to organic solvents such as benzene, toluene, and xylene, whereas B and null lymphocyte counts were not altered. Santesson et al. (9) and Lindblad and Lambert (10) have observed that the SCE frequency of T lymphocytes was about 2 times greater than that observed in B lymphocytes. Selective damage of T lymphocytes in vivo with a resultant lower

Tab. 2. Sister chromatid exchange frequencies and cell cycle kinetics in cultured lymphocytes from benzene-exposed workers.

Groups	No. of subjects	SCE[1]	Proportion of M2+ metaphases[2]	RI[3]
Workshop 1 workers	7	9.6±0.7*	0.52(0.40-0.68)	1.56
Workshop 2 workers	9	10.5±1.2	0.52(0.29-0.67)	1.58
Controls	7	11.4±1.1	0.51(0.40-0.66)	1.59

[1]mean ±S.D. [2]mean (range). [3]mean.

*p<0.01, as compared with control by t-test.

proportion of T lymphocytes among the analyzable mitotic cells in the in vitro phytohemagglutinin (PHA)-stimulated proliferating lymphocytes may account for the reduction in SCE frequency.

SCE STUDY IN STYRENE-EXPOSED WORKERS

The incidences of SCEs and the cell proliferation kinetics in cultured lymphocytes of styrene-exposed workers in 3 fiber reinforced plastics (FRP) boat factories and a polyester-resin board factory were investigated (Tabs. 3 and 4) (5,6). Exposure to styrene did not seem to affect the SCE frequencies in lymphocytes. Most investigators have also found that the frequency of SCEs among styrene-exposed workers was not different from control levels (11-14). However, Uggla et al. (15) and Camurri et al. (16) have shown that styrene-exposed workers had a slightly increased SCE frequency in their peripheral blood lymphocytes. In most of these studies the numbers of subjects examined were small, as in our study, and no detailed analysis of dose-effect relationships was performed. Improvement in analysis efficiency (17,18), including statistical methods, may be necessary.

CONFOUNDING FACTORS IN THE SCE TEST

The International Commission for Protection Against Environmental Mutagens and Carcinogens (ICPEMC) (19), has recommended that epidemiological studies on the possible effects of environmental mutagens should take into account the possible interfering effects of cigarette smoking. The smoking habit is considered an important confounding factor in various cytogenetic studies on environmental and occupational exposure. Cigarette smoking has been found to

Tab. 3. Sister chromatid exchange frequencies and cell cycle kinetics in cultured lymphocytes from styrene-exposed workers.

Groups	No. of subjects	SCE[1]	Proportion of M3 metaphases[2]	RI[3]
Workshop 3				
workers	9	7.8±1.6	0.40(0.27-0.51)	2.27
controls	5	7.6±1.2	0.36(0.30-0.44)	2.23
Workshop 4				
workers	7	6.7	0.37¶	2.28¶
nonsmokers	5	6.3±0.4	0.39(0.24-0.52)	2.30
smokers	2	7.7±0.8*	0.31(0.27-0.34)	2.23
controls	8	7.6	0.53	2.46
nonsmokers	3	6.7±1.2	0.56(0.37-0.70)	2.48
smokers	5	8.2±1.2	0.51(0.42-0.65)	2.45

[1]mean ±S.D. [2]mean (range). [3]mean.

*$p<0.05$, as compared with exposed nonsmokers.

¶$p<0.05$, as compared with controls.

Tab. 4. Sister chromatid exchange frequencies and cell cycle kinetics in cultured lymphocytes from styrene-exposed workers in FRP boat factory.

Groups	No. of subjects	SCE[1]	Proportion of M2 metaphases[2]	RI[3]
Workshops (5 and 6)				
workers	18	8.9	0.29¶	1.29¶
nonsmokers	8	7.9±5.8	0.27(0.09-0.53)**	1.27**
smokers	10	9.6±1.5†	0.31(0.14-0.56)**	1.30**
controls	6	8.5	0.41	1.42
nonsmokers	3	8.6±0.4	0.58(0.55-0.61)	1.58
smokers	3	8.4±0.5	0.25(0.06-0.37)*	1.25*

[1]mean ±S.D. [2]mean (range). [3]mean.

†$p<0.01$, as compared with styrene-exposed nonsmokers.

*$0.05<p<0.1$, **$p<0.05$, as compared with control nonsmokers.

¶$p<0.05$, as compared with controls.

increase SCE frequencies as well as to produce chromosomal aberrations (13,14,20-22). In Tab. 5, the literature on SCE frequencies in the lymphocytes of cigarette smokers is reviewed. Most, but not all, of the studies have reported an increase in SCE frequencies in lymphocytes of smokers as compared with those observed in the lymphocytes of nonsmokers. Recently, Livingston and Fineman (23) and Wulf et al. (24) have demonstrated a dose-related increase in SCEs as measured by the cumulative pack-years and tar contents. We have reported that the mean SCE frequency in cigarette-smoking university students was higher than that of nonsmoking students (Tab. 6) (25). This suggests that cigarette smoking for a short time can increase the SCE frequency significantly.

In an evaluation of the effects of styrene exposure, smoking habits were therefore taken into account. The subjects of each group were subdivided into smokers and nonsmokers (see Tabs. 3 and 4). Exposure to styrene did not induce a significant increase in SCEs. However, the SCE frequencies in lymphocytes of cigarette smokers were significantly higher than those of nonsmokers among styrene-exposed workers in workshops 4, 5, and 6. Among control subjects, no difference in SCE frequencies was observed between smokers and nonsmokers, although the sample size was rather small. This suggests that the SCE frequency in styrene-exposed workers is influenced by individual smoking habits and that both factors may act synergistically. In cytogenetic studies of occupationally lead-exposed workers in a smelter and ethylene oxide-exposed workers in a sterilization plant, a synergistic effect between smoking and these chemicals on SCE frequency has also been suggested (26,27). The possible combined effects of potential mutagens and cigarette smoking should be taken into consideration when using the SCE test system.

As to other confounding factors, alcohol drinking (28), drug intake (29), chemotherapy (30-32), infections (33,34) and vaccination (35-37) have been shown to influence SCE frequencies in humans. The influence of these factors on individual SCE frequencies within the population must be controlled in any cytogenetic monitoring study.

It has been suggested that sex and genetic factors are probably not a major source of subject variation in SCE mean values (38-41). However, a significant difference between younger and older subjects has been observed (40,42), and newborns had a significantly lower mean SCE values than their mothers (43,44). A control group with comparable personal and social characteristics to the exposed workers is advisable.

CELL CYCLE KINETICS

The differential staining technique of sister chromatids easily permits the identification of cell cycle kinetics without the

Tab. 5. SCE frequencies in cigarette smokers.

References	SCE frequencies		
	nonsmokers[a]	smokers[a]	results[b]
Hollander et al.[30]	11.9(9.8-15.0)	11.4 (8.4-15.3)	-
Lambert et al.[60]	13.1(7.8-19.6)	16.2 (9.2-24.2)	+
Meretoja et al.[11]	4.6	4.0	-
Murthy[71]	6.4	7.7	+
Valadaud-Barrieu and Izard[72]	4.7	5.8	+
Andersson et al.[13]	7.3	7.8	-
Ardito et al.[14]	8.1(5.6-10.3)	8.1 (6.6-10.1)	-
Crossen and Morgan[73]	10.0(7.0-14.0)	10.6 (8.3-16.2)	-
Hopkin and Evans[74] and Evans[75]	7.4(7.0- 8.5)	8.4 (6.5-12.0)	+
Husgafvel-Pursiainen et al.[14]	8.1(6.2-10.5)	9.6 (7.2-13.2)	+
Husum and Wulf[62]	c	c	-
Lambert and Lindblad[27]	14.0(7.8-20.2)	15.6 (9.2-24.2)	+
Mäki-Paakkanen et al.[56]	8.0(7.5- 8.2)	9.7 (7.2-12.2)	+
Carrano et al.[17]	c	c	+
Husum et al.[76,77]	9.4	10.6	+
Hüttner and Schöneich[21]	12.5	15.2	+
Mäki-Paakkanen et al.[26]	9.2(7.9-10.3)	10.4 (9.3-12.2)	+
Waksvik et al.[40]	7.4	7.3	-
Aronson et al.[31]	6.1(4.0- 7.7)	7.7 (6.3- 9.6)	+
Bauchinger et al.[55]	7.8(6.3- 9.2)	8.9 (6.5-10.3)	+
Bauchinger et al.[61]	7.6	8.9	+
Husum et al.[78]	8.3(6.1-11.2)	9.2 (6.8-14.1)	+

necessity of autoradiography (45). It is a sensitive method for detecting mitotic delay in human cultures. Cell cycle kinetics were examined by scoring the proportions of the cells at M1 (metaphases in the first replication cycle), M2 (metaphases in the second replication cycle) and M3 (metaphases in the third replication cycle) (46). A delay in cell proliferation is indicated by a decrease in the frequency of M2 and/or M3+ (metaphases in the third or subsequent replication cycle) cells and a corresponding increase in M1 replication cells, or as a decline in the replication index (RI) (47,48). The estimate of the RI is calculated according to the following formula:

$$RI = \frac{(M1 \times 1) + (M2 \times 2) + (M3 \times 3)}{100}$$

Tab. 5. (Continued) SCE frequencies in cigarette smokers.

References	SCE frequencies		
	nonsmokers	smokers	results
Meiying et al. [79]	4.4 (4.0- 5.8)	8.3 (6.9- 9.8)	+
Obe et al. [22]	4.8	5.8	+
Seshadri et al. [28]	11.4	11.9	-
Vijayalaxmi and Evans [80]	7.6 (7.5- 7.7)	11.6 (11.0-12.6)	+
Sorsa et al. [65]	8.4 (7.0- 9.1)	9.7 (6.6-12.2)	-
Lambert et al. [81]	14	15.7[d]	+
		18.3[e]	+
Watanabe et al. [25]	9.6 (7.6-11.7)	10.3 (8.1-11.9)	+
Hedner et al. [57]	8.9 (7.1-12.2)	9.1 (6.9-12.1)	-
Linnainmaa [68]	8.1 (7.4-10.2)	10.0 (8.1-12.2)	+
Lundberg and Linnainmaa [44]	9.0 (7.0-11.9)	11.4 (9.7-12.8)	+
Livingston and Fineman [23]	8.5	10.8	+
Livingston et al. [41]	9.6[f]	10.5	+
	8.7[g]	11.3	+
Wulf et al. [24]	8.3	9.0-10.5[h]	+

[a] mean (range).

[b] +: increased frequency of SCE. -: no change.

[c] no data given.

[d] less than 10 years of smoking.

[e] more than 10 years of smoking.

[f] male.

[g] female.

[h] cigarettes, cheroots or pipe tobacco.

where M1, M2, and M3 are expressed as percentage. In our observation in short-term cultured peripheral lymphocytes, there was no deviation in the percentages of M1, M2, or M3 cells in benzene- or tetrachloroethylene-exposed workers (Tab. 2) (3,4). However, delay of cell cycle was observed in styrene-exposed workers (workshops 4, 5, and 6). Cell proliferation of lymphocytes in vitro depends on the cellular response to mitogens such as PHA. Lymphocyte damage in the styrene-exposed individual may have manifested as a decreased response to mitogens. To our knowledge, little information has been obtained on the effects of occupational or environmental chemicals on cell cycle kinetics in exposed populations. It should be stressed that cell cycle kinetics analysis by differential chromatid staining provides valuable information as a biological indicator for the monitoring of workers exposed to xenobiotics.

Tab. 6. Sister chromatid exchange frequencies and cell cycle kinetics in cultured lymphocytes of cigarette smokers and nonsmokers.

Groups	No. of subjects	SCE[1]	Proportion of M2 metaphases[1]	RI[2]
Nonsmokers	18	9.6(7.6-11.7)	0.71(0.63-0.75)	1.80
Smokers	17	10.3(8.1-11.7)*	0.70(0.57-0.79)	1.85

[1]mean(range). [2]mean.

*0.05<p<0.10, as compared with nonsmokers by Mann-Whitney U test.

SAMPLE SIZE

In the setting of statistical testing, the significant level is referred to as the risk of erroneously rejecting a null hypothesis that is really true, which is called the α error (type I). A β error (type II) is defined as the chance of erroneously failing to reject a null hypothesis that is, in fact, false (49). In any statistical testing we should keep the value of the β error as low as possible. The value of β is a function of sample size, the magnitude of the difference between the two means, the variance of the means, and the level of significance (α) used. We calculated the sample size necessary to have a high probability of obtaining a statistical significance when a true difference exists between the exposed and control populations. Figure 1 demonstrates how the power ($1 - \beta$) of the statistical test increases with increasing sample size. In general, the power of 80-90% is recommended. For example, in order to achieve an 85% chance of detecting the difference of 1 SCE between the two means, at the significant level of 5%, we would need 17 subjects each for the exposed and the control groups. Conversely, in a case where data from only 9 workers and 9 controls are available, the power drops from 85% to 60% (Tab. 7). By an adequate setting of the number of examinees, monitoring would become more efficient in detecting an increased frequency of SCEs in the exposed population.

IMPROVING THE SCE TEST SYSTEM

Some investigators (50-52) have suggested that human blood lymphocytes are comprised of at least 2 different subpopulations with different sensitivities to DNA damage, SCE formation and cell proliferation. When lymphocytes are stimulated by PHA in whole blood

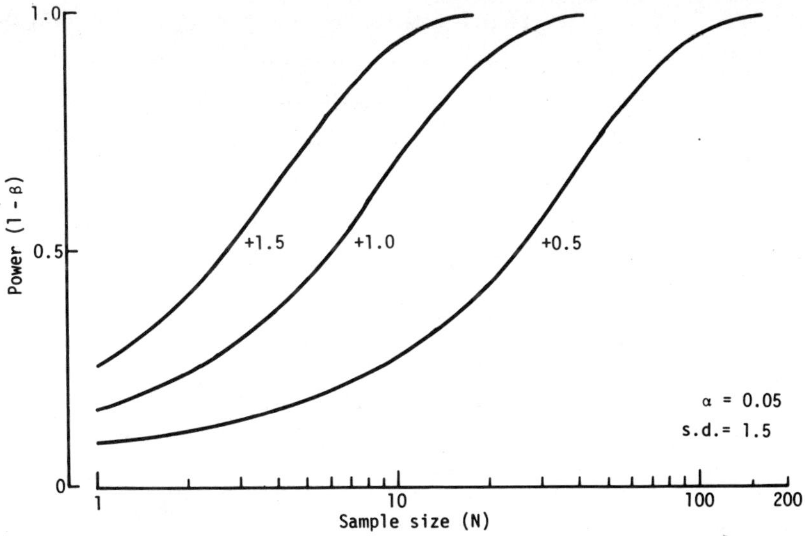

Fig. 1. Operating characteristic curve. The curves in the figure show the relation of sample size and power when the differences between the means are 0.5, 1.0, and 1.5. Standard deviation for the two means is assumed to be 1.5 each, and the statistical significance level (α) is set to be 0.05.

cultures, both T lymphocytes and B lymphocytes respond. The proportion of T cells and B cells in the peripheral blood has been reported to be 60% (29-75%) and 16% (9-25), respectively (10). Also, it has been reported that the SCE frequency of purified T lymphocytes is about 2 times greater than that of purified B lymphocytes, and that the latter cell type has a higher rate of proliferation than the former (9,10). Thus, the response to chemicals in each subpopulation may have to be measured separately. Schwartz and Gaulden (53) have recently reported that B cells were more sensitive to radiation than T cells.

Further studies will be required to improve the SCE test system which will enhance sensitivity in monitoring. Comparative analysis of SCE frequency between T lymphocytes and B lymphocytes in the exposed populations may be one of these needed studies. Also, cells dividing for the first, second, and third time in cultures can be distinguished by the differential staining method of sister chromatids. Chromosome analysis in the first replication cells will provide information on in vivo-occurring genetic damage.

Tab. 7. Sample size requirement for cytogenetic monitoring.

SCE increases	1 - β			
	0.90	0.85	0.80	0.60
0.5	78	65	56	33
1.0	20	17	14	9
1.5	9	8	7	4
2.0	5	5	4	3
3.0	3	2	2	1

α=0.05, s.d.=1.5.

SCE increases	1 - β			
	0.90	0.85	0.80	0.60
0.5	118	102	91	60
1.0	30	26	23	15
1.5	14	12	11	7
2.0	8	7	6	4
3.0	4	3	3	2

α=0.01, s.d.=1.5.

REFERENCES

1. Perry, P., and H.J. Evans (1975) Cytological detection of mutagen-carcinogen exposure by sister chromatid exchange. Nature (Lond.) 258:121-125.
2. Brøgger, A. (1982) Application of SCE to public health. In Sister Chromatid Exchange, A.A. Sandberg, ed. Alan R. Liss, New York, pp. 655-673.
3. Watanabe, T., A. Endo, Y. Kato, S. Shima, T. Watanabe, and M. Ikeda (1980) Cytogenetics and cytokinetics of cultured lymphocytes from benzene-exposed workers. Int. Arch. Occup. Environ. Health 46:31-41.
4. Ikeda, M., A. Koizumi, T. Watanabe, A. Endo, and K. Sato (1980) Cytogenetic and cytokinetic investigations on lymphocytes from workers occupationally exposed to tetrachloroethylene. Toxicol. Lett. 5:251-256.

5. Watanabe, T., A. Endo, K. Sato, T. Ohtsuki, M. Miyasaka, A. Koizumi, and M. Ikeda (1981) Mutagenic potential of styrene in man. Indust. Health 19:37-45.
6. Watanabe, T., A. Endo, M. Kumai, and M. Ikeda (1983) Chromosome aberrations and sister chromatid exchanges in styrene-exposed workers with reference to their smoking habits. Environ. Mutagen. 5:299-309.
7. Funes-Cravioto, F., C. Zapata-Gayon, B. Kolmodin-Hedman, B. Lambert, J. Lindsten, E. Norberg, M. Nordenskjöld, R. Olin, and Å. Swensson (1977) Chromosome aberrations and sister-chromatid exchange in workers in chemical laboratories and a rotoprinting factory and in children of women laboratory workers. Lancet ii:322-325.
8. Moszczynski, P. (1981) Organic solvents and T lymphocytes. Lancet i:438.
9. Santesson, B., K. Lindahl-Kiessling, and A. Mattsson (1979) SCE in B and T lymphocytes. Possible implications for Bloom's syndrome. Clin. Genet. 16:133-135.
10. Lindblad, A., and B. Lambert (1981) Relation between sister chromatid exchange, cell proliferation and proportion of B and T cells in human lymphocyte cultures. Human Genet. 57:31-34.
11. Meretoja, T., H. Järventaus, M. Sorsa, and H. Vainio (1978) Chromosome aberrations in lymphocytes of workers exposed to styrene. Scand. J. Work Environ. Health 4 (Suppl. 2):259-264.
12. Sorsa, M., M. Hyvönen, H. Järventaus, and H. Vainio (1979) Chromosomal aberrations and sister chromatid exchange in children of women in reinforced plastic industry. Sym. Toxicol. University of Turku, (Abstract) p. 16.
13. Andersson, H.C., E.Å. Tranberg, A.H. Uggla, and G. Zetterberg (1980) Chromosomal aberrations and sister-chromatid exchanges in lymphocytes of men occupationally exposed to styrene in a plastic-boat factory. Mutat. Res. 73:387-401.
14. Husgafvel-Pursiainen, K., J. Mäki-Paakkanen, H. Norppa, and M. Sorsa (1980) Smoking and sister chromatid exchange. Hereditas 92:247-250.
15. Uggla, A.H., H.C. Andersson, E.Å. Tranberg, and L.G. Zetterberg (1980) Correlation between exposure to styrene and the frequency of chromosomal aberrations and sister-chromatid exchanges in lymphocytes of workers in a plastic boat factory. Mutat. Res. 74:199 (Abstract).
16. Camurri, L., S. Codeluppi, C. Pedroni, and L. Scarduelli (1983) Chromosomal aberrations and sister-chromatid exchanges in workers exposed to styrene. Mutat. Res. 119:361-369.
17. Carrano, A.V., L.K. Ashworth, J.L. Minkler, and D.H. Moore, II (1981) Sister chromatid exchange frequencies in humans: The effect of smoking. Environ. Mutagen. 3:339 (Abstract).
18. Carrano, A.V., and D.H. Moore, II (1982) The rationale and methodology for quantifying sister chromatid exchange in humans. In Mutagenicity: New Horizons in Genetic Toxicology, J.A. Heddle, ed. Academic Press, New York, pp. 267-304.
19. Bridges, B.A., J. Clemmesen, and T. Sugimura (1979) Cigarette

smoking -- does it carry a genetic risk? ICPEMC publication no. 3. Mutat. Res. 65:71-81.
20. Obe, G., and J. Herha (1978) Chromosomal aberrations in heavy smokers. Human Genet. 41:259-263.
21. Hüttner, E., and J. Schöneich (1981) Effects of smoking on the frequencies of chromosomal aberrations and SCE in man. Mutat. Res. 85:255 (Abstract).
22. Obe, G., H.-J. Vogt, S. Madle, A. Fahning, and W.D. Heller (1982) Double-blind study on the effect of cigarette smoking on the chromosomes of human peripheral blood lymphocytes in vivo. Mutat. Res. 92:309-319.
23. Livingston, G.K., and R.M. Fineman (1983) Correlation of human lymphocyte SCE frequency with smoking history. Mutat. Res. 119:59-64.
24. Wulf, H.C., B. Husum, and E. Niebuhr (1983) Sister chromatid exchanges in smokers of high-tar cigarettes, low-tar cigarettes, cheroots and pipe tobacco. Hereditas 98:225-228.
25. Watanabe, T., T. Kasukawa, and A. Endo (1983) Sister chromatid exchanges in lymphocytes from cigarette smokers. Jpn. Human Genet. 28:152 (Abstract).
26. Mäki-Paakkanen, J., M. Sorsa, and H. Vainio (1981) Chromosome aberrations and sister chromatid exchanges in lead-exposed workers. Hereditas 94:269-275.
27. Lambert, B., and A. Lindblad (1980) Sister chromatid exchange and chromosome aberrations in lymphocytes of laboratory personnel. J. Toxicol. Environ. Health 6:1237-1243.
28. Seshadri, R., E. Baker, and G.R. Sutherland (1982) Sister-chromatid exchange (SCE) analysis in mothers exposed to DNA-damaging agents and their newborn infants. Mutat. Res. 97:139-146.
29. Waksvik, H., M. Boysen, A. Brøgger, and O. Klepp (1981) Chromosome aberrations and sister chromatid exchanges in persons occupationally exposed to mutagens/carcinogens. In Chromosome Damage and Repair, E. Seeberg and K. Kleppe, eds. Plenum Press, New York, pp. 563-566.
30. Hollander, D.H., M.S. Tockman, Y.W. Liang, D.S. Borgaonkar, and J.K. Frost (1978) Sister chromatid exchanges in the peripheral blood of cigarette smokers and in lung cancer patients; and the effect of chemotherapy. Human Genet. 44:165-171.
31. Aronson, M.M., P.C. Miller, R.B. Hill, W.W. Nichols, and A.T. Meadows (1982) Acute and long-term cytogenetic effects of treatment in childhood cancer: Sister-chromatid exchanges and chromosome aberrations. Mutat. Res. 92:291-307.
32. Crossen, P.E., and W.F. Morgan (1982) The effect of chlorpromazine on SCE frequency in human chromosomes. An in vitro and in vivo study. Mutat. Res. 96:225-232.
33. Ghosh, R., and P.K. Ghosh (1983) Sister-chromatid exchanges in herpes simplex infection. Mutat. Res. 119:303-308.
34. Kurvink, K., C.D. Bloomfield, and J. Cervenka (1978) Sister chromatid exchange in patients with viral disease. Exp. Cell Res. 113:450-453.

35. Lambert, B., A. Ehrnst, K. Hansson, A. Lindblad, M. Morad, and B. Werelius (1979) Sister chromatid exchange in peripheral lymphocytes of subjects vaccinated against measles. Human Genet. 50:291-296.
36. Knuutila, S., J. Mäki-Paakkanen, M. Kähkönen, and E. Hokkanen (1978) An increased frequency of chromosomal changes and SCE's in cultured blood lymphocytes of 12 subjects vaccinated against smallpox. Human Genet. 41:89-96.
37. Knuutila, S., A. Harkki, K. Ellimäki, and R. Salunen (1979) Decreased sister chromatid exchange in Down's syndrome after measles vaccination. Hereditas 90:147-149.
38. Morgan, W.F., and P.E. Crossen (1977) The incidence of sister chromatid exchanges in cultured human lymphocytes. Mutat. Res. 42:305-312.
39. Pedersen, C., E. Oláh, and U. Merrild (1979) Sister chromatid exchanges in cultured peripheral lymphocytes from twins. Human Genet. 52:281-294.
40. Waksvik, H., P. Magnus, and K. Berg (1981) Effects of age, sex and genes on sister chromatid exchange. Clin. Genet. 20:449-454.
41. Livingston, G.K., L.A. Cannon, D.T. Bishop, P. Johnson, and R.M. Fineman (1983) Sister chromatid exchange: Variation by age, sex, smoking, and breast cancer status. Cancer Genet. Cytogenet. 9:289-299.
42. de Arce, M.A. (1981) The effect of Donor sex and age on the number of sister chromatid exchanges in human lymphocytes growing in vitro. Human Genet. 57:83-85.
43. Ardito, G., L. Lamberti, E. Ansaldi, and P. Ponzetto (1980) Sister-chromatid exchanges in cigarette-smoking human females and their newborns. Mutat. Res. 78:209-212.
44. Lundberg, M.S., and G.K. Livingston (1983) Sister-chromatid exchange frequency in lymphocytes of smoking and nonsmoking mothers and their newborn infants. Mutat. Res. 121:241-246.
45. Tice, R., L. Schneider, and J.M. Rary (1976) The utilization of bromodeoxyuridine incorporation into DNA for the analysis of cellular kinetics. Exp. Cell Res. 102:232-236.
46. Crossen, P.E., and W.F. Morgan (1977) Analysis of human lymphocyte cell cycle time in culture measured by sister chromatid differential staining. Exp. Cell Res. 104:453-457.
47. Schneider, E.L., and J. Lewis (1981) Aging and sister chromatid exchange. VIII. Effect of the aging environment on sister chromatid exchange induction and cell cycle kinetics in Ehrlich ascites tumor cells. A brief note. Mech. Age. Develop. 17: 327-330.
48. Lamberti, L., P.B. Ponzetto, and G. Ardito (1983) Cell kinetics and sister-chromatid-exchange frequency in human lymphocytes. Mutat. Res. 120:193-199.
49. Colton, T. (1974) Statistics in Medicine, Little, Brown and Company, Boston.
50. Beek, B., and G. Obe (1974) The human leukocyte test system.

II. Different sensitivities of sub-populations to a chemical mutagen. Mutat. Res. 24:395-398.
51. Snope, A.J., and J.M. Rary (1979) Cell-cycle duration and sister-chromatid exchange frequency in cultured human lymphocytes. Mutat. Res. 63:345-349.
52. Morimoto, K., and S. Wolff (1980) Cell cycle kinetics in human lymphocyte cultures. Nature (Lond.) 288:604-606.
53. Schwartz, J.L., and M.E. Gaulden (1980) The relative contributions of B and T lymphocytes in the human peripheral blood mutagen test system as determined by cell survival, mitogenic stimulation, and induction of chromosome aberrations by radiation. Environ. Mutagen. 2:473-485.
54. Haglund, U., I. Lundberg, and L. Zech (1980) Chromosome aberrations and sister chromatid exchanges in Swedish paint industry workers. Scand. J. Work Environ. Health 6:291-298.
55. Bauchinger, M., E. Schmid, J. Dresp, J. Kolin-Gerresheim, R. Hauf, and E. Suhr (1982) Chromosome changes in lymphocytes after occupational exposure to toluene. Mutat. Res. 102:439-445.
56. Mäki-Paakkanen, J., K. Husgafvel-Pursiainen, P.-L. Kalliomäki, J. Tuominen, and M. Sorsa (1980) Toluene-exposed workers and chromosome aberrations. J. Toxicol. Environ. Health 6:775-781.
57. Hedner, K., B. Högstedt, A.-M. Kolnig, E. Mark-Vendel, B. Strömbeck, and F. Mitelman (1983) Sister chromatid exchanges and structural chromosome aberrations in relation to smoking in 91 individuals. Hereditas 98:77-81.
58. Kucerová, M., Z. Polívková, and J. Bátora (1979) Comparative evaluation of the frequency of chromosomal aberrations and the SCE numbers in peripheral lymphocytes of workers occupationally exposed to vinyl chloride monomer. Mutat. Res. 67:97-100.
59. Hansteen, I.-L., L. Hillestad, E. Thiis-Evensen, and S. Storetvedt Heldaas (1978) Effects of vinyl chloride in man. A cytogenetic follow-up study. Mutat. Res. 51:271-278.
60. Lambert, B., A. Lindblad, M. Nordenskjöld, and B. Werelius (1978) Increased frequency of sister chromatid exchanges in cigarette smokers. Hereditas 88:147-149.
61. Bauchinger, M., J. Dresp, E. Schmid, and R. Hauf (1982) Chromosome changes in lymphocytes after occupational exposure to pentachlorophenol (PCP). Mutat. Res. 102:83-88.
62. Husum, B., and H.C. Wulf (1980) Sister chromatid exchanges in lymphocytes in operating room personnel. Acta Anaesth. Scand. 24:22-24.
63. Husum, B., H.C. Wulf, and E. Niebuhr (1981) Sister chromatid exchanges in lymphocytes after anaesthesia with halothane or enflurane. Acta Anaesth. Scand. 25:97-98.
64. Mitelman, F., S. Fregert, K. Hedner, and K. Hillbertz-Nilsson (1980) Occupational exposure to epoxy resins has no cytogenetic effect. Mutat. Res. 77:345-348.
65. Sorsa, M., B. Kolmodin-Hedman, and H. Järventaus (1982) No effect of sulphur dioxide exposure, in aluminium industry, on chromosomal aberrations or sister chromatid exchanges. Hereditas 97:159-161.

66. Dalprà, L., M.G. Tibiletti, G. Nocera, P. Giulotto, L. Auriti, V. Carnelli, and G. Simoni (1983) SCE analysis in children exposed to lead emission from a smelting plant. Mutat. Res. 120:249-256.
67. Waksvik, H., and M. Boysen (1982) Cytogenetic analyses of lymphocytes from workers in a nickel refinery. Mutat. Res. 103: 185-190.
68. Linnainmaa, K. (1983) Sister chromatid exchanges among workers occupationally exposed to phenoxy acid herbicides 2,4-D and MCPA. Teratogen. Carcinogen. Mutagen. 3:269-279.
69. Brøgger, A., H. Waksvik, and P. Thune (1978) Psoralen/UVA treatment and chromosomes. II. Analysis of psoriasis patients. Arch. Dermatol. Res. 261:287-294.
70. Goh, K.-O. (1980) Sister chromatid exchange in normal adults long after thymus irradiation. Invest. Radiol. 15:332-334.
71. Murthy, P.B.K. (1979) Frequency of sister chromatid exchanges in cigarette smokers. Human Genet. 52:343-345.
72. Valadaud-Barrieu, D., and C. Izard (1979) Action de la phase gazeuse de fumeé de cigarette sur le taux d'échanges des chromatides-soeurs du lymphocyte humain in vitro. C.R. Acad. Sc. (Paris) 288:899-901.
73. Crossen, P.E., and W.F. Morgan (1980) Sister chromatid exchange in cigarette smokers. Human Genet. 53:425-426.
74. Hopkin, J.M., and H.J. Evans (1980) Cigarette smoke-induced DNA damage and lung cancer risks. Nature (Lond.) 283:388-390.
75. Evans, H.J. (1981) Cigarette smoke induced DNA damage in man. Prog. Mutat. Res. 2:111-128.
76. Husum, B., H.C. Wulf, and E. Niebuhr (1981) Sister chromatid exchanges in peripheral lymphocytes after preoperative mammography. Radiat. Res. 87:684-688.
77. Husum, B., H.C. Wulf, and E. Niebuhr (1981) Sister-chromatid exchanges in lymphocytes in women with cancer of the breast. Mutat. Res. 85:357-362.
78. Husum, B., H.C. Wulf, and E. Niebuhr (1982) Increased sister chromatid exchange frequency in lymphocytes in healthy cigarette smokers. Hereditas 96:85-88.
79. Meiying, C., X. Jiujin, and Z. Xianting (1982) Comparative studies on spontaneous and mitomycin-C-induced sister-chromatid exchanges in smokers and non-smokers. Mutat. Res. 105:195-200.
80. Vijayalaxmi, and H.J. Evans (1982) In vivo and in vitro effects of cigarette smoke on chromosomal damage and sister-chromatid exchange in human peripheral blood lymphocytes. Mutat. Res. 92:321-332.
81. Lambert, B., A. Bredberg, W. McKenzie, and M. Sten (1982) Sister chromatid exchange in human populations: The effect of smoking, drug treatment, and occupational exposure. Cytogenet. Cell Genet. 33:62-67.

SISTER CHROMATID EXCHANGES IN WORKERS EXPOSED TO LOW DOSES OF STYRENE

Lamberto Camurri, Susanna Codeluppi, Laura Scarduelli, and Silvia Candela[1]

Laboratory of Genetics
USL 9
Reggio Emilia, I 42100, Italy

INTRODUCTION

Several studies have been published on structural chromosomal aberrations (CAs) in the peripheral lymphocytes of workers in the reinforced plastic industry (1-9). A number of these studies have shown an association between styrene exposure and increased frequencies of CAs (1-4). Among the studies available on sister chromatid exchange (SCE) induction in the lymphocytes of styrene-exposed workers, only two studies report a slight increase in SCEs (1,2). The present study completes the work published by Camurri et al. (2) and attempts to describe the increase in SCEs at different styrene environmental exposures.

METHODS

Styrene Concentration in Air

The concentration of styrene in the air was evaluated by low-flow personal pumps and NIOSH-approved charcoal tubes.

Styrene Urinary Metabolites

The analysis of styrene urinary metabolites, mandelic acid (MA), and phenylglyoxylic acid (PGA), was carried out by gas-liquid chromatography. Friday morning urine samples were taken, and urinary metabolite levels were adjusted to a specific gravity of 1.024.

---

[1]Occupational Health Service, USL 12, Scandiano, I 42019, Italy.

Cytogenetics

Sister chromatid exchanges and CAs were analyzed in cells from cultures of peripheral lymphocytes of workers employed in 9 different plants manufacturing reinforced unsaturated polyester plastics. The control groups were selected on the basis of sex, age, and smoking habits, and tested at the same time. None of the subjects exposed or selected for the control group had recently had viral infections, vaccinations, or exposure to clastogenic agents.

Whole blood lymphocyte cultures (RPMI 1640 medium, 20% fetal calf serum, 0.2 ml PHA-M) from all individuals were incubated at 37°C for 50 hr, for the analysis of CAs, and for 72 hr in the presence of bromodeoxyuridine (BrdUrd) at 1 μg/ml for the analysis of SCE. All slides were coded; 100 metaphases per subject were analyzed for CAs, and about 50 harlequin-stained metaphases per subject were analyzed for SCE.

RESULTS AND DISCUSSION

In Tab. 1 the technological features and the styrene air concentrations of the 9 plants are summarized and combined with the urinary metabolites levels, CA frequencies, and SCE rates in the groups of analyzed workers. The statistical significance of the differences between each exposed group and its matched control was examined by Student's t-test.

Table 2 details the results of the CA and SCE analysis in the groups of workers tested in 1983. Chromosomal aberration mean values per plant were significantly higher ($p < 0.005$) in all of the cultures prepared from the styrene-exposed workers than in those of their matched controls.

A new attempt to relate CAs frequencies to exposure levels was made. Two exposure indices were considered, the styrene air concentration and the levels of styrene urinary metabolites. Figure 1 shows the relation between the mean values of CAs in the exposed workers and related controls and the styrene air concentration in each plant. When controls were subtracted, no significant increase related to the exposure level resulted. The individual CA rates were also correlated against urinary MA + PGA concentrations, again showing no significant linear regression.

The relation between SCEs (exposed workers and controls) and styrene air concentrations of the various plants is shown in Fig. 2. The SCE rates of the exposed groups were not significantly different from their matched controls at styrene air levels ranging from 30 to 200 mg/mc. In the exposed workers of plant No. 5, the SCE values,

Tab. 1. Technological features of the 9 plants, exposure parameters, CAs, and SCE mean values of the groups of tested workers.

Plant	Year of analysis	Category	Subjects	Styrene in the air (mg/mc)		Urinary metabolites		Chromosome aberrations					SCE			
				min	max	MA[c] (mg/l)	MA+PGA[d] (mg/mc)	Number of cells observed	Cells with chromatid type aberrations	Cells with chromosome type aberrations	Total (%)	p	Number of cells observed	SCE/cell Range	SCE/cell Mean	p
1	1982	Electrical	3	30	40	45-75	86-123	360	95	7	30±5	< 0.005	160	5-27	12.70 ±0.66	N.S.
2	1982	Small parts	4	70	100	65-133	160-222	502	108	7	23±3	< 0.001	222	6-23	12.72 ±0.36	< 0.05
3	1983	Helmets	4	100	150	170-694	379-1429	346	80	0	24±4	< 0.001	196	5-25	10.92 ±1.02	N.S.
4	1983	Tanks	5[a]	150	200	151-786	346-1168						173	2-25	10.26 ±0.93	N.S.
			9[b]			151-1083	346-1706	668	166	0	26±3	< 0.001				
5	1982	Tanks	6	200	250	340-671	656-1046	633	185	12	32±6	< 0.001	321	4-22	11.77 ±0.49	< 0.005
6	1982	Tanks	2	250	300	615-777	990-1336	182	61	10	39±1	< 0.005	101	9-35	19.62 ±5.2	N.S.
7	1982	Tanks (laminat.)	2[a]	300	350	489-828	878-1052						100	8-34	16.13 ±1.56	< 0.05
			3[b]			187-828	382-1052	289	102	6	37±3	< 0.005				
8	1983	Tanks	4	350	400	504-909	1465-2070	384	91	1	25±5	< 0.001	194	2-36	16.07 ±0.13	< 0.005
9	1982	Small cabins (laminat.)	7		400	389-1108	733-1781	390	175	2	44±7	< 0.001	279	4-29	15.14 ±0.46	< 0.001

a. Number of workers tested only for SCE analysis
b. Number of workers tested only for chromosomal aberration analysis
c. Mandelic acid, mg/l; spec.gr.,1.024
d. Mandelic acid + phenylglyoxylic acid, mg/l; spec. gr.,1.024

Tab. 2. Chromosomal aberrations and SCE in blood lymphocytes of styrene-exposed workers and controls belonging to the plants analyzed in the year 1983.

Plant	Subject N°	Sex	Age	Years of exposure	Smoke [a]	Urinary metabolites MA [b] (mg/l)	MA + PGA [c] (mg/l)	Chromosomal aberration analysis Number of cells observed	Chromatid type	Chromosome type	Total (%)	SCE analysis Number of cells observed	SCE/cell Range	SCE/cell Mean
3	1	M	16	1	+	386	699	56	16	0	28	60	7-17	10.55
	2	F	41	4	++	533	881	100	25	0	25	61	5-25	12.44
	3	F	19	2	-	170	379	101	19	0	19	53	5-18	10.53
	4	F	24	6	+	694	1429	89	20	0	22	22	5-14	10.18
								346	80	0	24±4	196		10.92 ±1.02
4	1	M	47	10	+++	786	1168	56	18	0	32	24	7-17	11.04
	2	F	36	12	+	263	660	72	20	0	28	41	2-25	9.54
	3	F	39	4	-	335	731	12	3	0	25			
	4	F	48	16	++	670	1498	37	14	0	38			
	5	M	19	4	+	151	346	32	7	0	22	27	5-16	10.55
	6	M	34	8	-	197	408	104	25	0	24			
	7	M	26	7	++	220	365	116	26	0	22	44	3-17	9.77
	8	F	58	13	-	449	899	54	13	0	24			
	9	F	45	8	++	1083	1706	185	40	0	22			
	10	M	39	13	+++							37	5-19	10.54
								668	166	0	26±3	173		10.26±0.93
8	1	M	26	5	-	465	1814	119	25	0	21	48	5-21	16.07
	2	M	42	20	+++	504	1465	52	17	0	33	55	2-36	15.94
	3	F	18	1	+	1440	3888	102	26	0	25	48	8-25	16.20
	4	F	18	1	-	909	2070	112	23	1	21	43	8-32	16.07
								384	91	1	25±5	194		16.07±0.13
(3,4)[d]	1	M	33		-			87	8	0	9	12	5-13	9.00
	2	F	37		-			20	1	0	5	23	3-14	9.17
	3	F	43		++			67	9	0	13			
	4	M	23		++			44	3	0	7	74	6-23	12.46
	5	F	25		++			62	5	0	8	25	5-16	9.84
	6	F	31		+							22	5-16	7.81
	7	F	34		++							53	3-19	9.84
								280	25	0	8.4±3	209		9.69±1.55
(8)[d]	1	M	26		-			102	2	0	2			
	2	M	43		-			41	1	0	2	48	5-14	9.33
	3	F	18		++			101	5	0	5	50	6-14	7.75
	4	F	18		-			100	7	0	7			
								344	15	0	4±2	98		8.54±1.12

a. +, 1-10 cigarettes/day; ++, 11-19 cigarettes/day; +++ 20 cigarettes/day
b. Mandelic acid, mg/l; spec. gr.,1.024
c. Mandelic acid + phenylglyoxylic acid, mg/l; spec. gr.,1.024
d. Control group: the number of the plant to wich the control group is related is indicated in parentheses.

although well above those of their matched controls, were not significantly different because of the low number of subjects involved (Tab. 1). A statistically significant increase in SCE rates existed at air styrene concentrations higher than 200 mg/mc, with a steep increase in the number of SCEs per cell occurring at about 250 mg/mc of styrene.

From the described trend of SCE rates, it could be inferred that a concentration of 250 mg/mc is the threshold value for styrene in air with regard to the induction of SCEs in peripheral lymphocytes of exposed individuals. Consequently, SCEs are not useful for demonstrating genetic damage caused by very low doses of styrene, at least in occupational monitoring studies.

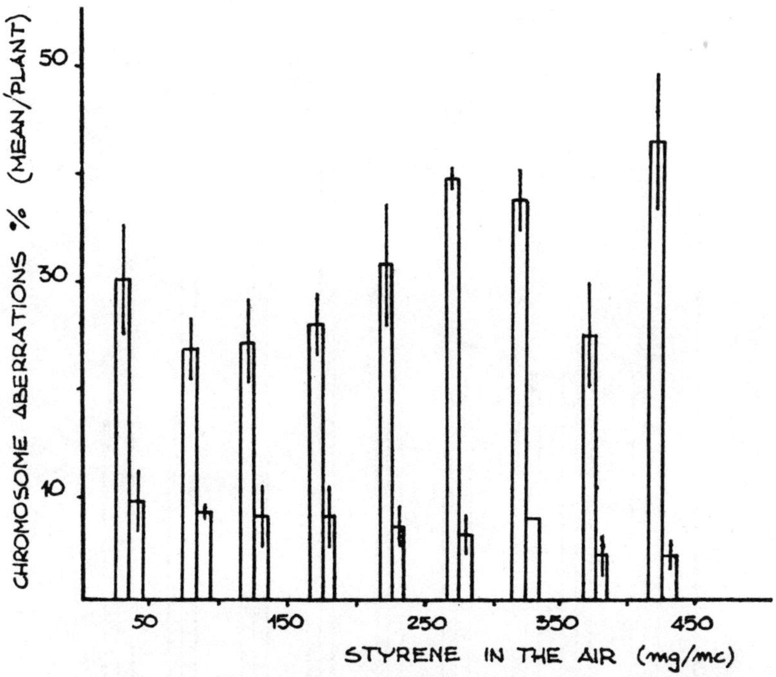

Fig. 1. Chromosomal aberrations mean values per plant versus styrene air concentrations. Left and right bars of each pair indicate exposed and control groups, respectively.

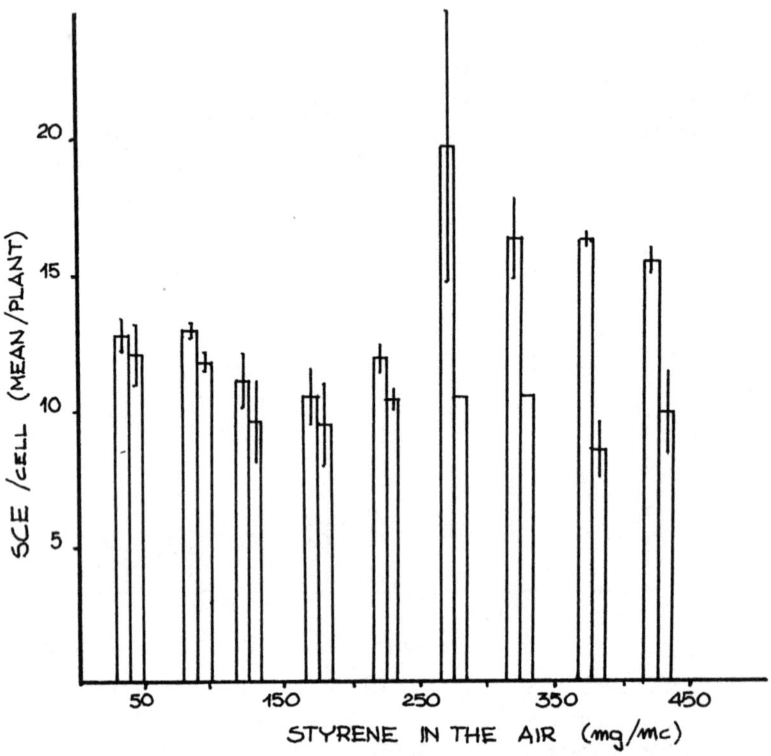

Fig. 2. Sister chromatid exchange mean values per plant versus styrene air concentration. Left and right bars of each pair indicate exposed and control groups, respectively.

REFERENCES

1. Andersson, H.C., E.A. Tranberg, A.H. Uggla, and G. Zetterberg (1980) Chromosomal aberrations and sister chromatid exchanges in lymphocytes of men occupationally exposed to styrene in a plastic boat factory. Mutat. Res. 73:387-401.
2. Camurri, L., S. Codeluppi, C. Pedroni, and L. Scarduelli (1983) Chromosomal aberrations and sister chromatid exchanges in workers exposed to styrene. Mutat. Res. 119:361-369.
3. Fleig, I., and A.M. Thiess (1978) Mutagenicity study of workers employed in the styrene and polystyrene processing and manufacturing industry. Scand. J. Work Environ. Health 4 (suppl. 2): 254-258.
4. Högstedt, B., K. Hedner, E. Mark-Vendel, F. Mitelman, A. Schütz, and S. Skerfving (1979) Increased frequency of chromosome aberrations in workers exposed to styrene. Scand. J. Work Environ. Health 5:333-335.
5. Meretoja, T., H. Järventaus, M. Sorsa, H. Vainio (1978) Chromosome aberrations in lymphocytes of workers exposed to styrene.
6. Meretoja, T., H. Vainio, M. Sorsa, and H. Härkönen (1977) Occupational styrene exposure and chromosomal aberrations. Mutat. Res. 56:193-197.
7. Sorsa, M., M. Hyvönen, H. Järventaus, and H. Vainio (1979) Chromosomal aberrations and sister chromatid exchanges in children of women in reinforced plastic industry. Symposium on Toxicology, Turku, Finland, May 29-30 (Abstracts: 16).
8. Thiess, A.M., H. Schwegler, and I. Fleig (1980) Chromosome investigations in lymphocytes of workers employed in areas in which styrene-containing unsaturated polyesther resins are manufactured. Amer. J. Ind. Med. 1:205-210.
9. Watanabe, T., A. Endo, K. Sato, T. Ohtsuki, M. Miyasaka, A. Koizumi, and M. Ikeda (1981) Mutagenic potential of styrene in man, Ind. Health, 19:37-45.

THE EFFECTS OF HYPOLIPIDEMIC PEROXISOME PROLIFERATORS ON THE INDUCTION OF SISTER CHROMATID EXCHANGES

Kaija Linnainmaa

Institute of Occupational Health
Department of Industrial Hygiene and Toxicology
Haartmaninkatu 1
SF-00290 Helsinki 29, Finland

INTRODUCTION

A novel class of suspected chemical carcinogens has been recently introduced by Reddy et al. (1). Characteristically, all the agents in the group possess hypolipidemic properties, induce proliferation of peroxisomes in the liver cells, and enhance the peroxisomal β-oxidation of fatty acids in the liver and kidneys of rodents (2-7). They have been shown to cause liver tumors in experimental animals (1,8-12), but, are not mutagenic in the Ames Salmonella mutagenicity assay (13,14). The induction of peroxisome proliferation was first introduced by the hypolipidemic drug, clofibrate (ethyl-α-p-chlorophenoxyisobutyrate), and some of its structural analogs (15). At present, however, the number of known peroxisome proliferators is about 20, and the group includes several structurally diverse compounds of great industrial importance (e.g., phthalates which are used as plasticiders in the plastic industry) (15). Quite recently, we have also shown that the widely used weed and brush killers, phenoxyacetic acid herbicides 2,4-dichlorophenoxyacetic acid (2,4-D) and 4-chloro-2-methylphenoxyacetic acid (MCPA) induce peroxisome proliferation and hypolipidemia in the liver cells of rodents (16,17). Similar to the previously known peroxisome proliferators, phenoxyacid herbicides are suspected carcinogens (18), even though they are not mutagenic in the Ames Salmonella assay (19). In addition to the suggestive data from experimental animal studies (18), epidemiological studies are indicative of increased risks of certain types of cancers among populations occupationally exposed to 2,4-D and MCPA (20,21). Because very little has been known about the induction of sister chromatid exchanges (SCEs) by hypolipidemic peroxisome proliferators, we have studied the induction of SCEs by 2,4-D, MCPA, and clofibrate (Fig. 1). Because of the widespread use of phenoxyacetic acid herbicides, and the epidemiological data which suggest that they have carcinogenic potency, a special emphasis has been given to the studies among populations occupationally exposed to phenoxyacid herbicides, 2,4-D, and MCPA, or their mixtures. However, experimental studies have also been carried out to find out whether clofibrate, 2,4-D, and MCPA are able to induce SCEs in cell cultures in vitro, or in laboratory animals in vivo.

SCEs IN CHINESE HAMSTER OVARY CELL CULTURES IN VITRO

The Chinese hamster ovary (CHO) cultures were treated with clofibrate, 2,4-D, and MCPA (22). Both commercial solutions and purified compounds of 2,4-D were used in the experiments. Each chemical was tested in 3 different subtoxic concentrations (10^{-5}, 10^{-4}, and 10^{-3} M, for 1 hr) and all the treatments were performed both in the presence and in the absence of the rat liver microsome activating system (S-9 mix). Benzo(a)pyrene was used as a positive control agent.

No clear dose-related increase of the SCE frequency was observed with any of the compounds in the study (Tabs. 1 and 2). On the other hand, all the compounds effected slight but consistant increases in the means of SCEs; these increases averaged 1-2 SCEs per cell. No differences in SCE induction efficiency were observed between the commercial herbicides and the corresponding pure phenoxyacetic acid. Further, the presence of S-9 mix did not affect the results. Thus, none of the compounds was found to be a potent SCE inducer although each gave slightly increased frequencies of SCEs.

SCEs IN EXPERIMENTAL ANIMALS IN VIVO

Rat Lymphocyte Cultures

The induction of SCEs has been studied in blood lymphocytes of rats treated intragastrically for 2 wk (100 mg/kg, 1 dose per da) with clofibrate, 2,4-D, and MCPA (22). No significant differences

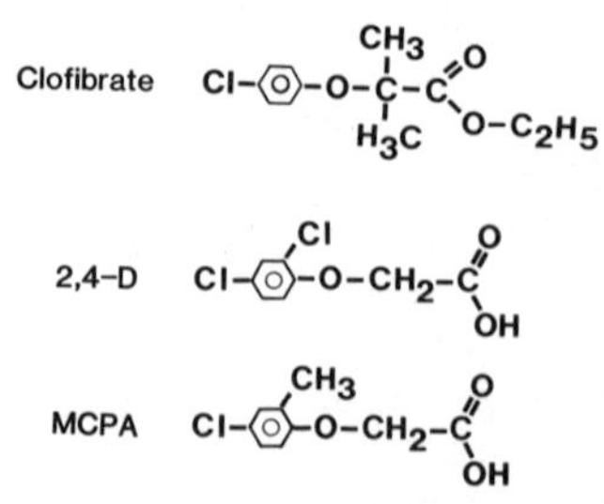

Fig. 1. Molecular structures of clofibrate, 2,4-D, and MCPA.

Tab. 1. Sister chromatid exchange in CHO cells treated for 1 hr either with pure 2,4-D and MCPA phenoxyacetic acids or commercial herbicide products.

Concentration	Mean SCE frequency/cell ± S.E.[a]			
	Pure compound		Commercial herbicide	
	(-S9)	(+S9)	(-S9)	(+S9)
CONTROLS				
-	8.6±0.4	7.9±0.4	7.8±0.4	8.6±0.4
BaP, (10^{-5}) M	7.7±0.6	16.5±0.6***		
2,4-D				
(10^{-5}) M	8.6±0.5	8.1±0.4	8.9±0.4*	9.0±0.4
(10^{-4}) M	9.0±0.4	9.1±0.4*	8.3±0.4	9.2±0.4
(10^{-3}) M	9.4±0.4	8.8±0.4	no growth	8.6±0.4
MCPA				
(10^{-5}) M	9.3±0.5	9.0±0.4*	7.8±0.4	9.1±0.4
(10^{-4}) M	9.7±0.4	9.1±0.4*	9.1±0.5*	9.3±0.5
(10^{-3}) M	9.2±0.7	9.4±0.4**	9.5±0.5**	9.8±0.4*

[a]The values are means ± standard errors from two experiments combined, and they represent the SCE frequency in 60 cells, 30 from each culture.

*Statistically significant difference from negative controls, $p < 0.05$.

**Statistically significant difference from negative controls, $p < 0.01$.

***Statistically significant difference from negative controls, $p < 0.001$.

From Linnainmaa, 1983 (ref. 22).

in the mean of the number of SCEs existed between the groups of treated animals and their respective control groups (Fig. 2). There was large interindividual variation in average SCE frequency among the animals. These variations were nevertheless equally prevalent within the control group and the groups exposed to 2,4-D, MCPA, and clofibrate. Furthermore, there were no significant delays in the cell cycle kinetics in the cultures from the treated animals compared to the controls; this is apparent from the frequencies of first-, second-, and third-division metaphase cells in the cultures (Tab. 3).

Tab. 2. Sister chromatid exchange in CHO cells treated with clofibrate for 1 hr.

Concentration	Mean SCE frequency/cell ±S.E.[a] (-S9)	(+S9)
CONTROLS		
- acetone (1 %)	9.6±0.5	10.5±0.7
- cyclophosphamide (10^{-5}) M	10.9±0.6*	19.6±1.0
Clofibrate		
(10^{-5}) M	10.2±0.5	9.6±0.4
(10^{-4}) M	10.6±0.6*	10.4±0.6
(10^{-3}) M	-[b]	11.1±0.5

[a]The values are means ± standard errors from two experiments combined, and they represent the SCE frequency in 60 cells, 30 from each culture.
[b]Only first division metaphases observable.
*Statistically significant difference acetone controls, $p < 0.05$.
From Linnainmaa, 1983 (ref. 22)

Bone Marrow Cells of Chinese Hamsters

The induction of SCEs has been studied in the bone marrow cells of Chinese hamsters after intragastric treatments with 2,4-D, MCPA, and clofibrate (treatments similar to those in the experiment with rats (22). No significant differences in the mean of SCE frequency were observed between the groups treated with clofibrate or 2,4-D and their respective controls, even though the test group means were slightly elevated (Fig. 3). The SCE levels in the bone marrow cells

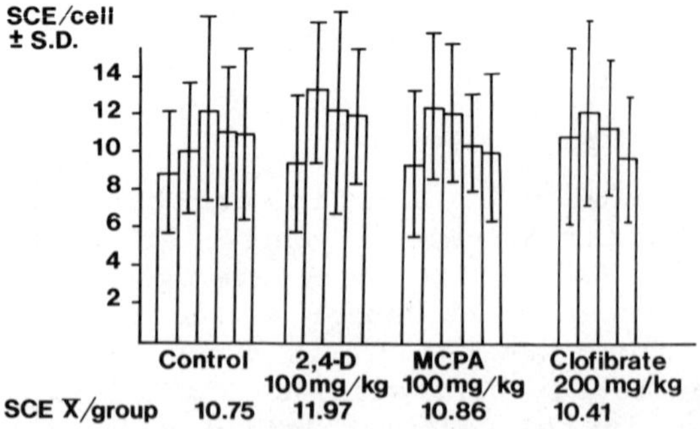

Fig. 2. SCEs in blood lymphocytes of rats after 2 wk treatments of the animals (intragastrically, 1 dose per da) with 2,4-D, MCPA, and clofibrate. The mean and standard deviation of 30 cells from each animal are shown.

Tab. 3. Percentages of 1st, 2nd, and 3rd division metaphase cells in blood lymphocyte cultures of groups of rats exposed to 2,4-D and MCPA herbicides, and clofibrate.

Treatment	No. animals	Percentage of cells + S.D.		
		1st division	2nd division	3rd division
Controls	5	60.2 ± 5.0	29.0 ± 2.1	10.7 ± 3.7
2,4-D, 100 mg/kg	4	58.0 ± 5.5	32.1 ± 4.1	9.9 ± 5.2
MCPA, 100 mg/kg	5	64.5 ± 9.2	25.9 ± 5.2	9.6 ± 5.7
Clofibrate, 100 mg/kg	4	61.9 ± 16.7	23.3 ± 7.4	14.9 ± 9.7

From Linnainmaa, 1983 (ref. 22)

of the Chinese hamsters treated with MCPA, on the other hand, differed significantly from those of the control animals; the mean in the treated group was 1.2 SCEs/cell higher than that in the control group.

SCEs AMONG WORKERS OCCUPATIONALLY EXPOSED TO 2,4-D AND MCPA

The induction of SCEs in peripheral lymphocytes of workers spraying foliage in forestry with 2,4-D and MCPA has been analyzed from 3 successive blood samples from 35 herbicide sprayers (23). The first sample was taken before, the second sample in the middle of, and the third sample at the end of the spraying season of July to October 1981. For the estimation of individual levels of exposure during spraying work, the concentrations of 2,4-D and MCPA

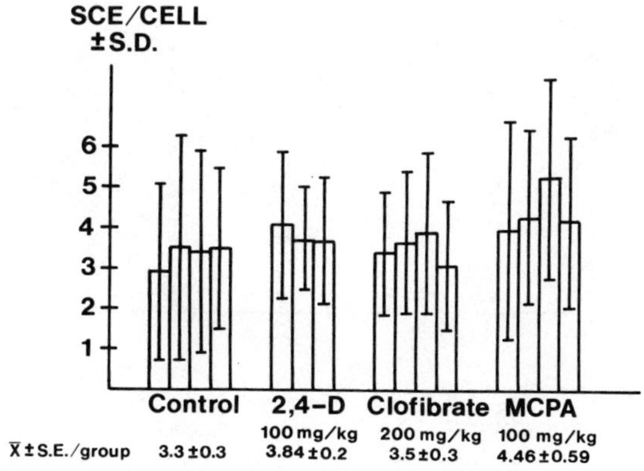

Fig. 3. SCEs in bone marrow cells of Chinese hamsters after 2 wk treatment of the animals (intragastrically, 1 dose per da) with 2,4-D, clofibrate, and MCPA. The mean and standard deviation of 30 cells from each animal are shown.

in urine samples, which were taken concurrently with the second blood sample, were also determined.

Urinary Concentrations of Phenoxyacetic Acids

The levels of phenoxy acids in the urine samples were found to vary highly between the individuals tested. In the group of nonsmokers, total values (2,4-D + MCPA) from 0.00 to 10.99 mg/l were observed, with a mean of 1.80 and a standard deviation of 2.85. For smokers, the values were from 0.00 to 10.93, the mean was 2.80, and the standard deviation was 3.73. In both groups several subjects had no detectable concentrations of 2,4-D or MCPA in their urine samples, a finding which indicates the proper use of protective equipment during the application of the herbicide.

SCEs in Peripheral Lymphocytes

No exposure-related differences in the means of the SCE frequencies in the lymphocytes of exposed workers were observed, either in the group level or in the individual level (Figs. 4, 5, and 6). Furthermore, the means of the SCE frequencies in the 15 unexposed control subjects fell in the same range as those of the exposed subjects. A slight difference in the means of SCEs was observed between smokers and nonsmokers, smokers having significantly higher mean values than nonsmokers.

DISCUSSION

The results of the present study support earlier data which indicated that 2,4-D, MCPA, and clofibrate do not act as agents

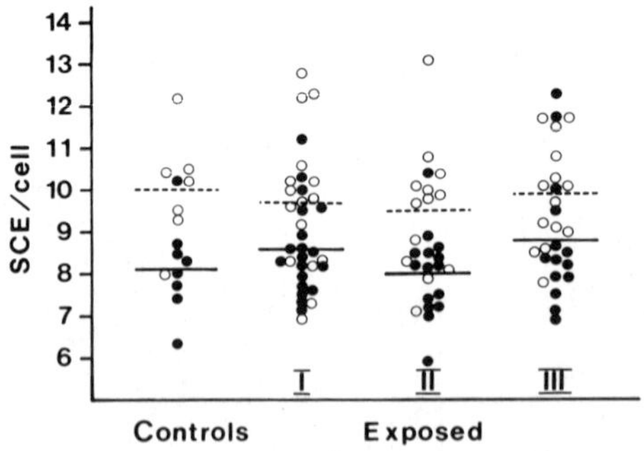

Fig. 4. The individual [(•), nonsmokers, (o), smokers] and group [(——), nonsmokers, (- - -), smokers] means of SCE frequencies in lymphocyte cultures of control subjects and in the cultures from the three successive blood samples. I. Samples taken before the exposure period; II. samples taken at the middle of the exposure period; III. samples taken at the end of the exposure period. From data Ref. 23.

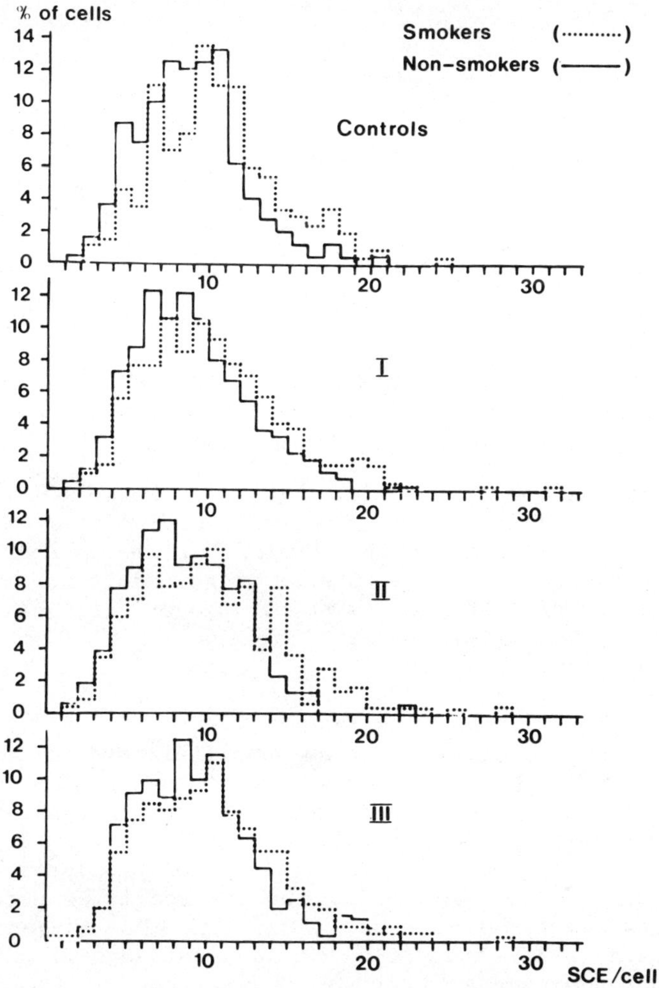

Fig. 5. Distributions of SCEs in individual cells in the lymphocyte cultures of control subjects and from the 3 successive blood samples of the herbicide workers. The number of nonsmoking and smoking individuals included in the distributions are: controls, 8 and 7, sample no. I: 19 and 16, no. II: 15 and 13, no III: 14 and 14. 30 cells from each culture were scored for SCEs. From data of Ref. 23.

which damage DNA directly (7,9). The data also are in accordance with the results of Abe and Sasaki (24) who found that a peroxisome proliferator di-(2-ethylhexyl)phthalate (DEHP) induced a non-dose-dependent, but significant increase in SCEs in CHO cells in vitro. Similar to the present results with MCPA in bone marrow cells of exposed Chinese hamsters, the studies of Lamb et al. (25) revealed slightly, although not significantly, higher frequencies of SCEs in the bone marrow cells of mice injected with 2,4-D.

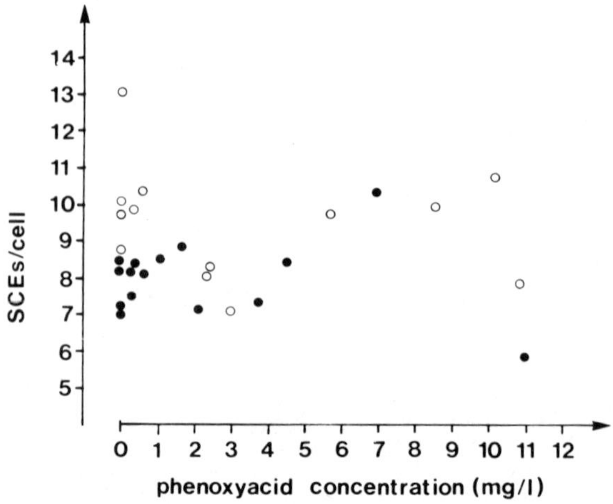

Fig. 6. Individual means of SCEs for nonsmokers (•) and smokers (o) in relation to the urinary concentrations of phenoxyacetic acid. The blood sample for the analysis of SCEs was taken at the same time as urine sample for the urinary concentration of phenoxy acid. Linear correlation coefficients for nonsmokers and smokers were 0.015 and 0.030, respectively.

The mechanism by which peroxisome proliferators initiate malignant transformation in the liver cells of rodents is not known. However, Reddy et al. (26) have pointed out that persistent proliferation of peroxisomes and the increase in peroxisomal β-oxidation systems may serve as an endogenous initiator of neoplastic transformation of liver cells by increasing the intracellular production of DNA-damaging hydrogen peroxide and other reactive oxygen radicals. The present data on SCE induction are in accordance with this hypothesis as genotoxic agents which are known to act indirectly via oxygen radicals have been shown to be rather poor SCE inducers (27,28). However, they typically have high clastogenic (chromosome-breaking) activity (29). For this type of genotoxic agent, the analysis of structural chromosome aberration may provide a more reliable method for monitoring occupational exposures than the analysis of SCEs. Structural chromosome aberrations are currently being studied for the peroxisome proliferators in the present study.

The number of recognized carcinogenic hypolipidemic peroxisome proliferators has been rapidly growing in recent years. Thus, if the hypothesis of the indirect genotoxic action of these compounds proves correct, hypolipidemic peroxisome proliferators may form a large and important group of carcinogens which cannot be detected by monitoring SCEs in the peripheral lymphocytes of the exposed subjects.

REFERENCES

1. Reddy, J.K., D.L. Azarnoff, C.E. Hignite (1980) Hypolipidaemic hepatic peroxisome proliferations form a novel class of chemical carcinogens. Nature 283:397-398.
2. Lazarow, P.B. (1978) Rat liver peroxisomes catalyze the β-oxidation of fatty acids. J. Biol. Chem. 253:1522-1528.
3. Lazarow, P.B., and C. de Duve (1976) A fatty acyl-CoA oxidizing system in rat liver peroxisomes; enhancement by clofibrate, hypolipidemic drug. Proc. Natl. Acad. Sci., USA 73:2043-2046.
4. Inestrosa, N.C., M. Bronfman, and F. Leighton (1979) Detection of peroxisomal fatty acyl-coentzyme A oxidase activity. Biochem. J. 182:779-778.
5. Lalwani, N.D., M.K. Reddy, M. Mangkornkanok-Mark, and J.K. Reddy (1981) Induction, immunochemical identity and immunofluorescence localization of peroxisome proliferation associated polypeptide (PPA-80) and peroxisomal enoyl-CoA hydratase of mouse liver and renal cortex. Biochem. J. 198:177-186.
6. Osumi, T., and T. Hashimoto (1978) Enhancement of fatty acyl-CoA oxidizing activity in rat liver peroxisomes by di-(2-ethylhexyl)phthalate. J. Biochem. 83:1361-1365.
7. Reddy, M.K., S.A. Qureshi, P.F. Hollenberg, and J.K. Reddy (1981) Immunological identity of peroxisomal enoyl-CoA hydratase with the peroxisome proliferation as associated 80 000 mol. wt. polypeptide in rat liver. J. Cell Biol. 89:406-417.
8. Reddy, J.K., and S.A. Qureshi (1979) Tumorgenicity of the hypolipidemic peroxisome proliferator ethyl-α-p-chloro-phenoxyisobutyrate (clofibrate) in rats. Br. J. Cancer 40:476-482.
9. Reddy, J.K., and M.S. Rao (1977) Malignant tumors in rats fed nafenopin, a hepatic peroxisome proliferator. J. Natl. Cancer Inst. 59:1645-1650.
10. Reddy, J.K., M.S. Rao, D.L. Azarnoff, and S. Sell (1979) Mitogenic and carcinogenic effects of hypolidemic peroxisome proliferator, (4-chloro-6-(2,3-xylidino)-2-pepimidinylthio)acetic acid (WY-14,643) in rat and mouse liver. Cancer Res. 39:152-161.
11. Reddy, J.K., M.S. Rao, and D.E. Moody (1976) Hepatocellular carcinomas in acatalasemic mice treated with nafenopin, a hypolipidemic peroxisome proliferator. Cancer Res. 36:1211-1217.
12. Svoboda, O.J., and D.L. Azarnoff (1979). Tumors in male rats fed ethyl chlorophenoxyisobutyrate, a hypolipidemic drug. Cancer Res. 39:3419-3428.
13. Warren, J.R., V.F. Simmon, J.K. Reddy (1980) Properties of hypolipidemic peroxisome proliferators in the lymphocyte ^{3}H-thymidine and Salmonella mutagenesis assays. Cancer Res. 40:36-41.
14. Kirby, P.E., R.F. Pizzarello, T.E. Lawlor, S.R. Haworth, and J.R. Hodgson (1983) Evaluation of di-(2-ethylhexyl)phthalate and its major metabolites in the Ames test and L5178Y Mouse Lymphoma Mutagenicity Assay. Environ. Mutag. 5:657-663.
15. Reddy, J.K., J.R. Warren, M.K. Reddy, and N.D. Lalwani (1982)

In Hepatic and Renal Effects of Peroxisome Proliferations: Biological Implications, K. Kindl and P.B. Lazarow, eds. The New York Academy of Sciences, New York, pp. 81-110.
16. Vainio, H., J. Nickels, and K. Linnainmaa (1982) Phenoxy acid herbicides cause peroxisome proliferation in Chinese hamsters. Scand. J. Work Environ. Health 8:70-73.
17. Vainio, H., K. Linnainmaa, M. Kähönen, J. Nickels, E. Hietanen, J. Marniemi, and P. Peltonen (1983) Hypolipidemia and peroxisome proliferation induced by phenoxyacetic acid herbicides in rats. Biochem. Pharmacol. 32:2775-2779.
18. IARC (International Agency for Research on Cancer) (1983) Monographs on the Evaluation of the Carcinogenic Risk of Chemicals to Humans. Miscellaneous Pesticides, Vol. 39, Lyon, IARC, pp. 137-147.
19. Seiler, J. (1978) The genetic toxicology of phenoxy acids other than 2,4,5-T. Mutat. Res. 55:197-226.
20. Eriksson, M., L. Hardell, N.O. Berg, T. Möller, and O. Axelson (1981) Soft-tissue sarcomas and exposure to chemical substances: a case referent study. Br. J. Ind. Med. 38:27-33.
21. Hardell, L., M. Eriksson, P. Lenner, and E. Lundgren (1981) Malignant lymphoma and exposure to chemicals, especially organic solvents, chlorophenols and phenoxy acids: a case control study. Br. J. Cancer 43:169-176.
22. Linnainmaa, K. (1983) Induction of sister chromatid exchanges by the peroxisome proliferators 2,4-D, MCPA, and clofibrate *in vivo* and *in vitro*. Carcinogenesis (in press).
23. Linnainmaa, K. (1983) Sister chromatid exchanges among workers occupationally exposed to phenoxy acid herbicides 2,4-D and MCPA. Terat. Carcinog. Mutag. 3:269-279.
24. Abe, S., and M. Sasaki (1977) Chromosome aberrations and sister chromatid exchanges in Chinese hamster cells exposed to various chemicals. J. Natl. Cancer Inst. 58:1635-1641.
25. Lamb, J.C., T.A. Marks, B.C. Gladen, J.W. Allen, and J.A. Moore (1981) Male fertility, sister chromatid exchange, and germ cell toxicity following exposure to mixtures of chlorinated phenoxyacids containing 2,3,7,8-tetrachlorodibenzo-p-dioxin. J. Toxicol. Environ. Health 8:825-834.
26. Reddy, J.K., N.D. Lalwani, M.K. Reddy, and S.A. Qureshi (1982) Excessive accumulation of autofluorescent lipofucin in the liver during hepatocarcinogenesis by methyl clofenapate and other hypolipidemic peroxisome proliferators. Cancer Res. 42: 259-266.
27. Wolff, S. (1982) Chromosome aberrations, sister chromatid exchanges, and the lesions that produce them. Sister Chromatid Exchange, S. Wolff, ed. John Wiley & Sons Inc., New York, pp. 41-57.
28. Emerit, I., and P.A. Cerutti (1981) Tumour promoter phorbol-12-myristate-13-acetate induces chromosomal damage via indirect action. Nature 293:144-146.
29. Kao, F., and T. Puck (1969) Genetics of somatic mammalian cells. IX. Quantitation mutagenesis by physical and chemical agents. J. Cell Physiol. 74:245-258.

ETHYLENE OXIDE AND SOME FACTORS AFFECTING THE MUTAGEN SENSITIVITY OF SISTER CHROMATID EXCHANGE IN HUMANS

V. F. Garry, J. K. Wiencke, and R. L. Nelson

University of Minnesota
Laboratory of Environmental Pathology
421 S. E. 29th Avenue
Minneapolis, Minnesota 55414

INTRODUCTION

In our laboratory, we consider mutagenesis and the factors surrounding a mutagenic event as a series of barrier systems (Fig. 1), any one of which can affect a final biological outcome in humans. For example, a mutagen can only exert an effect if it indeed has access to a portal of entry. Physical parameters such as solubility affect skin absorption (1). Particle size will determine whether a particulate is respirable (2). Taken together, the physical/chemical barriers at the portal of entry limit carcinogen access to an organism. Beyond this point, the extracellular fluid (e.g., blood plasma) may limit or augment a mutagenic effect. In our studies shown here and elsewhere in this volume, we (Wiencke, et al.) demonstrate that sister chromatid exchange (SCE) mutagen effects can be modulated by blood plasma (3). Indeed, Lambert et al. (4) have indicated this possibility earlier. The complex environment of peripheral blood presents a myriad of cell types (erythrocytes, other mononuclear cells, neutrophils, and platelets) which may alter the mutagen response in phytohemagglutinin (PHA)-stimulated lymphocytes. An elegant series of studies (5) shows quite clearly that erythrocytes activate at least 1 carcinogen, styrene, which otherwise is inactive in purified lymphocyte cultures, i.e., the target cell. Much remains to be done to describe the role of nontarget cells in the modulation of the mutagen response of a target cell population. Even the target cell, the T lymphocyte, for the SCE mutagen response may be modified by intercurrent disease. In another study (Fig. 2) (6), we showed in whole blood cultures that the lymphocytes from untreated patients with solid tumors, although uniformly respondent to the direct-acting mutagen diepoxybutane, varied in response to the mitomycin C

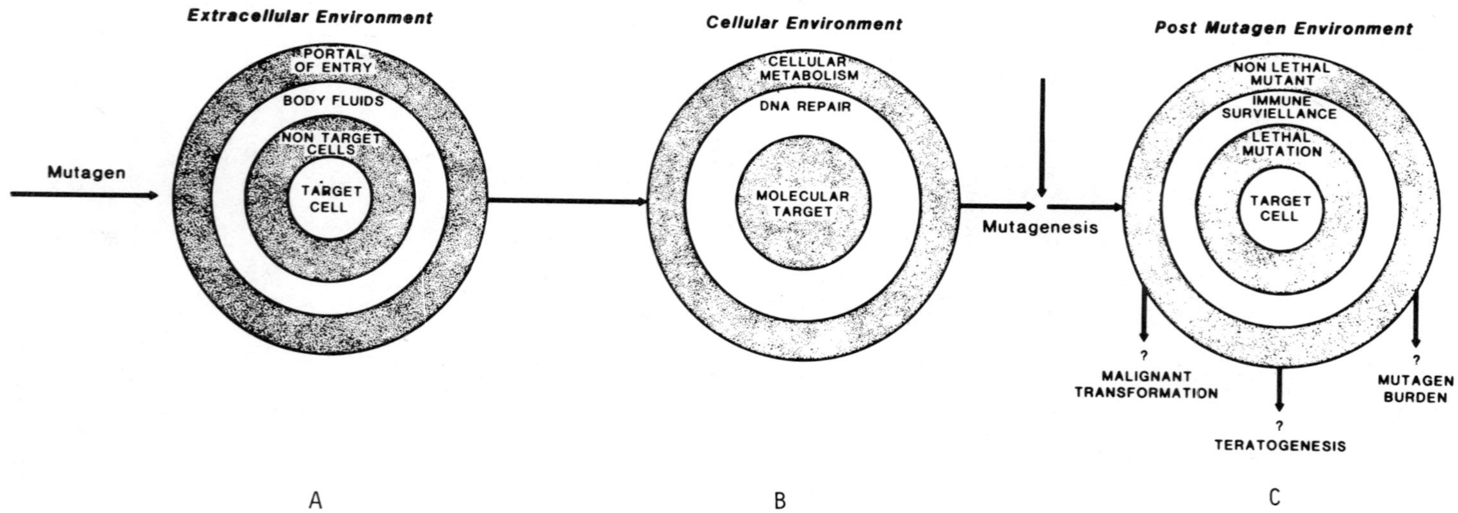

Fig. 1. The diagram presents the barrier system concept of the possible mutagen interactions in higher organisms. A) Environmental barrier to mutagen activities; B) cellular barriers limiting mutagen activity in a target cell; and C) the final barriers limiting health effect consequences given a mutagenic event.

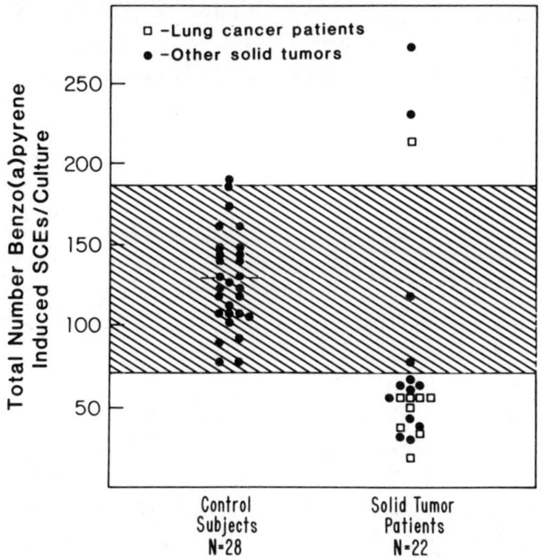

Fig. 2. The induction of SCEs by benzo(a)pyrene (BP) treatment (10 μg/ml) in lymphocyte cultures from untreated patients with solid tumors and controls is depicted. Shown on the left are SCE scores from 28 control subjects. The hatched area represents the total range of BP-induced SCEs in the control group. Shown on the right are data collected from 22 patients with solid tumors. □ = lung cancer patients; ● = all other patients (colon, testes, pancreas, esophagus, breast, bladder, and endometrium).

and were significantly divergent in response to the indirect-acting carcinogen, benzo(a)pyrene. This evidence suggests that modification of the target cell population can occur in chronic as well as in acute infectious states (7).

Once a mutagen has entered the target cell population, cellular metabolism may alter the mutagenic potential of the chemical. If a lesion has been introduced in DNA, it can be repaired. Should all of these fail-safe mechanisms function inadequately, then a mutagenic event will occur. A mutant cell thus born faces still other barriers to its survival (Fig. 1C). The mutation may be lethal and so may be suicidal for the cell. A mutation may render the cell so unique that it may be destroyed by immunological surveillance. Finally, a mutation may be silent with either no observable biological effect or with no real health effect consequence to the organism.

From these observations derived from the barrier systems concept, malignant transformation of a cell or the evolution of a neoplastic cell, and the occurrence of mutational teratogenic events, are rare final options with significant health effect consequences.

In the framework of the barrier systems view of mutagenesis, I will review some of our current findings and those of other workers regarding the effects of ethylene oxide (EO).

HUMAN STUDY

In early 1978, a group of hospital workers (8) engaged in the sterilization of supplies and periodically reported a number of health complaints (Fig. 3) ranging from upper respiratory irritation to neurological dysfunction. Review of the nursing staff's daily report indicated the periodic presence of respiratory symptoms that appeared to be related to efforts to repair the EO sterilization facility. Clearly, EO exhibited irritant effects on the respiratory tract (portal of entry) and systemic effects. This acute response suggested to us that EO, known to be a mutagen in may also have had an effect on the human genome. To test this hypothesis, we studied SCE frequencies in cultured peripheral blood cells. Several comparison groups between exposed persons and control subjects were evaluated. First, we studied persons with neurological symptoms (Tab. 1) because these were expected to have been exposed to the highest dose (11). Although ambient measurements of EO by Miran infrared spectroscopy indicated a level of 36 ppm, other measurements did, however, show a short burst of greater than 1,500 ppm at the site of release of EO to the drain during the

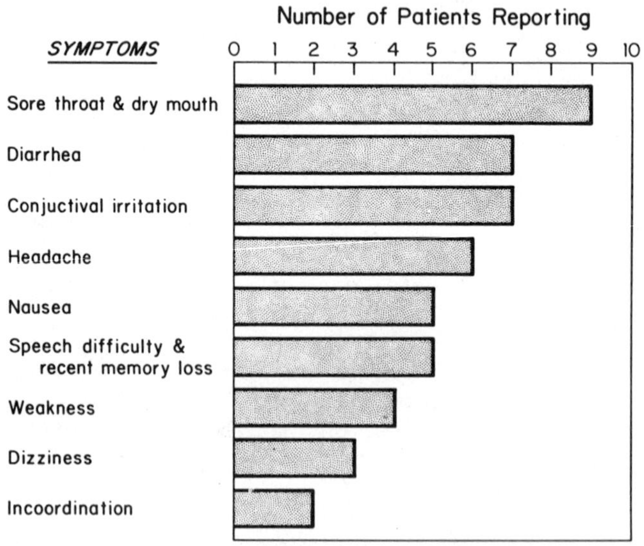

Fig. 3. All symptoms reported by 12 exposed individuals during April and May, 1978. Data were obtained from medical histories elicited during our investigation.

purge cycle of the sterilizer. A second exposed group of those persons who may have been incidentally exposed and were asymptomatic were also studied. Data from each exposed control group were compared to data from control subjects collected at the same time to minimize experimental variance. Persons with neurological symptoms (Tab. 1) showed increased SCE 3 wk after exposure to EO had been terminated. Most of these individuals showed persistent SCE elevation during the eighth wk. Baseline SCE was achieved in this population by the twenty-fourth wk after last exposure (Fig. 4). Persons with respiratory symptoms and incidentally exposed persons demonstrated increased SCE levels, measured between the seventh and ninth wk since their last exposure.

The data recounted in this study display two rather unusual findings in the SCE results. First, persistent elevation of SCE was noted. Other in vivo and in vitro human studies with cytostatic agents (12,13,14) show that transient increases in SCE usually occur after mutagen exposure. In the case of EO, it is likely that G_0 lymphocytes in the subset of long-lived lymphocytes are uniquely sensitive to EO. Other alternative explanations can also be considered. For example, it is also possible that the committed or the noncommitted hemopoietic stem cells have been altered by EO exposure leading to persistent SCE elevation. While these considerations are important for the overall outcome of human exposure to EO, a second unusual finding focused our interest: one individual exposed to 1,500 ppm EO for 5 min (Tab. 2) showed an increased SCE frequency some 8 wk after a single EO exposure. These data suggested the possibility that a short-term high dose exposure to the gas could increase the level of SCEs. If this hypothesis were correct, then long held occupational dicta

Tab. 1. Frequency of SCEs among individuals exposed to EO and unexposed volunteers. The data are recorded per metaphase; 20 metaphase cells were examined per individual culture. The difference in the exchange between exposed and unexposed subjects is significant at the $P < 0.005$ level (4 wk) and $P < 0.01$ (8 wk) by the Mann-Whitney test.

EXPOSED			CONTROLS	
Subject Number	SCE/ metaphase (3 weeks after last exposure)	SCE/ metaphase (8 weeks after last exposure)	Subject Number	SCE/ metaphase
			5	6.36
1	10.68	12.05	6	5.70
			7	5.85
2	8.85	10.45	8	5.70
3	9.75	12.15	9	4.90
			10	7.00
4	9.70	6.70	11	5.30
			12	7.05

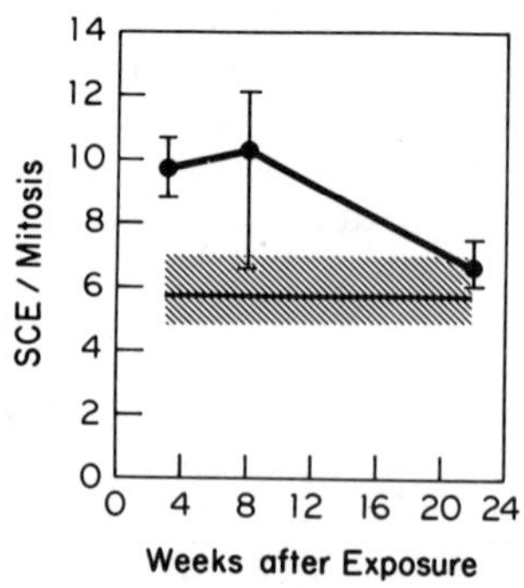

Fig. 4. The graph depicted in the figure follows the SCE frequencies of the same 4 individuals (Tab. 1) over a 24-wk period. SCE frequencies within the normal range were obtained 24 wk after the last EO exposure.

regarding limitation of exposure based on 8 hr, time-weighted average (15) may not be applicable to EO. Since the possibilities of exploring this idea were limited by ethical and other constraints, we decided to develop an in vitro system to test our hypothesis in a relatively limited manner.

MEMBRANE DOSIMETRY SYSTEM

To evaluate perturbation of the target cell in peripheral blood for SCEs by an EO burst, a quantitative gas delivery system was devised (16). The system consists of an infrared analyzer with

Tab. 2. Incidental/asymptomatic exposure to EO. Individuals incidentally exposed to EO were tested for SCE frequencies between the seventh and ninth wk after the last known exposure to EO. As a group, SCE rate was significantly elevated over controls (P < 0.02) by the Mann-Whitney test. The control group was the same source used in Tabs. 1 with 3 additional control individuals (average 6.37 ± 0.47).

SUBJECTS	SCE/METAPHASE	INCIDENT DESCRIPTION
1	11.6	Orderly, carries sterilized materials to operating room
2	9.5	Investigator, exposed to ETO during study
3	6.8	Part time (weekend) worker in sterilization facility
4	9.6	Investigator, exposed to approx. 1500 ppm ETO for 5 minutes

an injection port to measure ambient concentrations of the gas and an exposure chamber linked together forming a closed loop apparatus. The critical features of the system are the culture containers (Fig. 5), which are enclosed by porous diffusion membranes different in diameter for each of 4 separate cultures. Under nonequilibrium conditions, different amounts of EO are deposited into the culture media in a specified period of time. A cumulative media dose measured by gas chromatography greater than 10 μg delivered over 20 min was sufficient to alter SCE levels (Fig. 6). A linear dose response (EO vs. SCE) was observed.

These data suggest that a short, high-level burst of the gas is sufficient to induce SCEs in peripheral blood cells, at least in vitro. Moreover, the data show that the SCE assay system is a

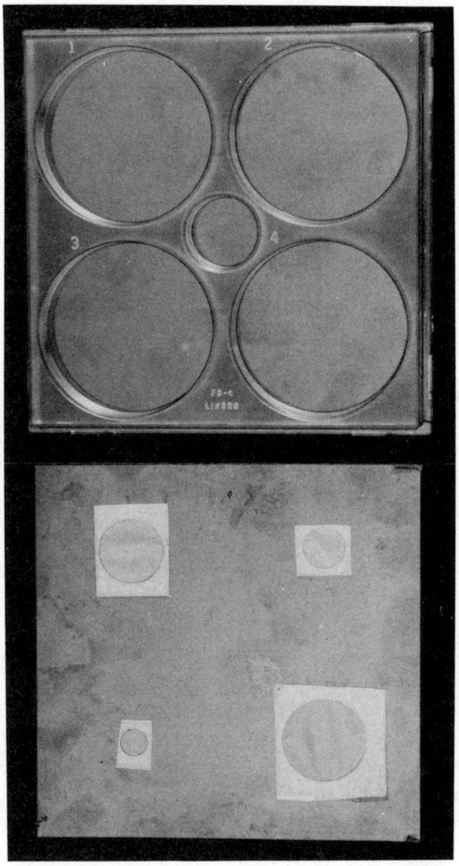

Fig. 5. The photograph presented pictures a 4-plex culture dish and a polycarbonate cover with 4 different-sized membrane inserts. When sealed in place over the individual cultures, delivery of different doses of EO to the cultures is achieved under nonequilibrium conditions.

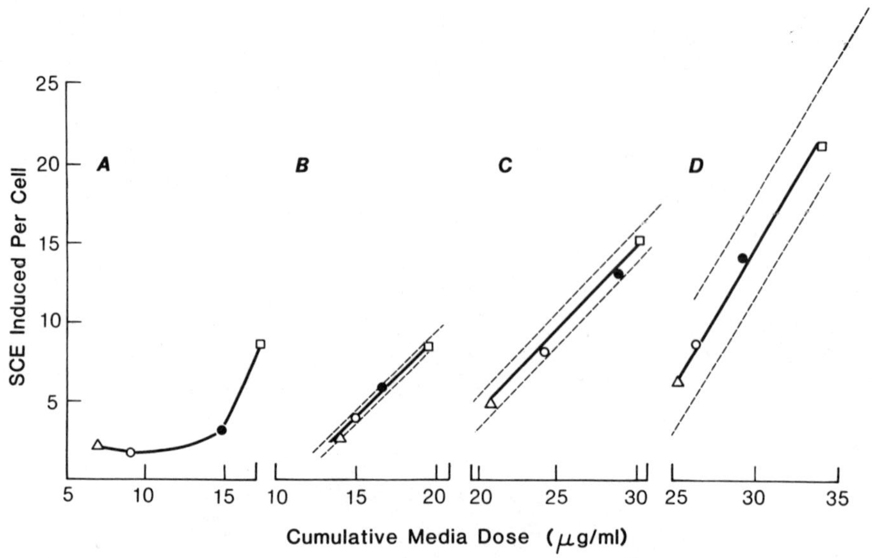

Fig. 6. The pooled data from EO-treated cultures show the relationship between EO-induced SCE frequencies and cumulative media dose. Each point on these graphs is the SCE score and individual EO determinations from separate cultures. The symbols identify the membrane area studied (same as Fig. 5). The ambient dose for each panel is indicated: A) 100, B) 131, C) 218, and D) 306 ppm, respectively. The values (± 16%) recorded in panel B are the mean of 3 samples from different blood sources. Above 100 ppm, the data follow first-order regression. The 95% confidence limits for the regression model (---) are delineated in the figure.

sensitive biological dosimeter for EO. As with any other assay system used to measure human effect in the clinical laboratory, sensitivity of the method is one aspect of the steps taken to demonstrate the validity of the assay. Since other mutagens yield a positive SCE response in vivo, the assay lacks specificity on these grounds. EO exposure is usually an isolated event; therefore, the number of SCE-positive mutagen interferences is limited. Smoking (17), for example, may make a minor contribution to the SCE response to EO. Vaccination (18) may interfere with interpretation of the SCE data. Aside from these and other expected variables, more subtle sources of variation are possible. Passage of a chemical mutagen from the environment through a portal of entry encounters a plasma milieu, one of the barriers referred to earlier in this paper.

PLASMA MODULATION OF SCE

To test the idea that plasma could effect the SCE response, we performed the following series of studies. In our laboratory, routine cultures of peripheral blood cells are ordinarily supplemented with 20% fetal calf serum. To explore the idea that plasma could affect mutagen-induced SCE, we cultured blood from EO-exposed workers supplemented with either 20% fetal calf serum or 20% autologous plasma. In this initial series of experiments (J. Hozier and V. F. Garry, unpublished data) shown in Fig. 7, the effects of autologous plasma from 4 different unexposed persons on SCE levels recorded are not significantly different from the same cells cultured in fetal calf serum. On the other hand, cells from persons who have been exposed to EO cultured in fetal calf serum show increased levels of SCEs compared to the same cells cultured in autologous human plasma. This interesting finding was further defined by J. Wiencke who, as a student in our laboratory, studied in detail the relationship between serum or plasma source and in vitro induction of SCE. From these data, it appears that supplementation of the blood cultures by serum or plasma alters the kinetics of cell proliferation. The expression of SCEs induced by some mutagens may be dependent on factors presented in these culture supplements. The data are discussed in depth elsewhere in this volume (3).

TARGET CELL ALTERATION

Another variable in the expression of SCEs in peripheral blood may be an alteration of the T lymphocyte target for an SCE event. From an immunological standpoint, subset populations of T lymphocytes can be altered in several disease states. In a

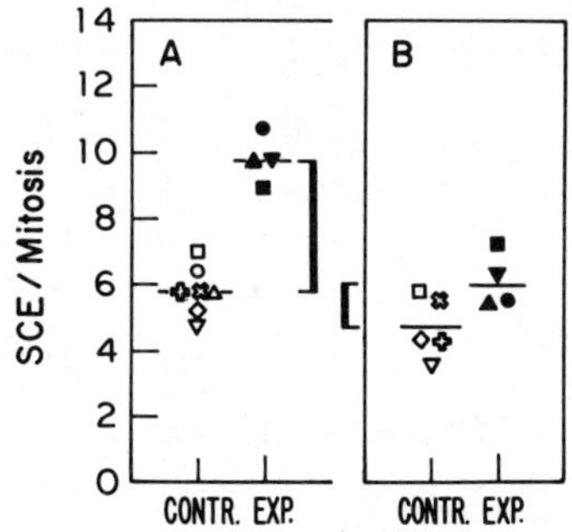

Fig. 7. In the figure, A) whole blood cultures were grown in the presence of 20% fetal calf serum. The SCE data with increased SCE levels B) occurred in EO-exposed persons compared to controls. Cultures from the same subjects grown in autologous human plasma show suppression of EO-induced SCE. Each symbol denotes a culture from a given subject.

similar sense, SCE-mutagen responsiveness could also be affected. At the moment, evidence for or against this notion is at best indirect (3,6).

CONCLUSIONS

The barrier systems concept reviewed here provides a perspective with which to explore the effects of mutagen exposure on SCE induction. It is apparent from the data we have gathered and from those of other workers (19,20,21), that EO has indeed crossed over many of the barriers and has a human effect with respect to SCE levels. Animal studies show that the chemical is an animal carcinogen (22,23). Inter-individual variance in the SCE-mutagen response may reflect an alteration of the microenvironment or of the T lymphocyte target cell.

REFERENCES

1. Bock, F. (1977) In Dermatotoxicology and Pharmacology, F. Marzullo and H. Majbach, eds. Wiley Publishing Co., New York, pp. 473-485.
2. Menzel, D., and R. McClellan (1980) In Toxicology, 2nd Ed, J. Doull, C. Klaasen, and M. Amdur, eds. MacMillan Pub. Co., New York, pp. 255-272.
3. Wiencke, J., K. Kelsey, R. Kreiger, and V. Garry (1984) Serum and plasma dependent variations in benzo(a)pyrene induced sister chromatid exchange in human lymphocytes. In Sister Chromatid Exchange, R. Tice and A. Hollaender, eds. Plenum Press, New York.
4. Lambert, B., A. Lindblad, K. Holmberg, and D. Francesconi (1982) The use of sister chromatid exchange to monitor human populations for exposure to toxicologically harmful agents. In Sister Chromatid Exchange, J. Wiley and Sons, New York, pp. 164-165.
5. Norppa, H., and J. Tursi (1984) Erythrocyte-mediated metabolic activation detected by SCE. In Sister Chromatid Exchange, 25 Years of Experimental Research, R. Tice and A. Hollaender, eds. Plenum Press, New York.
6. Wiencke, J., J. Vosika, P. Johnson, and V. Garry (1982) Differential induction of sister chromatid exchange by chemical carcinogens in lymphocytes cultured from patients with solid tumors. Pharmacology 24:67-73.
7. Kurvink, K., C. Bloomfield, and J. Cervenka (1978) Sister chromatid exchange in patients with viral disease. Exp. Cell Res. 113:450-453.
8. Garry, V., J. Hozier, D. Jacobs, R. Wade, and D. Gray (1979) Ethylene oxide: Evidence of human chromosomal effects. Environ. Mutagen. 1:375-382.

9. Glaser, Z. (1977) Use of ethylene oxide as a sterilant in medical facilities. DHEW (NIOSH) Publication No. 77-200, Government Printing Office, Washington, D.C.
10. Ehrenberg, L., and T. Hallston (1967) Haematologic studies on person occupationally exposed to ethylene oxide. International Agency Report, SM 92/26, pp. 327-334.
11. Gross, J., M. Haas, and T. Swift (1978) Ethylene oxide neuropathy. Neurology 28:355 (abstract).
12. Raposa, T. (1978) Sister chromatid exchange studies for monitoring DNA damage and repair capacity after cytostatics in vitro and in lymphocytes of leukemic patients under cytostatic therapy. Mutat. Res. 57:241-251.
13. Littlefield, L., S. Colyer, A. Sayer, and R. DuFrain (1979) Sister chromatid exchange in human lymphocytes exposed during G_0 to four classes of DNA-damaging chemicals. Mutat. Res. 67:259-269.
14. Kowalczyk, J. (1980) Sister chromatid exchanges in children treated with nalidixic acid. Mutat. Res. 77:371-375.
15. Djuric, D. (1983) Encyclopedia of Health and Safety, 3rd Ed; L. Parmeggiani, ed. International Labor Office, Geneva, Switzerland, pp. 812-818.
16. Garry, V., C. Opp, J. Wiencke, and D. Lakatua (1982) Ethylene oxide induced sister chromatid exchange in human lymphocytes using a membrane dosimetry system. Pharmacology 25:214-221.
17. Lambert, B., A. Lindblad, M. Nordenskjold, and B. Werelius (1978) Increased frequency of sister chromatid exchanges in cigarette smokers. Hereditas 88:147-149.
18. Knuutila, S., J. Maki-Pookkanen, M. Kahkonen, and E. Hokkanen (1978) An increased frequency of chromosomal changes and SCEs in cultured blood lymphocytes from 12 subjects vaccinated against smallpox. Human Gen. 41:89-96.
19. Yager, J., C. Hines, and R. Spear (1983) Exposure to ethylene oxide at work increases sister chromatid exchanges in human peripheral lymphocytes. Science 219:1221-1223.
20. Tan, E., R. Cummings, and A. Hsie (1981) Mutagenicity and cytotoxicity of ethylene oxide in the CHO/HGPRT system. Environ. Mutagen. 3:683-686.
21. Pero, R., B. Widegren, B. Hoegstedt, and F. Mitelman (1981) In vivo and in vitro ethylene oxide exposure of human lymphocytes assessed by chemical stimulation of unscheduled DNA synthesis. Mutat. Res. 83:271-290.
22. Dunkelberg, H. (1982) Carcinogenicity of ethylene oxide and 1,2 propylene oxide upon intragastric administration to rats. Br. J. Cancer 46:924-933.
23. Snellings, W., C. Weil, and R. Maronpot (1981) Final report on ethylene oxide two-year inhalation study on rats. Project Report 44-20, Bushy Run Research Center (formerly Carnegie-Mellon Institute of Research).

THE RELEVANCE OF SISTER CHROMATID EXCHANGE STUDIES TO PUBLIC HEALTH: PREVENTION AND INTERVENTION. INTRODUCTION TO A GENERAL DISCUSSION ON THE INTERPRETATION OF SISTER CHROMATID EXCHANGE DATA

T. A. Tsongas

Occupational Safety and Health Administration*
Washington, D.C.

The Occupational Safety and Health Administration (OSHA) is responsible for the development of standards to eliminate, to the extent that is feasible, health hazards in the occupational environment. Data from a variety of sources on the toxic effects of chemical or physical agents are reviewed in order to determine the level of exposure to these agents associated with potential adverse health effects in workers. Experimental bioassays, human case reports, and/or epidemiologic studies are evaluated.

Recently, the SCE assay has come under close scrutiny as a potentially useful short-term test to indicate exposure to chemical mutagens or carcinogens. Reviews and evaluations have included those by the U.S. Environmental Protection Agency's Gene-Tox Program (1) and the U.S. Congress' Office of Technology Assessment (2). An expert panel convened in 1981 stated that "... cytogenetic study of both chromosomal aberrations (CAs) and sister chromatid exchanges (SCEs) has a clear place in the evaluation of human populations exposed to known or suspected mutagens." Although cytogenetic changes cannot be used to predict specific health effects in an individual, they do give an estimate of the magnitude of an exposure that could increase the risk of disease in the exposed population (3).

Thus, evaluations of potential health hazards have recently included reports pertaining to elevations in numbers of SCEs observed

---

*The views expressed are those of the author and do not necessarily represent those of the Occupational Safety and Health Administration.

in experimental studies as well as in persons exposed to toxic chemicals in an occupational setting. Workers exposed to vinyl chloride monomer, ethylene oxide, antineoplastic agents, styrene, formaldehyde, benzene, and organic solvents have been evaluated to determine the effects of these chemicals on the frequency of SCEs. In addition, on the basis of SCE and aberration data, manufacturers have taken action to limit exposure of employees to chemical agents. Thus, results of the SCE assay are now being used for health intervention in the occupational setting.

During the course of this meeting a number of intrinsic factors affecting the outcome of SCE assays have been discussed. Some former concerns regarding the validity of SCE results have been laid to rest as a greater understanding of the mechanism of SCE formation and information on improved methodologies have become available. A number of questions regarding methodology and interpretation of data on SCEs have been raised in the evaluation of study results (see Tab. 1). Some of these questions may appear to oversimplify complex issues and may overlap, but they are intended to elicit discussion, to refine our use of this assay and interpretation of the results, and to point to the areas where recent data have substantially changed or augmented our understanding of the mechanism of SCE induction.

For example, the first question inquires about the nature of the information obtained by the SCE assay. How much does it tell us about exposure to health hazards? The presence of SCEs in a cell appears to indicate a change in the DNA of that cell. If that change is "error free" then one might assume that the exchange of material between sister chromatids will have no effect on the life or the activities of the cell. Indeed, SCEs do not seem to be associated with cell death until increases are considerable, perhaps reflecting other damage (than SCEs) from higher exposure concentrations or longer exposure time. The association between increases in SCEs and exposure to agents that are known to damage DNA is well supported by empirical evidence (2,4). Therefore, even if the exchange of portions of sister chromatids is not harmful in itself, it still indicates exposure to a harmful agent that may cause damage that will be manifested in other ways. Without seeing the effect of that damage directly, we still have a sensitive method for showing that the exposure has occurred and the risk of DNA damage is increased.

If the exchange of sister chromatids is itself a manifestation of damage to the DNA, that is, the exchange involves errors and thus possible mutations, then we have a direct indicator of an adverse effect of exposure to the SCE-producing agent. In either case, exposure to a hazardous substance has been indicated.

What correlations are there between SCEs and endpoints such as cancer, chromosomal aberrations, or mutations? At present, many known carcinogens have been found to produce SCEs, but no systematic sampling of chemical agents has been conducted to determine whether correlations for various classes of chemicals are truly predictive. A manageable beginning might be to randomly sample the chemicals tested in the National Toxicology Program (NTP) carcinogenesis bioassay program and test them for SCE induction. This sample of chemicals will admittedly still be biased by the selection criteria for the bioassay program but would be an improvement over selection of known carcinogens for testing since it will give us a better idea about SCE induction rates by chemicals that are both positive and negative in the carcinogenesis bioassay.

What sorts of adjustments of the data are biologically sound? What data are there to support these adjustments? We know that adjustments should be made for confounders such as cigarette smoking, age, and gender, and we know that these factors are best controlled by proper study design. Therefore, it is useful to know in advance what other potential confounding factors may also increase SCEs. For example, do commonly used substances such as alcohol, pharmaceutical drugs, or caffeine affect the frequency of SCEs?

Tab. 1. Questions on the interpretation of SCE data.

* WHAT IS THE RATIONALE FOR THE USE OF SCE ASSAYS FOR HEALTH HAZARD EXPOSURE MONITORING?

* WHAT DO INCREASES IN SCEs INDICATE ABOUT PRESENT OR FUTURE HEALTH RISKS?

* WHAT CORRELATIONS ARE THERE BETWEEN SCEs AND ENDPOINTS SUCH AS CANCER, CHROMOSOMAL ABERRATIONS, OR MUTATIONS?

* WHEN DO INCREASES IN SCEs INDICATE THAT HUMAN EXPOSURE TO CHEMICAL OR OTHER HAZARDS MUST BE REDUCED OR ELIMINATED?

* WHAT DO DATA ON SCEs IN OTHER SPECIES TELL US ABOUT HUMAN HEALTH RISKS?

* WHAT ARE THE STATISTICAL REQUIREMENTS FOR STUDIES OF SCEs IN THE OCCUPATIONAL OR ENVIRONMENTAL SETTING?

* WHAT CONFOUNDING FACTORS MAY ALSO INFLUENCE SCE LEVELS (E.G., AGE, GENDER, CIGARETTE SMOKING)?

SCE data are presently being used to contribute information on human health risks. The simple question that really underlies evaluation of these data is: When do increases in SCEs, alone or in combination with other data, indicate that human exposure to suspected chemicals or other hazards must be reduced or eliminated? In view of the importance of these data to public health intervention, we encourage the scientists who have developed and are using this assay to become involved in the development of regulations. Such participation is vital to assure that interpretations of SCE data for regulatory purposes are made on a reasonable scientific basis.

REFERENCES

1. Latt, S.A., J. Allen, S.E. Bloom, A.V. Carrano, E. Falke, D. Kram, E. Schneider, R. Schreck, R. Tice, B. Whitfield, and S. Wolff (1981) Sister-chromatid exchanges: A report of the Gene-Tox program. Mutat. Res. 87:17-62.
2. Office of Technology Assessment, Congress of the United States (1983) The Role of Genetic Testing in the Prevention of Occupational Disease. OTA-BA-194. U.S. Government Printing Office, Washington, D.C.
3. Bloom, A.D., ed. (1981) Guidelines for Studies of Human Populations Exposed to Mutagenic and Reproductive Hazards. March of Dimes Birth Defects Foundation, White Plains, N.Y.
4. Wolff, S. ed. (1982) Sister Chromatid Exchange. John Wiley & Sons, New York, NY.

SUMMARY OF GENERAL DISCUSSION

Raymond R. Tice

Medical Department
Brookhaven National Laboratory
Upton, LI, New York 11973

Dr. Tice began the General Discussion by introducing Dr. T. A. Tsongas (Occupational Safety and Health Administration, OSHA) and by indicating a desire to achieve an overall consensus among the European, American, and Japanese investigators participating in the symposium, with regard to the applicability, reliability, and utility of SCE analysis.

Dr. Tsongas indicated the need for scientists who are developing and using SCE assays to become involved in the interpretation of results because of their greater familiarity with the scientific bases, problems, and advantages of this assay (see Introduction). A set of questions had been developed and was presented as a focus for the General Discussion. These questions stem from the needs of regulatory agencies to interpret SCE data for public health intervention.

Dr. Tong-Man Ong (National Institute for Occupational Safety and Health) raised the following additional questions:

1. Is there any way that a standard protocol procedure can be established for SCE studies in the occupational setting? "Protocol" was defined as prearranged criteria for selection of workers for this type of study and for analysis of results.

2. Is there a good approach to getting cooperation from industry, not only management but also workers, to participate in SCE studies?

3. How are workers to be informed of the results of these studies? Who should tell them? How does one explain to an individual the significance of a high SCE frequency?

4. For follow-up studies, how long should SCE surveillance continue and at what time intervals? What health effects should we look for in the positive response group?

Because the ideas to be discussed were similar for both sets of questions, Dr. Tice proceeded with the question: WHAT IS THE RATIONALE FOR THE USE OF SCE ASSAYS FOR HEALTH HAZARD EXPOSURE MONITORING?

Dr. A. V. Carrano (Lawrence Livermore National Laboratory) responded by saying that there does not necessarily have to be a health hazard present to justify the use of SCEs in the workplace and that SCEs can be used as a biological indicator of exposure to a genotoxic agent(s). He stated that a health hazard can be defined as an exposure to a toxic agent or a substance that threatens bodily harm. There is irrefutable evidence that SCEs can occur only through the breakage and rejoining of four strands of DNA, a process which may or may not be error free. Geneticists would agree that there is a finite possibility that the process may not be error free. If so, then there is a threat of a health risk associated with the induction of SCEs. He suggested that SCEs can be used in occupational monitoring as an indicator of exposure and as an indicator of potential bodily harm.

Dr. J. Whang-Peng (National Cancer Institute) pointed out that we should be concerned about anything that is a hazard to DNA. However, SCEs are not a good indicator of all kinds of DNA damage and care should be taken to include other biological indicators, such as chromosomal aberrations, in biomonitoring studies.

Dr. M. Sorsa (Academy of Finland) concurred with this point and suggested that in discussing the possibilities for application of SCEs to routine occupational hygiene measurements, what the World Health Organization (WHO) refers to as biological monitoring, not only SCEs but cytogenetic monitoring as a whole should be discussed. The endpoints are different but the problems involved are the same. Biological monitoring is defined as using human samples to estimate internal exposures; for example, the detection of a chemical or its metabolites in urine or blood. The reason for conducting these studies is to prevent the potential ill effects of the exposures. The difficulty lies in accepting the use of "effect" monitoring. Cytogenetic findings, either SCEs or chromosomal aberrations, are biological effects in human cells. At present, enough is known to include effects measurements in biological monitoring and WHO recommends that such monitoring should comprise "regular" measurements with validated, specific methods. There are very few regular measuring activities using cytogenetic monitoring, with the exception of long term follow-up studies such as those of Dr. Linnainmaa and Dr. Galloway, and information on prospective types of studies is lacking.

Whether the methods are validated is another problem. This meeting has brought this goal closer to realization. What should be accomplished is to collaboratively construct a validated standardization for the methodologies: how a good cytogenetic study should be conducted, and what would be the recommended laboratory methods. Before entering into a cytogenetic monitoring study, experimental work should be conducted with the chemical in question to address questions of specificity, potential modulating influences, adaptation, etc. In practice, collaboration between different types of experts is important for conducting and interpreting a good study. For an occupational study, collaboration is needed between all interested parties, i.e., the workers, the employers, the on-site plant medical staff, etc. and more effort should go into the good design of a study than into the study itself.

A Nordic working group (involving, in addition to others, M. Sorsa, B. Lambert, and A. Brogger) has reviewed the literature on cytogenetic studies in occupational settings. There are quite a few studies, involving exposures to approximately 72 agents. The major weakness in these studies is that most totally lack information about the nature, duration, and magnitude of the exposure. These studies do not allow interpretation of their results. Among these 72 agents, the number of human clastogenic chemicals in occupational situations demonstrated by well-designed studies and confirmed by by several independent groups is very low: approximately 7 chemicals and ionizing radiation. However, from these studies, it was evident that an increase frequency of chromosomal aberrations did not necessarily correlate with an increased frequency of SCEs and vice versa. Thus, both endpoints need to be assessed to demonstrate a negative effect.

Dr. Sorsa further indicated that there are 2 stages for cytogenetic monitoring in occupational settings: (i) the identification or qualitative stage. This stage requires prior experimental data to justify the need for initiating a monitoring study. The results of a good study should be reported and upon detecting a significant increase in SCE and/or aberrations, the exposure should be minimized; and (ii) quantitative stage: estimation of risk. This stage can be applied only to well-documented exposure situations where mechanistic knowledge is available. Presently available information is probable too meager to allow stage ii studies.

Dr. B. Lambert (Karolinska Hospital) commented that monitoring studies should not be limited only to situations for which the nature and extent of exposure are well-documented. It would still be worthwhile to conduct cytogenetic monitoring in settings where such data are not readily obtainable, although such studies will place strong demands on the protocol.

The next question, WHAT DO INCREASES IN SCEs INDICATE ABOUT PRESENT OR FUTURE HEALTH RISKS? This questions is the most difficult one with which to deal because of a lack of knowledge about the significance of an elevated frequency of SCEs to an individual in terms of the future appearance of an adverse health outcome. Thus, Dr. Tice stated, in agreement with the majority participants that individual SCE frequencies, like individual aberration frequencies, can not currently be used to predict individual risk. However, elevated frequencies of either cytogenetic manifestation of in vivo genotoxic damage can and should be used to indicate increased risk to groups of individuals. More research on mechanisms is necessary and more human studies should be conducted to evaluate the magnitude and the individuality of the risks.

Dr. N. C. Popescu (National Cancer Institute) stated that he believes that there is a relationship between the induction of SCEs and the initiation of neoplastic development, and that the best approach for measuring this relationship currently is to use in vitro systems. There is a correlation between the induction of SCEs and transformation which is the best endpoint in vitro for measuring induction of neoplasia. Molecular and cellular evidence is available which implies that for certain carcinogens a common lesion is involved in SCE formation and transformation. Thus, exogenous factors which prevent or inhibit SCE formation should be identified and examined for their ability to prevent or inhibit transformation.

Dr. Lambert then commented that individuals with SCE frequencies much higher than is normally observed in control groups, i.e., outliers, should cause concern and be considered a medical problem. The outlier individual should probably be transferred from group monitoring to an individual patient-doctor relationship, and an effort should be made to discover the underlying cause. He also pointed out the ethical need to report unexpected karyotypic changes to individuals when they are found.

Dr. A. Brogger (Norsk Hydro's Institute for Cancer Research) continued the discussion by summarizing some of the ethical problems encountered in the Nordic countries by the Nordic working group. Scientific research is aimed at gaining knowledge. Public health work means the application of this knowledge for the benefit of the single individual, as well as for the population as a whole. However, public health work cannot be performed without the certain knowledge of the health situation of single individuals. Conflict may arise between the health authorities' demand for knowledge about their clients and these clients' demands for their privacy. An ethical dilemma will arise when different human values are not obtained at the same time. An open discussion of such ethical questions between these different parties may solve many problems.

Chromosome studies in occupational health are aimed at revealing the possible occurrence of genotoxic agents in the work environment. The problems that arise involve reporting results to individuals and in providing meaningful explanations of the results. Personal knowledge of individual cytogenetic data can cause anxiety, which should be controlled to the extent possible. Unless specifically requested, group data only should be provided along with an appropriate interpretation. Dr. Brogger stressed that solving the exposure problem by moving individuals with high frequencies of SCEs or chromosome damage to other areas was not appropriate. Rather, measures must be taken to reduce exposure. He also reiterated the importance of informing individuals of constitutive chromosomal alterations.

Dr. K. Morimoto (University of Tokyo) made a general comment that it is easier to quantify SCEs than to quantify an adverse health response and that great care must be taken in any study to collect adequate data on the health status of each individual involved.

Dr. R. D. Benz (Brookhaven National Laboratory) indicated the need for caution in interpreting data from these studies, holding the opinion that more mechanistic studies are required to demonstrate the usefulness of SCEs as a good indicator of threat of bodily harm and a good indicator of somatic cell damage.

One participant asked how high an increase in SCEs a cell could survive, and suggested that a cell-tolerated range of SCEs might be developed to indicate the number to be tolerated by an individual without future adverse health outcome.

Dr. Tice responded by stating that the maximum frequency of SCEs in a cell dependents on cell type and on the genotoxic nature of the compound. The process of SCE formation may only relate to toxicity in that the lesion(s) causing one event may also cause the other. Because aberrations are much more likely a cause of cell death than are SCEs, the relative ability of an agent to induce one effect versus the other must be examined. Thus, with regard to toxicity, he felt it unlikely that a standard or unsafe level of SCEs could be determined.

Dr. P. J. O'Neill (University of Vermont) suggested that the ability of SCEs to indicate future adverse health outcomes will not be resolved without prospective follow-up studies in humans and indicated the importance of conducting a worst-case population study. One example of such a study would involve assessing SCEs in cancer patients being treated with chemotherapeutic agents, where a high risk for secondary tumors is well documented.

Dr. Tice agreed with the need for a prospective study of this kind and mentioned Dr. Raposa's data, presented at this symposium, on SCE frequencies in chemotherapy-treated cancer patients where the ability of a chemotherapeutic regime to induce SCEs correlated with the risk of secondary tumors.

Dr. R. Becher (West Germany Tumor Center) commented that secondary leukemias are quite different from other leukemias and that the difference is reflected in a unique chromosomal aberration pattern. This difference would suggest that this population may not be representative of cancer induction in general. This does not mean, however, that such a study should not be conducted, but, rather, that the results should be kept in proper perspective.

Dr. Tice introduced the next question: WHAT CORRELATIONS ARE THERE BETWEEN SCEs AND ENDPOINTS SUCH AS CANCER, CHROMOSOMAL ABBERATIONS, AND MUTATIONS?

Dr. S. Parodi (Scientific Institute for the Study and the Cure of Tumors) stated that many factors are involved in determining the degree of predictivity of short-term tests, including the SCE assay. These factors include metabolic activation, target organ specificity, and endpoint. For many short-term tests the relationship between the test response and carcinogenicity contains more "noise" than information. Sister chromatid exchanges do not indicate the specific nature of the lesion responsible for their induction. However, SCEs do indicate an interaction between the DNA and the test agent. Their predictive value for carcinogenicity is no worse than any other short-term test, including preneoplastic liver nodules or in vitro transformation, and is better than most. It may well be that within certain classes or chemicals, SCEs are an extremely good or relatively poor predictor of carcinogenicity. Not enough information is currently available to decide upon the global predictivity of SCEs. However, it must be remembered that marginally weak SCE responses may result from other than genotoxic damage and may not indicate a carcinogenic interaction. Thus, risk assessment involving SCE induction should be conservative.

Dr. J. H. Taylor (Florida State University) commented that SCEs are a very sensitive test for DNA damage, even if the response is not specific for all kinds of genotoxic insult. Using that knowledge it is possible to decide whether or not a potential health risk exists.

Dr. F. B. Oleson (Bristol-Myers Co.) agreed and suggested the need for increased research to elucidate the mechanisms involved in SCE formation. One specifically interesting line of investigation would involve cell mutants which lack an ability to form SCEs or have them at a constitutively elevated level. Also, when thinking about SCEs and DNA damage, and the correlation between SCEs and cancer, he raised a question about the meaning of fluctuations in SCE frequencies between individuals.

Dr. J. Schwartz (Harvard School of Public Health) stated that the induction of SCEs is not just the consequence of, or a measure of, DNA damage but is also a measure of a cellular effect. Because adverse health effects are not necessarily always the result of DNA damage, this fact improves the usefulness of SCE assays.

Dr. Tice noted that there had been many presentations dealing with the relationship between SCEs and mutations, between SCEs and specific adducts, and the correlation between SCE-inducing potency and classes of compounds. The data presented to date indicate that one-to-one correlations do not exist and that many kinds of DNA damage are capable of inducing an SCE response. Some of the same lesions responsible for inducing SCEs are also responsible for inducing mutations and transformation. Regarding chromosomal aberrations, Dr. Tice stated that the differing correlations with SCEs for different types of clastogens, i.e., S-dependent versus S-independent, provides information about the mechanism(s) involved in SCE formation. He then proceeded to the next question: WHEN DOES AN INCREASE IN SCEs INDICATE THAT HUMAN EXPOSURE TO SUSPECTED CHEMICAL OR OTHER HAZARDS MUST BE REDUCED OR STOPPED?

Dr. Sorsa replied that one should never act on the basis of SCE data alone. Experimental information, exposure measurements, other cytogenetic information such as chromosomal aberrations, and some idea of epidemiologic endpoints should all be considered. However, where substitutions of chemicals are feasible, then another, non-SCE-inducing chemical should be used.

Dr. Ong stated that if an SCE effect cannot be related to a health problem then it will be difficult to ask industries to change their operations based on the results of SCE studies.

Dr. W. F. Morgan (University of California, San Francisco) agreed that an effect based on an SCE response alone would be difficult to interpret. He then went on to state, however, that SCEs are more than just an exchange between chromatid arms. As more information on oncogenes becomes available, the possibility that promoting and enhancing sequences are being shifted around and therefore becoming localized next to one of these oncogenes is something that should be considered. So, although SCEs cannot be correlated currently with adverse health outcomes because of the variation between people, between different cell lines, etc., current research in these areas will form the basis on which to make such correlations in the future.

Dr. D. W. Matheson (Stauffer Chemical Co.) commented that when regulating or setting dose levels or tolerances, the most sensitive indicator of an effect is often used. SCEs seem to be very sensitive indicators of an effect. Therefore, if they can be detected, then there is an obligation to reduce the exposure to the chemical as much as possible.

Dr. Schwartz commented, using the film badge analogy, that if the toxic agent and the dose-response curve are known, then SCE data for a person could be used to indicate too great an exposure.

Dr. R. J. DuFrain (Oak Ridge Associated Universities) cautioned that the assay as an exposure film badge may be misused to deprive people of their livelihood.

Dr. Carrano agreed and stated that a major problem is a legalistic one involving tort liability.

Dr. Tice responded to the next question: WHAT DO DATA ON SCEs IN OTHER SPECIES TELL US ABOUT HUMAN HEALTH RISKS?, by stating his belief that a positive SCE response in an animal test system provides a qualitative estimation of risk. This qualitative estimation can become quantitative only when the identity of the SCE-inducing agent, i.e., parent compound vs. metabolite(s), and the metabolic, pharmacokinetic, and DNA-repair differences between man and the test species are fully understood.

Dr. Tice asked Dr. Whorton to comment on the next question: WHAT ARE THE STATISTICAL REQUIREMENTS FOR STUDIES OF SCEs IN THE OCCUPATIONAL OR ENVIRONMENTAL SETTING?

Dr. E. B. Whorton (University of Texas Medical Branch) commented that this is more than a sample size question. In agreement with an earlier comment by Dr. Sorsa, he also suggested that 90% of the time should go into planning of the experiment, controlling for as many factors in the planning, before the experiment, as are relevant. Collaborative effort is necessary to determine the ramifications of the inferences and there are no simple answers. However, several presentations at this symposium have discussed statistical methods for dealing with SCE data and for constructing meaningful sample sizes of cells within individuals and numbers of individuals within groups. What is needed is a data base of an adequate size with which to compare these methodological approaches for differences in sensitivity and reliability.

Dr. Tice responded to the next question: WHAT CONFOUNDING FACTORS MAY ALSO INFLUENCE SCE LEVELS (E.G., AGE, GENDER, CIGARETTE SMOKING)? He listed the factors that had been discussed throughout the meeting: experimental variables, including the number of cells, BrdUrd concentration, type of serum, temperature, and medium. In addition, some human population studies have shown a gender difference in SCE frequency; smoking and presumably other personal habits are also confounding variables. Age differences in SCE frequency have been seen in animal populations. Dr. Whorton agreed that these important variables should be considered prior to the study.

Dr. Tice asked for comments on the question raised by Dr. Ong concerning a standardized protocol. The discussion that followed reflected a consensus that each laboratory should have a completely standardized protocol but that a standardized protocol for use everywhere was neither necessary, wise, nor feasible. The importance of running controls in each laboratory was emphasized. Moreover, many laboratories have well-developed protocols and can act as an information resource for other laboratories. It was also pointed out that a standardized protocol for all laboratories would cause the loss of much information such as has been obtained on mechanisms of SCE formation and confounders.

Dr. Tice asked the audience to discuss one other question asked by Dr. Ong which dealt with post-exposure surveillance intervals and the significance of induced SCE frequencies that quickly or slowly decline with time.

Dr. Parodi raised the problem of lymphocyte turnover and dilution, which may appear to cause a decline in SCE frequency. Dr. Tice agreed and suggested that the repair competency of individual cells combined with the nature and duration of the original genotoxic insult would also significantly contribute to the rate of decline detected in peripheral blood lymphocytes. However, it is not possible to conclude whether an individual with an elevated level of SCEs which is persistent across time is more at risk than an individual who exhibited the same initial elevation but which disappeared quickly. It was pointed out that studies are in progress, especially in treated cancer patients undergoing chemotherapy, which may help to resolve this question. Moreover, it should be remembered that the cell population in which the SCEs are being analyzed, i.e., mature lymphocytes, are not a cell type at risk of tumor formation.

Dr. Tice closed the general discussion with a few summary statements. These statements included the following points:

1. The induction of SCEs is an manifestation of DNA damage and, as such, can be used to indicate human exposure to genotoxic agents.

2. Thus, SCEs provide a useful tool for biological effect monitoring studies, especially when used in conjunction with other biological effect indicators of genotoxic exposure and where prior experimental data indicate that potential risk exists.

3. However, SCEs can also be induced by indirect mechanisms, including altered nucleotide pools, and awareness of these interactions should be included in any evaluation of SCE data.

4. It is not currently possible to quantify the health risk associated with an increased frequency of SCEs for a group

of exposed individuals, and even more so, for the individual. Longitudinal, prospective studies should be conducted on a worst-case population to help resolve this issue.

5. SCEs can be correlated with mutations, transformation, and cancer but the correlations are not constant and depend on the spectrum of lesions induced by a specific agent, some of which will be unique to one or the other endpoint, and some of which will be common to more than one endpoint.

6. Statistically correct approaches for analyzing SCE data are available. However, to define the most sensitive method for SCE analysis under different conditions, adequately large data bases must be made generally accessible. Efforts to locate and make available a suitable, large SCE data base which will include data from in vitro systems, in vivo systems, and human biomonitoring studies are in progress. (Information on this subject can be obtained from the editors.)

7. In any monitoring study, potentially confounding variables, i.e., differences in age, gender, smoking history, drug usage, etc., should be controlled for in the design of the study.

8. More basic research is necessary to increase our understanding of how SCEs occur, whether their formation is error-free or error-prone and the limitations involved in their application.

9. It is hoped that in 4 or 5 years a second international symposium on SCEs would be held to address advances in SCE research.

Dr. A. Hollaender (Council for Research Planning In Biological Sciences, Inc.), offered his appreciation to the various individuals involved in making the Symposium successful, and made a plea for more basic research on SCE formation--research that should involve the new molecular biology methods in genetic engineering. The efforts of Drs. Lambert, Morimoto, and Tice in developing the program especially were acknowledged.

PARTICIPANTS AND SPEAKERS

ABE, S.	Chromosome Research Unit, Faculty of Science, Hokkaido University, Sapporo 060, Japan
ALHADEFF, B.	Sloan Kettering Institute of Cancer Research, New York, NY 10021, USA
*ALLEN, James W.	U.S. Environmental Protection Agency, Genetic Toxicology Division, Research Triangle Park, NC 27711, USA
ALLEN, Jane S.	American Cyanamid Company, Agricultural Research Division, P.O. Box 400, Princeton, NJ 08540, USA
ANDERSON, H. Christen	Department of Genetics, University of Uppsala, P.O. Box 7003, S-750 07 Uppsala, Sweden
ARUNDEL, Carla M.	Brown University, Laboratory of Radiation Biology, Box G, Providence, RI 02912, USA
AUERBACH, Arleen D.	Rockefeller University, 1230 York Avenue, New York, NY 10021, USA
*AWA, Akio A.	Cytogenetics Section, Radiation Effects Research Foundation, Hijiyama Park, Hiroshima 730, Japan
BACKER, Lorraine C.	Division of Biological Sciences, University of Kansas, Show Hall, Lawrence, KS 66044, USA
BAN, Sadayuki	Biology Department, Brookhaven National Laboratory, Upton, NY 11973, USA
BASLER, Armin	Bundesgesundheitsamt, Postfach 33 00 13, D-1000 Berlin 33, FRG
BECHER, Reinhard	Innere Universitatsklinik (Tumorforschung), Westdeutsches Tumorzentrum, Hafelandstr 55, D-4300 Essen 1, FRG

BECKER, Doris	INBIFO Institut fur Biologische Forschung, Fuggerstr. 3, 5000 Cologne 90, FRG
*BENZ, Daniel R.	Medical Department, Brookhaven National Laboratory, Upton, NY 11973, USA
BLAKEY, David	Canadian Department of Health and Welfare, Environmental Health Center, Ottawa K1A 0L2, Canada
BLANK, Charles	American Petroleum Institute, 1220 L Street, NW, Washington, DC 20005, USA
BLOCK, Anne Marie W.	Department of Genetics and Endocrinology, Roswell Park Memorial Institute, 666 Elm Street, Buffalo, NY 14263, USA
*BLOOM, Arthur D.	Columbia University, College of Physicians and Surgeons, 630 West 168th Street, New York, NY 10032, USA
*BLOOM, Stephen E.	Department of Poultry and Avian Sciences, Cornell University, Ithaca, NY 14853, USA
BODELL, William	Brain Tumor Research Center, University of California, San Francisco, CA 94143, USA
BOND, Victor P.	Director's Office, Brookhaven National Laboratory, Upton, NY 11973, USA
BORG, Donald C.	Medical Department, Brookhaven National Laboratory, Upton, NY 11973, USA
*BROGGER, A.	Norsk Hydro's Institute for Cancer Research, Norwegian Radium Hospital, Montebello, Oslo 3, Norway
BROOKS, Anthony L.	Inhalation Toxicology Research Institute, P.O. Box 5890, Albuquerque, NM 87117, USA
CAMURRI, Lamberto	Laboratory of Genetics USL 9, Via Dante Alighieri 11 - 42100 Reggio Emilia, Italy
CAO, En-Hua	Biology Department, Brookhaven National Laboratory, Upton, NY 11973, USA
*CARRANO, Anthony V.	Biomedical Sciences Division, Lawrence Livermore National Laboratory, P.O. Box 5507, Livermore, CA 94550, USA

CARSTEN, A. L.	Medical Department, Brookhaven National Laboratory, Upton, NY 11973, USA
CASCIANO, Daniel A.	Division of Mutagenesis Research, National Center for Toxicological Research, Jefferson, AR 72079, USA
CHROMEY, Nancy C.	Haskell Laboratory, E.I. DuPont de Nemours, P.O. Box 50, Elkton Road, Newark, DE 19711, USA
CIARAVINO, Victor	Department of Radiation Biology and Biophysics, University of Rochester, School of Medicine and Dentistry, Rochester, NY 14642, USA
CONAWAY, Clifford C.	Texaco Inc., P.O. Box 509, Beacon, NY 12508, USA
CONNELL, Jane R.	ARC Institute of Animal Physiology, Cambridge CB2 4AT, UK
*CONNER, Mary K.	University of Pittsburgh, Graduate School of Public Health, Pittsburgh, PA 15261, USA
CREMER, Thomas	Institute of Anthropology and Human Genetics, University of Heidelberg, 6900 Heidelberg, FRG
DEARLOVE, G.	DuPont Pharmaceuticals, 100 Stewart Avenue, Garden City, NY 11530, USA
DEKNUDT, Gh.	Laboratory of Mammalian Genetics, Department of Radiobiology, C.E.N/S.C.K., B-2400 Mol, Belgium
DELIHAS, Neva C.	Biology Department, Brookhaven National Laboratory, Upton, NY 11973, USA
DE MARCO, Charles	Xerox Corporation, Building 317, 800 Phillips Road, Webster, NY 14580, USA
DILLEHAY, Larry E.	Department of Radiology, George Washington University Medical Center, 2300 I Street, NW, Washington, DC 20037, USA
DI LORENZO, Ann Marie	Biology Department, Montclair State College, Upper Montclair, NJ 07043, USA
DORIGAN, Janet V.	Office of Energy Research, U.S. Department of Energy, ER-75 GTN, Washington, DC 20545, USA
*DU FRAIN, Russell J.	Medical and Health Sciences Division, Oak Ridge Associated Universities, P.O. Box 117, Oak Ridge, TN 37831, USA

DUNBAR, Virginia G.	Human and Behavioral Genetics Research Laboratory, Georgia Mental Health Institute, Atlanta, GA 30306, USA
EMERIT, Ingrid	CNRS, 15, rue de l' Ecole de Medecine, Paris VI, France
FISCHMAN, Ken	New York State Psychiatric Institute, 722 West 168th Street, New York, NY 10032, USA
FOX, Donald P.	Genetics Department, University of Aberdeen, 2 Tillydrone Avenue, Aberdeen AB9 2TN, UK
*FUJIWARA, Y.	Department of Radiation Biophysics, Kobe University School of Medicine, Kunsunoki-cho 7-5-1, Chuo-ku, Kobe 650, Japan
*GALLOWAY, Sheila M.	Litton Bionetics Inc., 5516 Nicholson Lane, Kensington, MD 20895, USA
GARRY, Vincent F.	University of Minnesota Medical School, Stone Lab. I., Room 110, 421 S.E. 29th Avenue, Minneapolis, MN 55414, USA
GEARD, Charles R.	Columbia University, College of Physicians and Surgeons, 630 West 168th Street, New York, NY 10032, USA
*GEBHART, Erich	Institute of Human Genetics, University of Erlangen-Nurnberg, Bismarckstr. 10, D-8520 Erlangen, FRG
GENTIL, A.	Institute de Recherches Scientifique sur le Cancer, 7, rue Guy Mocquet, 94802, Villejuif Cedex, France
GHIDONI, Achille	Dipartimento di Biologia, Universita di Milano, via Celoria 26, Milano 20133, Italy
GLATT, H. R.	Institute of Toxicology of the University, Obere Zahlbacher Strasse 67, D-6500 Mainz, FRG
GRANT, William F.	Genetics Laboratory, McGill University, Ste. Anne de Bellevue, Quebec, Canada H9X 1CO
GUTIERREZ, Crisanto	Instituto de Biologia Celular, Velazquez, 144, Madrid-6, Spain

HARRIS, Clifton R.	Department of Radiology, University of Texas Health Science Center, 7703 Floyd Curl Drive, San Antonio, TX 78284, USA
HASHIMOTO, Tomoko	Department of Genetics, Hyogo College of Medicine, Mukogawa, Nishinomiya, Hyogo 663, Japan
HATCHER, Norma H.	New York State Department of Health, Center for Labs and Research, 120 New Scotland Avenue, Albany, NY 12201, USA
HEARTLEIN, Michael W.	Oak Ridge Graduate School of Biomedical Sciences, P.O. Box Y, Oak Ridge, TN 37830, USA
HEEREMA, Nyla A.	Department of Medical Genetics, Indiana University School of Medicine, Indianapolis, IN 46223, USA
HEIDEMANN, Albrecht	Technische Hochschule Darmstadt, Institut fur Zoologie, Schnittspahnstr. 3, D-1600 Darmstadt, FRG
HIRSCH, Betsy	Division of Cytogenetics and Genetics, University of Minnesota, Minnapolis, MN 55455, USA
HOLLAENDER, Alexander	Council for Research Planning in Biological Sciences, Inc., 1717 Massachusetts Avenue, NW, Washington, DC 20036-2077, USA
HOOK, Graham	Medical Department, Brookhaven National Laboratory, Upton, NY 11973, USA
HOPKIN, Julian	Department of Medicine, Queen Elizabeth Hospital, University of Birmingham, Birmingham, UK
HORI, Tada-aki	Division of Genetics, National Institute of Radiological Sciences, Anagawa, Chiba 260, Japan
HUANG, Chester C.	Roswell Park Memorial Institute, 666 Elm Street, Buffalo, NY 14263, USA
HUGHES, Pete	Medical Department, Brookhaven National Laboratory, Upton, NY 11973, USA
IKUSHIMA, Takaji	Research Reactor Institute, Kyoto University, Kumatori, Sennan-Gun, Osaka 590-04, Japan
*ISHII, Yutaka	Department of Fundamental Radiology, Faculty of Medicine, Osaka University, Osaka 530, Japan

*IVETT, James L.	Medical Department, Brookhaven National Laboratory, Upton, NY 11973, USA
JABLONKA, Eileen	Biology Department, Brookhaven National Laboratory, Upton, NY 11973, USA
JABLONKA, Paul	University of Arizona, Tucson, AZ 85719, USA
JACKMAN, Joany	Department of Nutrition and Food Science, Massachusetts Institute of Technology, 20-C-129, 18 Vassar Street, Cambridge, MA 02139, USA
JORDAN, Diane K.	S-374 Hospital School, Department of Pediatrics, Iowa Hospital and Clinics, Iowa City, IA 52242, USA
KALE, P.	Safety and Environmental Protection Division, Brookhaven National Laboratory, Upton, NY 11973, USA
KANO, Yoshio	Laboratory of Radiobiology, Harvard School of Public Health, Boston, MA 02115, USA
KAO-SHAN, Chien Song	National Cancer Institute, National Institutes of Health, Building 10, 12C 112, Bethesda, MD 20205, USA
KELSEY, Karl	Department of Occupational Medicine, Harvard School of Public Health, Boston, MA 02115, USA
KLIGERMAN, Andrew D.	CIIT, P.O. Box 12137, Research Triangle Park, NC 27709, USA
KNUUTILA, Sakari	Department of Medical Genetics, University of Helsinki, Haartmaninkatu 3, SF-00290 Helsinki 29, Finland
KOSZ-VNENCHAK, Magdalena	Microbiology Department, University of Massachusetts, Amherst, MA 01002, USA
*LAMBERT, Bo	Department of Clinical Genetics, Karolinska Hospital, 10401 Stockholm 60, Sweden
LATT, Samuel A.	Genetics Division, Children's Hospital, 300 Longwood Avenue, Boston, MA 02115, USA
LAURENT, Christian P.A.	Genetics Department, C.H.U. Sart Tilman - Pathology (B23), B-4000 Liege, Belgium

LINDAHL-KIESSLING, Kerstin Institute of Zoophysiology, University of Uppsala, Box 560, 75722 Uppsala, Sweden

LINNAINMAA, Kaija Institute of Occupational Health, Haartmaninkatu 1, SF-00290 Helsinki 29, Finland

*LITTLEFIELD, Gayle L. Oak Ridge Universities, P.O. Box 117, Oak Ridge, TN 37831, USA

LOCK, Joe Wayne Genetics, Parkview Hospital, 2200 Randalia Drive, Fort Wayne, IN 46805, USA

LUKE, Carol A. Medical Department, Brookhaven National Laboratory, Upton, NY 11973, USA

LUNDGREN, Karsten National Institute of Environmental Health Sciences, Research Triangle Park, NC 27709, USA

MACKAY, James Morrison Genetics Department, University of Aberdeen, 2 Tillydrone Avenue, Aberdeen AB9 2TN, UK

MAHONEY, Eileen Xerox Corporation, 800 Philips Road, Building 843, Rochester, NY 14580, USA

MATHESON, Dale W. Stauffer Chemical Company, 400 Farmington Avenue, Farmington, CT 06032, USA

MC FEE, Alfred, Oak Ridge Associated Universities, P.O. Box 117, Oak Ridge, TN 37831, USA

MILLER, Valerie M. Medical Department, Brookhaven National Laboratory, Upton, NY 11973, USA

MIURA, Kunihiko Department of Public Health, Faculty of Medicine, University of Tokyo, Tokyo 113, Japan

MOORE, Dan Biomedical Sciences Division, Lawrence Livermore National Laboratory L-452, Livermore, CA 94550, USA

*MORALES-RAMIREZ, P. Departmento de Radiobiologia y Genetica, Instituto Nacional de Investigaciones Nucleares, C.P. 06140 Mexico City, Mexico

MORGAN, William F. Laboratory of Radiobiology and Environmental Health, University of California, San Francisco, CA 94143, USA

*MORIMOTO, Kanehisa	Department of Public Health, Faculty of Medicine, University of Tokyo, Tokyo 113, Japan
MORRIS, Suzanne M.	Division of Mutagenesis, National Center for Toxicological Research, Jefferson, AK 72079, USA
MOURELATOS, Denis	Department of General Biology Faculty of Medicine, Aristotelian University, Thessaloniki, Greece
NAGASAWA, Hatsumi	Laboratory of Radiobiology, Harvard School of Public Health, Boston, MA 02115, USA
NAISMITH, Robert	Pharmakon Research International, Inc., P.O. Box 313, Waverly, PA 18471, USA
NAKANO, Mimako	Radiation Effects Research Foundation, 5-2 Hijiyama Park, Minami-Ward, Hiroshima 730, Japan
NAWROCKY, Marta M.	Medical Department, Brookhaven National Laboratory, Upton, NY 11973, USA
NEAL, S. B.	Toxicology Division, Lilly Research Laboratories, Greenfield, IN 46140, USA
NEFT, Robin E.	Graduate School of Public Health, University of Pittsburgh, Pittsburgh, PA 15217, USA
NISHI, Yoshisuke	Section of Cell Biology, Biological Research Center, Japan Tobacco and Salt Public Corporation, Nakagi Hatano, Kanagawa 257, Japan
NORPPA, Hannu	Institute of Occupational Health, Haartmaninkatu 1, SF-00290 Helsinki 29, Finland
OCKEY, Charles H.	Paterson Laboratories, Christie Hospital, Withington, Manchester M20 9BX, UK
*OIKAWA, Atsushi	Department of Pharmacology, Research Institute for Tuberculosis and Cancer, Tohoku University Sendai, Miyagi 980, Japan
OLESON, Frederick B.	Pharmaceutical Research and Development Division, Bristol-Myers Company, P.O. Box 4755, Thompson Road, Syracuse, NY 13221, USA

O'LONE, Susan — Stauffer Chemical Company, 400 Farmington Avenue, Farmington, CT 06032, USA

O'NEILL, Patrick J. — E-311 Given Building, College of Medicine, University of Vermont, Burlington, VT 05405, USA

ONG, Tong-Man — National Institute of Occupational Safety and Health, Morgantown, WV 26505, USA

OSTROSKY, Patricia — Instituto de Investigaciones Biomedicas, UNAM Apt Postal 70228, Mexico DF, Mexico 20

PALITTI, Fabrizio — Consiglio Nazionale delle Ricerche, Centro di Studio di Genetica Evoluzionistica del CNR, Istituto di Genetica (Scienze), Citta Universitaria, I-00185 Rome, Italy

PARODI, Silvio — Istituto Nazionale per la Ricerca sul Cancro, Viale Benedetto XV n. 10, 16132 Genova, Italy

*PERRY, Paul E. — MRC Clinical and Population Cytogenetics Unit, Western General Hospital, Crewe Road, Edinburgh EH4 2XU, UK

PESCH, Gerald — Environmental Research Laboratory, U.S. Environmental Protection Agency, South Ferry Road, Narragansett, RI 02882, USA

POET, Lauren — Haskell Laboratory, E. I. Du Pont de Nemours Elkton Road, Newark, DE 19711, USA

POPESCU, N. C. — Laboratory of Biology, Building 37, Division of Cancer Cause and Prevention, National Cancer Institute, N.I.H., Bethesda, MD 20205, USA

QIN, Shizhen — Roswell Park Memorial Institute, 666 Elm Steet, Buffalo, NY 14263, USA

*RAPOSA, Tibor — Department of Medicine, Semmelweis University Medical School, Eotvos u. 12, 1121 Budapest, Hungary

ROGERS, J. Felix — Department of Pathology/Cytogenetics, Mercer University School of Medicine, Macom, GA 31207, USA

RUBINSON, Harriet F.	Graduate School of Public Health, University of Pittsburgh, Pittsburgh, PA 15260, USA
RUDENKO, Larisa	Department of Pharmacological Sciences, State University of New York, Stony Brook, NY 11794, USA
*RUDIGER, H. W.	Unit of Hereditary and Constitutional Diseases, Universitat Hamburg, Universitats-Krankenhous Eppendorf, Martinistr. 52, 2000 Hamburg 20, FRG
SALAMONE, Michael	Ontario Ministry of Environment, Biohazard Unit, P.O. 213, Resources Road, Toronto (Rexdale), Ontario M9W SL1 Canada
*SANDBERG, Avery A.	Department of Genetics and Endocrinology, Roswell Park Memorial Institute, 666 Elm Street, Buffalo, NY 14263, USA
SAN SEBASTIAN, Juan R.	Pharmakon Research International, Inc., P.O. Box 313, Waverly, PA 18471, USA
*SCHVARTZMAN, J. B.	Instituto de Biologia Celular, Velazques 144, Madrid-6, Spain
*SCHWARTZ, Jeffrey	Harvard School of Public Health, 665 Huntington Avenue, Boston, MA 02115, USA
SETLOW, Richard B.	Biology Department, Brookhaven National Laboratory, Upton, NY 11973, USA
SHAFER, David A.	Emory University and Georgia Medical Mental Health Institute, Genetics, GMHI, 1256 Briarcliff Road, Atlanta, GA 30306, USA
SHARMA, Rajendar K.	Bioassay Systems Corporation, 225 Wildwood Avenue, Woburn, MA 01801, USA
*SHIRAISHI, Yukimasa	Department of Anatomy, Kochi Medical School, Kochi 781-51, Japan
SHIRYOA, Tsugio	Faculty of Science, University of Tokyo, Tokyo 113, Japan
SINGH, Narendra P.	Department of Radiology, Ohio State University, Columbus, OH 43210, USA
SINISCALCO, M.	Sloan Kettering Institute of Cancer Research, New York, NY 10021, USA
SIRLIN, J. L.	Cornell University Medical College, 515 E. 71st Street (S-400), New York, NY 10021, USA

*SORSA, Marza	Institute of Occupational Health, Academy of Finland, Haartmaninkatu 1, SF-00290 Helsinki 29, Finland
*SPEIT, Gunter	Abteilung Klinische Genetik der Universitat Ulm, Oberer Eselsberg, D-7900 Ulm, FRG
SUN, Clare	University of Albuquerque, St. Joseph Place, NW, Albuquerque, NM 87140, USA
TAKESHITA, Tatsuya	Industrial Environmental Health Sciences, Graduate School of Public Health, University of Pittsburgh, Pittsburgh, PA 15261, USA
TANZARELLA, Caterina	Centro Genetica Evoluzionistica, Dipartmento di Genetica e Biologia Molecolare, Piazzale Aldo Moro, 5, 00185 Roma, Italy
TAYLOR, J. Herbert	Department of Biological Sciences, Institute of Molecular Biophysics, Florida State University, Tallahassee, FL 32306, USA
TEPPERBERG, James H.	Department of Pediatrics, West Virginia Genetics Center, West Virginia University Hospital, Morgantown, WV 26506, USA
*TICE, Raymond	Medical Department, Brookhaven National Laboratory, Upton, NY 11973, USA
TOLER, Douglas L.	Department of Pediatrics, West Virginia Genetics Center, West Virginia University Hospital, Morgantown, WV 26506, USA
TSONGAS, Theodora A.	Occupational Safety and Health Administration, Room N3718, 200 Constitution Avenue, NW, Washington, DC 20210, USA
TSUJI, Hideo	Division of Genetics, National Institute of Radiological Sciences, Anagawa, Chiba 260, Japan
TUCKER, James	National Institute of Occupational Safety and Health, Morgantown, WV 26505, USA
*UTAKOJI, T.	Department of Cell Biology, Japanese Foundation for Cancer Research, Kami-Ikebukuro, Tokyo 170, Japan

VAN'T HOF, Jack	Biology Department, Brookhaven National Laboratory, Upton, NY 11973, USA
VAN STADEN, J. L.	Department of Physiology, Faculty of Medicine, University of Orange Free State, Bloemfontein, Republic of South Africa
VED BRAT, S.	American Health Foundation, 1 Dana Road, Valhalla, NY 10595, USA
*VERCAUTEREN, Philip	Division of Human Genetics, Minderbroeder Street 12, 3000 Leuven, Belgium
VILIETINCK, Robert	Division of Human Genetics, Minderbroeder Street 12, 3000 Leuven, Belgium
WANG, Ju-jun	Biology Department, Brookhaven National Laboratory, Upton, NY 11973, USA
WATANABE, Toshiaki	Department of Hygiene and Preventive Medicine, Yamagata University, Zao-lida, Yamagata 990-23, Japan
WENGER, Sharon, L.	Children's Hospital of Pittsburgh, 125 DeSoto Street, Pittsburgh, PA 15213, USA
WHANG-PENG, Jacqueline	National Cancer Institute, 900 Rockville Pike, Building 31, Bethesda, MD 20205, USA
*WHORTON, Elbert B. Jr.	Division of Biometry, Department of Preventive Medicine and Community Health, University of Texas Medical School, Galveston, TX 77550, USA
WIENCKE, J. K.	Laboratory of Radiobiology, University of California, San Francisco, CA 94115, USA
WILEY, John E.	Cytogenetics Unit, State Laboratory of Hygiene, 465 Henry Mall, University of Wisconsin, Madison, Madison, WI 53706, USA
WILMER, James L.	Department of Genetic Toxicology, Chemical Industry Institute of Toxicology, P.O. Box 12137, Research Triangle Park, NC 27709, USA
*WOLFF, Sheldon	Laboratory of Radiobiology and Environmental Health, University of Calfornia, San Francisco, CA 94143, USA

WOODS, Philip S.	Biology Department, Queens College, 65-30 Kissena Blvd., Flushing, NY 11367, USA
WULF, Hans Christian	Department of Dermatology, The Finsen Institute, Strandboulevarden 49, DK-2100 Copenhagen Ø, Denmark
YAGER, Janice W.	Department of Biomedical and Environmental Health Sciences, Warren Hall, Room 317, University of California, Berkeley, CA 94720, USA
ZAKOUR, Helen R.	School of Fisheries, SH-10, University of Washington, Seattle, WA 98195, USA
ZWANENBURG, T. S. B.	Department of Radiation Genetics and Chemical Mutagenesis, Sylvius Laboratory, State University of Leiden, Wassenaarseweg 72, AL-2333 Leiden, The Netherlands

*Session chairperson.

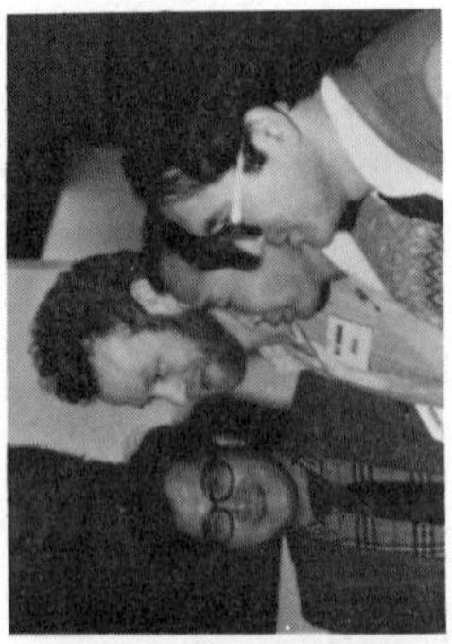

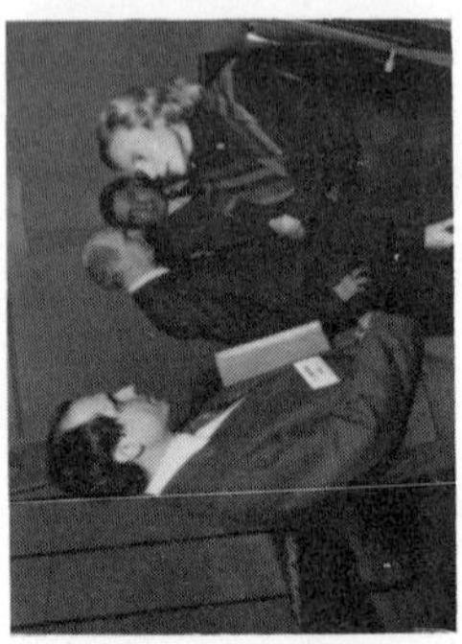

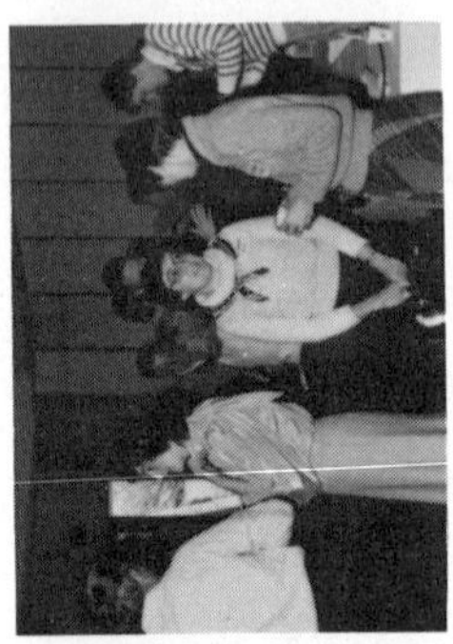
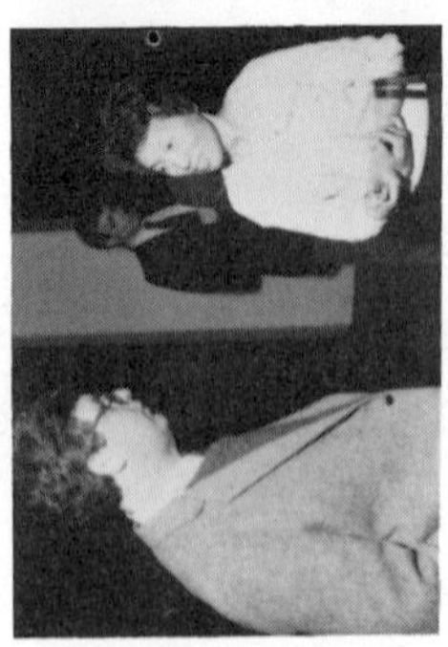

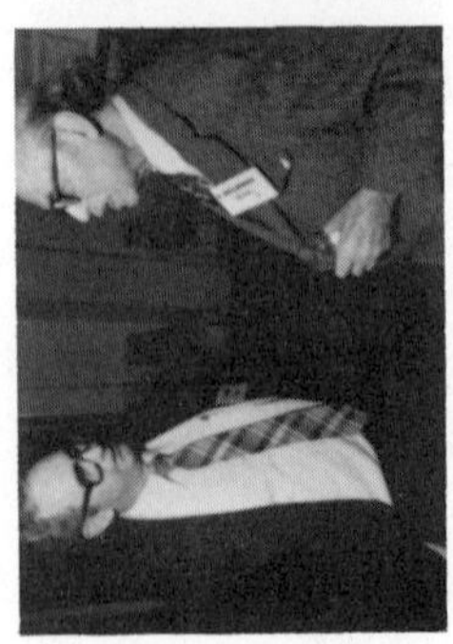

INDEX